Modern Semiconductor Physics and Device Applications

Modern Semiconductor Physics and Device Applications

Vitalii K. Dugaev and Vladimir I. Litvinov

CRC Press is an imprint of the
Taylor & Francis Group, an informa business

First edition published 2022
by CRC Press
6000 Broken Sound Parkway NW, Suite 300, Boca Raton, FL 33487-2742

and by CRC Press
2 Park Square, Milton Park, Abingdon, Oxon, OX14 4RN

Library of Congress Cataloging-in-Publication Data
Names: Dugaev, Vitalii, 1945- author.
Title: Modern semiconductor physics and device applications / Vitalii Dugaev and Litvinov Vladimir.
Description: First edition. | Boca Raton : CRC Press, [2022] | Includes bibliographical references and index.
Identifiers: LCCN 2021023385 | ISBN 9780367250805 (paperback) | ISBN 9780367250829 (hardback) |
ISBN 9780429285929 (ebook)
Subjects: LCSH: Semiconductors. | Solid-state physics.
Classification: LCC QC611 .D84 2022 | DDC 537.6/22—dc23
LC record available at https://lccn.loc.gov/2021023385

ISBN: 978-0-367-25082-9 (hbk)
ISBN: 978-0-367-25080-5 (pbk)
ISBN: 978-0-429-28592-9 (ebk)

DOI: 10.1201/9780429285929

Typeset in Minion Pro
by codeMantra

Access the Support Material: www.routledge.com/9780367250829

Contents

3 Impurities and Disorder in Semiconductors

4 Statistics of Electrons in Semiconductors

5 Electrons in a Magnetic Field

6 Phonons and Electron–Phonon Interaction

Preface

This textbook on semiconductor physics begins with the fundamentals of semiconductors and then includes the modern methods of the condensed matter theory such as Green functions, Feynman diagrams, functional integration, Keldysh technique, as well as the Berry phase and topology considerations. The approach helps to understand the physics of various quantum phenomena: the Hall conductivity, anomalous Hall and spin Hall effects as well as their quantized versions.

Particular attention is paid to electron transport in disordered crystals and localization effects. The book discusses specific features of topological insulators, graphene, and related materials.

A large part of the text is intended for graduate students in theoretical and experimental physics who have studied electrodynamics, statistical physics, and quantum mechanics. Therefore, in this part of the book, we use the advanced methods of quantum mechanics and field theory.

The textbook relates the solid-state physics fundamentals to the semiconductor device applications and, wherever necessary, includes the auxiliary results from mathematics and quantum mechanics, thus making the book useful also for graduate students in electrical engineering and material science.

Recent development in solid-state physics has resulted in the discovery of new materials which have become application-ready. To explore them in full, it is necessary to use specific theoretical methods. As planned, the textbook describes many of these methods and repeatedly discusses physical phenomena using different levels of presentation making the book accessible for a wide range of prospective readers. Some chapters require just general physics and mathematics courses as a prerequisite, so that part of the book is recommended for undergraduate students in physics and electrical engineering.

Vitalii K. Dugaev
Vladimir I. Litvinov

Authors

Vitalii K. Dugaev, PhD, is a professor at the Department of Physics and Medical Engineering in Rzeszow University of Technology (Poland) since 2006. He earned his MS degree in electrical engineering at Lviv Technical University (Ukraine), PhD and Doctor of Science in physics at Chernivtsi University (Ukraine). He worked as a research fellow for nearly 30 years in the Institute of Materials Science, National Academy of Sciences of Ukraine, Technical University of Lisbon (Portugal), Max Planck Institute for Microstructure Physics (Germany) and Néel Institute (France), and also spent one year as a visiting scientist at the Landau Institute for Theoretical Physics (Russia). He has taught Physics I and II and Solid State Physics at both undergraduate and graduate levels. His main scientific interests are mostly related to the electronic structure of semiconductors and low dimensional structures, in addition to the transport properties of semiconductors and spin-resolved transport.

Vladimir I. Litvinov, PhD, is a principal scientist at the Sierra Nevada Corporation, Irvine, California since 1999. He holds PhD and Doctor of Science degrees in physics from Chernivtsi National University (Ukraine) and Institute of Physics Estonian Academy of Sciences (now the Institute of Physics, University of Tartu, Estonia), respectively. From 1978 to 1995, he was a member and subsequently the head of the theoretical lab at the Institute of Material Science, National Academy of Science of Ukraine. From 1996 to 1999, he was a senior research associate at the Center of Quantum Devices at Northwestern University, Evanston, Illinois, USA. His research interests include solid-state and semiconductor physics, semiconductor spintronics, topological insulators, optoelectronic devices, and millimeter-wave scanning antennas.

Authors

[illegible]

[illegible]

1

Quantum Electron States and Energy Bands

1.1 Crystal Structures

In a crystal, atoms are located at certain positions, forming a lattice-structure characterized by spatial periodicity. Some spatial translations of lattice do not change the crystal structure (if we consider an "infinite crystal"), which means *translational symmetry.* One can present any vector of translation $\boldsymbol{R}$ preserving this symmetry by using the set of *basis vectors* $\boldsymbol{a}$, $\boldsymbol{b}$, $\boldsymbol{c}$:

$$\boldsymbol{R} = n\boldsymbol{a} + m\boldsymbol{b} + l\boldsymbol{c}, \tag{1.1}$$

where n, m, l are arbitrary integer numbers. The set of numbers (nml) specifies the translation vector $\boldsymbol{R}$. The cell of volume $\Omega_0 = \boldsymbol{a} \cdot (\boldsymbol{b} \times \boldsymbol{c})$, formed by the basis vectors, is called the *elementary cell.*

Besides the translational symmetry, there is a point symmetry, which is the group of operations that leaves at least one point unchanged. Point symmetry includes rotations, reflections, inversions, or any combinations of these. Symmetry operations may also include a combination of translation with any of the point transformations. A full set of symmetry operations characterizes the crystal structures from the symmetry standpoint. There is a large number of possible crystal structures that correspond to 230 *spatial groups.*

1.2 Bravais Lattices

Translation invariance allows describing real solid using the *crystal lattice*, or *Bravais lattice* – a set of nodes generated by vectors $\boldsymbol{R}$. It is a geometrical construction, each node of which does not necessarily coincide with the position of a real atom in a solid. One may create a real solid by all possible translations of the set of atoms called the *base* – group of atoms attached to every node of a crystal lattice. In other words, a crystal lattice becomes a real structure once we assign the base to each node. In metals like Cu, Fe, or Al, the base is a single atom, while in elemental semiconductors Si or Ge, it comprises two identical atoms. In III-V semiconductors, the base is also diatomic but including two different atoms, for example, Ga and As.

The analysis of the symmetry types of crystal lattices is much simpler than the classification of real crystalline structures. The symmetry of crystal lattices includes seven crystalline classes, and each of them has several types of Bravais lattices. Each type of Bravais lattice has all the symmetries of the crystal lattice. Fourteen *Bravais cells* present all possible types, some of which one finds in Figure 1.1. For more information, see Ref. [1,2].

One can choose basis vectors $\boldsymbol{a}, \boldsymbol{b}, \boldsymbol{c}$ in different ways. Correspondingly, an elementary cell can have a different shape but the same volume. One of the ways to choose the elementary cell is as follows.

DOI: 10.1201/9780429285929-1

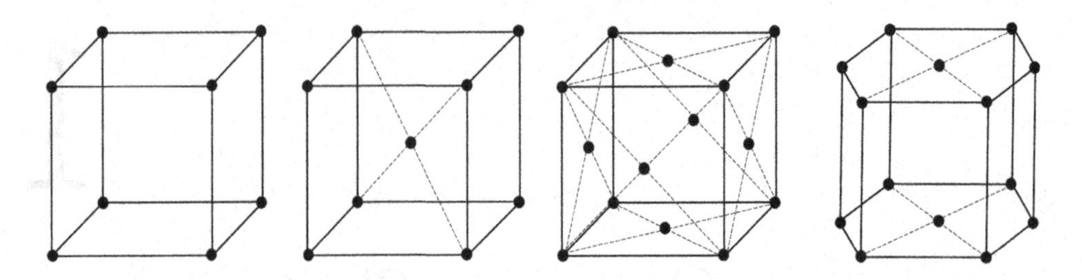

FIGURE 1.1 Several types of Bravais lattices: simple cubic, body-centered cubic (bcc), face-centered cubic (fcc), and hexagonal lattice.

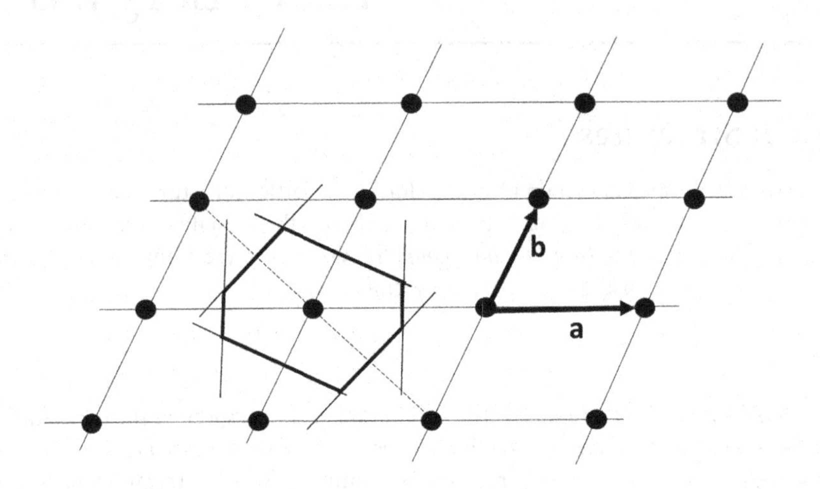

FIGURE 1.2 Wigner–Seitz elementary cell.

Choosing an arbitrary site in the crystal lattice, one connects it to all neighbors by straight lines, then draws the planes perpendicular to each of these lines dividing the line by half. As a result, one obtains the bounded-by-planes cell called *Wigner-Seitz elementary cell* (see Figure 1.2).

1.3 Schrödinger Equation for an Electron in a Periodic Crystal Lattice Potential

To find the wave function which determines the state and corresponding electron energy, we use the wave equation of quantum mechanics describing a nonrelativistic electron in the periodic electric field of ions in a crystal,

$$\left(i\hbar\frac{\partial}{\partial t}-H\right)\Psi(\boldsymbol{r},t)=0, \tag{1.2}$$

where $\Psi(\boldsymbol{r},t)$ is the wave function, and H is the Hamiltonian of an electron in the field of all ions in the lattice,

$$H=-\frac{\hbar^2\Delta}{2m_0}+V(\boldsymbol{r}). \tag{1.3}$$

Here $\Delta=\partial^2/\partial x^2+\partial^2/\partial y^2+\partial^2/\partial z^2$ is the Laplace operator, m_0 is the mass of a free electron, $V(\boldsymbol{r})$ is the potential field of the crystal lattice

$$V(\boldsymbol{r}) = \sum_i \mathrm{v}(\boldsymbol{r} - \boldsymbol{R}_i) \tag{1.4}$$

and $\mathrm{v}(\boldsymbol{r} - \boldsymbol{R}_i)$ is the potential created by a single ion located at the ith site $\mathbf{R}_i$ of the crystal lattice.

The periodicity of the crystalline field means $V(\boldsymbol{r}) = V(\boldsymbol{r} + \boldsymbol{R})$, where $\boldsymbol{R}$ is the lattice translation vector.

The Schödinger equation for the *stationary state* $\psi(\boldsymbol{r})$ of an electron with certain energy ε follows from Eq. (1.2) after substitution $\Psi(\boldsymbol{r},t) = \exp(-i\varepsilon t/\hbar)\psi(\boldsymbol{r})$,

$$(H - \varepsilon)\,\psi(\boldsymbol{r}) = 0. \tag{1.5}$$

Due to the periodicity of function $V(\boldsymbol{r})$, the solution to Eq. (1.5) has the form of Bloch function normalized on the crystal volume:

$$\psi_{\boldsymbol{k}}(\boldsymbol{r}) = e^{i\boldsymbol{k}\cdot\boldsymbol{r}}\, u_{\boldsymbol{k}}(\boldsymbol{r}), \tag{1.6}$$

where Bloch amplitude $u_{\boldsymbol{k}}(\boldsymbol{r})$ has lattice periodicity, and discrete wavevector $\boldsymbol{k}$ numerates the electron states and varies from $-\infty$ to $+\infty$". The lattice periodicity requires vector $\boldsymbol{k}$ to be discrete, and that discerns electron states in a crystal as compared to a free electron – the subject to be discussed below. Wave function (1.6) describes periodically modulated wave propagating in the $\boldsymbol{k}$-direction.

Substituting (1.6) in (1.5), we find that $\psi_{\boldsymbol{k}}(\boldsymbol{r})$ is an eigenfunction of Hamiltonian H as long as the Bloch amplitude $u_{\boldsymbol{k}}(\boldsymbol{r})$ is the solution to the equation,

$$\left(-\frac{\hbar^2\Delta}{2m_0} - \frac{i\hbar^2}{m_0}\boldsymbol{k}\cdot\nabla + \frac{\hbar^2 k^2}{2m_0} + V(\boldsymbol{r}) - \varepsilon\right) u_{\boldsymbol{k}}(\boldsymbol{r}) = 0. \tag{1.7}$$

As potential $V(\boldsymbol{r})$ is lattice-periodic, Eq. (1.7) is invariant to translations $(\boldsymbol{r} \to \boldsymbol{r} + \boldsymbol{R})$ if $u_{\boldsymbol{k}}(\boldsymbol{r})$ is also periodic. The solution to Eq. (1.7) gives us $u_{\boldsymbol{k}}(\boldsymbol{r})$ and corresponding eigenenergy $\varepsilon(\boldsymbol{k})$, which is the energy of an electron in the state described by wavefunction $\psi_{\boldsymbol{k}}(\boldsymbol{r})$. Then the dependence $\varepsilon(\boldsymbol{k})$ is the *energy spectrum* of an electron in the periodic potential $V(\boldsymbol{r})$.

It should be noted that the nonrelativistic Hamiltonian (1.3) does not include operators acting on an electron spin. The spin operators appear in the relativistic quantum mechanics described by the Dirac Hamiltonian. Since the velocity of electrons in semiconductors is much smaller than the speed of light, the spin operators appear in the Hamiltonian in the form of small *spin-orbit corrections*. These corrections determine the electronic structure of some semiconductors made of heavy atoms (like, e.g., Pb atoms in PbTe). They are also at the origin of spin-dependent transport properties like the *anomalous Hall effect* and *spin Hall effect*, the problems to be discussed later in the text.

1.3.1 Symmetry and Classification of Electron Energy States

One can formulate translational invariance in a more general form. If Hamiltonian H is invariant to transformation with unitary operator $\hat{T}$: $\hat{T}H\hat{T}^\dagger = H$, one can use this transformation in (1.5) by applying operator $\hat{T}$ from the left:

$$\hat{T}(H - \varepsilon)\psi = 0, \tag{1.8}$$

$$\hat{T}(H - \varepsilon)\hat{T}^\dagger\hat{T}\psi = 0, \tag{1.9}$$

$$(H - \varepsilon)\hat{T}\psi = 0, \tag{1.10}$$

where we use the property of *unitary operators*, $\hat{T}^\dagger\hat{T} = \hat{T}\hat{T}^\dagger = 1$. Thus, if $\psi(\boldsymbol{r})$ is an eigenfunction of H, then $\hat{T}\psi(\boldsymbol{r})$ is also the eigenfunction of H with the same energy ε.

Multiplying equation $\hat{T}H\hat{T}^{\dagger} = H$ by $\hat{T}$ from the right we get $\left[\hat{T}, H\right] = 0$, where $\left[\hat{T}, H\right] \equiv \hat{T}H - H\hat{T}$ is the commutator of operators H and $\hat{T}$. It means that operators $\hat{T}$ and H have the same eigenfunctions. Thus, the invariance of Hamiltonian to unitary transformation $\hat{T}$ means that $\hat{T}$ commutes with H, or one can say that since H and $\hat{T}$ commute they have the same eigenfunctions.

In particular, the Bloch theorem is related to the translational invariance of the Hamiltonian. Indeed, the Bloch function (1.6) is the eigenfunction of the translation operator $\hat{R}$ that generates shift $\boldsymbol{r} \to \boldsymbol{r} + \boldsymbol{R}$,

$$\hat{R}\,\psi_{\boldsymbol{k}}(\boldsymbol{r}) \equiv \psi_{\boldsymbol{k}}(\boldsymbol{r} + \boldsymbol{R}) = e^{i\boldsymbol{k}\cdot\boldsymbol{R}}\,\psi_{\boldsymbol{k}}(\boldsymbol{r}). \tag{1.11}$$

At the same time, operator $\hat{R}$ commutes with the Hamiltonian H. Therefore, $\psi_{\boldsymbol{k}}(\boldsymbol{r})$ is also the eigenfunction of H.

1.3.2 Reciprocal Lattice and Brillouin Zone

Condition (1.11) follows from the translational symmetry only and defines the wavevector through the eigenvalue of the translation operator. One may handle the wavevector in a crystal in much the same way as it was in free space. However, there are two main differences between a wavevector in free space and the one in a crystal. First, in a crystal, $\boldsymbol{k}$ is a discrete variable. Second, the periodic crystal field introduces an ambiguity in that the wavevector is defined up to the reciprocal lattice vector $\boldsymbol{K}$. Below we consider these differences in more detail.

Since we are dealing with a crystal of finite size, we have to impose boundary conditions on the wave function. There are two options for applying boundary conditions. First, we may equate the wavefunction to zero outside the boundaries of the crystal. That would correspond to taking the surface effects into account. Another option is to consider surface effects negligible and assume that crystal comprises of an infinite number of the periodically repeated parts of volume Ω (volume of a crystal). Then impose the Born–Karman cyclic boundary conditions $\psi_{\boldsymbol{k}}(\boldsymbol{r} + \boldsymbol{L}_i) = \psi_{\boldsymbol{k}}(\boldsymbol{r})$, where $\boldsymbol{L}_i = N_i \boldsymbol{b}_i$ is the linear size of the crystal in the direction of basic Bravais vector $\boldsymbol{b}_i$, N is the number of Bravais cells. Using (1.11), we express the boundary conditions as $e^{i\boldsymbol{k}\cdot\boldsymbol{L}_i} = 1$, and thus the wavevector takes discrete values $\boldsymbol{k}_i = 2\pi m_i / \boldsymbol{L}_i, m_i = 0, \pm 1 \ldots$.

In bulk materials $(L_i \to \infty)$ the wave vector is a quasi-continuous variable, so the exact summation over $\boldsymbol{k}$ can be replaced by the integral: $\sum_{\boldsymbol{k}}(\ldots) \to \Omega \int (\ldots) d^3k/(2\pi)^3$. Formally, to define the reciprocal space, we expand arbitrary lattice periodic function into the Fourier series:

$$\varrho(\boldsymbol{r}) = \sum_{\boldsymbol{K}} \varrho(\boldsymbol{K}) \exp(i\boldsymbol{K}\cdot\boldsymbol{r}) \tag{1.12}$$

Let us expand $\varrho(\boldsymbol{r})$ displaced on the lattice vector $\boldsymbol{R}$:

$$\varrho(\boldsymbol{r} + \boldsymbol{R}) = \sum_{\boldsymbol{K}} \varrho(\boldsymbol{K}) \exp\left(i\boldsymbol{K}\cdot(\boldsymbol{r} + \boldsymbol{R})\right) \tag{1.13}$$

As the translation $\boldsymbol{R}$ cannot change $\varrho(\boldsymbol{r})$ due to the lattice periodicity, the condition $\varrho(\boldsymbol{r}) = \varrho(\boldsymbol{r} + \boldsymbol{R})$ holds

$$\sum_{\boldsymbol{K}} \varrho(\boldsymbol{K}) \exp(i\boldsymbol{K}\cdot\boldsymbol{r}) = \sum_{\boldsymbol{K}} \varrho(\boldsymbol{K}) \exp\left(i\boldsymbol{K}\cdot(\boldsymbol{r} + \boldsymbol{R})\right). \tag{1.14}$$

It follows from Eq. (1.14)

$$\exp(i\boldsymbol{K}\cdot\boldsymbol{R}) = 1. \tag{1.15}$$

According to (1.15), real lattice defined by vectors $\boldsymbol{R}$ generates the *reciprocal lattice* (or dual lattice) defined by vectors $\boldsymbol{K}$. From Eqs. (1.11) and (1.15) one concludes that replacement $\boldsymbol{k} \to \boldsymbol{k} + \boldsymbol{K}$ does not change the Bloch wave function (1.6), so $\boldsymbol{k}$ and $\boldsymbol{k}+\boldsymbol{K}$ are physically equivalent. So, handling electron kinematics, one may use wavevectors in a part of the reciprocal space – the first Brillouin zone.

Equation (1.15) dictates the relation between reciprocal lattice vectors and basic vectors of the crystal lattice as follows:

$$\boldsymbol{K}\cdot\boldsymbol{R}=\boldsymbol{K}\cdot(n\boldsymbol{a}+m\boldsymbol{b}+l\boldsymbol{c})=2\pi(n_1+m_1+l_1), \tag{1.16}$$

$$\boldsymbol{K}\cdot\boldsymbol{a}=2\pi n',\quad \boldsymbol{K}\cdot\boldsymbol{b}=2\pi m',\quad \boldsymbol{K}\cdot\boldsymbol{c}=2\pi l', \tag{1.17}$$

where all n,m,l are arbitrary integers.

Wavevector $\boldsymbol{K}$ corresponds to the wavefunctions with the wavelength λ equal to the *lattice constant*, which is the linear size of the *elementary cell* of the crystal lattice.[1] Thus one can expect the diffraction of corresponding electron waves at the crystal lattice. We will show later that due to diffraction, the dependence $\varepsilon(\boldsymbol{k})$ has discontinuities at wavevectors $\boldsymbol{k}=\boldsymbol{K}$, where $\boldsymbol{K}$ is solutions to (1.17):

$$\boldsymbol{K}=n'\boldsymbol{A}+m'\boldsymbol{B}+l'\boldsymbol{C}, \tag{1.18}$$

where

$$\boldsymbol{A}=2\pi\frac{\boldsymbol{b}\times\boldsymbol{c}}{\Omega_0},\quad \boldsymbol{B}=2\pi\frac{\boldsymbol{c}\times\boldsymbol{a}}{\Omega_0},\quad \boldsymbol{C}=2\pi\frac{\boldsymbol{a}\times\boldsymbol{b}}{\Omega_0} \tag{1.19}$$

are the *basis vectors of the reciprocal lattice,* $\Omega_0=\boldsymbol{a}\cdot(\boldsymbol{b}\times\boldsymbol{c})$ is the volume of the elementary cell, and n',m',l' are the arbitrary integer numbers. Basis vectors $\boldsymbol{A}$, $\boldsymbol{B}$, $\boldsymbol{C}$ generate the reciprocal lattice the same way vectors $\boldsymbol{a}$, $\boldsymbol{b}$, $\boldsymbol{c}$ generate the crystal (direct) lattice. The symmetry of reciprocal lattice does not necessarily coincide with that of the crystal lattice. Choosing an elementary cell in a reciprocal lattice by the Wigner-Seitz method, one obtains the first *Brillouin zone* or just the Brillouin zone. At the edges of the Brillouin zone separated by $\boldsymbol{K}$, the energy spectrum $\varepsilon(\boldsymbol{k})$ experiences discontinuity.

The Bloch electron states (1.6), numbered by vectors $\boldsymbol{k}$, span the whole $\boldsymbol{k}$-space. As mentioned after Eq. (1.15), due to ambiguity in a wavevector, it is enough to enumerate all states within the first Brillouin zone. After reducing the $\boldsymbol{k}$-space to the first Brillouin zone, the energy spectrum $\varepsilon(\boldsymbol{k})$ folds into a set of the *energy bands* $\varepsilon_n(\boldsymbol{k})$, which correspond to Bloch states $\psi_{nk}(\boldsymbol{r})$ defined in the first Brillouin zone. Energy bands are shown in fig.3.

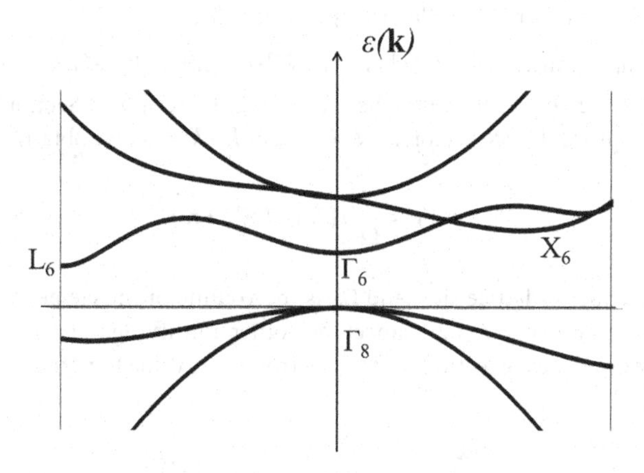

FIGURE 1.3 Example semiconductor band structure with conduction and valence bands separated by the energy gap E_g.

[1] The wavelength λ is related to wavevector k by $\lambda=2\pi/k$.

1.4 The Nearly-Free Electron Approximation

There are several toy models to describe the electron states and to understand better the formation of the energy bands in crystals. One of them is the model of the quasi-free electron. Namely, suppose that the potential $V(\boldsymbol{r})$ is so small that we can use the perturbation theory of quantum mechanics to solve the Schrödinger equation (1.5). To simplify the problem, we assume that the crystalline structure consists of ions of the same type located in the lattice nodes. Also, we consider a finite crystal of volume Ω.

If $V(\boldsymbol{r})=0$ (free electron), the solution of the Schrödinger equation is the plane wave $\psi_{0\mathbf{k}}(\boldsymbol{r})=\exp(i\boldsymbol{k}\cdot\boldsymbol{r})/\sqrt{\Omega}$, and the corresponding eigenvalue is

$$\varepsilon_0(\boldsymbol{k})=\frac{\hbar^2k^2}{2m_0}, \tag{1.20}$$

where m_0 is the free electron mass.

Following the perturbation theory method, we calculate matrix elements of potential $V(\boldsymbol{r})$ using the eigenfunctions of a free electron,

$$\begin{aligned}
V_{kk'} &= \frac{1}{\Omega}\int_{\Omega} d^3r\, e^{-i(k-k')r}\, V(\boldsymbol{r}) = \frac{1}{\Omega}\sum_{R}\int_{\Omega} d^3r\, e^{-i(k-k')r}\, \mathrm{v}(\boldsymbol{r}-\boldsymbol{R}) \\
&= \frac{1}{\Omega}\sum_{R} e^{-i(k-k')R}\int_{\Omega} d^3r\, e^{-i(k-k')(r-R)}\, \mathrm{v}(\boldsymbol{r}-\boldsymbol{R}) \\
&= \frac{1}{\Omega}\sum_{R} e^{-i(k-k')R}\int_{\Omega} d^3r'\, e^{-i(k-k')r'}\, \mathrm{v}(\boldsymbol{r}') \\
&= \frac{\mathrm{v}(\boldsymbol{k}-\boldsymbol{k}')}{\Omega}\sum_{R} e^{-i(k-k')R} \\
&= \frac{\mathrm{v}(\boldsymbol{k}-\boldsymbol{k}')}{\Omega}\sum_{n} e^{-in(k-k')a}\sum_{m} e^{-im(k-k')b}\sum_{l} e^{-il(k-k')c},
\end{aligned} \tag{1.21}$$

where $\mathrm{v}(\boldsymbol{q})$ is the Fourier transform of single-atom potential $\mathrm{v}(\boldsymbol{r})$: $\mathrm{v}(\boldsymbol{q})=\int d^3r\,\mathrm{v}(\boldsymbol{r})\exp(-i\boldsymbol{q}\cdot\boldsymbol{r})$. In the points where $(\boldsymbol{k}-\boldsymbol{k}')\cdot\boldsymbol{a}=2\pi n'$, $(\boldsymbol{k}-\boldsymbol{k}')\cdot b=2\pi m'$, $(\boldsymbol{k}-\boldsymbol{k}')\cdot c=2\pi l'$ all the exponents in (1.21) are equal to unity. Then following the results of Section 2.2, in these points, $\boldsymbol{k}-\boldsymbol{k}'$ is equal to the reciprocal lattice vector: $\boldsymbol{k}-\boldsymbol{k}'=\boldsymbol{K}$. For $\boldsymbol{k}-\boldsymbol{k}'=\boldsymbol{K}$, we obtain

$$V_{kk'}\equiv V_K=\frac{N}{\Omega}\mathrm{v}(\boldsymbol{K})=\Omega_0^{-1}\,\mathrm{v}(\boldsymbol{K}), \tag{1.22}$$

where N is the total number of lattice sites and Ω_0 is the volume of the elementary cell. If $\boldsymbol{k}-\boldsymbol{k}'\neq\boldsymbol{K}$, a large number of oscillating terms in (1.21) cancel each other, nullifying the matrix element: $V_{kk'}\simeq 0$.

Perturbation theory gives the correction to free-electron energy due to perturbation $V(\boldsymbol{r})$ as

$$\Delta\varepsilon(\boldsymbol{k})=\Omega\int\frac{d^3k'}{(2\pi)^3}\frac{|V_{kk'}|^2}{\varepsilon_0(\boldsymbol{k})-\varepsilon_0(\boldsymbol{k}')}, \tag{1.23}$$

where integration runs over the whole reciprocal space. This formula is valid if the correction (1.23) is relatively small, $\Delta\varepsilon(\mathbf{k}) \ll |\varepsilon_0(\mathbf{k})|$. As follows from (1.23), $\Delta\varepsilon(\mathbf{k})$ is not always small as $V_{\mathbf{kk}'} \neq 0$ in the points $\mathbf{k}' = \mathbf{k} + \mathbf{K}$, where the spectrum may be degenerate, $\varepsilon_0(\mathbf{k}) \approx \varepsilon_0(\mathbf{k}')$, in which case simple perturbation approach fails. It happens if $\mathbf{K} \cdot (\mathbf{K}/2 + \mathbf{k}) \approx 0$, meaning that potential $V(\mathbf{r})$ strongly couples states $\mathbf{k} \approx -\mathbf{K}/2$ and $\mathbf{k}' = \mathbf{k} + \mathbf{K} \approx \mathbf{K}/2$ (opposite momenta at the edges of BZ).

To solve the problem, we use the perturbation theory for degenerate levels. Coupling between wavefunctions in points $\mathbf{k}$, in the vicinity of the surface of Brillouin zone, and $\mathbf{k}' = \mathbf{k} - \mathbf{K}$, should be explicitly taken into account by constructing 2×2 matrix representation of the Hamiltonian H on the two-state basis, $\psi_{0\mathbf{k}}(\mathbf{r})$ and $\psi_{0\mathbf{k}'}(\mathbf{r})$:

$$H_{\mathbf{kk}'} = \begin{pmatrix} \varepsilon_0(\mathbf{k}) & V_K \\ V_K^* & \varepsilon_0(\mathbf{k}') \end{pmatrix}. \tag{1.24}$$

Eigenvalues ε of this matrix can be found from $\det(H_{\mathbf{kk}'} - \varepsilon) = 0$, which gives us

$$\varepsilon^2 - \varepsilon\left[\varepsilon_0(\mathbf{k}) + \varepsilon_0(\mathbf{k}')\right] + \varepsilon_0(\mathbf{k})\varepsilon_0(\mathbf{k}') - |V_K|^2 = 0. \tag{1.25}$$

Solving this equation, we obtain

$$\varepsilon_{1,2} = \frac{\varepsilon_0(\mathbf{k}) + \varepsilon_0(\mathbf{k}')}{2} \pm \sqrt{\left(\frac{\varepsilon_0(\mathbf{k}) - \varepsilon_0(\mathbf{k}')}{2}\right)^2 + |V_K|^2}. \tag{1.26}$$

If $|\varepsilon_0(\mathbf{k}) - \varepsilon_0(\mathbf{k}')| \gg 2|V_K|$ then

$$\varepsilon_{1,2} \approx \frac{\varepsilon_0(\mathbf{k}) + \varepsilon_0(\mathbf{k}')}{2} \pm \left(\frac{\varepsilon_0(\mathbf{k}) - \varepsilon_0(\mathbf{k}')}{2} + \frac{|V_K|^2}{\varepsilon_0(\mathbf{k}) - \varepsilon_0(\mathbf{k}')}\right) \tag{1.27}$$

or

$$\varepsilon(\mathbf{k}) \simeq \varepsilon_0(\mathbf{k}) + \frac{|V_K|^2}{\varepsilon_0(\mathbf{k}) - \varepsilon_0(\mathbf{k}')},$$

$$\varepsilon(\mathbf{k}') \simeq \varepsilon_0(\mathbf{k}') - \frac{|V_K|^2}{\varepsilon_0(\mathbf{k}) - \varepsilon_0(\mathbf{k}')}. \tag{1.28}$$

In the opposite limit of $|\varepsilon_0(\mathbf{k}) - \varepsilon_0(\mathbf{k}')| \ll 2|V_K|$, we obtain

$$\varepsilon_{1,2} \simeq \frac{\varepsilon_0(\mathbf{k}) + \varepsilon_0(\mathbf{k}')}{2} \pm \left(|V_K| + \frac{\left[\varepsilon_0(\mathbf{k}) - \varepsilon_0(\mathbf{k}')\right]^2}{4|V_K|}\right). \tag{1.29}$$

Discontinuities $2|V_K|$ in $\varepsilon(\mathbf{k})$ correspond to energy gaps, which are related to the interference of two waves of wavelength $\lambda = 2a$ propagating in the opposite directions, a is the lattice constant. The dependence $\varepsilon(\mathbf{k})$ is shown in Figure 1.4.

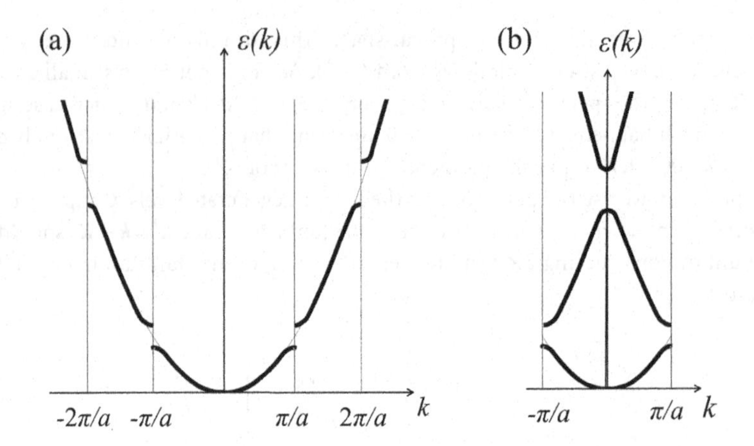

FIGURE 1.4 (a) Energy bands in the quasi-free electron model. (b) Same energy bands folded into the first Brillouin zone.

1.5 The Kronig-Penney Model

To better understand the formation of electron bands in crystals, one can consider the one-dimensional model of an electron in a periodic potential $V(x)$ formed by a series of rectangular wells, as shown in Figure 1.5. Here, the periodicity of potential is a, the width of each well is b, and its depth is V_0. This model can be solved exactly by the standard methods of quantum mechanics.

If the distance between the potential wells is very high ($a \to \infty$), then the electron states are mostly localized within every single well, and an energy spectrum is a set of discrete levels for $\varepsilon < 0$. For finite a, there is tunneling of an electron between neighboring wells, forming the energy bands from discrete levels.

The Hamiltonian of the Kronig-Penney model is

$$H = -\frac{\hbar^2}{2m}\frac{d^2}{dx^2} + V(x). \tag{1.30}$$

Within each period, the solution of the Schrödinger equation, $(H - \varepsilon)\psi(x) = 0$, can be easily found as the function $V(x)$ is constant within the segment. Within the n-th period, the wave function is ($-V_0 < \varepsilon < 0$):

$$\psi_n(x) = \begin{cases} A_n e^{\kappa x} + B_n e^{-\kappa x}, & na + b < x < (n+1)a, \\ C_n e^{ik_1 x} + D_n e^{-ik_1 x}, & na < x < na + b, \end{cases} \tag{1.31}$$

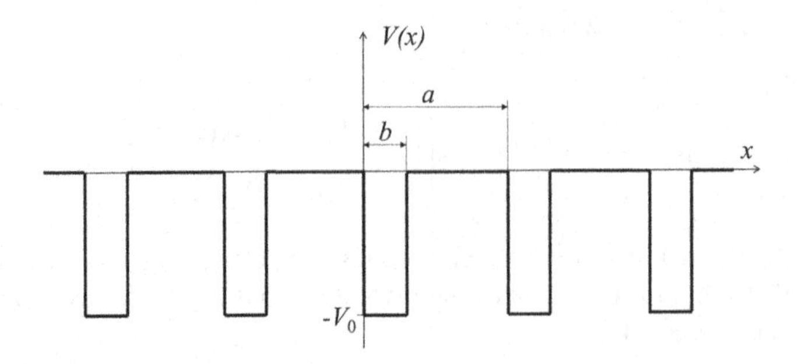

FIGURE 1.5 Potential profile in the Kronig–Penney model.

where

$$\kappa = \sqrt{-2m\varepsilon}/\hbar, \quad k_1 = \sqrt{2m(\varepsilon + V_0)}/\hbar. \tag{1.32}$$

The continuity of wave function and its derivative at $x = na$ and $x = na + b$ gives the following relations between the coefficients:

$$A_{n-1}e^{\kappa na} + B_{n-1}e^{-\kappa na} = C_n e^{ik_1 na} + D_n e^{-ik_1 na}, \tag{1.33}$$

$$A_{n-1}\kappa e^{\kappa na} - B_{n-1}\kappa e^{-\kappa na} = iC_n k_1 e^{ik_1 na} - iD_n k_1 e^{-ik_1 na}, \tag{1.34}$$

$$A_n e^{\kappa(na+b)} + B_n e^{-\kappa(na+b)} = C_n e^{ik_1(na+b)} + D_n e^{-ik_1(na+b)} \tag{1.35}$$

$$A_n \kappa e^{\kappa(na+b)} - B_n \kappa e^{-\kappa(na+b)} = iC_n k_1 e^{ik_1(na+b)} - iD_n k_1 e^{-ik_1(na+b)}. \tag{1.36}$$

Bloch theorem requires wave function to be in the form,

$$\psi_n(x) = e^{ikx} u_n(x), \tag{1.37}$$

where the Bloch amplitude $u_n(x)$ has the periodicity of potential $V(x)$, i.e. $u_{n-1}(x-a) = u_n(x)$. Using (1.31) and (1.37) we obtain the following equations:

$$A_{n-1}e^{\kappa(x-a)+ika} + B_{n-1}e^{-\kappa(x-a)+ika} = A_n e^{\kappa x} + B_n e^{-\kappa x}, \quad na+b < x < (n+1)a, \tag{1.38}$$

$$C_{n-1}e^{ik_1(x-a)+ika} + D_{n-1}e^{-ik_1(x-a)+ika} = C_n e^{ik_1 x} + D_n e^{-ik_1 x}, \quad na < x < na+b, \tag{1.39}$$

which are satisfied if

$$A_{n-1}e^{(ik-\kappa)a} = A_n, \tag{1.40}$$

$$B_{n-1}e^{(ik+\kappa)a} = B_n, \tag{1.41}$$

$$C_{n-1}e^{i(k-k_1)a} = C_n, \tag{1.42}$$

$$D_{n-1}e^{i(k+k_1)a} = D_n. \tag{1.43}$$

Using (1.40) and (1.41) we can present (1.33)–(1.36) as a set of equations for the coefficients A_n, B_n, C_n, D_n. This system has a nonzero solution upon the condition:

$$\det\begin{pmatrix} e^{-ika+\kappa(n+1)a} & e^{-ika-\kappa(n+1)a} & -e^{ik_1 na} & -e^{-ik_1 na} \\ \kappa e^{-ika+\kappa(n+1)a} & -\kappa e^{-ika-\kappa(n+1)a} & -ik_1 e^{ik_1 na} & ik_1 e^{-ik_1 na} \\ e^{\kappa(na+b)} & e^{-\kappa(na+b)} & -e^{ik_1(na+b)} & -e^{-ik_1(na+b)} \\ \kappa e^{\kappa(na+b)} & -\kappa e^{-\kappa(na+b)} & -ik_1 e^{ik_1(na+b)} & ik_1 e^{-ik_1(na+b)} \end{pmatrix} = 0 \tag{1.44}$$

Note that the determinant in (1.44) does not depend on n.

Since k_1 and κ are known functions of energy (1.32), the equation (1.44) determines $\varepsilon(k)$. The numerical solution to (1.44) is shown in Figure 1.6. The coefficients A_n, B_n, C_n, D_n can be determined by using Eqs. (1.33)–(1.36) and (1.40)–(1.43) along with the normalization condition for the wave function.

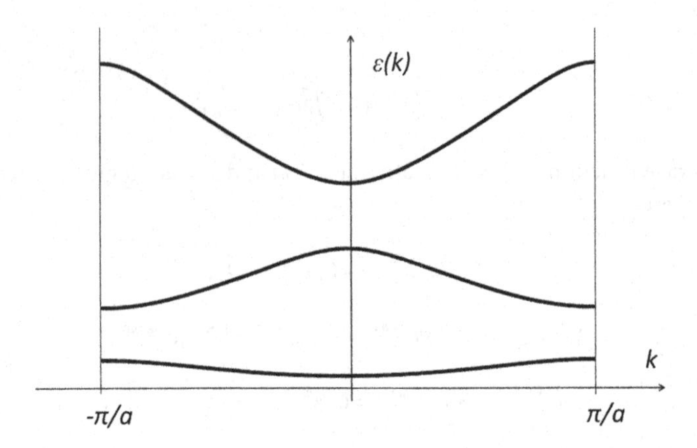

FIGURE 1.6 Energy bands in the Kronig-Penney model.

The model we considered here is the electron propagation in a periodic chain of quantum wells, which helps to understand the origin of energy bands in crystals. The problem can also be seen as the tunneling of an electron through an infinite set of barriers. Any shape of the finite barrier does not prevent the propagation of electron waves if the barriers are arranged periodically.

1.6 The Tight-Binding Approximation

Another approach to electron energy spectrum in solids uses the electronic structure of constituent atoms periodically arranged in the crystal, meaning that it relies on wave functions localized around atomic sites, and in that sense, it is "opposite" to the free-electron model. The Hamiltonian of a single atom located at the point $\boldsymbol{R}_n$ is

$$H_n = -\frac{\hbar^2 \Delta}{2m_0} + \mathrm{v}(\boldsymbol{r} - \boldsymbol{R}_n). \tag{1.45}$$

Let us assume that we know the exact wave functions (atomic orbitals) $\psi_{i,n}(\boldsymbol{r})$ (in other notations, $|i,n\rangle$) and corresponding eigenenergies $\varepsilon_{i,n}$ of the Hamiltonian H_n. Then we can use functions $\psi_{i,n}$ as a basis to find matrix elements of Hamiltonian (1.2). Since the wave functions belonging to different sites are not orthogonal to each other, we should use the basis of orthonormalized functions[2] (see Ref. [3] for details). The main approximation of the method is to use only the matrix elements with the functions of neighboring atoms n, n',

$$\langle i,n|H|j,n'\rangle \equiv \int d^3r\, \psi^*_{i,n}(\boldsymbol{r}) H \psi_{j,n'}(\boldsymbol{r}), \tag{1.46}$$

neglecting overlap of wave functions of the atoms at longer distances.

As an example, we consider a simple two-dimensional model, in which identical atoms reside in the sites of the regular lattice shown in Figure 1.7. Each atom has many-electron states (orbitals) with energies ε_i. The orbital is localized, but tunneling between the atoms let the electron moving across the lattice. The simplest model includes the possibility of tunneling between the neighboring atoms only. Corresponding Hamiltonian in the second quantization representation has the form (for details of the second quantization method, see Appendix 1), https://www.routledge.com/Modern-Semiconductor-Physics-and-Device-Applications/Dugaev-Litvinov/p/book/9780367250829#

---

[2] For example, for slightly overlapping states $|i,n\rangle$ and $|i,n'\rangle$, orthogonalized to $|i,n'\rangle$ state has the form, $|i,n\rangle_{ort} = |i,n\rangle - \frac{1}{2}|i,n'\rangle\langle i,n'|i,n\rangle$. In the following, we imply orthogonalization performed.

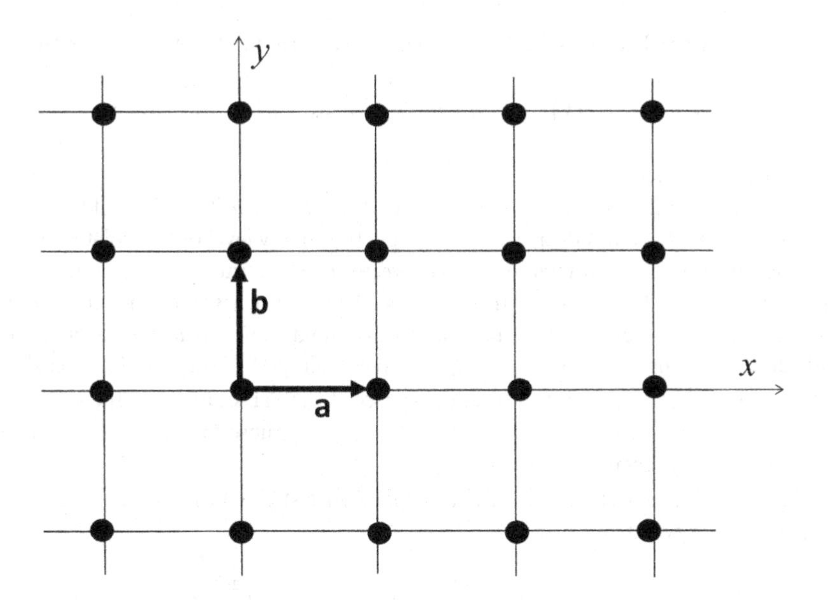

FIGURE 1.7 Two-dimensional regular lattice.

$$H=\sum_{nm}\left[\varepsilon_i\, c^{\dagger}_{nm}c_{nm}-t_i\left(c^{\dagger}_{nm}c_{n+1,m}+c^{\dagger}_{n+1,m}c_{nm}+c^{\dagger}_{nm}c_{n,m+1}+c^{\dagger}_{n,m+1}c_{nm}\right)\right], \quad (1.47)$$

where $c^{\dagger}_{nm}$ and c_{nm} are the operators of creation and annihilation of an electron in ith atomic orbital belonging to the site (nm) of the crystal lattice shown in Figure 1.7 (n and m enumerate the atoms along axes x and y). The first term in (1.47) corresponds to the energy of an electron at the ith orbital; the electron hopping integral t_i appears due to the overlap between ith- and neighboring orbitals.

To determine the eigenvalues, we diagonalize the Hamiltonian (1.47) by using Fourier transforms

$$c_{nm}=\Omega\int\frac{d^2k}{(2\pi)^2}c_k\, e^{ik\cdot R_{nm}}=\Omega\int\frac{d^2k}{(2\pi)^2}c_k\, e^{ik\cdot(na+mb)},$$

$$c^{\dagger}_{nm}=\Omega\int\frac{d^2k}{(2\pi)^2}c^{\dagger}_k\, e^{-ik\cdot R_{nm}}=\Omega\int\frac{d^2k}{(2\pi)^2}c^{\dagger}_k\, e^{-ik\cdot(na+mb)}, \quad (1.48)$$

In k-representation, the Hamiltonian is diagonal:

$$H=NS\int\frac{d^2k}{(2\pi)^2}\left[\varepsilon_i-t_i\left(e^{ik\cdot a}+e^{-ik\cdot a}+e^{ik\cdot b}+e^{-ik\cdot b}\right)\right]c^{\dagger}_{k}c_{k}$$

$$=NS\int\frac{d^2k}{(2\pi)^2}\left\{\varepsilon_i-2t_i\left[\cos(\boldsymbol{k}\cdot\boldsymbol{a})+\cos(\boldsymbol{k}\cdot\boldsymbol{b})\right]\right\}c^{\dagger}_{k}c_{k}, \quad (1.49)$$

where N is the total number of sites in the lattice and S is the area of the elementary cell. Here we used relation

$$\sum_{nm}\exp\left(in(\boldsymbol{k}-\boldsymbol{k}')\cdot\boldsymbol{a}+im(\boldsymbol{k}-\boldsymbol{k}')\cdot\boldsymbol{b}\right)=\begin{cases}N, & \text{if } \boldsymbol{k}'=\boldsymbol{k}+\boldsymbol{K},\\ 0, & \text{if } \boldsymbol{k}'\neq\boldsymbol{k}+\boldsymbol{K}.\end{cases}$$

The expression in curly brackets of Eq. (1.49) is the energy spectrum of the electron system

$$\varepsilon(\boldsymbol{k})=\varepsilon_i-2t_i\left[\cos(k_x a)+\cos(k_y a)\right], \tag{1.50}$$

where a is the lattice parameter.

The Brillouin zone is limited by $-\pi/a\le k_x\le\pi/a$ and $-\pi/a\le k_y\le\pi/a$. The width of the energy band is $8t_i$. In the limit of $t_i\to 0$, the width decreases, and we get the energy level of the isolated atom.

If $t_i\neq 0$, the electron state is characterized by wavevector $\boldsymbol{k}$, which means that the electron is delocalized, and it can move like a free electron. This result is related to the assumed model, in which we consider only one electron in the periodic crystal field (this is called *one-particle approximation*).

The band edges are at the energies $\varepsilon=\varepsilon_i\mp 4t_i$. For these values of energy $\cos(k_x a)+\cos(k_y a)=\pm 2$. It corresponds, respectively, to the point $(0,0)$ at the center of the Brillouin zone (the so-called Γ point) and four points $(\pi/a,\pi/a)$, $(-\pi/a,\pi/a)$, $(\pi/a,-\pi/a)$, $(-\pi/a,-\pi/a)$ (these four points are *equivalent* since they differ by the vectors of reciprocal lattice).

Near the Γ point, at $(k_{x,y}a)\ll 1$, one can use the expansion $\cos(k_{x,y}a)\simeq 1-(k_{x,y}a)^2/2$. From Eq. (1.50), we obtain the low-energy electron spectrum,

$$\varepsilon(\boldsymbol{k})\simeq\varepsilon_i-4t_i+2t_i a^2k^2=\varepsilon_i-4t_i+\frac{\hbar^2k^2}{2m^*}, \tag{1.51}$$

where

$$m^*=\frac{\hbar^2}{4t_i a^2} \tag{1.52}$$

presents the electron *effective mass* near the band edge. The effective mass depends on the lattice parameter explicitly as well as on the hopping matrix element t_i, which drops exponentially when a grows. So, the effective mass tends to infinity when a increases.

The calculations with the Hamiltonian (1.47) can be generalized to take into account the formation of energy bands from the other atomic orbitals. Since the wave functions corresponding to inner atomic orbitals have a small spatial size, their overlap is small, meaning the low-energy bands are narrow. The width of high-energy bands is relatively large, and the bands stemmed from different orbitals may overlap.

In a more sophisticated version of the tight-binding model, one could take into account the matrix elements of Hamiltonian between different orbitals at the nearest-neighbor sites. It turns out that the results obtained by the method can be quite a good approximation for the band structure of real semiconductors.

1.7 The *k-p* Method

The *k-p* method allows us to find the electron wave functions and energy spectrum in the vicinity of a point in the Brillouin zone in which solutions are known.

We start with the Schrödinger equation for Bloch functions,

$$H_0\psi_{n\boldsymbol{k}}(\boldsymbol{r})=E_n(\boldsymbol{k})\psi_{n\boldsymbol{k}}(\boldsymbol{r}),\quad H_0=\frac{p^2}{2m_0}+V(\boldsymbol{r}), \tag{1.53}$$

where m_0 is the free electron mass, $\boldsymbol{p}=-i\hbar\nabla$ is the momentum operator, and $V(\boldsymbol{r})$ is the electron potential energy in the periodic crystal field. Using $\psi_{n\boldsymbol{k}}(\boldsymbol{r})=e^{i\boldsymbol{k}\cdot\boldsymbol{r}}\,u_{n\boldsymbol{k}}(\boldsymbol{r})$ in (1.53), we obtain the *k-p Hamiltonian* and Schrödinger equation in terms of Bloch amplitudes:

$$H_k u_{nk} = E_{nk} u_{nk},$$
$$H_k = H_0 + \frac{\hbar^2 k^2}{2m_0} + \frac{\hbar \mathbf{k} \cdot \mathbf{p}}{m_0}. \tag{1.54}$$

To find the eigenvalues E_{nk}, one has to choose the full set of known orthonormalized functions that create the initial basis on which we can expand the unknown amplitudes $u_{nk}(\mathbf{r})$. If we are looking for the spectrum in the vicinity of point $\mathbf{k}_0$, the set of known Bloch amplitudes $u_{nk_0}(\mathbf{r})$ can serve as the basis wave functions (*Luttinger-Kohn representation*). To make a transition to this representation, we present Bloch amplitude as a linear combination of basis functions,

$$u_{nk}(\mathbf{r}) = \sum_m c_m u_{mk_0}(\mathbf{r}). \tag{1.55}$$

Substitution (1.55) into Eq. (1.54) gives matrix Hamiltonian in Luttinger-Kohn representation, $H_{nn'}(\mathbf{k}) = \int d^3r\, u^*_{nk_0}(\mathbf{r}) H_k u_{n'k_0}(\mathbf{r})$. The matrix dimension is equal to the number of bands in sum (1.55). Within *kp-perturbation theory*, the third term in (1.54) is being treated as a perturbation. The diagonal elements of this matrix are the values ε_{nk_0}. In the off-diagonal matrix elements, one uses the expansion in $(\mathbf{k} - \mathbf{k}_0)$ limiting consideration by the low-order (e.g., linear) terms.

Usually, we can restrict ourselves by a relatively small number of basis functions corresponding to the bands energetically close to the Fermi level as only these bands determine the electrical, transport, and magnetic properties of semiconductors. For example, the 6-band Kane model presents the 6×6 matrix $H_{nn'}(\mathbf{k})$ and describes the electron states in the valence band of III–V semiconductors like GaAs or InSb. For the calculation of matrix elements, one uses the symmetry of the crystal. The elements of crystal symmetry are translations, rotations around crystal axes, reflections in planes (mirror symmetry), and combined symmetry elements like, for example, the rotation with translation. Mathematically, these operations are forming the symmetry *group*.[3] Under any symmetry transformation of the group, an arbitrary function of $\mathbf{r}$ transforms as $\psi(\mathbf{r}) \to \hat{T}\psi(\mathbf{r})$, where $\hat{T}$ is the operator of a symmetry transformation. Operators of rotation and reflection in a plane are forming a subgroup of transformations, which do not change at least one point, *the point group*.

Let us consider a function ψ_{nk_0}, which is the solution of the Schrödinger equation with the Hamiltonian $H_{nn'}(\mathbf{k})$ at $\mathbf{k} = \mathbf{k}_0$. Acting on this function with operators from the point group results in a set of functions $\{\psi_{nk_0}\}$. All of these functions are the solutions of the Schrödinger equation with the same energy as the corresponding operators of symmetry transformation commute with the Hamiltonian. We usually say that the set of functions $\{\psi_{nk_0}\}$ realizes the *representation of the point group of crystal symmetry*. It means that each element of the point group can be represented by the matrix acting on basis functions $\{\psi_{nk_0}\}$.

If we are looking for the spectrum near the Γ-point in the Brillouin zone, the right choice for Bloch amplitudes in $\mathbf{k}_0 = 0$ is *s*- and *p*-atomic orbitals. They are the spherical harmonics $Y_{lm}(\theta, \varphi)$, which are the eigenfunctions of the angular momentum operator $\mathbf{L}$ (where l is the orbital quantum number, m is the magnetic quantum number, θ and φ are Euler angles):

$$u_{n0}(\mathbf{r}) \to \psi_{n,lm}(r, \theta, \varphi) = f_n(r)\, Y_{lm}(\theta, \varphi), \tag{1.56}$$

where $f_n(r)$ is the spherically-symmetric function. This form of wavefunction allows us to determine nonzero matrix elements $H_{nn'}(\mathbf{k})$.

In Section 1.12, we briefly discuss the results of this method in group-IV and III–V semiconductors. Also, in Appendix 2, https://www.routledge.com/Modern-Semiconductor-Physics-and-Device-Applications/Dugaev-Litvinov/p/book/9780367250829# we present details of *k-p* Hamiltonian for wide bandgap (InAlGa)N semiconductors.

[3] In mathematics, the elements $a, b, c, \ldots$ form the group if we can introduce the operation of group multiplication, $a * b = c$; establish the element 1 (group unity), $1 * a = a * 1 = a$; and identify the inverted to a element a^{-1}, so that $a^{-1} * a = a * a^{-1} = 1$.

1.8 Effective Mass of an Electron

In the vicinity of band extremum, one can approximate the energy spectrum by a parabolic function of the wavevector $\boldsymbol{k}$. As an example, for electrons in the conduction band near a minimum at the Γ-point, one can take,

$$\varepsilon(\boldsymbol{k})=\varepsilon_c+\frac{\hbar^2}{2}\left(m^{-1}\right)_{ij}k_ik_j,\quad i,j=x,y,z, \tag{1.57}$$

where ε_c is the band edge energy, and $(m^{-1})_{ij}$ is the *effective mass tensor.* The expression (1.57) is just the expansion of $\varepsilon(\boldsymbol{k})$ near $\boldsymbol{k}=0$.

Note that the energy spectrum (1.57) is the eigenvalue of the *effective Hamiltonian* describing electrons near the minimum of the conduction band,

$$H_{eff}=\varepsilon_c-\frac{\hbar^2}{2}(m^{-1})_{ij}\nabla_i\nabla_j. \tag{1.58}$$

Various properties of semiconductors can be studied using effective Hamiltonians that describe electrons near extremum points in the Brillouin zone. For crystals in an external field, the effective Hamiltonian would contain only the external perturbation with periodic field already included in the effective mass tensor. This approach simplifies the eigenvalue problem and is called the *effective mass theory.*

As follows from (1.57), tensor $(m^{-1})_{ij}$ is symmetric. By the proper choice of the coordinate system, it can be made diagonal.[4] The "proper" choice of the coordinate system means that the axes k_x, k_y, and k_z should be some symmetry axes of the system. Thus, we can present the electron energy spectrum near extremum as

$$\varepsilon(\boldsymbol{k})=\varepsilon_c+\frac{\hbar^2k_x^2}{2m_x^e}+\frac{\hbar^2k_y^2}{2m_y^e}+\frac{\hbar^2k_z^2}{2m_z^e}, \tag{1.59}$$

where m_x^e, m_y^e, m_z^e are the components of electron effective mass in k_x, k_y, k_z directions, respectively.

Similarly, near the edge of the valence band in the Γ point, we obtain,

$$\varepsilon(\boldsymbol{k})=\varepsilon_v-\frac{\hbar^2k_x^2}{2m_x^h}-\frac{\hbar^2k_y^2}{2m_y^h}-\frac{\hbar^2k_z^2}{2m_z^h} \tag{1.60}$$

where ε_v is the energy of the band edge, m_x^h, m_y^h, m_z^h are the components of *hole effective mass* in k_x, k_y, k_z directions.

The symmetry of crystal lattice to certain rotations and reflections makes some components of the effective mass tensor equal. For example, symmetry to $\pi/2$-rotations around x,y,z axes (it corresponds to the point group C_4) gives $m_x=m_y=m_z$. That is the case of a crystal of cubic symmetry, which gives isotropic Hamiltonian of electrons and holes in the vicinity of the Γ-point.

1.9 Electron Density of States

The states of an electron, which confines in a cube of dimensions $L\times L\times L$, are quantized due to zero boundary conditions: the wavefunction equals zero at the boundary. In an infinite crystal, the quantization stems from the *Born-von Karman boundary condition* which imply that an infinite crystal is the

[4] The symmetric tensor has three diagonal and three off-diagonal components. Rotation of the coordinate system implies a proper choice of three angles so one can make zero three off-diagonal components.

periodically repeated 3D-region of volume Ω: $\psi(x,y,z)=\psi(x+L_x,y,z)$, $\psi(x,y,z)=\psi(x,y+L_y,z)$, $\psi(x,y,z)=\psi(x,y,z+L_z)$, where $\Omega = L_x L_y L_z$ is called crystal volume. After applying the boundary conditions, wavevector components k_x, k_y, k_z become discrete (quantized): $k_x = 2\pi n / L_x$, $k_y = 2\pi m / L_y$, $k_x = 2\pi l / L_z$, where n,m,l are integers.

One can calculate the *density of states* of electrons (DOS) in the conduction band (1.59): a number of electron states per unit energy per unit volume,

$$\rho_e(\varepsilon) = \frac{dn_e(\varepsilon)}{d\varepsilon}, \tag{1.61}$$

where $n_e(\varepsilon)$ is the total number of electron states per unit volume in the energy range from ε_c to ε

$$n_e(\varepsilon) = \frac{2}{\Omega}\sum_{\boldsymbol{k}} \theta[\varepsilon - \varepsilon(\boldsymbol{k})] = 2\int \frac{d^3k}{(2\pi)^3}\theta[\varepsilon - \varepsilon(\boldsymbol{k})], \tag{1.62}$$

where we denote $d^3k \equiv dk_x dk_y dk_z$, and the Heaviside function $\theta(x)$ is zero for $x<0$ and 1 if $x>0$. The factor 2 is due to the electron spin – each state with definite $\boldsymbol{k}$ is double-degenerate by spin directions.

With spectrum (1.59), the integral (1.62) is the volume of paraboloid in the k-space divided by $4\pi^3$. Let us measure the energy of electrons in the conduction band from the band edge, $\varepsilon_c = 0$. The axes of paraboloid are $k_{x,\max} = \sqrt{2m_x\varepsilon}/\hbar$, $k_{y,\max} = \sqrt{2m_y\varepsilon}/\hbar$, $k_{z,\max} = \sqrt{2m_z\varepsilon}/\hbar$. If we rescale the axes, introducing $k_i = k_i'\sqrt{m_i/m_c}$, Eq. (1.62) takes the form,

$$n_e(\varepsilon) = 2\int \frac{d^3k'}{(2\pi)^3}\theta[\varepsilon - \varepsilon(\boldsymbol{k}')], \tag{1.63}$$

where $m_c = (m_x m_y m_z)^{1/3}$ is called the *density-of-states effective mass* and $\varepsilon(\boldsymbol{k}') = \hbar^2 k'^2/2m_c$.

Now the integral (1.63) is proportional to the volume of a sphere of the radius $k_{\max} = \sqrt{2m_c\varepsilon}/\hbar$. Correspondingly, we obtain

$$n_e(\varepsilon) = \frac{(2m_c\varepsilon)^{3/2}}{3\pi^2\hbar^3}. \tag{1.64}$$

Calculating the derivative (1.61), we find

$$\rho_e(\varepsilon) = \frac{(2m_c)^{3/2}\varepsilon^{1/2}}{2\pi^2\hbar^3}. \tag{1.65}$$

That is the electron density of states near the edge of the conduction band of electrons with a parabolic energy spectrum. Similarly, in 2D and 1D semiconductors, one finds $\rho_e^{(2D)}(\varepsilon) = m_c/\pi\hbar^2$ and $\rho_e^{(1D)}(\varepsilon) = \sqrt{m_c}/\pi\hbar\sqrt{2\varepsilon}$, respectively.

In a general case of the nonparabolic energy spectrum, one can use the following expression:

$$\rho(\varepsilon) = 2\int \frac{d^3k}{(2\pi)^3}\,\delta[\varepsilon - \varepsilon(\boldsymbol{k})], \tag{1.66}$$

where $\delta(x)$ is the Dirac delta-function.[5] This expression follows from (1.62) as an ε-derivative of the θ-function.

---

[5] The Dirac delta-function $\delta(x)$ is defined as follows: $\delta(x)$ is zero when $x \neq 0$ and ∞ at $x=0$ so that $\int_{-\infty}^{\infty}\delta(x)dx = 1$. For any function $f(x)$ we get $\int_{-\infty}^{\infty} f(x)\delta(x)dx = f(0)$.

1.10 Korringa-Kohn-Rostoker and Ab-Initio Methods

The most reliable and commonly recognized methods of calculation of the electron band structure of semiconductors are *ab-initio* calculations. There are several different methods known as ab initio. One of the most popular is the Korringa-Kohn-Rostoker approach we outline below. Instead of exactly solving the Schrödinger equation (1.5), one can formulate a variational approximation for the functional Λ (a real-valued function depending on the function as an argument):

$$\Lambda = \int d^3 r \psi^*(\boldsymbol{r})\left[\varepsilon - H_0 - V(\boldsymbol{r})\right]\psi(\boldsymbol{r}), \tag{1.67}$$

where $H_0 = -\hbar^2 \Delta / 2m_0$ is the Hamiltonian of free electron and $\psi(\boldsymbol{r})$ is a variational function, which minimizes the value of Λ for electron energy ε.

One can introduce the Green's function operator,

$$G_0(\varepsilon) = (\varepsilon - H_0)^{-1}. \tag{1.68}$$

It corresponds to the Green's function of free electron since it uses the Hamiltonian H_0. Then we can write

$$\begin{aligned}\Lambda &= \int d^3 r \psi^*(\boldsymbol{r})\left[G_0^{-1} - V(\boldsymbol{r})\right]\psi(\boldsymbol{r}) \\ &= \int d^3 r \psi^*(\boldsymbol{r})\left[V(\boldsymbol{r}) - V(\boldsymbol{r})G_0 V(\boldsymbol{r})\right]V^{-1}(\boldsymbol{r})G_0^{-1}\psi(\boldsymbol{r}).\end{aligned} \tag{1.69}$$

The variation of Λ over $\psi^*(\boldsymbol{r})$ gives

$$\frac{\delta\Lambda}{\delta\psi^*(\boldsymbol{r})} = \left[V(\boldsymbol{r}) - V(\boldsymbol{r})G_0 V(\boldsymbol{r})\right]V^{-1}(\boldsymbol{r})G_0^{-1}\psi(\boldsymbol{r}) = 0. \tag{1.70}$$

This operator equation can be presented in a matrix form using the basis comprising localized orbitals,

$$\psi(\boldsymbol{r}) = \sum_{il} c_{il}\,\phi_l(\boldsymbol{r} - \boldsymbol{R}_i) = \sum_{ilk} c_{il}\,\phi_{lk}(\boldsymbol{r})e^{i\boldsymbol{k}\cdot\boldsymbol{R}_i} = \sum_{lk} c_{kl}\,\phi_{kl}(\boldsymbol{r}), \tag{1.71}$$

where index i corresponds to the site location $\boldsymbol{R}_i$, index l identifies the atomic orbital (it includes the set of quantum numbers other than orbital ones), and we denote

$$c_{kl} = \sum_i c_{il}\,e^{i\boldsymbol{k}\cdot\boldsymbol{R}_i}. \tag{1.72}$$

Using the basis of function $\phi_{lk}(\boldsymbol{r})$ Eq. (1.70) can be represented as a set of homogeneous linear equations for the coefficients c_{kl}:

$$\sum_{k'l'}\left\{\left[V(\boldsymbol{r}) - V(\boldsymbol{r})G_0\,V(\boldsymbol{r})\right]V^{-1}(\boldsymbol{r})G_0^{-1}\right\}_{kl,k'l'} c_{k'l'} = 0. \tag{1.73}$$

This set of equations has a nonzero solution if the determinant of the matrix in braces is equal to zero. As the determinant of the product of two matrices equals the product of determinants, the nonzero solution to Eq. (1.73) exists if $\det\Lambda_0 = 0$, where matrix Λ_0 has the form,

$$\left[\Lambda_0\right]_{kl,k'l'} = \left[V(\boldsymbol{r}) - V(\boldsymbol{r})G_0\,V(\boldsymbol{r})\right]_{kl,k'l'}, \tag{1.74}$$

or alternatively

$$[\Lambda_0]_{kl,k'l'} = \left[V(\mathbf{r}) - \int d^3r' V(\mathbf{r}) G_0(\mathbf{r}-\mathbf{r}') V(\mathbf{r}')\right]_{kl,k'l'}, \tag{1.75}$$

where the Green's function $G_0(\mathbf{r}-\mathbf{r}')$ is the solution to the equation

$$G_0^{-1} G_0(\mathbf{r}-\mathbf{r}') = \delta(\mathbf{r}-\mathbf{r}'). \tag{1.76}$$

Using the Fourier transform of operator G_0 defined in Eq. (1.68), one can find the Fourier transform of function $G_0(\mathbf{r})$, i.e., $G_0(\mathbf{k}) = \int d^3r\, e^{-i\mathbf{k}\cdot\mathbf{r}} G_0(\mathbf{r})$. Indeed, we get

$$(\varepsilon - H_0)\, G_0(\mathbf{r}-\mathbf{r}') = \delta(\mathbf{r}-\mathbf{r}'), \tag{1.77}$$

or after Fourier transformation,

$$(\varepsilon - \varepsilon_{\mathbf{k}}) G_0(\mathbf{k}) = 1, \tag{1.78}$$

which gives

$$G_0(\mathbf{k}) = \frac{1}{\varepsilon - \varepsilon_{\mathbf{k}} + i\delta\, sign(\varepsilon - \mu)}, \tag{1.79}$$

where $\varepsilon_{\mathbf{k}} = \hbar^2 k^2 / 2m_0$ is the free-electron spectrum, μ is the chemical potential of electrons, and $i\delta$ is the infinitesimal shift of the pole to the complex plane that guarantees (1.79) is a *causal* Green's function.

The inverse Fourier transform is written as

$$G_0(\mathbf{r}) = \int \frac{d^3k}{(2\pi)^3} \frac{e^{i\mathbf{k}\cdot\mathbf{r}}}{\varepsilon - \varepsilon_{\mathbf{k}} + i\delta\, sign(\varepsilon - \mu)}. \tag{1.80}$$

The solution of equation $\det \Lambda_0 = 0$ defines the electron energy spectrum. In practical calculation, one uses the Green's function (1.80) and the crystal lattice potential $V(\mathbf{r})$ in *muffin-tin* form, which approximates the real potential created by atoms in a crystal.

1.11 Spin–Orbit Interaction

Schrödinger equation (1.5) describes electron states and the energy spectrum in a nonrelativistic approximation, which is well justified due to the relatively small velocity of electrons in solids ($v \approx 10^6$ m/s) as compared to the speed of light, $v \ll c$.

Electron in an eigenstate of the Hamiltonian (1.3) can be in state spin-up or spin-down along the arbitrarily chosen quantization axis. Nonrelativistic approximation ignores the interaction between spin and orbital momenta of an electron and thus cannot explain various magnetic and spin-dependent transport phenomena observed in semiconductors. Taking into account relativistic correction terms, which originate from the Dirac theory of relativistic electron in the limit of $v \ll c$, we find the free-electron-like Hamiltonian that includes the spin-orbit interaction,

$$H = H_{non-rel} - \frac{i\lambda_0^2}{4}\left[\boldsymbol{\sigma} \times (\nabla V)\right] \cdot \nabla, \tag{1.81}$$

where $V(\boldsymbol{r})$ is the periodic lattice potential, $\lambda_0 = \hbar/m_0 c$ and $\boldsymbol{\sigma}$ is the vector with Pauli matrix components $\sigma_x,\sigma_y,\sigma_z$ acting on the spinor wave function of the electron.[6] Constant λ_0 in (1.81) has the dimension of length, and the value $\lambda_c = 2\pi\lambda_0$ is the Compton wavelength of an electron. The spin-orbit interaction (the second term in Eq. (1.81)) mixes pure up- and down-spin states.

In a crystal, the periodic potential makes the electron spectrum a set of the energy bands $\varepsilon_n(\boldsymbol{k})$, where n is the band number. Besides, in the vicinity of the conduction and valence band edges the Hamiltonian looks like that of a free electron where electron mass m_0 is replaced with the effective mass tensor m_{ij} (see Eq. (1.58)). Similarly, the periodic potential transforms the free-electron spin-orbit constant λ_0 into that depending on the magnitude of crystalline potential.

An important example is a two-dimensional *Rashba* Hamiltonian describing the electron gas (electrons in conduction band near the band edge) in a thin semiconducting film on a substrate,

$$H = -\frac{\hbar^2\left(\nabla_x^2 + \nabla_y^2\right)}{2m^*} - i\alpha\left(\sigma_x\nabla_y - \sigma_y\nabla_x\right), \tag{1.82}$$

where the first term is 2D Hamiltonian of a free electron with isotropic effective mass m^* and the second term looks like in Eq. (1.81) with a potential gradient $(\nabla V)_z$ created by the substrate in the perpendicular to film direction. The magnitude of *Rashba constant* α depends on material parameters, and in many cases can be much larger than $\lambda_0^2(\nabla V)_z/4$ in (1.81), and allows engineering of spin effects in artificial semiconductor structures such as quantum wells and heterostructures, creating the base for semiconductor spintronics.

The Rashba Hamiltonian describes the 2D system, in which the symmetry $z \to -z$ is broken. If this symmetry holds, for example, in symmetric III–V quantum wells, the spin–orbit interaction is given by the *Dresselhaus spin-orbit* term, which can be presented as

$$\hat{H} = -\frac{\hbar^2\left(\nabla_x^2 + \nabla_y^2\right)}{2m^*} - i\beta\left(\sigma_x\nabla_x + \sigma_y\nabla_y\right), \tag{1.83}$$

and comes from a small deviation of electron dispersion from the parabolic form.

1.12 Luttinger-Kohn Band Structure: Si, Ge, and GaAs

In the vicinity of the band edge, the energy dispersion is approximately parabolic. The most important examples are Si, Ge, and GaAs semiconductors.

In Ge, the minimum of the conduction band is in point L along the axis [111] of the Brillouin zone shown in Figure 1.3. The energy spectrum in this band referenced to the valence band maximum is

$$\varepsilon_c(\boldsymbol{k}) = E_g + \frac{\hbar^2\left(k_x^2 + k_y^2\right)}{2m_t^e} + \frac{\hbar^2 k_z^2}{2m_l^e}, \tag{1.84}$$

---

[6] The spinor wave functions have two components usually written as a vector $\begin{pmatrix} \varphi \\ \chi \end{pmatrix}$, where $\varphi(\boldsymbol{r})$ and $\chi(\boldsymbol{r})$ are the amplitudes of spin up and spin down states, respectively. The Pauli matrices act on the spinor functions: $\sigma_x = \begin{pmatrix} 0 & 1 \\ 1 & 0 \end{pmatrix}$, $\sigma_y = \begin{pmatrix} 0 & -i \\ i & 0 \end{pmatrix}$, $\sigma_z = \begin{pmatrix} 1 & 0 \\ 0 & -1 \end{pmatrix}$.

where E_g, m_t^e, and m_l^e are the bandgap, transverse, and longitudinal with respect to the [111] direction effective masses, respectively. The valence band consists of *light-hole, heavy-hole,* and *spin-orbit split* subbands. The maximum is in the Γ-point, and near the maximum, the electron dispersion has the form,

$$\varepsilon_{lh}(\boldsymbol{k}) = -\frac{\hbar^2 k^2}{2m_l^h}, \quad \varepsilon_{hh}(\boldsymbol{k}) = -\frac{\hbar^2 k^2}{2m_h^h}, \quad \varepsilon_{so}(\boldsymbol{k}) = -\Delta - \frac{\hbar^2 k^2}{2m_{so}^h}, \tag{1.85}$$

where Δ is the spin-orbit splitting energy.

For Si, Ge, and GaAs, the dispersions are similar. The difference for Si is that the conduction band minimum is near point X along [100] direction, which discerns the transverse and longitudinal effective masses. In GaAs, the conduction band minimum lies Γ-point and thus $m_l^e = m_t^e = m^e$.

If one neglects the spin-orbit split valence band, the structure of degenerate in Γ-point valence bands is described by the *Luttinger model.* This model is based on k-p approximation and includes the basis of four wave functions corresponding to electron states in Γ-point. These states originate from atomic p-orbitals of constituent atoms which carry orbital momentum $L=1$ and spin $S=1/2$. Thus, our basis functions should be eigenfunctions of the total angular momentum $J=3/2$ or $J=1/2$. Four valence bands degenerate in Γ- point, correspond to $J=3/2$, and eigenfunctions $Y_{J,J_z}(\boldsymbol{r})$ correspond to four projections of the total moment on the quantization axis: $J_z = \pm 3/2, \pm 1/2$. The model describes the light- and heavy-hole bands in semiconductors Ge, Si, and III–V family. In the representation which uses four k-p basis functions

$$\psi_{\boldsymbol{k},J,J_z}(\boldsymbol{r}) = \frac{e^{i\boldsymbol{k}\cdot\boldsymbol{r}}}{\sqrt{\Omega}} Y_{J,J_z}(\boldsymbol{r}), \tag{1.86}$$

the *Luttinger Hamiltonian* is a 4×4 matrix:

$$H_{\boldsymbol{k}} = \frac{\hbar^2 k^2}{2m_0}\left(\gamma_1 + \frac{5}{2}\gamma_2\right) - \frac{\hbar^2 \gamma_2}{m_0}(\boldsymbol{J}\cdot\boldsymbol{k})^2, \tag{1.87}$$

where m_0 is the free electron mass, $\gamma_{1,2}$ are the *Luttinger parameters,* and $\boldsymbol{J}$ is the vector which components are 4×4 matrices of angular momentum $J=3/2$:

$$J_x = \begin{pmatrix} 0 & \frac{i\sqrt{3}}{2} & 0 & 0 \\ -\frac{i\sqrt{3}}{2} & 0 & i & 0 \\ 0 & -i & 0 & \frac{i\sqrt{3}}{2} \\ 0 & 0 & -\frac{i\sqrt{3}}{2} & 0 \end{pmatrix}, \quad J_y = \begin{pmatrix} 0 & \frac{\sqrt{3}}{2} & 0 & 0 \\ \frac{\sqrt{3}}{2} & 0 & 1 & 0 \\ 0 & 1 & 0 & \frac{\sqrt{3}}{2} \\ 0 & 0 & \frac{\sqrt{3}}{2} & 0 \end{pmatrix},$$

$$J_z = \begin{pmatrix} 3/2 & 0 & 0 & 0 \\ 0 & 1/2 & 0 & 0 \\ 0 & 0 & -1/2 & 0 \\ 0 & 0 & 0 & -3/2 \end{pmatrix}, \tag{1.88}$$

Hamiltonian (1.87) is spherically symmetric. The spherical model is an approximation, which does not account for a small warping of valence bands in real semiconductors.

If one expands the basis to include the spin-orbit split valence band, the additional two functions are eigenfunctions of total angular momentum $J=1/2$. Then, the basis of six functions generates a 6×6 matrix Hamiltonian of *Luttinger-Kohn model*: the full description of the valence band. Further expanding the basis to include two more functions for the conduction band results in 8×8 matrix *Kane* Hamiltonian [4]:

$$H_k \begin{pmatrix} E_g+\varepsilon_c & 0 & \frac{pk_-}{\sqrt{2}} & \frac{pk_+}{\sqrt{6}} & -\frac{i\sqrt{2}pk_z}{\sqrt{3}} & 0 & \frac{pk_z}{\sqrt{3}} & -\frac{pk_+}{\sqrt{6}} \\ 0 & E_g+\varepsilon_c & 0 & \frac{\sqrt{2}pk_z}{\sqrt{3}} & \frac{ipk_-}{\sqrt{6}} & \frac{ipk_+}{\sqrt{2}} & \frac{pk_-}{\sqrt{6}} & \frac{ipk_z}{\sqrt{3}} \\ \frac{pk_+}{\sqrt{2}} & 0 & F & H & I & 0 & \frac{iH}{\sqrt{2}} & -i\sqrt{2}I \\ \frac{pk_-}{\sqrt{6}} & \frac{\sqrt{2}pk_z}{\sqrt{3}} & H^* & G & 0 & I & \frac{i(G-F)}{\sqrt{2}} & \frac{i\sqrt{3}H}{\sqrt{2}} \\ \frac{i\sqrt{2}pk_z}{sqrt3} & -\frac{ipk_+}{\sqrt{6}} & I^* & 0 & G & -H & -\frac{i\sqrt{3}H^*}{\sqrt{2}} & \frac{i(G-F)}{\sqrt{2}} \\ 0 & -\frac{ipk_-}{\sqrt{2}} & 0 & I^* & -H^* & F & -i\sqrt{2}I^* & -\frac{iH^*}{\sqrt{2}} \\ \frac{pk_z}{\sqrt{3}} & \frac{pk_+}{\sqrt{6}} & -\frac{iH^*}{\sqrt{2}} & -\frac{i(G-F)}{\sqrt{2}} & \frac{i\sqrt{3}H}{\sqrt{2}} & i\sqrt{2}I & -\Delta+\frac{F+G}{2} & 0 \\ \frac{ipk_-}{\sqrt{6}} & -\frac{ipk_z}{\sqrt{3}} & i\sqrt{2}I^* & -\frac{i\sqrt{3}H^*}{\sqrt{2}} & -\frac{i(G-F)}{\sqrt{2}} & \frac{iH}{\sqrt{2}} & 0 & -\Delta+\frac{F+G}{2} \end{pmatrix} \tag{1.89}$$

where

$$\varepsilon_c=\frac{\hbar^2k^2}{2m_c},\quad F=\frac{(L+M)\left(k_x^2+k_y^2\right)}{2}+Mk_z^2,$$

$$G=\frac{F}{3}+\frac{2}{3}\left[M\left(k_x^2+k_y^2\right)+Lk_z^2\right],\quad H=-\frac{N}{\sqrt{3}}\left(k_yk_z+ik_xk_z\right),$$

$$I=\frac{1}{\sqrt{12}}\left[(L-M)\left(k_x^2-k_y^2\right)-2iNk_xk_y\right],\quad k_\pm=k_x+ik_y. \tag{1.90}$$

Parameters m_c, Δ, L, M, N are the matrix elements of the initial Hamiltonian (1.3) on eight basis functions. They can be estimated fitting theory to experimental measurements. The magnitude of Δ is related to the strength of spin-orbit interaction.

Further basis expansion may include remote conduction and valence bands and results in a Hamiltonian matrix of higher dimension. The conduction and valence bands in zinc-blende III–V semiconductors are illustrated in Figure 1.8b.

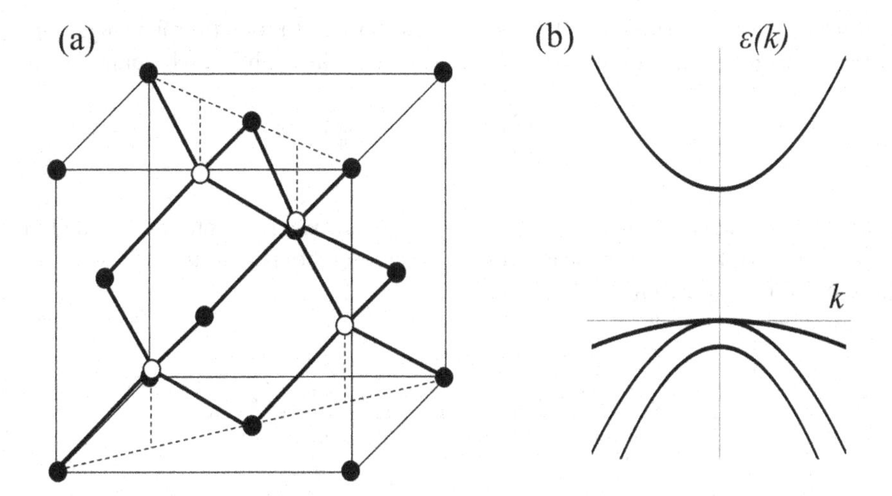

FIGURE 1.8 (a) Crystal lattice and (b) energy bands of GaAs in the vicinity of Γ point.

1.13 Energy Bands in Graphene

Graphene is a two-dimensional honeycomb structure of carbon atoms shown in Figure 1.9. The crystal lattice has hexagonal symmetry, and the elementary cell includes two carbon atoms.

In tight-binding approximation, the Hamiltonian has the form,

$$H=\varepsilon_0\sum_i a_i^\dagger a_i+t\sum_{\langle ij\rangle}a_i^\dagger a_j, \tag{1.91}$$

where ε_0 is the energy of electron localized at the carbon atom, t is the hopping energy describing jumps between neighboring sites, $a_i^\dagger$ and a_i are the creation and annihilation operators, respectively, of an

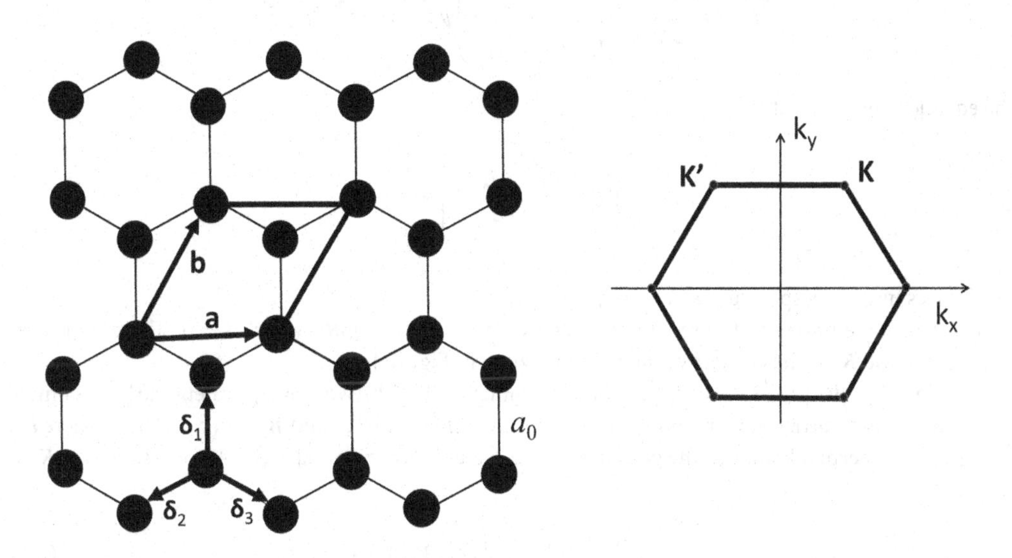

FIGURE 1.9 Two-dimensional crystal lattice and the Brillouin zone of graphene.

electron on site i. The double sum in (1.91) runs over neighboring pairs. In the following, we put $\varepsilon_0 = 0$, meaning the energy reference is from the energy level on site i. The double sum becomes,

$$H = t\sum_{nms}\left(a_{nm}^{\dagger}a_{nms} + a_{nms}^{\dagger}a_{nm}\right), \tag{1.92}$$

where (nm) identifies the site located at $R_{nm} = n\boldsymbol{a} + m\boldsymbol{b}$, whereas (nms) is the site at the second sublattice, i.e., $R_{nms} = n\boldsymbol{a} + m\boldsymbol{b} + \boldsymbol{\delta}_s$. By $\boldsymbol{a}$ and $\boldsymbol{b}$, we denote the basis vectors of the crystal lattice, n and m are integers, and vectors $\boldsymbol{\delta}_s$ $(s = 1,2,3)$ are shown in Figure 1.3.

Making use of Fourier transforms

$$a_{nm} = \Omega\int \frac{d^2k}{(2\pi)^2}\, e^{ik\cdot R_{nm}}\, a_k, \quad a_{nms} = \Omega\int \frac{d^2k}{(2\pi)^2}\, e^{ik\cdot R_{nms}}\, b_k \tag{1.93}$$

in (1.92) gives

$$H = tS\sum_{s}\int \frac{d^2k}{(2\pi)^2}\left(e^{-ik\cdot\delta_s}\, a_k^{\dagger} b_k + e^{ik\cdot\delta_s}\, b_k^{\dagger} a_k\right). \tag{1.94}$$

where S is the area of the unit cell. As defined in (1.93), operators a_k and b_k belong to different sublattices. It is convenient to represent the operators in spinor form,

$$\psi_k = \begin{pmatrix} a_k \\ b_k \end{pmatrix} \tag{1.95}$$

with components corresponding to different sublattices. Hamiltonian (1.94) is a matrix in pseudospin space,

$$H = S\int \frac{d^2k}{(2\pi)^2}\, \psi_k^{\dagger}\begin{pmatrix} 0 & f_k \\ f_k^{*} & 0 \end{pmatrix}\psi_k, \quad f_k = t\sum_{s} e^{-ik\cdot\delta_s}. \tag{1.96}$$

The equation for eigenvalues:

$$\det\begin{pmatrix} -\varepsilon & f_k \\ f_k^{*} & -\varepsilon \end{pmatrix} = 0, \tag{1.97}$$

which gives us energy spectrum $\varepsilon(\boldsymbol{k}) = \pm|f_k|$.

There are some points in the reciprocal space $\boldsymbol{k}$, called *Dirac points*, where $f_k = 0$. These points are labeled as $\boldsymbol{K}$ and $\boldsymbol{K}'$ at the corners of the Brillouin zone in Figure 1.3.

Indeed, as $|\boldsymbol{a}| = |\boldsymbol{b}| = a_0\sqrt{3}$, the area of the elementary cell with two nonequivalent carbon atoms is $S = 3a_0^2$. Reciprocal lattice vectors follow from (1.19): $\boldsymbol{A} = 2\pi\left(\boldsymbol{b}\times\hat{\boldsymbol{c}}\right)/\Omega_0$ and $\boldsymbol{B} = 2\pi\left(\hat{\boldsymbol{c}}\times\boldsymbol{a}\right)/\Omega_0$, where $\hat{\boldsymbol{c}}$ is the unit vector perpendicular to the graphene plane. Calculation gives $|\boldsymbol{A}| = |\boldsymbol{B}| = 4\pi/3a_0$, so for $\boldsymbol{k} = \boldsymbol{K}$, f_k turns zero:

$$f_K = e^{-2\pi i/3} + e^{2\pi i/3} + 1 = 2\cos(2\pi/3) + 1 = 0. \tag{1.98}$$

The same result holds for point $\boldsymbol{K}'$.

Expanding f_k up to linear in $\boldsymbol{k}$ terms near Dirac points $\boldsymbol{K}$ and $\boldsymbol{K}'$ we obtain the Hamiltonian of Dirac electrons,

$$H_k = \begin{pmatrix} 0 & \mathrm{v}\left(\pm k_x - ik_y\right) \\ \mathrm{v}\left(\pm k_x + ik_y\right) & 0 \end{pmatrix}, \tag{1.99}$$

where signs ± correspond to $\boldsymbol{K}$ and $\boldsymbol{K}'$ points, respectively, and $\boldsymbol{k}$ is measured from the corresponding Dirac point. Parameter v is related to the electron velocity as $\mathrm{v}_F = \mathrm{v}/\hbar$.

Eigenvalues of Hamiltonian gives the electron energy spectrum $\varepsilon_{1,2}(k) = \pm \mathrm{v}k$. They are usually called the *Dirac cones*. Since there are six corners in the Brillouin zone in Figure 1.9, we get two Dirac cones belonging to the first Brillouin zone.

1.14 Narrow-Gap Semiconductors

In narrow-gap semiconductors, the energy gap is small that requires both conduction and valence states to be in the set of basis functions for spectrum calculation. That makes the problem similar to the case of nearly degenerated states in quantum mechanics. A good approximation for the band structure calculation is the k-p model.

There are two main classes of narrow-gap semiconductors like $Hg_{1-x}Cd_xTe$ alloys and IV–VI semiconductors like $Pb_{1-x}Sn_xTe$ alloys.

Figure 1.10 shows the crystal lattice, Brillouin zone, and the energy bands of PbTe in the vicinity of point L, the location of the bandgap. The basis functions are the states at the edges of conduction and valence bands in L-point multiplied by $\exp(i\boldsymbol{k}\cdot\boldsymbol{r})$. Due to the spin-orbit interaction, the basis consists of four functions corresponding to two bands and two spin states. The Hamiltonian for electrons is a 4×4 matrix that looks like the relativistic Dirac model[7]:

$$H_k = \begin{pmatrix} \Delta & \mathrm{v}_t\left(\sigma_x p_x + \sigma_y p_y\right) + \mathrm{v}_z\sigma_z p_z \\ \mathrm{v}_t\left(\sigma_x p_x + \sigma_y p_y\right) + \mathrm{v}_z\sigma_z p_z & -\Delta \end{pmatrix} \tag{1.100}$$

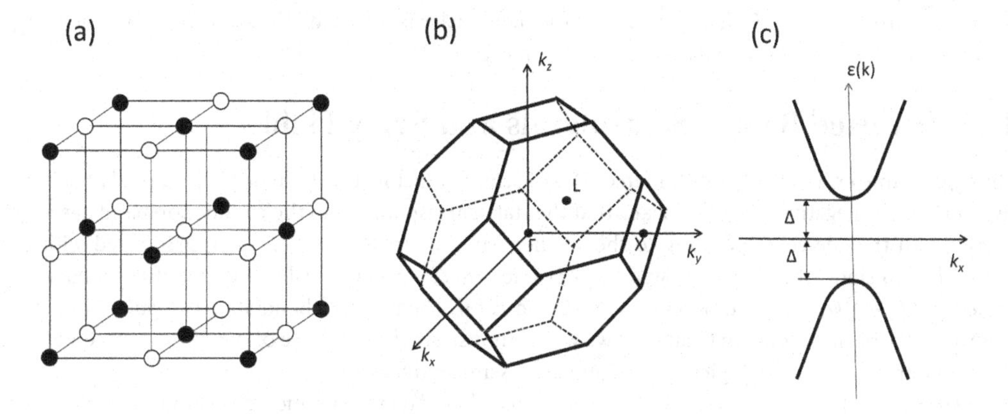

FIGURE 1.10 (a–c) Crystal structure, Brillouin zone, and the band structure of IV–VI semiconductor near the L point.

[7] In quantum electrodynamics, the Dirac Hamiltonian contains $\Delta = mc^2$ and $\mathrm{v}_t = \mathrm{v}_z = c$, where c is the speed of light in vacuum.

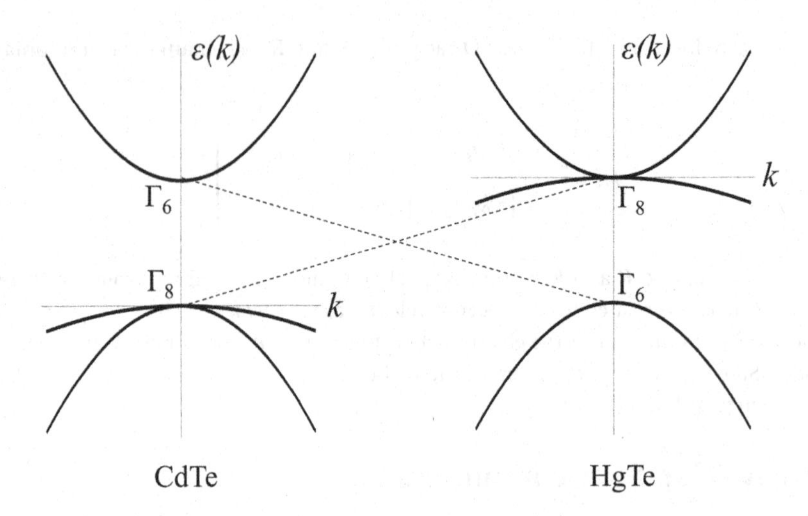

FIGURE 1.11 Energy spectrum in II–VI alloy: normal band position of CdTe vs. inverted bands in HgTe.

where σ_i are Pauli matrices, $\boldsymbol{p}=\hbar\boldsymbol{k}$, Δ is the gap parameter, the axis z is along the [111] direction (in direction to point L), and the parameters v_t and v_z correspond to transversal and longitudinal motion of electron with respect to the z-axis. As illustrated in Figure 1.10, there are eight equivalent L-points and, correspondingly, four valleys since only half of each valley at L belongs to the first Brillouin zone.

The Schrödinger equation is $(H_{\boldsymbol{k}}-\varepsilon)\psi_{\boldsymbol{k}}=0$, where $\psi_{\boldsymbol{k}}$ is the bispinor wavefunction. In (1.100), we restricted ourselves by linear in k terms. The energy spectrum, the following from (1.100) two double spin-degenerate branches

$$\varepsilon_{1,2}(\boldsymbol{p})=\pm\sqrt{\Delta^2+\mathrm{v}_t^2\left(p_x^2+p_y^2\right)+\mathrm{v}_z^2p_z^2}, \tag{1.101}$$

are shown in Figure 1.10.

The typical II–VI narrow gap semiconductors are $Hg_{1-x}Cd_xTe$ alloys, which can be described by the six-band Kane model that includes the conduction band and two valence bands: for the energy spectrum, see Figure 1.11. The variation of alloy composition x changes the energy gap E_g resulting in $E_g\simeq 0$ at $x=x_0\simeq 0.16$. At $x<x_0$, E_g is negative, and the band order is inverted. The transformation of energy bands with a decrease of Cd content is indicated in Figure 1.11 by a dotted line.

1.15 Inverted Bands. Semimetals and Berry Field

The gap parameter Δ in the Hamiltonian of IV–VI semiconductor (1.101) can be positive (PbTe) or negative (SnTe). The negative sign of Δ means that the state corresponding to the conduction band has lower energy than the one corresponding to the valence band. Therefore, negative Δ is associated with the *inverted band* state. The change of sign (Δ) does not affect eigenenergy, but the eigenfunctions determined by $(H-\varepsilon)\psi_{\boldsymbol{k}}=0$ are different for $\Delta>0$ and $\Delta<0$. The inverted band state is *topologically nonequivalent* to the direct one. In Chapter 2 we will demonstrate the existence of specific interface states at the boundary between topologically nonequivalent semiconductors.

The topological classification of electronic systems is based on the symmetry of Hamiltonian and corresponding transformation properties of wave functions. In a general case of matrix Hamiltonian $H_{\boldsymbol{k}}$, one can consider the transformation of eigenfunction $\psi_{\boldsymbol{k}}$ related to the variation of $\boldsymbol{k}$ (i.e., to the motion in the reciprocal space). The corresponding *Berry connection* defines as

$$\boldsymbol{A}(\boldsymbol{k})=-i\left\langle\psi_{\boldsymbol{k}}^{\dagger}\middle|\nabla_{\boldsymbol{k}}\middle|\psi_{\boldsymbol{k}}\right\rangle. \tag{1.102}$$

The integration of the Berry connection on an arbitrary closed path defines the *Berry phase*. Namely, for the circulation on closed contour C, we obtain the following value of the Berry phase,

$$\gamma_C = \oint_C d\mathbf{s} \cdot \mathbf{A}(\mathbf{k}) = \int_{S_C} d\mathbf{S} \cdot \mathbf{B}(\mathbf{k}), \tag{1.103}$$

where $\mathbf{B}(\mathbf{k}) = \nabla_k \times \mathbf{A}(\mathbf{k})$ is the *Berry curvature*, and we used the Stokes theorem to transform the integral on closed contour C to the integration over surface S_C bounded by C. The Berry phase, acquired by the circular electron motion in real space, will be discussed in Chapter 5 concerning the Aharonov–Bohm effect.

The Berry curvature $\mathbf{B}(\mathbf{k})$ characterizes the electron state with given $\mathbf{k}$. As follows from definition, the Berry curvature is associated with the transformation properties of the corresponding wave function ψ_k in the reciprocal space. Berry curvature naturally appears when the wave packet makes a circular motion in coordinate space (see Chapter 7 for details). Using the Berry curvature, one can also define the *monopole density* in the reciprocal space,

$$\rho(\mathbf{k}) = \frac{1}{2\pi} \nabla_k \cdot \mathbf{B}(\mathbf{k}) \tag{1.104}$$

Formally, the Berry connection $\mathbf{A}(\mathbf{k})$ and Berry curvature $\mathbf{B}(\mathbf{k})$ look similar to the usual vector potential $\mathbf{A}(\mathbf{r})$ and magnetic field $\mathbf{B}(\mathbf{r}) = rot\, \mathbf{A}(\mathbf{r})$ in the real space. Unlike a magnetic field existing without magnetic charges (monopoles), the Berry field generates monopoles. The monopoles reveal themselves in the quantum Hall effect.

References

1. N.W. Ashcroft and N.D. Mermin, *Solid State Physics* (Harcourt College, New York, 1976).
2. M.P. Marder, *Condensed Matter Physics* (Willey, New York, 2000).
3. W.A. Harrison, *Electronic Structure and the Properties of Solids* (Dover, New York, 1989).
4. G.L. Bir and G.E. Pikus, *Symmetry and Strain-Induced Effects in Semiconductors* (Wiley, New York, 1974).

2

Quantum Confinement in Semiconductors

2.1 Size Quantization

Low-energy electrons in a single conduction band of a bulk isotropic semiconductor are described by the effective mass Hamiltonian (see Chapter 1):

$$H_{eff} = -\frac{\hbar^2 \Delta}{2m^*}, \tag{2.1}$$

where Δ is the Laplace operator. The spectrum is parabolic, $\varepsilon(\boldsymbol{k}) = \hbar^2 k^2 / 2m^*$, and the eigenfunction has a free-electron form, $f(\boldsymbol{r}) \sim \exp(i\boldsymbol{k} \cdot \boldsymbol{r})$. For small $k \ll |\boldsymbol{K}|$, $\boldsymbol{K}$ being the reciprocal lattice vector, $f(\boldsymbol{r})$ is the long-periodic function of distance, and thus called envelope function. To obtain the Bloch function, one has to multiply $f(\boldsymbol{r})$ by the unit-cell-periodic Bloch amplitude: $\psi(\boldsymbol{r}) = u_0(\boldsymbol{r})\, f(\boldsymbol{r})$. Here, we use Bloch amplitude $u_0(\boldsymbol{r}) \equiv u_{k=0}(\boldsymbol{r})$ as a basis function in the Luttinger-Kohn approach to k-p theory (Chapter 1, Section 1.7). So, H_{eff} is called effective Hamiltonian in the sense that the corresponding Schrödinger equation $(H_{eff} - \varepsilon) f(\boldsymbol{r}) = 0$ is the equation for the envelope function. The solution to the equation with Hamiltonian (2.1) has the form of a free-electron state, in which effective mass m^* carries information on lattice-periodic potential. For electrons in a crystal subjected to external fields, the envelope function $f(\boldsymbol{r})$ is determined by the Schrödinger equation that now includes external field and excludes periodic lattice potential. A practically important instance of an electric field is the built-in potential in bulk semiconductors and heterojunctions. Built-in fields appear in inhomogeneously doped semiconductors and heterojunctions. A heterojunction comprises semiconductor layers of different bandgaps, thus limiting electron motion in the direction perpendicular to layers (growth direction) and giving rise to quantum confinement effects. Below we consider model envelope functions for electrons in various low-dimensional semiconductor structures.

Within the k-p perturbation theory, we account for many bands by expressing total wave function as $\psi(\boldsymbol{r}) = \sum_j u_{j0}(\boldsymbol{r}) f_j(\boldsymbol{r})$, where index j enumerates bands included in the basis set.

Let us consider a semiconductor that is infinite in x and y directions but has a finite thickness d in the z-direction. This model *quasi-2D semiconductor* describes a thin semiconducting film (or a quantum well). The electron wave function is the solution of the Schrödinger equation $(H_{eff} - \varepsilon) f(\boldsymbol{r}, z) = 0$ ($\boldsymbol{r}$ is the 2D vector in the x–y plane) with some boundary conditions at $z = 0, d$. One can assume zero boundary conditions which correspond to infinite barriers for electrons at both surfaces of the film:

$$f(\boldsymbol{r}, z = 0) = f(\boldsymbol{r}, z = d) = 0. \tag{2.2}$$

DOI: 10.1201/9780429285929-2

The solution to the Schrödinger equation with boundary conditions (2.2) is

$$f_{k_z}(\boldsymbol{r})=\sqrt{\frac{2}{Sd}}\exp(i\boldsymbol{k}\cdot\boldsymbol{r})\sin(k_z z), \tag{2.3}$$

where S is the film area, $\boldsymbol{k}$ is the two-dimensional wave vector, and $k_z = n\pi/d$, where $n=1,2,\ldots$ is an integer. The corresponding energy spectrum is a series of parabolic *minibands* labeled by n:

$$\varepsilon_n(\boldsymbol{k})=\frac{\pi^2\hbar^2 n^2}{2m^* d^2}+\frac{\hbar^2 k^2}{2m^*}. \tag{2.4}$$

Since k_z acquires discrete values, the motion of an electron in the z-direction is quantized. That is the *size quantization* effect in a quantum well. If the electron motion is limited in both z and y directions, we obtain the solution

$$f_{k_y,k_z}(\boldsymbol{r})=\frac{2\exp(ik_x x)}{\sqrt{Ld_y d_z}}\sin(k_y y)\sin(k_z z), \tag{2.5}$$

where L is the wire length, $k_y = \pi n_y/d_y$, $k_z = \pi n_z/d_z$, and $n_y, n_z = 1,2,3\ldots$ are the quantum numbers that numerate minibands. The energy spectrum is

$$\varepsilon_{n_y,n_z}(k_x)=\frac{\pi^2\hbar^2 n_y^2}{2m^* d_y^2}+\frac{\pi^2\hbar^2 n_z^2}{2m^* d_z^2}+\frac{\hbar^2 k_x^2}{2m^*}. \tag{2.6}$$

The electron motion quantized in y and z directions models the behavior of electrons in a rectangular quantum wire with infinitely high barriers.

Finally, in the case of a rectangular quantum dot (quantum box) with 3D confinement, we obtain the wave function and spectrum as

$$f_{k_x,k_y,k_z}(\boldsymbol{r})=\frac{2\sqrt{2}}{\sqrt{d_x d_y d_z}}\sin(k_x x)\sin(k_y y)\sin(k_z z), \tag{2.7}$$

$$\varepsilon_{n_x,n_y,n_z}=\frac{\pi^2\hbar^2 n_x^2}{2m^* d_x^2}+\frac{\pi^2\hbar^2 n_y^2}{2m^* d_y^2}+\frac{\pi^2\hbar^2 n_z^2}{2m^* d_z^2}. \tag{2.8}$$

The spectrum consists of discrete energy levels with quantum numbers (n_x, n_y, n_z).

2.1.1 Envelope Functions and Effective Hamiltonian

Using the recipe discussed in the previous section, one may construct the Bloch functions in a quantum well as follows:

$$\psi_{k_z}(\boldsymbol{r})=f_{k_z}(\boldsymbol{r})u_0(\boldsymbol{r})=\frac{\sqrt{2}\exp(i\boldsymbol{k}\cdot\boldsymbol{r})}{\sqrt{Sd}}\sin(k_z z)u_0(\boldsymbol{r}), \tag{2.9}$$

where ***k* is the inplane vector.** Analogously, in quantum wire, where electron motion restricted in both y and z dimensions, the electron wave function is expressed as

$$\psi_{k_y,k_z}(\boldsymbol{r})=\frac{2e^{ik_x x}}{\sqrt{Sd_y d_z}}\sin(k_y y)\sin(k_z z)u_0(x,y,z). \tag{2.10}$$

In the following, we consider some examples for the electrons states in finite-barrier quantum wells.

2.2 Electrons in Quantum Wells

Example quantum well comprises a thin layer of GaAs sandwiched by two thick layers of $Al_xGa_{1-x}As$ alloy. The bandgap of GaAs is about 1.42 eV, whereas in $Al_xGa_{1-x}As$ it is 1.67 eV for $x = 0.2$ (at the temperature of 300 K). The conduction band offset $V_0 \approx 0.15$ eV at $x = 0.2$ (valence band offset at $x = 0.2$ has about the same magnitude) [1]. The band's edge profile in the growth direction z is shown schematically in Figure 2.1.

Conduction and valence band offsets follow either from experiments or from first-principle numerical calculations [2–4].

The effective Hamiltonian for conduction band electrons has the form,

$$H_{eff} = -\frac{\hbar^2 \Delta}{2m^*} + V_c(z), \tag{2.11}$$

where $V_c(z) = 0$ for $0 < z < d$ and $V(z) = V_0$ otherwise, d is the width of the GaAs well. We assume that the effective mass is the same in GaAs and $Al_{0.2}Ga_{0.8}As$ and also that the structure is large in the $x-y$ plane (i.e., the size in x, y directions is much larger than the wavelength of electrons). The in-plane motion ($x-y$ plane) is like the motion of free electrons with effective mass m^*, whereas in the z-direction, electrons dwell in a quantum well of finite height, so their energy is quantized. The energy spectrum has the form

$$\varepsilon_n(\boldsymbol{k}) = \frac{\hbar^2 \left(k_x^2 + k_y^2 + k_z^2\right)}{2m^*} \tag{2.12}$$

where in-plane wavevectors k_x and k_y have arbitrary values in BZ, while k_z is discrete due to the spatial confinement. The value of k_z can be found from equations [5]:

$$\cos\left(\frac{k_z d}{2}\right) = \pm\gamma\left(\frac{k_z d}{2}\right) \quad \text{and} \quad tg\left(k_z d/2\right) > 0 \quad \text{– for } n \text{ odd}, \tag{2.13}$$

$$\sin\left(\frac{k_z d}{2}\right) = \pm\gamma\left(\frac{k_z d}{2}\right) \quad \text{and} \quad tg\left(k_z d/2\right) < 0 \quad \text{– for } n \text{ even}, \tag{2.14}$$

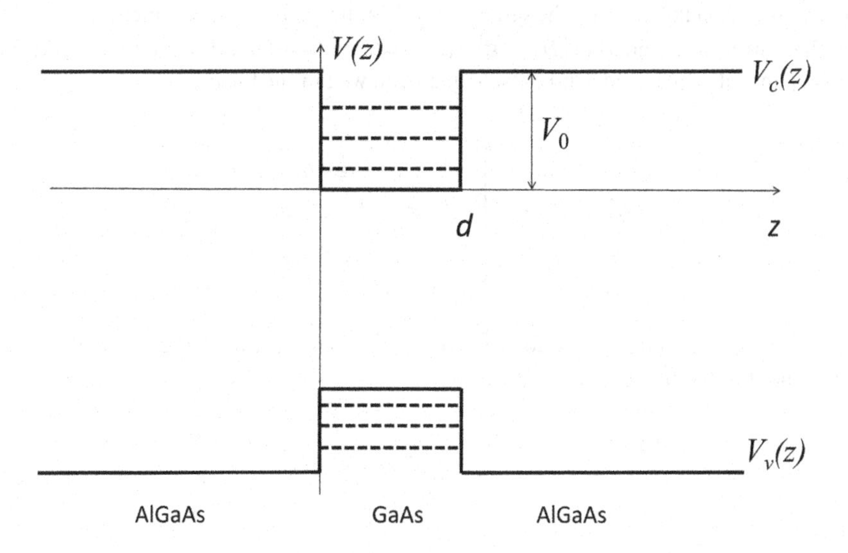

FIGURE 2.1 Conduction and valence bands profiles in AlGaAs/GaAs/AlGaAs quantum well. The dashed lines correspond to the edges of electron minibands.

where $n = 1,2,3,\ldots$ and $\gamma = (\hbar/d)(2/m^* V_0)^{1/2}$. These equations originate from matching the envelope functions $\psi(z)$ and $(1/m^*)\, d\psi(z)/dz$ on both sides of interfaces $x = 0$ and $z = d$. These boundary conditions follow from the integration of the Schrödinger equation in the close vicinity of the interface. If masses are equal on both sides, it reduces to the continuity of the wave function derivative.

2.3 Quantum Well States in the Valence Band

Heavy and light holes in III-V semiconductors are degenerate in the Γ-point, so we have to use the Luttinger Hamiltonian describing *electron states* in two valence bands (see Chapter 1, Section 1.12). We present it here in the form of a 4×4 matrix

$$H_{eff} = -\beta_1 \Delta - \beta_2 (\boldsymbol{J} \cdot \nabla)^2 + V_v(z), \tag{2.15}$$

where $\beta_1 = \hbar^2 (\gamma_1 + 5\gamma_2/2)/2m_0$ and $\beta_2 = -\hbar^2 \gamma_2/m_0$, $V_v(z)$ is the electron potential energy in the valence band. The matrices $\boldsymbol{J}$ of angular momentum $J = 3/2$ are defined in Chapter 1, Eq. (1.88).

For holes, the energy reference is the top of the valence band. Besides, since a hole state is an unoccupied electron state, the hole excitation (i.e., creation of an unoccupied electron state in the valence band) means the transition of an electron to the state with higher energy. In other words, we measure the hole energy in the direction opposite to electrons: positive energy is going downward in the valence band. Then the Hamiltonian of holes in GaAs quantum well can be expressed as

$$H_{eff}^h = \beta_1 \Delta + \beta_2 (\boldsymbol{J} \cdot \nabla)^2 + V_v^h(z), \tag{2.16}$$

where $V_v^h(z) = 0$ for $0 < z < d$ and $V_v^h(z) = V_0^h$ otherwise. In other words, $V_0^h > 0$ is a barrier for holes that prevents their penetration to the AlGaAs -barrier from GaAs quantum well.

The problem reduces to calculating the eigenvalues and eigenvectors of a matrix whose dimension is equal to the number of basis wave functions. One may simplify the problem to avoid numerical calculations. Below we outline successive steps to analytically obtain the energy spectrum and wave functions of holes in a quantum well. Note that the same method applies to H_{eff}^h of any dimension.

To solve the Schrödinger equation $(H_{eff}^h - \varepsilon)\psi(\boldsymbol{r}, z) = 0$ for 4×4 Hamiltonian (light- and heavy-hole bands), we look for solutions inside and outside quantum well in the form,

$$\psi_{\boldsymbol{k}}(\boldsymbol{r}, z) \sim e^{i\boldsymbol{k}\cdot\boldsymbol{r}} \left\{ \begin{array}{ll} e^{\kappa z} \chi_\kappa, & z < 0, \\ e^{ik_z z} \chi_{k_z}, & 0 < z < a, \\ e^{\kappa z} \chi_\kappa, & z > a, \end{array} \right\} \tag{2.17}$$

where χ_κ, χ_{k_z} are the 4-vectors to be found. For $z < 0$, after substituting (2.17) in the Schrödinger equation, we find the equation for χ_κ:

$$\hat{A}_1 \chi_\kappa = 0, \tag{2.18}$$

where

$$\hat{A}_1 = \begin{pmatrix} a_1 - \varepsilon & \sqrt{3}\beta_2 \kappa k_- & \frac{\sqrt{3}\beta_2 k_-^2}{2} & 0 \\ -\sqrt{3}\beta_2 \kappa k_+ & b_1 - \varepsilon & 0 & \frac{\sqrt{3}\beta_2 k_-^2}{2} \\ \frac{\sqrt{3}\beta_2 k_+^2}{2} & 0 & b_1 - \varepsilon & -\sqrt{3}\beta_2 \kappa k_- \\ 0 & \frac{\sqrt{3}\beta_2 k_+^2}{2} & \sqrt{3}\beta_2 \kappa k_+ & a_1 - \varepsilon \end{pmatrix} \tag{2.19}$$

$$a_1 = \beta_1 \left(\kappa^2 - k^2\right) - \frac{3\beta_2 \left(k^2 - 3\kappa^2\right)}{4} + V_0^h, \tag{2.20}$$

$$b_1 = \beta_1 \left(\kappa^2 - k^2\right) - \frac{\beta_2 \left(7k^2 - \kappa^2\right)}{4} + V_0^h, \tag{2.21}$$

and $k_\pm = k_x \pm ik_y$. The set of linear equations (2.18) for the components of 4-vector χ_κ has a nontrivial solution if $\det \hat{A}_1 = 0$. It determines parameter κ as a function of ε and $\boldsymbol{k}$. Equation $\det \hat{A}_1 = 0$ is the 8th order algebraic equation for κ. In the region $0 < \varepsilon < V_0^h$, the equation has eight real roots we denote as $\pm\kappa_1, \pm\kappa_2, \pm\kappa_3, \pm\kappa_4$. For $z < 0$, we choose positive values of κ to guarantee the decay of the wavefunction $\psi_{\boldsymbol{k}}(\boldsymbol{r}, z) \to 0$ at $z \to -\infty$.

The eigenvectors χ_κ follow from (2.18), and proper normalization of $\psi_{\boldsymbol{k}}(\boldsymbol{r}, z)$ requires $\chi_\kappa^\dagger \chi_\kappa = 1$. Then the general solution for $\psi_{\boldsymbol{k}}(\boldsymbol{r}, z)$ at $z < 0$ is

$$\psi_{\boldsymbol{k}}(\boldsymbol{r}, z) = e^{i\boldsymbol{k}\cdot\boldsymbol{r}} \left[C_1 e^{\kappa_1 z} \chi_{\kappa_1} + C_2 e^{\kappa_2 z} \chi_{\kappa_2} + C_3 e^{\kappa_3 z} \chi_{\kappa_3} + C_4 e^{\kappa_4 z} \chi_{\kappa_4} \right], \quad z < 0, \tag{2.22}$$

where C_1, C_2, C_3, C_4 are arbitrary coefficients. Similarly, one can find general solution at $0 < z < d$:

$$\begin{aligned} \psi_{\boldsymbol{k}}(\boldsymbol{r}, z) = e^{i\boldsymbol{k}\cdot\boldsymbol{r}} \Big[& D_1 e^{ik_{z1} z} \chi_{k_{z1}} + D_2 e^{ik_{z2} z} \chi_{k_{z2}} + D_3 e^{ik_{z3} z} \chi_{k_{z3}} + D_4 e^{ik_{z4} z} \chi_{k_{z4}} + D_5 e^{-ik_{z1} z} \chi_{-k_{z1}} \\ & + D_6 e^{-ik_{z2} z} \chi_{-k_{z2}} + D_7 e^{-ik_{z3} z} \chi_{-k_{z3}} + D_8 e^{-ik_{z4} z} \chi_{-k_{z4}} \Big], \quad 0 < z < d, \end{aligned} \tag{2.23}$$

where $\pm k_{z1}, \pm k_{z2}, \pm k_{z3}, \pm k_{z4}$ are the solutions to $\det \hat{A}_2(k_z) = 0$:

$$\hat{A}_2 = \begin{pmatrix} a_2 - \varepsilon & i\sqrt{3}\beta_2 k_z k_- & \frac{\sqrt{3}\beta_2 k_-^2}{2} & 0 \\ -i\sqrt{3}\beta_2 k_z k_+ & b_2 - \varepsilon & 0 & \frac{\sqrt{3}\beta_2 k_-^2}{2} \\ \frac{\sqrt{3}\beta_2 k_+^2}{2} & 0 & b_2 - \varepsilon & -i\sqrt{3}\beta_2 k_z k_- \\ 0 & \frac{\sqrt{3}\beta_2 k_+^2}{2} & i\sqrt{3}\beta_2 k_z k_+ & a_2 - \varepsilon \end{pmatrix} \tag{2.24}$$

$$a_2 = -\beta_1\left(k_z^2 + k^2\right) - \frac{3\beta_2\left(k^2 + 3k_z^2\right)}{4}, \tag{2.25}$$

$$b_2 = -\beta_1\left(k_z^2 + k^2\right) - \frac{\beta_2\left(7k^2 + k_z^2\right)}{4}. \tag{2.26}$$

For $z > d$,

$$\psi_k(\boldsymbol{r}, z) = e^{i\boldsymbol{k}\cdot\boldsymbol{r}}\left[F_1 e^{-\kappa_1 z}\chi_{-\kappa_1} + F_2 e^{-\kappa_2 z}\chi_{-\kappa_2} + F_3 e^{-\kappa_3 z}\chi_{-\kappa_3} + F_4 e^{-\kappa_4 z}\chi_{-\kappa_4}\right], \quad z > d, \tag{2.27}$$

with arbitrary coefficients F_1, F_2, F_3, F_4. As a result, we have 16 unknown coefficients $C_1, \ldots C_4$, $D_1, \ldots D_8$, $F_1, \ldots F_4$. They follow from the continuity of wavefunction $\psi_k(\boldsymbol{r}, z)$ and derivative $\partial\psi_k(\boldsymbol{r}, z)/\partial z$ at $z = 0, d$. As each vector χ_κ or χ_{k_z} has 4 components, we come to a system of 16 linear equations

$$C_1\chi_{\kappa_1} + C_2\chi_{\kappa_2} + C_3\chi_{\kappa_3} + C_4\chi_{\kappa_4} = D_1\chi_{k_{z1}} + D_2\chi_{k_{z2}} + D_3\chi_{k_{z3}} + D_4\chi_{k_{z4}} + D_5\chi_{-k_{z1}} + D_6\chi_{-k_{z2}} + D_7\chi_{-k_{z3}} + D_8\chi_{-k_{z4}} \tag{2.28}$$

$$\begin{aligned} C_1\kappa_1\chi_{\kappa_1} + C_2\kappa_2\chi_{\kappa_2} + C_3\kappa_3\chi_{\kappa_3} + C_4\kappa\chi_{\kappa_4} &= iD_1k_{z1}\chi_{k_{z1}} + iD_2k_{z2}\chi_{k_{z2}} + iD_3k_{z3}\chi_{k_{z3}} + iD_4k_{z4}\chi_{k_{z4}} - iD_5k_{z1}\chi_{-k_{z1}} \\ &\quad - iD_6k_{z2}\chi_{-k_{z2}} - iD_7k_{z3}\chi_{-k_{z3}} - iD_8k_{z4}\chi_{-k_{z4}} \end{aligned} \tag{2.29}$$

$$\begin{aligned} &D_1e^{ik_{z1}d}\chi_{k_{z1}} + D_2e^{ik_{z2}d}\chi_{k_{z2}} + D_3e^{ik_{z3}d}\chi_{k_{z3}} + D_4e^{ik_{z4}d}\chi_{k_{z4}} + D_5e^{-ik_{z1}d}\chi_{-k_{z1}} + D_6e^{-ik_{z2}d}\chi_{-k_{z2}} \\ &+ D_7e^{-k_{z3}d}\chi_{-k_{z3}} + D_8e^{-ik_{z4}d}\chi_{-k_{z4}} = F_1e^{-\kappa_1 d}\chi_{-\kappa_1} + F_2e^{-\kappa_2 d}\chi_{-\kappa_2} + F_3e^{-\kappa_3 d}\chi_{-\kappa_3} + F_4e^{-\kappa_4 d}\chi_{-\kappa_4} \end{aligned} \tag{2.30}$$

$$\begin{aligned} &iD_1k_{z1}e^{ik_{z1}d}\chi_{k_{z1}} + iD_2k_{z2}e^{ik_{z2}d}\chi_{k_{z2}} + iD_3k_{z3}e^{ik_{z3}d}\chi_{k_{z3}} + iD_4k_{z4}e^{ik_{z4}d}\chi_{k_{z4}} - iD_5k_{z1}e^{-ik_{z1}d}\chi_{-k_{z1}} - iD_6k_{z2}e^{-ik_{z2}d}\chi_{-k_{z2}} \\ &- iD_7k_{z3}e^{-k_{z3}d}\chi_{-k_{z3}} - iD_8k_{z4}e^{-ik_{z4}d}\chi_{-k_{z4}} = -F_1\kappa_1e^{-\kappa_1 d}\chi_{-\kappa_1} - F_2\kappa_2e^{-\kappa_2 d}\chi_{-\kappa_2} - F_3\kappa_3e^{-\kappa_3 d}\chi_{-\kappa_3} - F_4\kappa_4e^{-\kappa_4 d}\chi_{-\kappa_4} \end{aligned} \tag{2.31}$$

The determinant of this system of 16 linear algebraic equations should be zero, and this gives us the equation for the energy spectrum of holes in the quantum well. Equations (2.28)–(2.31) along with normalization condition

$$\int d^2r \int_{-\infty}^{\infty} dz\, \psi_k^\dagger(\boldsymbol{r}, z)\psi_{k'}(\boldsymbol{r}, z) = \delta(\boldsymbol{k} - \boldsymbol{k}') \tag{2.32}$$

determine all 16 coefficients and, therefore, the wavefunctions of all hole states in the quantum well.

2.4 2D-Semiconductors

An example of a two-dimensional semiconductor is a thin semiconductor film. Using the simplest quantization model (2.4), one finds that at low temperature, electrons occupy the lowest energy level and thus present 2D electron gas if $(k_BT, E_F) < 3\pi^2\hbar^2/2m^*d^2$.

At $T = 0$, in terms of Fermi wavelength, the 2D character of electrons starts prevailing when $d < \lambda_F \equiv \pi\hbar\sqrt{2}/\sqrt{m^*E_F}$.

Practically, such a 2D system can be grown as a semiconductor film on a substrate or as an asymmetric quantum well sandwiched between two barriers of different heights and thicknesses. In both instances, the built-in potential breaks spatial inversion symmetry, and, if the spin-orbit interaction is essential in the quantum well material, the inversion asymmetry gives rise to the spin-splitting in an electron spectrum (see Chapter 1 Eq. (1.11)). Then 2D electrons can be described by Rashba Hamiltonian,

$$H_k = \frac{\hbar^2 k^2}{2m^*} + \alpha\left(\sigma_x k_y - \sigma_y k_x\right) = \begin{pmatrix} \varepsilon_k & i\alpha k_- \\ -i\alpha k_+ & \varepsilon_k \end{pmatrix}, \tag{2.33}$$

where α is the constant of Rashba spin-orbit interaction, $k_\pm = k_x \pm i k_y$, and $\varepsilon_k = \hbar^2 k^2 / 2m^*$. The equation for eigenvalues follows:

$$\det\begin{pmatrix} \varepsilon_k - \varepsilon & i\alpha k_- \\ -i\alpha k_+ & \varepsilon_k - \varepsilon \end{pmatrix} = 0, \tag{2.34}$$

and gives

$$\varepsilon_{1,2}(k) = \varepsilon_k \pm \alpha k. \tag{2.35}$$

Thus, the spectrum consists of two energy bands shown in Figure 2.2. The picture demonstrates $\boldsymbol{k}$-dependent *spin splitting* due to spin–orbit interaction.

The corresponding eigenvectors of Hamiltonian H_k are the solution to the Schrödinger equation

$$\begin{pmatrix} \varepsilon_k - \varepsilon & i\alpha k_- \\ -i\alpha k_+ & \varepsilon_k - \varepsilon \end{pmatrix}\begin{pmatrix} \varphi_k \\ \chi_k \end{pmatrix} = 0, \tag{2.36}$$

where the wave function ψ_k has the spinor form

$$\psi_k = \begin{pmatrix} \varphi_k \\ \chi_k \end{pmatrix}. \tag{2.37}$$

Matrix equation (2.36) can be written as two equations for φ_k and χ_k:

$$\begin{aligned} &(\varepsilon_k - \varepsilon)\varphi_k + i\alpha k_- \chi_k = 0, \\ &-i\alpha k_+ \varphi_k + (\varepsilon_k - \varepsilon)\chi_k = 0. \end{aligned} \tag{2.38}$$

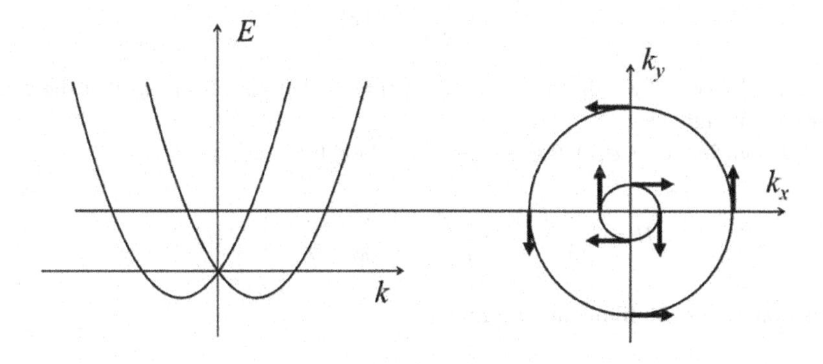

FIGURE 2.2 Electron energy spectrum in a 2D semiconductor with Rashba spin-orbit interaction. The right panel shows how the spin orientation of electrons is related to $\boldsymbol{k}$ in spin-split energy bands.

From (2.38), we have

$$\chi_k = \frac{i\alpha k_+}{\varepsilon_k - \varepsilon}\varphi_k, \tag{2.39}$$

And, after using (2.35), obtain

$$\chi_k^{(1,2)} = \mp\frac{ik_+}{k}\varphi_k = \mp\, i\varphi_k e^{i\phi}, \tag{2.40}$$

where in polar coordinates $k_\pm = k\exp(\pm i\phi)$, $tg\,\phi = k_y/k_x$. Finally, the normalized eigenfunctions of Hamiltonian (2.33) corresponding to eigenvalues $\varepsilon_{1,2}(k)$ are

$$\psi_k^{(1,2)} = \frac{1}{\sqrt{2}}\begin{pmatrix} 1 \\ \mp i\exp(i\phi) \end{pmatrix}. \tag{2.41}$$

Let us consider the state $|1,\boldsymbol{k}\rangle$ with $k = (k,0)$. For this state $\phi = 0$ and the wave function is

$$\psi_k^{(1)} = \frac{1}{\sqrt{2}}\begin{pmatrix} 1 \\ -i \end{pmatrix}. \tag{2.42}$$

This spinor is the eigenfunction of σ_y operator with $S_y = -1$

$$\sigma_y\psi_k^{(1)} = \frac{1}{\sqrt{2}}\begin{pmatrix} 0 & -i \\ i & 0 \end{pmatrix}\begin{pmatrix} 1 \\ -i \end{pmatrix} = -\psi_k^{(1)}. \tag{2.43}$$

Thus, the electron in this state has spin $S_y = -1$ (in units of $\hbar$). Similarly, the state $|2,\boldsymbol{k}\rangle$ with $k = (k,0)$ is the eigenfunction of σ_y, corresponding to $S_y = 1$. The orientation of axes can be chosen arbitrarily, as long as the system has circular symmetry. Therefore, any state $|i,\boldsymbol{k}\rangle$, where $i = 1,2$, can be characterized by spin perpendicular to vector $\boldsymbol{k}$, as illustrated in Figure 2.2.

2.5 Quantum Wires and 1D-Semiconductors

Now we consider the energy spectrum of electrons in a thin cylindric semiconducting wire of radius R. The effective Hamiltonian has the form,

$$H = -\frac{\hbar^2\Delta}{2m^*} + V(\boldsymbol{r}), \tag{2.44}$$

where the potential $V(\boldsymbol{r}) = 0$ within the wire and $V(\boldsymbol{r}) = \infty$ otherwise. It means that the electrons are locked in the wire of radius R.

In cylindrical coordinates (r,ϕ,z), the Hamiltonian (2.44) has the form:

$$H = -\frac{\hbar^2}{2m^*}\left(\frac{\partial^2}{\partial r^2} + \frac{1}{r}\frac{\partial}{\partial r} + \frac{1}{r^2}\frac{\partial^2}{\partial\phi^2} + \frac{\partial^2}{\partial z^2}\right) + V(r). \tag{2.45}$$

The eigenfunction of this Hamiltonian is given by

$$\psi(r,\phi,z) = \frac{e^{ik_z z}}{\sqrt{2\pi}}\frac{e^{im\phi}}{2\pi}f(r), \tag{2.46}$$

where m (azimuthal quantum number) is an integer.

Substituting (2.46) into the Schrödinger equation with Hamiltonian (2.44), we obtain an equation for the radial function $f(r)$:

$$\left[-\frac{\hbar^2}{2m^*}\left(\frac{d^2}{dr^2}+\frac{1}{r}\frac{d}{dr}-\frac{m^2}{r^2}-k_z^2\right)+V(r)-\varepsilon\right]f(r)=0. \tag{2.47}$$

For $r < R$, we get

$$\left(\frac{d^2}{dx^2}+\frac{1}{x}\frac{d}{dx}-\frac{m^2}{x^2}+1\right)f(x)=0, \tag{2.48}$$

where we introduced the dimensionless variable $x = kr$ with $k=\sqrt{2m^*\varepsilon/\hbar^2-k_z^2}$ (assuming $\varepsilon > \hbar^2 k_z^2/2m^*$). One can present (2.48) in the standard form of the Bessel equation:

$$x^2 f'' + xf' + \left(x^2 - m^2\right)f = 0. \tag{2.49}$$

In other words, we have found the solution $f(r) = J_m(kr)$, where $J_m(kr)$ is the Bessel function, which is not divergent at $r \to 0$.[1]

The energy spectrum of electrons in the wire follows from zeroes of the oscillating function $J_m(kr)$ with $r = R$:

$$J_m\left(R\sqrt{\frac{2m^*\varepsilon}{\hbar^2}-k_z^2}\right)=0, \tag{2.50}$$

That is the zero boundary condition on the surface of the cylinder, meaning that electrons are locked in the wire. For each m, the equation has many solutions $\varepsilon_{nm}(k_z)$, where n numerates size-quantized energy levels.

2.6 3D-Confinement: Quantum Dots

By method discussed in the previous section, we solve the Schrödinger equation in a spherical *quantum dot*, which confines electrons in three dimensions. Naturally, the quantization can be of importance when the quantum dot size is of the order of electron wavelength. The 3D effective Hamiltonian is given by

$$H = -\frac{\hbar^2}{2m^*}\left(\frac{\partial^2}{\partial r^2}+\frac{2}{r}\frac{\partial}{\partial r}-\frac{\hat{l}^2}{r^2}\right)+V(r), \tag{2.51}$$

where we used spherical coordinates (r,θ,ϕ), spherically symmetric potential $V(r)$, and $\hat{l}^2$ is the operator of the square of angular momentum,

$$\hat{l}^2 = -\frac{1}{\sin\theta}\frac{\partial}{\partial\theta}\left(\sin\theta\frac{\partial}{\partial\theta}\right)-\frac{1}{\sin^2\theta}\frac{\partial^2}{\partial\phi^2}. \tag{2.52}$$

[1] Bessel function for $m \geq 0$ and small argument $J_m(x) \approx (x/2)^m/\Gamma(m+1)$, where $\Gamma(m)$ is the Gamma function. For negative order m one can use the relation $J_{-m} = (-1)^m J_m(x)$.

As $\hat{l}^2$ commutes with angular momentum $\hat{l}$, they have common eigenfunctions. Due to the spherical symmetry of our problem, as an angular part of the solution to the Schrödinger equation $\psi(r,\theta,\phi)$, we take spherical functions, the eigenfunctions of the angular momentum operator:

$$\hat{l}^2 Y_{lm}(\theta,\phi) = l(l+1) Y_{lm}(\theta,\phi),$$

$$\psi(r,\theta,\phi) = Y_{lm}(\theta,\phi) f(r). \tag{2.53}$$

Substituting (2.53) to the Schrödinger equation with Hamiltonian (2.51), we obtain the equation for the radial function $f(r)$:

$$\left[-\frac{\hbar^2}{2m^*}\left(\frac{d^2}{dr^2} + \frac{2}{r}\frac{d}{dr} - \frac{l(l+1)}{r^2}\right) + V(r) - \varepsilon\right] f(r) = 0. \tag{2.54}$$

We assume that the potential $V(r)$ forms the spherical quantum well with an infinite potential barrier,

$$V(r) = \begin{cases} 0, & r < R, \\ \infty, & r \geq R. \end{cases} \tag{2.55}$$

Then Eq. (2.54) for $r < R$ is presented in the form,

$$x^2 f'' + 2xf' + \left[x^2 - l(l+1)\right] f = 0, \tag{2.56}$$

where $x = kr$ and $k = \sqrt{2m^*\varepsilon/\hbar}$. Equation (2.56) determines the spherical Bessel function $f(r) = j_l(kr)$.[2] Zeroes of this function at $r = R$ (boundary condition for infinite barrier) determine the energy levels of the localized states

$$j_l\left(\frac{R\sqrt{2m^*\varepsilon}}{\hbar}\right) = 0. \tag{2.57}$$

Numerical solutions ε_{nl} to Eq. (2.57) for each l give the energy spectrum of the quantum dot. Note that each level is degenerate because all the states with the same values of (n, l), but different angular momentum m, have the same energy. The degeneracy is the consequence of the spherical symmetry.

2.7 Electrons States on Surfaces and Interfaces: Tamm States

Formally, one could view a surface of a semiconductor as a defect (perturbation) added to the Hamiltonian of the ideal bulk crystal. This perturbation may create surface electron states stemming from the disruption of chemical bonds on the surface. For example, in bulk silicon, each atom is bound to four nearest neighbors by covalent bonds formed with four valence electrons in atomic configuration $3s^2 3p^2$. At the surface, bonds break, and there appear "excess" electrons not bound to Si atoms. It means that the energy $\Delta\varepsilon_{ex}$, needed to excite these electrons to the conduction band, is much smaller than the activation energy of valence electrons. In other words, the band edge for electrons at the surface is at the energy level $\varepsilon_s = \varepsilon_c - \Delta\varepsilon_{ex}$, where ε_c is the conduction band edge. If this level falls in the bandgap, it corresponds to electrons moving along the surface and localized in the perpendicular direction.

When we use the effective Hamiltonian for bulk electrons, one can form crystal termination by introducing the surface potential that attracts electrons and thus creates localized states.

[2] This Bessel function is not divergent at $x \to 0$ for $l \geq 0$, namely $j_l(x) \sim x^l$ when $x \to 0$.

The one-dimensional model of semiconductor surface at $z = 0$ describes the motion of electrons in the z-direction:

$$H = \begin{cases} -\frac{\hbar^2}{2m}\frac{d^2}{dz^2} + V(z), & z < 0, \\ -\frac{\hbar^2}{2m}\frac{d^2}{dz^2} + V_0, & z \geq 0, \end{cases} \tag{2.58}$$

where m is the free-electron mass, $V(z)$ is the periodic lattice potential, and V_0 is the potential barrier for electrons in the crystal. Consider one of the solutions to the Schrödinger equation at $z < 0$ corresponding to incoming to surface Bloch wave propagating inside semiconductor:

$$\psi_k(z) = a_k u_k(z)\exp(ikz), \quad z < 0, \tag{2.59}$$

where $u_k(z)$ is the Bloch amplitude, and a_k is a constant. Outside the crystal, the wavefunction is the decaying function of z:

$$\psi_k(z) = b_k \exp(-\kappa z), \quad z > 0, \tag{2.60}$$

where $\kappa > 0$. Constants a_k and b_k follow from the conditions of continuity for the wave function and its derivative at $z = 0$.

If the energy of the electron state, described by (2.59), (2.60), falls in the conduction or valence band, then the wavenumber k is real. Let us assume that (2.59) and (2.60) also describe the surface states in the bandgap. For these states, k is a complex number: $k_r - ik_i$, with real k_r and k_i. For $k_i > 0$, the wave function (2.59) decays when $z \to -\infty$.

The boundary conditions at $z = 0$ have the form:

$$a_k u_k(0) = b_k, \tag{2.61}$$

$$(ik_r + k_i)a_k u_k(0) + a_k u'_k(0) = -\kappa b_k, \tag{2.62}$$

where $u'_k(0) \equiv du_k(z)/dz\big|_{z\to 0}$. Substituting b_k from the first equation to the second one, we obtain

$$(ik_r + k_i + \kappa)u_k(0) + u'_k(0) = 0,$$

or

$$ik_r + k_i + \kappa = -\frac{d\log u_k(z)}{dz}\bigg|_{z\to 0} \equiv \alpha + i\beta, \tag{2.63}$$

where α, β are the real and imaginary parts of the logarithmic derivative. As follows from the Schrödinger equation (2.58) at $z > 0$, κ is related to the electron energy by $\varepsilon = V_0 - \hbar^2\kappa^2/2m$. Following (2.63), we find $\kappa = \alpha + i\beta - k_i - ik_r$. As κ is real, the allowed energy level in the bandgap (where $k_i \neq 0$) is $\varepsilon = V_0 - \hbar^2(\alpha - k_i)^2/2m$. Wave functions pin to the surface:

$$\psi_k(z) = \begin{cases} b_k \exp(-\kappa z), & z > 0, \\ u_k(z)\exp(ik_r z + k_i z), & z < 0. \end{cases} \tag{2.64}$$

The existence of surface states in the bandgap was theoretically predicted in 1932 by Igor Tamm. Surface states described by this model are called *Tamm surface states*.

2.8 Topological Insulator and Surface States. Gapless Dirac Fermions

As was discussed in Chapter 1, the energy gap parameter in a semiconductor can be positive or negative, not affecting the electron energy spectrum. For example, 2D Dirac Hamiltonian for electrons in thin IV-VI semiconductor layers is a 4×4 block-diagonal matrix:

$$H_k \equiv H_k^{(1)} \oplus H_k^{(2)} = \begin{pmatrix} \Delta & vk_- \\ vk_+ & -\Delta \end{pmatrix} \oplus \begin{pmatrix} -\Delta & vk_- \\ vk_+ & \Delta \end{pmatrix}, \tag{2.65}$$

where the symbol $\oplus$ means the direct sum, 2Δ is the bandgap. Eigenvalues give a double-spin-degenerate two-band energy spectrum $\varepsilon_{1,2}(k) = \pm\left(\Delta^2 + v^2k^2\right)^{1/2}$, which does not depend on the sign of Δ. When two semiconductors, one with $\Delta > 0$ (ordinary insulator) and another one with $\Delta < 0$ (band-inverted insulator), are brought into contact, the gapless electron states exist at the interface. To justify the statement, we consider the model with the interface at $x = 0$:

$$\Delta(x) = \begin{cases} \Delta = -\Delta_0, & x < 0, \\ \Delta = \Delta_0, & x > 0 \end{cases}, \tag{2.66}$$

where $\Delta_0 = \text{const} > 0$. Now the Hamiltonian $H^{(1)}$ takes the form,

$$H_{k_y}^{(1)} = \begin{pmatrix} \Delta(x) & v\left(-i\nabla_x - ik_y\right) \\ v\left(-i\nabla_x + ik_y\right) & -\Delta(x) \end{pmatrix}, \tag{2.67}$$

where we substituted k_x by the operator $k_x \to -i\nabla_x$. The Schrödinger equation for $\left(\varphi_{k_y}, \chi_{k_y}\right)$ components of the wave function $\psi_{k_y}(x)$ reads as

$$(\Delta - \varepsilon)\varphi_{k_y} + v\left(-i\nabla_x - ik_y\right)\chi_{k_y} = 0,$$

$$v\left(-i\nabla_x + ik_y\right)\varphi_{k_y} - (\Delta + \varepsilon)\chi_{k_y} = 0. \tag{2.68}$$

In the region $x < 0$, we find from (2.68) that there is decaying at $x \to -\infty$ solution

$$\psi_1(x) = C_1 \begin{pmatrix} 1 \\ \dfrac{iv\left(\kappa - k_y\right)}{\Delta_0 - \varepsilon} \end{pmatrix} \exp(\kappa x), \quad x < 0,$$

$$\kappa = \frac{1}{v}\sqrt{\Delta_0^2 + v^2k_y^2 - \varepsilon^2}, \tag{2.69}$$

where C_1 is an arbitrary constant. Similarly, for $x > 0$, we obtain from (2.68) the solution decaying at $x \to \infty$

$$\psi_1(x) = C_2 \begin{pmatrix} 1 \\ \dfrac{iv\left(\kappa + k_y\right)}{\Delta_0 + \varepsilon} \end{pmatrix} \exp(-\kappa x), \quad x > 0 \tag{2.70}$$

with the same κ and arbitrary C_2. Matching these solutions at $x = 0$, one gets the energy spectrum of electrons at the interface:

$$\varepsilon_1(k_y) = k_y \Delta_0 / \kappa \tag{2.71}$$

and $C_1 = C_2$. Hence, substituting $\varepsilon_1(k_y)$ in (2.69) and (2.68), we obtain localized at interface solution,

$$\psi_1(x) = C \begin{pmatrix} \Delta_0 \\ iv\kappa \end{pmatrix} \exp(-\kappa|x|). \tag{2.72}$$

Constant C follows from the normalization condition for $\psi_1(x)$. Then, we find another solution using the Hamiltonian $H^{(2)}$, which differs from $H^{(1)}$ by the sign of Δ. The solution

$$\psi_2(x) = C \begin{pmatrix} -\Delta_0 \\ iv\kappa \end{pmatrix} \exp(-\kappa|x|) \tag{2.73}$$

corresponds to

$$\varepsilon_2(k_y) = -k_y \Delta_0 / \kappa. \tag{2.74}$$

So, by making the junction of two insulators with opposite signs of the energy gap parameter, we come to the localized at surface electron gas with a linear-k energy spectrum. It turns out that the absence of the gap at the surface and the linearity of energy spectrum $\varepsilon_{1,2}(k_y)$ near the point $\varepsilon = 0$ is robust to any perturbations which do not break the symmetry of the initial Hamiltonian H.

In the example discussed, the effective Hamiltonian describes translationally symmetric two-dimensional IV-VI semiconductors. Now we consider the surface states in the inverted-band semiconductor heterojunction HgTe/CdTe. As one needs to account for band inversion, the k-p Hamiltonian is to be the Kane model, which includes one conduction and three valence bands. With the spin variable in mind, the effective Hamiltonian is the 8×8 matrix given in Chapter 1 [see Eq. (1.89)]. In a band-inverted HgTe, the gap parameter $E_g < 0$, whereas in normal (noninverted) semiconductor CdTe, $E_g > 0$. Here we restrict ourselves to the 6×6 model, which does not include the spin-split band:

$$H_k = \begin{pmatrix} E_g + \varepsilon_c & 0 & \frac{pk_-}{\sqrt{2}} & \frac{pk_+}{\sqrt{6}} & -\frac{i\sqrt{2}pk_z}{\sqrt{3}} & 0 \\ 0 & E_g + \varepsilon_c & 0 & \frac{\sqrt{2}pk_z}{\sqrt{3}} & \frac{ipk_-}{\sqrt{6}} & \frac{ipk_+}{\sqrt{2}} \\ \frac{pk_+}{\sqrt{2}} & 0 & F & H & I & 0 \\ \frac{pk_-}{\sqrt{6}} & \frac{\sqrt{2}pk_z}{\sqrt{3}} & H^* & G & 0 & I \\ \frac{i\sqrt{2}pk_z}{\sqrt{3}} & -\frac{ipk_+}{\sqrt{6}} & I^* & 0 & G & -H \\ 0 & -\frac{ipk_-}{\sqrt{2}} & 0 & I^* & -H^* & F \end{pmatrix} \tag{2.75}$$

We assume the interface between HgTe and CdTe is at $z = 0$, and

$$E_g(z) = \begin{cases} -E_0, & z < 0, \\ E_0, & z > 0, \end{cases} \tag{2.76}$$

where $E_0 > 0$.

At $k_x = k_y = 0$, Equation (2.75) can be split into two independent blocks, one is 4×4 matrix in the basis of states $(1/2, 1/2)$, $(1/2,-1/2)$, $(3/2,1/2)$, $(3/2,-1/2)$, and another one 2×2, in the heavy holes states $(3/2,\pm 3/2)$. In the Γ-point, the heavy hole band is decoupled from the others. So, the band inversion concerns only the light hole and conduction bands which form the 4×4 effective Hamiltonian:

$$H_{k_x,k_y=0} = \begin{pmatrix} E_g(z) - \frac{\hbar^2\nabla_z^2}{2m_c} & 0 & 0 & -\frac{\sqrt{2}p\nabla_z}{\sqrt{3}} \\ 0 & E_g(z) - \frac{\hbar^2\nabla_z^2}{2m_c} & \frac{-i\sqrt{2}p\nabla_z}{\sqrt{3}} & 0 \\ 0 & \frac{-i\sqrt{2}p\nabla_z}{\sqrt{3}} & -\frac{(L+M)\nabla_z^2}{3} & 0 \\ \frac{\sqrt{2}p\nabla_z}{\sqrt{3}} & 0 & 0 & -\frac{(L+M)\nabla_z^2}{3} \end{pmatrix} \tag{2.77}$$

where we replaced k_z by the operator, $k_z \to -i\nabla_z$.

In its turn, Hamiltonian (2.77) also comprises two independent 2×2 blocks, each corresponding to coupled $(1/2,1/2)$-$(3/2,-1/2)$ and $(1/2,-1/2)$-$(3/2,1/2)$ states. Two eigenfunctions $\psi_{10}(z)$, $\psi_{20}(z)$ are the degenerate at energy $\varepsilon = 0$ states localized near the interface $z = 0$ (surface states). In the spirit of *k-p* approximation, one can use these functions to construct a new basis:

$$\psi_{1k}(\boldsymbol{r},z) = \frac{1}{\sqrt{\Omega}}\exp(i\boldsymbol{k}\cdot\boldsymbol{r})\,\psi_{10}(z), \quad \psi_{2k}(\boldsymbol{r},z) = \frac{1}{\sqrt{\Omega}}\exp(i\boldsymbol{k}\cdot\boldsymbol{r})\,\psi_{20}(z), \tag{2.78}$$

where $\boldsymbol{r}$ is the 2D in-plane vector at the interface. In the limit of $k_x, k_y \to 0$, the problem reduces to the block-diagonal 2D Dirac model (2.65). At the interface, $E_g(z) = 0$, and the model becomes the Dirac 2D gapless Hamiltonian $H = v\,\boldsymbol{\tau}\cdot\boldsymbol{k}$, where $\boldsymbol{\tau} = (\tau_x, \tau_y)$ are the Pauli matrices in the space of basis functions (2.69) and v is the material parameter determined by initial effective Hamiltonian (2.75). The main result of this example is that we obtain a 2D gas of electrons localized at the interface. The low-energy spectrum of interface electrons, $\varepsilon_{1,2}(k) = \pm vk$, is called the *Dirac cone*. As demonstrated within various models, the solution of this type appears at the interface in an inverted-band heterojunction. Also, localized states of this type exist near the surface that separates an inverted-band semiconductor and a vacuum, as a vacuum can be considered a regular insulator of a large energy gap. In modern language, the regular and inverted-band semiconductors belong to different *topological classes*. The topological class of a semiconductor is determined by invariants associated with the symmetry of the Hamiltonian. Gapless surface (edge) states that appear when 3D (2D) inverted-band semiconductor in contact to an ordinary insulator or vacuum, is the specific feature of *topological insulator*: it has an energy gap in the bulk (interior) and gapless, thus conductive, states on the boundary.

An example of the topological insulator is graphene with *intrinsic* spin–orbit interaction. The electron spectrum in graphene is discussed in Chapter 1. Spin–orbit interaction creates the energy gap in the interior, and gapless (linear in momentum) edge modes are spin-polarized: spin-up and spin-down modes propagate in opposite directions. The reason for that is the time-inversion symmetry. If the time-inversion symmetry holds, the Kramers theorem dictates states $|k,\uparrow\rangle$ and $|-k,\downarrow\rangle$ (k is the wavevector along the edge) to have the same energy. Moreover, any perturbations which do not break the time-inversion symmetry (for example, nonmagnetic impurities) cannot provide a gap in the energy spectrum of edge electrons. Since in graphene, the intrinsic spin-orbit coupling and thus the energy gap is small, the topological modes are observable only in clean samples at low temperatures. The gapless spectrum *protected by the time-inversion symmetry* is a general property of topological insulators and takes place not only in graphene.

Another example of a 2D-topological insulator is the CdTe/HgTe/CdTe quantum well, where *c*- and v-bands position in HgTe remain inverted if the well is thick enough [6]. In bulk semiconductors HgTe or Bi_2Te_3 (examples of 3D-topological insulators), the surface states are also gapless and protected by time-inversion symmetry [7].

One more class of semiconductors with gapless surface states is the inverted-band IV-VI semiconductor alloys SnTe–PbTe. The interface with regular semiconductor PbTe carries 2D Dirac electrons [8]. The bandgap (direct and inverted) in IV–VI semiconductors originates from the crystal lattice potential, and the time-inversion symmetry is of no importance. Nevertheless, there exists a crystalline symmetry responsible for symmetry-protected gapless surface states. Correspondingly, the inverted-band IV–VI semiconductors are called *crystalline topological insulators.*

References

1. J. Batey and S.L. Wright, "Energy band alignment in GaAs:(Al,Ga)As heterostructures: the dependence on alloy composition", *J. Appl. Phys.* **59**, 200 (1986).
2. S. Adachi, *Physical Properties of III-V Semiconductor Compounds* (John Wiley & Sons, New York, 1992).
3. H. Wei and A. Zunger, "Calculated natural band offsets of all II–VI and III–V semiconductors: chemical trends and the role of cation *d* orbitals", *Appl. Phys. Lett.* **72**, 2011 (1998)
4. M.P.C.M. Krijn, "Heterojunction band offsets and effective masses in III-V quaternary alloys", *Semicond. Sci. Technol.* **6**, 27 (1991).
5. L.D. Landau and E.M. Lifshitz, *Quantum Mechanics*, Chapter 3 (Pergamon, New York, 1977).
6. B.A. Bernevig, T.L. Hughes, and S.-C. Zhang, "Quantum spin Hall effect and topological phase transition in HgTe quantum wells", *Science.* **314**, 1757 (2006).
7. M.Z. Hasan and C.L. Kane, Colloquium: topological insulators", *Rev. Mod. Phys.* **82**, 3045–3067 (2010).
8. B.A. Volkov and O.A. Pankratov, "Two-dimensional massless electrons in an inverted contact", *JETP Lett.* **42**, 178–181 (1985).

3

Impurities and Disorder in Semiconductors

3.1 Single Impurity: Donors and Acceptors

Doping controls the electrical properties of semiconductors. Even a small number of impurities may substantially change the observable properties such as conductivity and magnetic behavior. As an example, we consider a single atom of group-V (let say As) in silicon. Four valence electrons of Si atoms form covalent bonds with the four surrounding atoms. The As atom has five valence electrons, four of which go to saturation the bonds while the excess electron remains weakly bonded to the As atom, which means that a small amount of energy is enough to excite the electron from the localized As-state to the conduction band. Thus, As impurity donates an electron to the conduction band, so is called a *donor* in Si. On the contrary, an impurity atom of group III (for example, In) has three valence electrons and needs one more electron to saturate four covalent bonds to surrounding Si atoms. The impurity may accept an electron from the valence band of Si, so it is an *acceptor* or impurity atom with a bound hole. The excitation of that hole to the valence band means excitation of an electron from the valence band to the impurity level. For weakly bound holes, the process does not require high energy. So, the group-III acceptors supply holes to the valence band. For any semiconductor material, one can identify the type of impurities, which serve as donors or acceptors. In this chapter, we consider several models for impurity states in semiconductors.

3.2 Screening of Impurity Potential by Free Electrons

In a vacuum, electron energy in the field of charge e has the form of attracting Coulomb interaction $V_0(\boldsymbol{r}) = -e^2/4\pi\varepsilon_0 r$, ε_0 is the vacuum permittivity. In semiconductors, the effect of lattice polarization comes to the static screening of the potential by substitution $\varepsilon_0 \to \varepsilon_0\varepsilon$, where ε is the static dielectric constant (for details on static polarization, see Chapter 12). One more source is dynamic screening by free carriers. Even in an intrinsic semiconductor, i.e., semiconductor without doping, there is some density of thermally excited carriers (electrons and holes). Similar to dielectrics, free electrons present the medium polarization of which results in screening. In other words, excitations in the electron system transform the bare Coulomb force into screened interaction. The problem has found a solution by the method of quantum field theory. The process of screening by electron excitations is represented graphically by Feynman diagrams in Figure 3.1.

Diagrams illustrate how bare interaction becomes renormalized (screened) due to free carriers [1]. Graphic notations in Figure 3.1 are as follows: a thin dashed line with star – the bare interaction with impurity $V_0(\boldsymbol{r}-\boldsymbol{r}')$, a thick dashed line with star – the renormalized interaction with impurity $V(\boldsymbol{r}-\boldsymbol{r}')$, another thin and thick dashed lines – the bare $\mathrm{v}_0(\boldsymbol{r}-\boldsymbol{r}')$ and renormalized $\mathrm{v}(\boldsymbol{r}-\boldsymbol{r}')$ electron–electron interactions, respectively, and the solid line – the Green's function of a free electron, G_ε^0, which is the

DOI: 10.1201/9780429285929-3

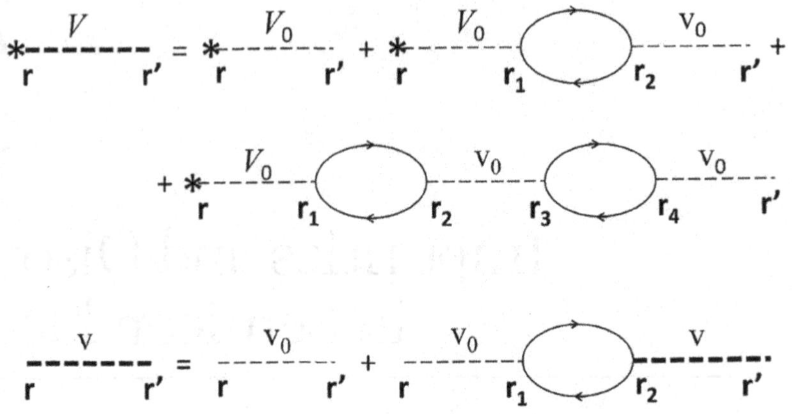

FIGURE 3.1 Feynman diagrams for renormalization of the Coulomb interaction due to the screening by electrons.

solution to equation $\left(\varepsilon+\hbar^2\Delta/2m\right)G_\varepsilon^0(\boldsymbol{r}-\boldsymbol{r}')=\delta(\boldsymbol{r}-\boldsymbol{r}')$, where Δ is the Laplace operator. Note that the repulsive electron–electron Coulomb interaction is positive, $\mathrm{v}_0(\boldsymbol{r})=e^2/4\pi\varepsilon\varepsilon_0 r$.

The last row in Figure 3.1 is the integral equation for electron-electron interaction, which is obtained by summation of interaction diagrams (factor 2 is due to spin):

$$\mathrm{v}(\boldsymbol{r}-\boldsymbol{r}')=\mathrm{v}_0(\boldsymbol{r}-\boldsymbol{r}')+2i\int\frac{d\varepsilon}{2\pi}d^3\boldsymbol{r}_1\,d^3\boldsymbol{r}_2\,\mathrm{v}_0(\boldsymbol{r}-\boldsymbol{r}_1)G_\varepsilon(\boldsymbol{r}_1-\boldsymbol{r}_2)G_\varepsilon(\boldsymbol{r}_2-\boldsymbol{r}_1)\mathrm{v}(\boldsymbol{r}_2-\boldsymbol{r}'). \tag{3.1}$$

After Fourier transformation, we obtain

$$V(\boldsymbol{r})=\frac{1}{(2\pi)^3}\int d^3q\,e^{i\boldsymbol{q}\cdot\boldsymbol{r}}\,V(\boldsymbol{q}),$$

$$\mathrm{v}(\boldsymbol{r})=\frac{1}{(2\pi)^3}\int d^3q\,e^{i\boldsymbol{q}\cdot\boldsymbol{r}}\,\mathrm{v}(\boldsymbol{q}),$$

$$V(\boldsymbol{q})=V_0(\boldsymbol{q})+2iV_0(\boldsymbol{q})\int\frac{d\varepsilon}{2\pi}\frac{d^3k}{(2\pi)^3}G_\varepsilon(\boldsymbol{k}+\boldsymbol{q})G_\varepsilon(\boldsymbol{k})\mathrm{v}(\boldsymbol{q}),$$

$$\mathrm{v}(\boldsymbol{q})=\mathrm{v}_0(\boldsymbol{q})+2i\mathrm{v}_0(\boldsymbol{q})\int\frac{d\varepsilon}{2\pi}\frac{d^3k}{(2\pi)^3}G_\varepsilon(\boldsymbol{k}+q)G_\varepsilon(\boldsymbol{k})\mathrm{v}(\boldsymbol{q}), \tag{3.2}$$

where $V_0(\boldsymbol{q})=-e^2/\varepsilon_0\varepsilon q^2$ and $\mathrm{v}_0(\boldsymbol{q})=e^2/\varepsilon_0\varepsilon q^2$. It can be also presented as

$$V(\boldsymbol{q})=V_0(\boldsymbol{q})-V_0(\boldsymbol{q})\Pi(\boldsymbol{q})\mathrm{v}(\boldsymbol{q}),$$

$$\mathrm{v}(\boldsymbol{q})=\mathrm{v}_0(\boldsymbol{q})-\mathrm{v}_0(\boldsymbol{q})\Pi(\boldsymbol{q})\mathrm{v}(\boldsymbol{q}), \tag{3.3}$$

where

$$\Pi_0(q) = -2i \int \frac{d\varepsilon}{2\pi} \frac{d^3k}{(2\pi)^3} G_\varepsilon(\boldsymbol{k}+\boldsymbol{q}) G_\varepsilon(\boldsymbol{k}) \tag{3.4}$$

is the *polarization operator* of the electron gas. The solution to Eq. (3.3) for $V(\boldsymbol{q})$ is the *renormalized Coulomb interaction*,

$$V(\boldsymbol{q}) = \frac{V_0(\boldsymbol{q})}{1 - v_0(\boldsymbol{q})\Pi_0(\boldsymbol{q})} = -\frac{e^2}{\varepsilon\varepsilon_0 q^2 - e^2 \Pi_0(\boldsymbol{q})}. \tag{3.5}$$

The renormalized interaction includes the effect of screening of the Coulomb interaction by free electrons. In particular, for small $q \to 0$, we obtain

$$V(\boldsymbol{q}) \simeq -\frac{e^2}{\varepsilon\varepsilon_0\left(q^2 - \frac{e^2\Pi_0(0)}{\varepsilon\varepsilon_0}\right)} = -\frac{e^2}{\varepsilon\varepsilon_0\left(q^2 + \alpha^2\right)}, \tag{3.6}$$

where α is the *inverse screening length*:

$$\alpha^2 = -\frac{e^2\Pi_0(0)}{\varepsilon\varepsilon_0}. \tag{3.7}$$

As we show below, $\Pi_0(0) < 0$, so α is real.

The inverse Fourier transformation of the interaction (3.6) gives us

$$V(\boldsymbol{r}) = \int \frac{d^3q}{(2\pi)^3} e^{-i\boldsymbol{q}\cdot\boldsymbol{r}}\, V(\boldsymbol{q}) = -\frac{e^2}{4\pi\varepsilon\varepsilon_0 r} \exp(-\alpha r). \tag{3.8}$$

As we see from this expression, free electrons cause an exponential decay of the screened potential $V(\boldsymbol{r})$ at large distances.

Now we calculate the polarization operator $\Pi_0(\boldsymbol{q})$ using the free-electron Green's function with the parabolic energy spectrum in the vicinity of the band edge

$$G_E(\boldsymbol{k}) = \frac{1}{E - \varepsilon_k + \mu + i\delta\, sign\, E}, \tag{3.9}$$

where $\varepsilon_k = \hbar^2 k^2/2m$ is the energy spectrum of electrons near the band edge, μ is the electron chemical potential, and the electron energy E counts from the chemical potential.

Let us assume that the chemical potential is inside the conduction band (degenerate semiconductor, $\mu > 0$). The Green's function (3.9) with the imaginary shift of pole $i\delta\, sign\, E$ is called the *causal* Green's function.

Using Eqs. (3.4) and (3.9) we find

$$\Pi_0(\boldsymbol{q}) = -2i \int \frac{dE}{2\pi} \frac{d^3k}{(2\pi)^3} \frac{1}{\left[E - \varepsilon_{k+q} + \mu + i\delta\, sign\, E\right]\left[E - \varepsilon_k + \mu + i\delta\, sign\, E\right]} \tag{3.10}$$

The integral over E runs from $-\infty$ to $+\infty$ along the real axis in the complex ε-plane. We calculate the integral by closing the integration path in the complex plane since the contribution from semicircle $|E| \to \infty$ is zero (at large $|E|$ the integrand decays as $1/E^2$). Then the integration along the closed contour gives[1]

$$\Pi_0(q) = 2\int \frac{d^3k}{(2\pi)^3}\left(\frac{f(\varepsilon_{k+q})}{\varepsilon_{k+q}-\varepsilon_k} + \frac{f(\varepsilon_k)}{\varepsilon_k - \varepsilon_{k+q}}\right) = 2\int \frac{d^3k}{(2\pi)^3}\frac{f(\varepsilon_{k+q}) - f(\varepsilon_k)}{\varepsilon_{k+q}-\varepsilon_k}, \tag{3.11}$$

where we introduced the Fermi-Dirac function $f(E) = \left\{\exp\left[(E-\mu)/kT\right]+1\right\}^{-1}$ at $T \to 0$, which equals 1 when $E < \mu$. We performed the calculation using the Green's function formalism at $T = 0$. However, with Fermi-Dirac functions introduced, the result (3.11) is also valid at $T \neq 0$ that can be proved by similar calculations within the Matsubara technique [2,3].

In the limit $q \to 0$, we obtain from (3.11):

$$\Pi_0(0) = -2\int \frac{d^3k}{(2\pi)^3}\left(-\frac{\partial f}{\partial \varepsilon_k}\right) = -\int \rho(E)\,dE\left(-\frac{\partial f}{\partial \varepsilon}\right), \tag{3.12}$$

where $\rho(E) = \sqrt{8m^3E}/2\pi^2\hbar^3$ is the electron density of states in the conduction band.

Since we assumed the chemical potential μ is in the conduction band, at $T \to 0$ $\partial f/\partial E \simeq -\delta(E-\mu)$. Then we get $\Pi_0(0) = -\rho(\mu)$, and thus the inverse square of screening length (3.7) in degenerate semiconductor becomes $\alpha^2 = e^2\rho(\mu)/\varepsilon\varepsilon_0$.

In a nondegenerate semiconductor, $\mu < 0$ and $|\mu| \gg k_BT$, one can use Eq. (3.12) with the classical approximation to distribution function, $f(E) = \exp\left[(-E-\mu)/k_BT\right]$ (see Chapter 4 for details). Then we obtain

$$\Pi_0(0) = -\frac{(2m)^{3/2}\exp(\mu/k_BT)}{2\pi^2\hbar^3 k_BT}\int_0^\infty \sqrt{E}\exp(-E/k_BT)\,dE$$

$$= -\frac{(2m)^{3/2}\sqrt{k_BT}}{4\sqrt{\pi^3}\hbar^3}\exp\left(\frac{\mu}{k_BT}\right), \tag{3.13}$$

where we used $\int_0^\infty \sqrt{x}\exp(-x)\,dx = \Gamma(3/2) = \sqrt{\pi}/2$ and $\Gamma(x)$ is the Gamma-function. Hence, for the non-degenerate semiconductor, we obtain

$$\alpha^2 = \frac{e^2(2m)^{3/2}\sqrt{k_BT}}{4\pi^{3/2}\hbar^3\varepsilon\varepsilon_0}\exp\left(\frac{\mu}{k_BT}\right). \tag{3.14}$$

Physically, the screening of Coulomb potential originates from the spatial redistribution of free electrons in the vicinity of impurity. In particular, an attractive (repulsive) potential increases (decreases) the electron density near impurity that causes screening at large distances.

[1] We omit the imaginary part of $\Pi_0(q)$ as it is not related to the screening.

3.3 Electron in the Field of Impurity Potential

The attractive potential of a donor creates a bound state for electrons. We estimate the bound state energy by using a simplified model to describe the donor impurity in Si. Neglecting the screening of the impurity potential by free electrons, we deal with the Hamiltonian,

$$H = -\frac{\hbar^2 \Delta}{2m} - \frac{e^2}{4\pi\varepsilon_0 \varepsilon r}, \tag{3.15}$$

where m is the effective mass of an electron in the conduction band. Hamiltonian (3.15) describes band electrons in an attractive potential of ionized donor impurity. The model can be used at low enough donor density when screening by free carriers is negligible.

Energy and wave functions of bound states follow from quantum mechanical calculations:

$$E_n = -\frac{me^4}{32\varepsilon_0^2 \varepsilon^2 \hbar^2}\frac{1}{n^2}, \tag{3.16}$$

where $n = 1, 2, \ldots$ is the main quantum number. By using typical parameters $m = 0.4m_0$ (m_0 is the free-electron mass) and $\varepsilon = 10$, we find the energy of the ground state $E_1 \simeq 0.05$ eV.

From quantum mechanics, the ground state wave function is

$$\psi_1(r) = \frac{1}{\sqrt{\pi}\, a^{3/2}} \exp\left(-\frac{r}{a}\right), \tag{3.17}$$

where $a = 4\pi\varepsilon\varepsilon_0 \hbar^2 / me^2$ is the *effective Bohr radius* of the electron. In silicon, $a \simeq 5.3 \times 10^{-11}$ m, which is ten times the lattice constant.

3.4 Model of Short-Range Impurity Potential

If a single impurity (or a defect such as a vacancy) is neutral, it can be modeled by short-range potential so that the Hamiltonian has the following form:

$$H = -\frac{\hbar^2 \Delta}{2m} - V_0\, \delta(\mathbf{r}), \tag{3.18}$$

where V_0 is the strength of perturbation, and we assumed potential is attractive.

The δ-function impurity potential in Eq. (3.18) does not correspond to real defects in semiconductors. In reality, the size of such a defect can be of the order of lattice constant. This model is justified if the size of the wave function of a localized state (effective Bohr radius) is much larger than the lattice constant, a_0. Hard-sphere approximation then implies that a_0 is the radius of electron–impurity interaction. For spherically-symmetric perturbation, the Hamiltonian written in spherical coordinates (r, θ, φ) has the form:

$$H = -\frac{\hbar^2}{2m}\left[\frac{\partial^2}{\partial r^2} + \frac{2}{r}\frac{\partial}{\partial r} + \frac{1}{r^2 \sin\theta}\left(\sin\theta \frac{\partial}{\partial \theta}\right) + \frac{1}{r^2 \sin^2\theta}\frac{\partial^2}{\partial \varphi^2}\right] - \tilde{V}(r), \tag{3.19}$$

where $\tilde{V}(r) = 3V_0 / 4\pi a_0^3$ for $r < a_0$ and $\tilde{V}(r) = 0$ otherwise. We assume a_0 is small compared to the characteristic size of the wave function chosen as

$$\psi(r, \theta, \varphi) = R_l(r) Y_{lm}(\theta, \varphi), \tag{3.20}$$

where $Y_{lm}(\theta,\varphi)$ is the spherical function. The radial function $R_l(r)$ obeys the equation:

$$R_l''+\frac{2R_l'}{r}-\frac{l(l+1)R_l}{r^2}+\frac{2m}{\hbar^2}\tilde{V}(r)R_l-\kappa^2R_l=0, \tag{3.21}$$

where $\kappa^2=-2m\varepsilon/\hbar^2$ implies the energy of bound state $\varepsilon<0$. If the angular momentum $l=0$,

$$R_0''+\frac{2R_0'}{r}+w(r)R_0-\kappa^2R_0=0,$$

$$w(r)=\begin{cases} w_0=\dfrac{3mV_0}{2\pi\hbar^2a_0^3}>0, & r<a_0, \\ 0, & r\geq a_0. \end{cases} \tag{3.22}$$

The general solution to Eq. (3.22) is

$$R_0(r)=\frac{A\sin(kr)}{r}, \quad r<a_0, \tag{3.23}$$

where $k=\sqrt{w_0-\kappa^2}$, A is a constant. The solution exists for $w_0>\kappa^2$.

For $r>a_0$, $w(r)=0$. Hence, the solution finite at $r\to\infty$ is

$$R_0(r)=\frac{B}{r}e^{-\kappa r}, \quad r>a_0. \tag{3.24}$$

The condition of continuity for the function $R_0(r)$ and its derivative at $r=a_0$ gives two equations for A and B

$$A\sin(ka_0)=Be^{-\kappa a_0}, \tag{3.25}$$

$$A\left[ka_0\cos(ka_0)-\sin(ka_0)\right]=-Be^{-\kappa a_0}(\kappa a_0+1). \tag{3.26}$$

The condition of small a_0 is $\kappa a_0\ll 1$. In this limit from (3.25), (3.26) follows $B=A\sin(ka_0)$ and $\cos(ka_0)=0$, which gives an equation for the energies of localized spherically symmetric states ($l=0$):

$$\varepsilon_n=-\frac{\hbar^2}{2m}\left(w_0-\frac{n^2\pi^2}{4a_0^2}\right)=-\frac{3}{4\pi a_0^3}\left(V_0-V_0^{cr}n^2\right),$$

$$V_0^{cr}=\frac{\pi^3\hbar^2a_0}{6m}, \tag{3.27}$$

where $n=1,2,\ldots$ and the ground state corresponds to $n=1$.

3.5 T-Matrix Approximation

Instead of solving the Schrödinger equation, the alternative approach uses Green's functions. The Schrödinger equation for an electron in the conduction band with parabolic energy spectrum $\varepsilon_k=\hbar^2k^2/2m$ can be written as

$$\left[\hat{G}_0^{-1} - V(\boldsymbol{r})\right]\psi(\boldsymbol{r}) = 0, \tag{3.28}$$

where $\hat{G}_0$ is the free-electron Green's function operator (see Chapter 1),

$$\hat{G}_0 = \left(E + \frac{\hbar^2 \Delta}{2m}\right)^{-1}. \tag{3.29}$$

Instead of operator $\hat{G}_0$, one can use Green's function in the "matrix" form of $G_0(E;\boldsymbol{r},\boldsymbol{r}')$ defined by

$$\hat{G}_0^{-1} G_0(E;\boldsymbol{r},\boldsymbol{r}') = \delta(\boldsymbol{r} - \boldsymbol{r}'). \tag{3.30}$$

The free-electron Green's function is spatially homogeneous, which means $G_0(E;\boldsymbol{r},\boldsymbol{r}') \equiv G_0(E;\boldsymbol{r} - \boldsymbol{r}')$.

The solution of Eq. (3.28) with $V(\boldsymbol{r}) = 0$ is the wave function of a free electron $\psi_0(\boldsymbol{r})$. When the perturbation potential is present in (3.28), the Schrödinger equation is equivalent to the integral equation

$$\psi(\boldsymbol{r}) = \psi_0(\boldsymbol{r}) + \int d^3r'\, G_0(E;\boldsymbol{r},\boldsymbol{r}') V(\boldsymbol{r}') \psi(\boldsymbol{r}'). \tag{3.31}$$

Equivalence between (3.31) and (3.28) can be checked by acting with operator $\hat{G}_0^{-1}$ from the left and using the definition (3.30). To find the solution by a perturbation method, one can write the series

$$\begin{aligned}\psi(\boldsymbol{r}) &= \psi_0(\boldsymbol{r}) + \int d^3r'\, G_0(E;\boldsymbol{r},\boldsymbol{r}') V(\boldsymbol{r}') \psi_0(\boldsymbol{r}') \\ &\quad + \int d^3r' d^3r''\, G_0(E;\boldsymbol{r},\boldsymbol{r}') V(\boldsymbol{r}') \psi_0(\boldsymbol{r}') G_0(E;\boldsymbol{r}',\boldsymbol{r}'') V(\boldsymbol{r}'') \psi_0(\boldsymbol{r}'') + \ldots \\ &\equiv \psi_0(\boldsymbol{r}) + \int d^3r' d^3r''\, G_0(E;\boldsymbol{r},\boldsymbol{r}') T(E;\boldsymbol{r}',\boldsymbol{r}'') \psi(\boldsymbol{r}''),\end{aligned} \tag{3.32}$$

where the *T-matrix* is introduced as

$$\begin{aligned}T(\varepsilon;\boldsymbol{r},\boldsymbol{r}') &= V(\boldsymbol{r})\delta(\boldsymbol{r} - \boldsymbol{r}') + V(\boldsymbol{r}) G_0(E;\boldsymbol{r},\boldsymbol{r}') V(\boldsymbol{r}') \\ &\quad + \int d^3r_1\, V(\boldsymbol{r}) G_0(E;\boldsymbol{r},\boldsymbol{r}_1) V(\boldsymbol{r}_1) G_0(E;\boldsymbol{r}_1,\boldsymbol{r}') V(\boldsymbol{r}') + \ldots,\end{aligned} \tag{3.33}$$

which formally includes all orders of the perturbation related to impurity potential. Equation (3.33) is presented graphically in Figure 3.2:

After Fourier transformation, Eq. (3.33) becomes

$$T_{\boldsymbol{kk}'}(E) = V_{\boldsymbol{kk}'} + \int \frac{d^3k_1}{(2\pi)^3} V_{\boldsymbol{kk}_1} G_0(E,\boldsymbol{k}_1) V_{\boldsymbol{k}_1\boldsymbol{k}'} + \int \frac{d^3k_1}{(2\pi)^3} \frac{d^3k_2}{(2\pi)^3} V_{\boldsymbol{kk}_1} G_0(E,\boldsymbol{k}_1) V_{\boldsymbol{k}_1\boldsymbol{k}_2} G_0(E,\boldsymbol{k}_2) V_{\boldsymbol{k}_2\boldsymbol{k}'} + \ldots, \tag{3.34}$$

where we use notations

$$V_{\boldsymbol{kk}'} = \int d^3r\, e^{-i(\boldsymbol{k}-\boldsymbol{k}')\boldsymbol{r}}\, V(\boldsymbol{r}), \quad G_0(E,\boldsymbol{k}) = \int d^3(\boldsymbol{r} - \boldsymbol{r}')\, e^{-i\boldsymbol{k}\cdot(\boldsymbol{r}-\boldsymbol{r}')} G_0(E;\boldsymbol{r},\boldsymbol{r}').$$

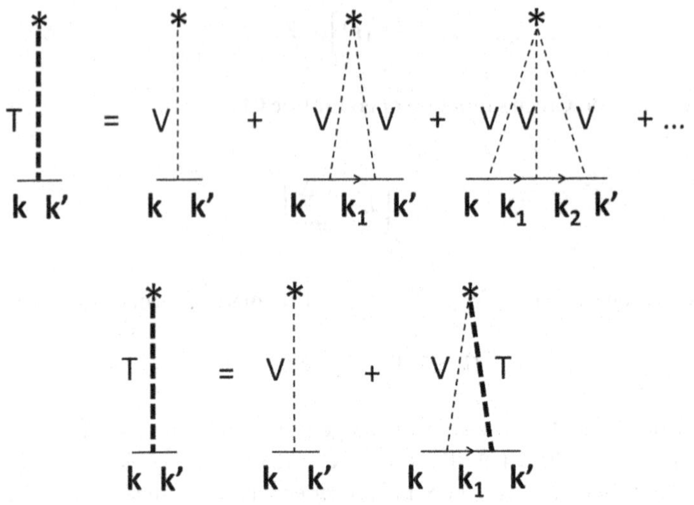

FIGURE 3.2 Feynman diagrams for the T-matrix (upper row) and Eq. (3.35) (lower row).

Summation in (3.34) leads to the equation for T-matrix:

$$T_{kk'}(E) = V_{kk'} + \int \frac{d^3k_1}{(2\pi)^3} V_{kk_1} G_0(E,\boldsymbol{k}_1) T_{k_1k'}(E). \tag{3.35}$$

Let us assume that the impurity potential is short-ranged so that $V(\boldsymbol{r}) = -V_0\,\delta(\boldsymbol{r})$. In this case, we get $V_{kk'} = -V_0$, Eq. (3.35) simplifies, and the solution has the form:

$$T(E) = -\frac{V_0}{1 + V_0\,F(E)}, \tag{3.36}$$

where

$$F(E) = \int \frac{d^3k}{(2\pi)^3} G_0(E,\boldsymbol{k}) \tag{3.37}$$

is the *Hilbert transform* of Green's function. As follows from (3.30), in the model of parabolic energy band the Green's function is[2]

$$G_0(E,\boldsymbol{k}) = \frac{1}{E - \varepsilon_k + i\delta\, sign\, E}. \tag{3.38}$$

Substituting (3.38) to (3.37), we obtain

$$F(E) = \int \frac{\rho(x)dx}{E - x + i\delta\, sign\, E} \tag{3.39}$$

where $\rho(x) = (2m)^{3/2}\sqrt{x}/4\pi^2\hbar^3$ is the electron density of states per spin. Integral (3.39) formally diverges at large x. Therefore, we have to introduce energy cutoff E_{max} of the order of the width of the conduction band.[3]

[2] For causal Green's function, we have to shift the pole from the real axis to the complex plane E.

[3] In 1D-model, one can use delta-potential without introducing cutoff as the integral over $\boldsymbol{k}$ converges. In the 2D-case, the divergence is logarithmic.

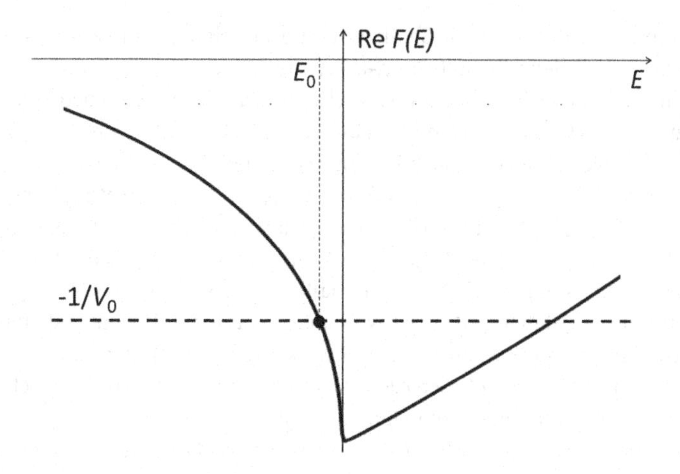

FIGURE 3.3 The graphic solution of Eq. (3.40) for the energy of the impurity state in the model of delta-potential.

The T-matrix (3.36) has simple poles at energies of bound states. The equation for poles is

$$\operatorname{Re} F(E) = -\frac{1}{V_0}. \tag{3.40}$$

The graphic solution to (3.40) is illustrated in Figure 3.3.

There are two solutions for bound states. As the energy reference point is at the edge of the conduction band, solution $E_0 < 0$ corresponds to the localized impurity state that may fall into a bandgap. The energy of the other level is positive, so it is a *resonant state* in the conduction band. The electron in this state has a finite lifetime due to the possibility to escape into a continuum of conduction band states. Note that there is a critical value for impurity potential to trap an electron: if $V_0 < V_0^{cr}$, an impurity does not form a localized state. Taking $E = 0$ in (3.39) and calculating the integral, we find

$$V_0^{(cr)} = \frac{2\pi^2\hbar^3}{(2m)^{3/2}\sqrt{E_{\max}}}. \tag{3.41}$$

Parameter $E_{\max}$ is explicitly introduced to cut the integral at high energy. That is the property of delta-function potential, for which the creation of localized level involves all possible conduction band states up to infinite energy. On the contrary, when discussing the same problem in the previous section, we used finite impurity potential (hard-sphere approximation). There was no need for the explicitly introduced cutoff at large k as it arose automatically as $k_{\max} \approx \pi/a_0$ (see Eq. (3.27)). Assuming $E_{\max} = \hbar^2 k_{\max}^2/2m$, we see a match between (3.27) and (3.41) up to a numerical factor of the order of unity. The numerical difference stems from the difference in initial models.

3.6 Electron States of an Impurity Atom in the Crystalline Lattice: Tight-Binding Approach

In previous sections, we discussed the mechanism of the formation of impurity levels in a semiconductor bandgap. It happens if the impurity potential is strong enough for electron states to split from the conduction band and to form donor levels. The binding energy of these levels (energy distance to the edge of the conduction band) is small as compared to the bandgap.[4] These levels belong to so-called

[4] The typical excitation energy of donors in Si is about 10^{-2} eV, whereas the bandgap is about 1 eV.

shallow donors, for instance, P atoms in Si. Shallow acceptor levels appear in a similar way – by splitting the electron states from the valence band (group-III atoms in Si).

Another type of impurities, such as transition metal atoms, cannot be described by a simple model presented above because the perturbation introduced by them affects the electron states of both conduction and valence bands. The levels created by such impurities can be located anywhere in the bandgap (they are called *deep levels*), as well as in the conduction or valence band, thus creating impurity resonances.

The theory of deep levels deals with impurity and neighboring atoms as a cluster whose electron states one determines in the tight-binding approximation. As an example, we consider transition metal atoms (for example, Mn) in the crystal lattice of GaAs. Usually, Mn replaces Ga (Ga is the positively charged cation), where it has four neighboring As (negatively charged anions). The outer shell of isolated Ga has three electrons in configuration s^2p, whereas As has five electrons in configuration s^2p^3. In a lattice, these eight electrons create four s-p bonds between each Ga and four neighboring As. The electron state in the bond is a superposition of s- and p-states of neighboring atoms.

Isolated Mn has electron configuration $3d^5 4s^2$. When Mn replaces Ga, it has to create new bonds with anions generating impurity states in GaAs. The spherical symmetry of isolated Mn requires five d-electrons to be in the five-fold degenerate state of azimuthal quantum number $l=2$ and magnetic quantum numbers $m=0,\pm1,\pm2$. In the lattice, Mn experiences the crystal field, which lowers the symmetry from spherical to tetrahedral. The crystal field splits d-level into the doublet e_g ($d_{x^2-y^2}$ and $d_{r^2-3z^2}$ orbitals) and triplet t_{2g} (d_{xy}, d_{yz}, d_{zx} orbitals). Positions of levels relative to energy bands one can determine from the calculation of the electronic states of a cluster comprising Mn and four neighboring As. The chemical bonds form due to the overlap of wave functions of atoms Mn and As.

Let us consider the cluster consisting of one Mn atom at point $\boldsymbol{r}=0$ and four nearest atoms of As at the points $\boldsymbol{r}=\boldsymbol{R}_j$ ($j=1,\ldots4$ labels As atoms). Model Hamiltonian of the cluster includes the coupling of all d-orbitals of the atom Mn at $\boldsymbol{r}=0$ to the sp^3 states of cations located at point $\boldsymbol{r}=\boldsymbol{R}_j$. The coupling in each pair of the states Mn and As can be presented as

$$H_{ij}=a_i^\dagger \varepsilon_{di} a_i + b_j^\dagger \varepsilon_{cj} b_j + \left(a_i^\dagger t_{ij} b_j + b_j^\dagger t_{ji}^* a_i\right), \tag{3.42}$$

where $a_i^\dagger$, a_i are the operators of creation and annihilation of an electron in state d_i at Mn atom, $b_j^\dagger$, b_j are the operators of creation and annihilation of state c_j on As-Mn bond located in $\boldsymbol{r}=\boldsymbol{R}_j$ (rendered as a double line in the picture) (Figure 3.4)

The hopping integral is

$$t_{ij}=\int d^3r\psi_{di}^*(\boldsymbol{r})U(\boldsymbol{r})\,\psi_{cj}(\boldsymbol{r}-\boldsymbol{R}_j), \tag{3.43}$$

where $\psi_{di}(\boldsymbol{r})$ is the wave function of d_i state, $\psi_{cj}(\boldsymbol{r})$ is the wave function of sp^3 orbital (denoted as c_j-state), and $U(\boldsymbol{r})$ is the perturbation related to the substitution of atom *Ga* by impurity atom. The state

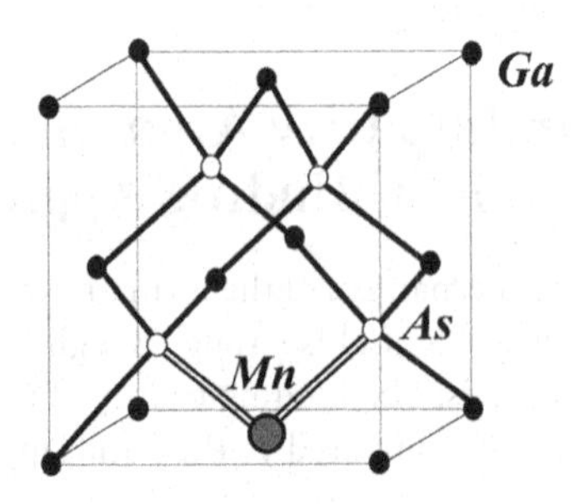

FIGURE 3.4 An atom of Mn in a crystal cell of GaAs.

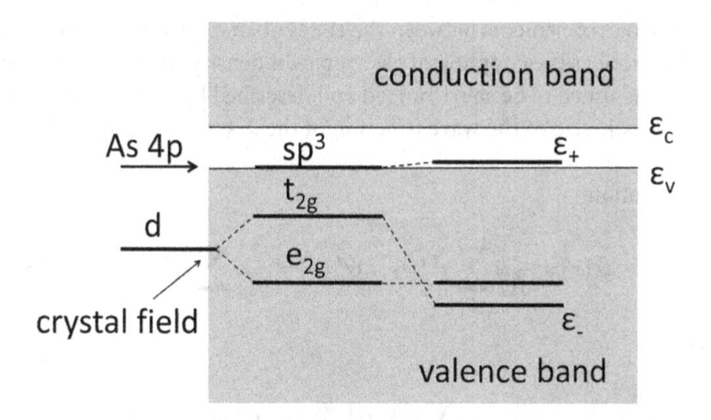

FIGURE 3.5 The genesis of the states of transition-metal impurity in GaAs.

d_i can be one of the states of doublet e_{2g} or triplet t_{2g}. Hopping integral (3.43) for triplet states is much larger than for the doublet ones, meaning that the level e_{2g} is not affected by the coupling to cations. The reason is that wavefunctions $\psi_{e_{2g}}(\boldsymbol{r})$ are elongated in [100] directions and do not overlap to the sp^3 states pointed in [111] directions.

The eigenenergies of Hamiltonian (3.42) are

$$\varepsilon_{\pm} = \frac{1}{2}\left(\varepsilon_{di} + \varepsilon_{cj}\right) \pm \frac{1}{2}\sqrt{\left(\varepsilon_{di} - \varepsilon_{cj}\right)^2 + 4\left|t_{ij}\right|^2}. \tag{3.44}$$

The genesis of the states appearing in GaAs with Mn impurity is shown schematically in Figure 3.5. Note that one of the levels falls in the bandgap near the valence band edge. It plays the role of the acceptor level so that GaAs(Mn) is a p-type semiconductor.

3.7 Magnetic Impurity

The atoms of magnetic impurities in semiconductors usually preserve their magnetic properties even though the electronic structure of the embedded atom changes, as discussed in the previous section. Isolated Mn has five d-electrons arranged among five $(2l+1, l=2)$ states. According to Hund's rule, all electron spins being uncompensated give a total spin of 5/2 . In GaAs host, Mn preserves spin 5/2.

The interaction of magnetic impurity with free electrons includes spin-independent and magnetic parts. The first one is the Coulomb or hard-sphere scattering discussed in the previous sections. The magnetic interaction depends on the impurity spin $\boldsymbol{S}_i$ located at $\boldsymbol{r} = \boldsymbol{R}_i$ and electron spin $\boldsymbol{\sigma}$. It is usually assumed that the impurity spin $\boldsymbol{S}_i$ is classical because its value for magnetic atoms like Mn is relatively large, $S = 5/2$. Hence, the local spin-electron interaction reads as

$$H_{int} = J\int d^3r\, \hat{\psi}^{\dagger}(\boldsymbol{r})(\boldsymbol{\sigma}\cdot\boldsymbol{S}_i)\hat{\psi}(\boldsymbol{r})\delta(\boldsymbol{r}-\boldsymbol{R}_i), \tag{3.45}$$

where $\hat{\psi}^{\dagger}(\boldsymbol{r})$ and $\hat{\psi}(\boldsymbol{r})$ are the spinor operators of creation and annihilation of electrons in the point $\boldsymbol{r}$, respectively, $\sigma_x, \sigma_y, \sigma_z$ are Pauli matrices, constant J is the exchange part of Coulomb interaction between free electron and electron in d-state. Two components $\varphi^{\dagger}$ and $\chi^{\dagger}$ of spinor operator $\hat{\psi}^{\dagger} \equiv (\varphi^{\dagger}, \chi^{\dagger})$ refer to the creation of spin-up and spin-down electrons, respectively. Correspondingly, the second quantization representation for electron spin density operator looks as follows:

$$\hat{s}(\boldsymbol{r}) = \hat{\psi}^{\dagger}(\boldsymbol{r})\boldsymbol{\sigma}\hat{\psi}(\boldsymbol{r}). \tag{3.46}$$

The operator (3.46) has matrix elements between the states of an electron of definite spin and located at $\boldsymbol{r}$. Hamiltonian (3.45) describes the coupling of electron spin density $\boldsymbol{s}(\boldsymbol{r})$ to the impurity spin $\boldsymbol{S}_i$ located at $\boldsymbol{R}_i$. The interaction is assumed to be short-ranged and described by delta-function $\delta(\boldsymbol{r}-\boldsymbol{R}_i)$, which is justified if the characteristic size of the wave function of the d-state is much smaller than the electron wavelength.

Using the $\boldsymbol{k}$-representation

$$\hat{\psi}(\boldsymbol{r})=\frac{1}{\sqrt{\Omega}}\sum_{\boldsymbol{k}}e^{i\boldsymbol{k}\cdot\boldsymbol{r}}\hat{\psi}_{\boldsymbol{k}},\quad \hat{\psi}^{\dagger}(\boldsymbol{r})=\frac{1}{\sqrt{\Omega}}\sum_{\boldsymbol{k}}e^{-i\boldsymbol{k}\cdot\boldsymbol{r}}\hat{\psi}^{\dagger}_{\boldsymbol{k}}, \tag{3.47}$$

we transform (3.45) to

$$H_{int}=J\sum_{\boldsymbol{k}\boldsymbol{k}'}e^{i(\boldsymbol{k}'-\boldsymbol{k})\cdot\boldsymbol{R}_i}\,\hat{\psi}^{\dagger}_{\boldsymbol{k}}\left(\boldsymbol{\sigma}\cdot\boldsymbol{S}_i\right)\hat{\psi}_{\boldsymbol{k}'}. \tag{3.48}$$

As follows from (3.45), scattering of electrons on magnetic impurity may accompany with reversing its spin (*spin-flip*). The impurity spin also changes its orientation so that the total angular moment of impurity and electron is conserved.

Electron spin-flip scattering is at the origin of the Kondo effect in metals at low temperatures: the logarithmic increase in electrical resistivity when temperature decreases.

3.8 Anderson Model for Magnetic Impurity

In the previous section, one assumes that d-electrons are tightly bound to Mn, thus preserving spin $\boldsymbol{S}_i$ of an isolated magnetic atom. If the unfilled atom state of the impurity is not deep enough to prevent electron transitions between localized on the atom and delocalized band states, the magnetic moment, not given a priori, may form. The Anderson model shows how the localized spin forms by taking into account the electrons hopping between the atom state and the delocalized energy band as well as Coulomb interaction.

The Hamiltonian of this model in the second-quantization representation reads

$$H=E_0\sum_{\sigma}a^{\dagger}_{\sigma}a_{\sigma}+\sum_{k\sigma}\varepsilon_k c^{\dagger}_{k\sigma}c_{k\sigma}+\sum_{k\sigma}\left(V_k c^{\dagger}_{k\sigma}a_{\sigma}+V^{*}_k a^{\dagger}_{\sigma}c_{k\sigma}\right), \tag{3.49}$$

where E_0 is the energy level on the impurity atom, ε_k is the conduction energy band, and V_k is the hybridization energy between $\boldsymbol{k}$-states and localized states at the impurity.

In the absence of hybridization, the unperturbed Green's functions of the localized and band electrons have the form,

$$G_{i0}(E)=\frac{1}{E-E_0+i\delta\,sign(E-\mu)}, \tag{3.50}$$

$$G_{k0}(E)=\frac{1}{E-\varepsilon_k+i\delta\;sign(E-\mu)}, \tag{3.51}$$

where μ is the chemical potential. To take into account the hybridization, we use diagrams in Figure 3.6, where thin lines denote the bare (unperturbed) Green's functions, thick lines – perturbed Green's functions. Dashed and solid lines refer to the Green's function of localized and band electrons, respectively.

Analytical expressions for full Green's functions corresponding to series in Figure 3.6 have the form,

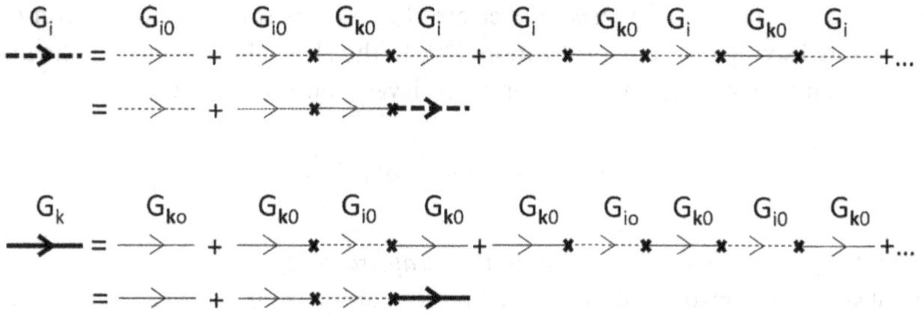

G_i G_{i0} G_{i0} G_{k0} G_i G_i G_{k0} G_i G_{k0} G_i

G_k G_{ko} G_{k0} G_{i0} G_{k0} G_{k0} G_{io} G_{k0} G_{i0} G_{k0}

FIGURE 3.6 Diagrams for the Green's function of the band and localized electrons in the Anderson model of magnetic impurity.

$$G_i(E) = G_{0i}(E) + G_{i0}(E)\Sigma_i(E)G_i(E), \tag{3.52}$$

$$G_k(E) = G_{k0}(E) + G_{k0}(E)\Sigma_k(E)G_k(E), \tag{3.53}$$

where we define the self energies of localized and band electrons, respectively:

$$\Sigma_i(E) = \sum_k |V_k|^2 G_{k0}(E), \tag{3.54}$$

$$\Sigma_k(E) = |V_k|^2 G_{i0}(E) \tag{3.55}$$

Using (3.53) and (3.55), we find

$$G_{k0}(E) = \frac{E - E_0}{(E-\varepsilon_k)(E-\varepsilon_0) - |V_k|^2 + i\delta\, sign(E-\mu)}. \tag{3.56}$$

The poles of the Green's function determine the energy spectrum of electrons hybridized with localized energy level:

$$E_\pm(\boldsymbol{k}) = \frac{\varepsilon_k + E_0}{2} \pm \frac{1}{2}\sqrt{(\varepsilon_k - E_0)^2 + 4|V_k|^2}. \tag{3.57}$$

The real part of the self-energy of localized electrons (3.52) defines the shift of E_0, while the imaginary part describes the finite width of the level caused by hybridization:

$$\mathrm{Im}\,\Sigma_i(E) = -\pi\, sign(E-\mu)\sum_k |V_k|^2 \delta(E-\varepsilon_k) = -\pi|V_E|^2 \rho(E)\, sign(E-\mu) \tag{3.58}$$

where $\rho(E)$ is the electron density of states in unperturbed energy band, and $V_E = V_k\,|_{\varepsilon_k = E}$.

Using Eqs. (3.50), (3.52), and (3.58), we find

$$G_i(E) \simeq \left[E - E_0 + \frac{i\hbar}{2\tau_i} sign(E_0 - \mu)\right]^{-1}, \tag{3.59}$$

where

$$\frac{\hbar}{\tau_i} = 2\pi|V_{E_0}|^2 \rho(E_0). \tag{3.60}$$

We have introduced the *lifetime of quasibound* electron τ_i. That is the time an electron resides on an impurity atom before leaving due to hybridization. Due to the Heisenberg principle, the uncertainty in time casts to uncertainty in energy, in other words, the level acquires finite width,

$$\Gamma_i \equiv \frac{\hbar}{2\tau_i} = \pi |V_{E_0}|^2 \rho(E_0), \tag{3.61}$$

Γ_i is the *width of the impurity level* that relates to the *escape rate* $1/\tau_i$.

The density of states of electrons, bound to the impurity atom, is expressed as

$$\rho_i(E) = -\frac{1}{\pi} \operatorname{Im} G_i^R(E), \tag{3.62}$$

where

$$G_i^R(E) = \frac{1}{E - E_0 + i\Gamma_i} \tag{3.63}$$

is retarded Green's function of the bound electron. Using Eq. (3.63), we find

$$\rho_i(E) = \frac{1}{\pi} \frac{\Gamma_i}{(E - E_0)^2 + \Gamma_i^2}, \tag{3.64}$$

which is the Lorentz function centered at $E = E_0$ of the width Γ_i.

The occupation number of bound electrons on the impurity level of finite width:

$$n_i = \int_{-\infty}^{\mu} \rho_i(E)\, dE = \frac{\Gamma_i}{\pi} \int_{-\infty}^{\mu} \frac{dE}{(E - E_0)^2 + \Gamma_i^2} = \frac{1}{\pi}\left(arctg \frac{\mu - E_0}{\Gamma_i} + \frac{\pi}{2} \right) \tag{3.65}$$

In particular, for $E_0 = \mu$, one gets $n_i = \frac{1}{2}$. This result is the same for the spin up and down electrons, $n_{i\uparrow} = n_{i\downarrow}$, as we assumed all states are spin-degenerate and omitted the spin index after Eq. (3.49). The spin balance means that the total impurity spin is zero, and impurity is nonmagnetic.

The model with hybridization can explain the formation of the magnetic impurity. For this purpose, Hamiltonian (3.49) should include the Coulomb repulsion between two electrons when they occupy the impurity level.[5] The Coulomb repulsion on an impurity site, the *Hubbard interaction* term, is

$$H_{int} = U \hat{n}_{i\uparrow} \hat{n}_{i\downarrow}, \tag{3.66}$$

where $\hat{n}_{i\sigma} = a_\sigma^\dagger a_\sigma$ is the number operator of localized electrons of spin $\sigma = \uparrow, \downarrow$ and U is the Hubbard constant, which determines the Coulomb interaction of two electrons at a distance of the characteristic size of d-state.

The model (3.49) with Hubbard interaction (3.66) included is called the *Anderson model for magnetic impurities*. To solve this model, we use the *mean-field approximation*. Namely, we assume that there is a mean number of spin-up, $n_{i\uparrow}$, and spin-down, $n_{i\downarrow}$, electrons localized at the impurity. It allows us to present the Hamiltonian as a sum for spin-up and spin-down electrons, $H = H_\uparrow + H_\downarrow$, where

$$H_\uparrow = (\varepsilon_0 + U n_\downarrow) a_\uparrow^\dagger a_\uparrow + \sum_k \varepsilon_k c_{k\uparrow}^\dagger c_{k\uparrow} + \sum_k \left(V_k c_{k\uparrow}^\dagger a_\uparrow + V_k^* a_\uparrow^\dagger c_{k\uparrow} \right). \tag{3.67}$$

[5] The Pauli principle allows two electrons to be with the same quantum numbers if they have opposite spins.

$$H_{\downarrow} = \left(\varepsilon_0 + Un_{\uparrow}\right) a_{\downarrow}^{\dagger} a_{\downarrow} + \sum_k \varepsilon_k c_{k\downarrow}^{\dagger} c_{k\downarrow} + \sum_k \left(V_k c_{k\downarrow}^{\dagger} a_{\downarrow} + V_k^{*} a_{\downarrow}^{\dagger} c_{k\downarrow}\right). \tag{3.68}$$

In other words, we approximate H as $\hat{n}_{i\uparrow}\hat{n}_{i\downarrow} \to \hat{n}_{i\uparrow} n_{i\downarrow}$ and $\hat{n}_{i\uparrow}\hat{n}_{i\downarrow} \to \hat{n}_{i\downarrow} n_{i\uparrow}$ in $H_{\uparrow}$ and $H_{\downarrow}$, respectively. Physically, it means that the mean number of spin-down electrons creates an effective potential $Un_{i\downarrow}$ for spin-up electrons, and the mean number of spin-up electrons creates a potential $Un_{\uparrow}$ for spin-down electrons. In this approximation, we can substitute $E_0 \to E_0 + Un_{i\sigma}$ in Eqs. (3.50) to (3.65) and find

$$n_{i\sigma} = \frac{1}{\pi}\left[arctg\left(\frac{\mu - E_{0\sigma}}{\Gamma_{i\sigma}}\right) + \frac{\pi}{2}\right], \tag{3.69}$$

where $E_{i\uparrow} = E_0 + Un_{i\downarrow}$, $E_{i\downarrow} = E_0 + Un_{i\uparrow}$, and $\Gamma_{i\sigma} = \pi\left|V_{E_{0\sigma}}\right|^2 \rho\left(E_{0\sigma}\right)$. These equations have solution $n_{i\uparrow} \neq n_{i\downarrow}$, which means the magnetization of impurity. Indeed, if $E_{0\downarrow} < E_{0\uparrow}$, from (3.69) follows $n_{i\downarrow} > n_{i\uparrow}$. In particular, if $\Gamma_{i\sigma} \ll U$ and $E_{0\downarrow} < \mu < E_{0\uparrow}$, we get $n_{i\downarrow} \simeq 1$ and $n_{i\uparrow} \simeq 0$.

So, if the localized state energy is close to the Fermi level, the hybridization and on-site Coulomb interaction might create the magnetic moment of an impurity.

3.9 Heavily Doped Semiconductors: Disorder Potential

The previous discussion of the formation of an impurity level and magnetic moment concerns one impurity. This approach is valid if the impurity density is low enough to prevent the overlap of localized states. Usually, the concentration of donors N_i in doped Si is below 10^{18} cm^{-3}. At these concentrations, the overlap of the wavefunctions localized at impurities at an average distance $d \sim N_i^{-1/3}$ is small, so the probability of electron hopping between impurity sites is negligible. When concentration grows, the effects of hopping and interaction between localized electrons become essential. It turns out that the properties of such *heavily doped semiconductors* with $a \geq d$ (here a is the radius of impurity state) are substantially different as compared to weakly doped semiconductors. The difference is associated with the impurity-induced random field that introduces disorder to the crystal lattice. The impurity potential is

$$V(\boldsymbol{r}) = \sum_i \mathrm{v}_i\left(\boldsymbol{r} - \boldsymbol{R}_i\right), \tag{3.70}$$

where $\mathrm{v}_i\left(\boldsymbol{r} - \boldsymbol{R}_i\right)$ is the potential of a single impurity localized at $\boldsymbol{r} = \boldsymbol{R}_i$, and the sum runs over all impurity sites. When all impurities are identical, functions $\mathrm{v}_i\left(\boldsymbol{r} - \boldsymbol{R}_i\right)$ differ only by their position in a crystal lattice.

If locations of impurities in the lattice are accidental and not correlated, each vector $\boldsymbol{R}_i$ is an independent random variable. Since we believe that the physical properties (e.g., electrical resistance) of a large crystal do not depend on specific positions of impurities, the macroscopic characteristics of the sample are an average over the random variables.

Another model of impurity disorder introduces random field $V(\boldsymbol{r})$ with a specific type of correlation. For example, *Gaussian disorder* is given by random field $V(\boldsymbol{r})$ with the correlator,

$$\left\langle V(\boldsymbol{r})\right\rangle = 0, \quad \left\langle V(\boldsymbol{r})V(\boldsymbol{r}')\right\rangle = \gamma\,\delta\left(\boldsymbol{r} - \boldsymbol{r}'\right), \tag{3.71}$$

where we neglect all higher correlators, γ is the magnitude of disorder, $\langle \ldots \rangle$ means an average over all possible realizations of the random field. The Gaussian disorder usually applies to heavily doped semiconductors where wavefunctions localized at neighboring impurities overlap.

3.10 Impurity Bands. Impurity Band Tail

Using the Gaussian disorder potential, one can calculate the density of states in a heavily doped semiconductor. For this purpose, we consider the Hamiltonian of electrons with a parabolic energy spectrum:

$$H = -\frac{\hbar^2 \Delta}{2m} + V(\boldsymbol{r}), \tag{3.72}$$

where $V(\boldsymbol{r})$ is the realization of the random field. The electron density of states is expressed by average over possible configurations of the random field,

$$\rho(E) = -\frac{1}{\pi} Tr\left\langle \mathrm{Im}\hat{G}^R(E)\right\rangle = -\frac{1}{\pi}\int d^3r \left\langle \mathrm{Im}G^R(E;\boldsymbol{r},\boldsymbol{r})\right\rangle, \tag{3.73}$$

where averaging $\langle\ldots\rangle$ to be explained later and Green's function $G^R(E;\boldsymbol{r},\boldsymbol{r}')$ is the solution to the equation

$$(E - H + i\delta)G^R(E;\boldsymbol{r},\boldsymbol{r}') = \delta(\boldsymbol{r}-\boldsymbol{r}'). \tag{3.74}$$

Term $i\delta$ defines $G^R(E;\boldsymbol{r},\boldsymbol{r}')$ as *retarded* Green's function. If $\varphi_n(\boldsymbol{r})$ is the eigenfunction of Schrödinger equation $(H - \varepsilon_n)\,\varphi_n = 0$ with the Hamiltonian (3.72), the solution to (3.74) has the form,

$$G^R(E;\boldsymbol{r},\boldsymbol{r}') = \sum_n G_n^R(E)\varphi_n(\boldsymbol{r})\varphi_n^*(\boldsymbol{r}'), \tag{3.75}$$

where $G_n^R(E)$ is the Green's function in the representation of eigenfunctions of operator H:

$$G_n^R(E) = \frac{1}{E - \varepsilon_n + i\delta}. \tag{3.76}$$

By substituting (3.75) into (3.74) we get $(E - H + i\delta)G^R(E;\boldsymbol{r},\boldsymbol{r}') = \sum_n G_n^R(E)(E - \varepsilon_n + i\delta)\varphi_n(\boldsymbol{r})\varphi_n^*(\boldsymbol{r}') = \delta(\boldsymbol{r}-\boldsymbol{r}')$, where we used the orthogonality condition $\sum_n \varphi_n(\boldsymbol{r})\varphi_n^*(\boldsymbol{r}') = \delta(\boldsymbol{r}-\boldsymbol{r}')$. Thus, (3.75) is the solution to (3.74), where both ε_n and $\varphi_n(\boldsymbol{r})$ are determined assuming a particular configuration of $V(\boldsymbol{r})$.

To find the density of states, we substitute (3.75) in (3.73) and get:

$$\rho(E) = \sum_n \langle \delta(E - \varepsilon_n)\rangle \tag{3.77}$$

From (3.73), the density of states is the averaged Green's function at $\boldsymbol{r} = \boldsymbol{r}'$. The averaging procedure implies integration over all realizations of $V(\boldsymbol{r})$ with the Gaussian distribution function:[6]

$$\begin{aligned}\left\langle G^R(\varepsilon;\boldsymbol{r},\boldsymbol{r})\right\rangle &= \frac{1}{A}\int \mathcal{D}V(\boldsymbol{r})\exp\left[-(2\gamma)^{-1}\int d^3r\,V^2(\boldsymbol{r})\right]G^R(\varepsilon;\boldsymbol{r},\boldsymbol{r}) \\ &= \frac{1}{A}\int \mathcal{D}V(\boldsymbol{r})\exp\left\{-\int d^3r\frac{V^2(\boldsymbol{r})}{2\gamma} - \log G^R(\varepsilon;\boldsymbol{r},\boldsymbol{r})\right\},\end{aligned} \tag{3.78}$$

[6] The Gaussian distribution for simple variable x reads $\rho(x) = \exp(-x^2/2\gamma)/A$ where $A = \sqrt{2\pi\gamma}$ is the normalization constant so that $\int_{-\infty}^{\infty}\rho(x)dx = 1$.

where $A=\int \mathcal{D}V(\mathbf{r})\exp\left[-(2\gamma)^{-1}\int d^3r V^2(\mathbf{r})\right]$ is the normalization factor in Gaussian probability distribution. Equation (3.78) explicitly determines the averaging procedure. It includes the *functional* (or *path) integral* $\int \mathcal{D}V(\mathbf{r})\ldots$ which is an infinite number of independent Gaussian integrals over V in each point of space $\mathbf{r}$: $\int \mathcal{D}V(\mathbf{r})\equiv\int\ldots\int\prod_{\mathbf{r}} dV(\mathbf{r})$. The main contribution to functional integral (3.78) comes from the potential profile $V(\mathbf{r})$, for which the function in curvy brackets of the exponent has the minimum. The corresponding *saddle point* potential $V_{sp}(\mathbf{r})$ is the solution to the equation,

$$\frac{\delta}{\delta V(\mathbf{r})}\left[\frac{V^2(\mathbf{r})}{2\gamma}-\log G^R(E;\mathbf{r},\mathbf{r})\right]=0, \tag{3.79}$$

where $\delta/\delta V$ is the functional derivative.

To find the derivative of the second term we present $\left(G^R\right)^{-1}=\left(G_0^R\right)^{-1}-V$, where G_0^R corresponds to $V=0$. Then we get $\delta\log G^R/\delta V=-\delta\log\left(G^R\right)^{-1}/\delta V=G^R$. As a result, we find the solution of (3.79) in the form:

$$V_{sp}(\mathbf{r})=\gamma G^R(E;\mathbf{r},\mathbf{r})=\gamma\sum_n\frac{|\varphi_n(\mathbf{r})|^2}{E-\varepsilon_n+i\delta}, \tag{3.80}$$

where ε_n and $\varphi_n(\mathbf{r})$ correspond to potential $V_{sp}(\mathbf{r})$. Function $V_{sp}(\mathbf{r})$ – *optimal trajectory* – determines the main contribution to functional integral.

In the optimal trajectory method, one assumes that at a given E, the ground state $\varphi_0(\mathbf{r})$ of Schrödinger equation with $V_{sp}(\mathbf{r})$ determined by (3.80) has energy E_0 close to E. Having said in other words, the density of states at energy E gets the main contribution from that realization of random potential, in which the energy of localized state $E_0\simeq E$. Thus, we can approximate the relation (3.80) by

$$V_{sp}\simeq\frac{\gamma|\varphi_0(\mathbf{r})|^2}{E-E_0} \tag{3.81}$$

and substitute it to the Schrödinger equation for $\varphi_0(\mathbf{r})$:

$$\left(-\frac{\hbar^2\Delta}{2m}+\frac{\gamma|\varphi_0(\mathbf{r})|^2}{E-E_0}-E_0\right)\varphi_0(\mathbf{r})=0. \tag{3.82}$$

That is a nonlinear equation to be solved numerically. However, using the dimensional analysis, one can estimate the solution without calculations. First, we present (3.82) in the form,

$$\left(-\Delta+\frac{2m\gamma|\varphi_0(\mathbf{r})|^2}{\hbar^2(E-E_0)}+\kappa^2\right)\varphi_0(\mathbf{r})=0, \tag{3.83}$$

where $\kappa=\sqrt{2m|E_0|}/\hbar$ and $E_0<0$ falls into the bandgap as the energy of the localized state. Parameter κ determines the radius of state $\varphi_0(\mathbf{r})$, $r_c\simeq\kappa^{-1}$, i.e., $\varphi_0(\mathbf{r})\sim\exp(-\kappa r)$. Indeed, in (3.83), the first term $\sim\kappa^2$ is of the order of the third one. The second term should also be of the same order of magnitude. The characteristic value of $\varphi_0(r)$, $\varphi_0(0)\sim\kappa^{3/2}$ follows from the normalization condition $\int d^3r\varphi_0^2(r)\simeq\varphi_0^2(0)r_c^3=1$. Using relation (we omit numerical coefficients)

$$\frac{m\gamma\kappa^3}{\hbar^2|E-E_0|}\sim\kappa^2, \tag{3.84}$$

we find that $|E-E_0|\sim\frac{\gamma m^{3/2}\sqrt{|E_0|}}{\hbar^3}\ll|E_0|$, provided that $|E_0|\gg\frac{\gamma^2m^3}{\hbar^6}$. So our approach is valid at large $|E|$. Now, the saddle-point potential (3.81) takes the form,

$$V_{sp}(\boldsymbol{r})\sim-\frac{\hbar^2\kappa^2}{m}e^{-\kappa r}. \tag{3.85}$$

Correspondingly, the probability of realization of $V_{sp}(\boldsymbol{r})$ is

$$P\left[V_{sp}(\boldsymbol{r})\right]=\frac{1}{A}\exp\left[-(2\gamma)^{-1}\int d^3\boldsymbol{r}\,V_{sp}^2(\boldsymbol{r})\right]\sim\frac{1}{A}\exp\left(-\frac{\hbar^3\sqrt{|E|}}{\gamma m^{3/2}}\right). \tag{3.86}$$

Assuming that E_0, the ground level in the saddle-point potential, determines the density of states (DOS) at energy E, we conclude that $\rho(E)\sim\exp\left(-\hbar^3\sqrt{|E|}/\gamma m^{3/2}\right)$ for large $|E|$. The dependence $\rho(E)$ exponentially decaying with energy inside the bandgap is called the *tail of the density of states.* The nonzero DOS at $E<0$ is due to a large number of discrete energy levels created by potential wells of random magnitude, size, and shape. If the impurity density is not too high, DOS has a maximum at the energy of a single impurity level. The width of the DOS peak is of the order of hopping integral at the mean distance between impurities. The DOS peak of finite width is called the *impurity band.*

Forming an impurity band from impurity levels ε_i with increasing density N_i can be understood as the "repulsion" of these levels caused by the overlap of impurity states placed at small distances, $d\sim a$ (a is the radius of a localized impurity state with energy ε_i). For a pair of close impurities, an electron hopping creates collective states $\varepsilon_{1,2}$, which are superpositions of single-level states in that pair. When d decreases, the splitting between ε_1 and ε_2 increases exponentially. Thus, at a small impurity concentration, the density of states is forming from a large number of impurity pairs at random distances, having random splitting of energy levels. This picture remains the same if one considers a cluster of three impurities. Since distances between the three are different, the two nearest would have superposition states, and the third one remains as an isolated state.

When we study the formation of the impurity band, the Coulomb interaction comes into play. Calculating the donor (acceptor) states in semiconductors, we assume the localized state occupied by one electron (hole). It is because the two-electron state has much larger energy, ε_i+U, U is the on-site Coulomb interaction between two electrons with opposite spins (not prohibited by the Pauli principle). In the impurity band, the electron hopping decreases the role of electron interaction: if hopping integral W is large, $W>U$, two-electron states are to be taken into account. The transition from the one-electron to the two-electron state due to Coulomb interaction is called the *Mott transition.* In a periodic crystal, the nonzero probability of a two-electron state on the same lattice site means the possibility of translation motion, so Mott transition models the transition metal-insulator. The electron–electron interactions affects the density of states in doped semiconductors as schematically illustrated in Figure 3.7.

The impurity band has two peaks with a quasi-gap between them caused by Coulomb interaction. With increasing impurity density, the width of each subband grows, and the gap disappears that corresponds to the Mott dielectric-to-metal transition in the impurity band [4].

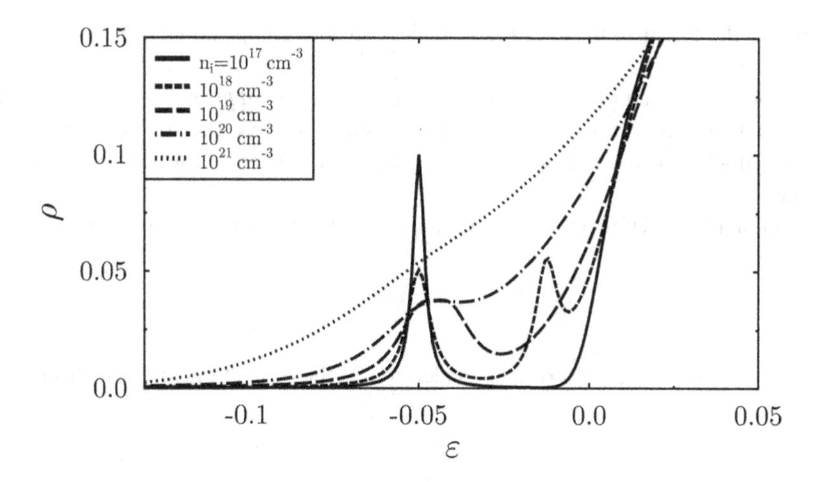

FIGURE 3.7 Variation of the density of state with increasing concentration of donor impurities in Si.

3.11 Semiconductor Alloys

Ternary semiconductor alloy is a binary semiconductor matrix where atoms in one of the sublattices are replaced randomly by other atoms of the same group. Example alloy is $Ga_{1-x}Al_xAs$ where Al replaces Ga. Alloying presents a specific type of disorder. The atomic electronic structures of Ga and Al are similar so that the perturbation potential is weak and, unlike doping, does not create localized states. On the other hand, the parameters of electron Hamiltonian vary when we increase Al content x so that at 100% substitution ($x = 1$), we come to the Hamiltonian with parameters of AlAs.

In contrast to the relatively small density of doping impurities,[7] the alloy content may vary in the whole range from 0 to 1.

3.11.1 Virtual Crystal and CPA Approximation

We start with the model Hamiltonian for electrons in the conduction band with a parabolic energy spectrum in the field of randomly distributed substitution atoms, which create short-range potential in points $\boldsymbol{R}_i$

$$H = E_0 - \frac{\hbar^2 \Delta}{2m} + V_0 \sum_{i=1}^{N_{imp}} \delta(\boldsymbol{r} - \boldsymbol{R}_i), \tag{3.87}$$

where Ω is the crystal volume, N_{imp} is the number of impurity atoms, $\boldsymbol{R}_i$ is the random variable, and E_0 is the conduction band edge.

The first term of this Hamiltonian describes the semiconductor compound without substitutional atoms ($x = 0$). One may express alloy composition as $x = n_i/n_0$, where $n_i = N_{imp}/\Omega$ is the density of substitutional atoms, and $n_0 = N_0/\Omega$ is the density of sites in the host sublattice.

In the *virtual crystal* approximation, we replace the random potential with its mean value:

$$V_0 \left\langle \sum_i \delta(\boldsymbol{r} - \boldsymbol{R}_i) \right\rangle_0 = \frac{V_0}{\Omega} \sum_{i=1}^{N_{imp}} \int_{\Omega} d^3 R_i \, \delta(\boldsymbol{r} - \boldsymbol{R}_i) = x n_0 V_0. \tag{3.88}$$

[7] Chemical solubility limits the densities of donors and acceptors in the material. When a density exceeds a solubility limit, clusters and inclusions of different phases may appear.

Term (3.88) makes the conduction band edge dependent on alloy composition: E_0 at $x = 0$ to $E = E_0 + n_0V_0$ at $x = 1$.

In a more sophisticated approach, known as *coherent potential approximation* or CPA, the average of the random potential depends on x nonlinearly. Besides, in CPA, the coherent potential can be a complex-valued function with an imaginary part that causes disorder-induced decay of plane-wave electronic states.

The coherent potential is defined formally as a model potential which turns zero the average T-matrix:

$$\langle T_j(E)\rangle = 0, \tag{3.89}$$

where $T_j(E)$ is the T-matrix of scattering of an electron in the coherent field $V_{ch}(E)$ from site j. In the equation for T-matrix,

$$T_j(E) = V_{0j} + V_{0j}\int \frac{d^3k}{(2\pi)^3} G_{\mathbf{k}}(E) T_j(E). \tag{3.90}$$

Green's function includes $V_{ch}(E)$ explicitly:

$$G_{\mathbf{k}}(E) = \frac{1}{E - \varepsilon_{\mathbf{k}} - V_{ch}(E) + i\delta\, sign(E - \mu)} \tag{3.91}$$

where $\varepsilon_{\mathbf{k}} = E_0 + \hbar^2k^2/2m$ is the energy spectrum at small k. As introduced in Eqs. (3.89) to (3.91), the coherent potential $V_{ch}(E)$ presents the mean-field potential for electrons in a disordered system.

In a binary alloy $A_{1-x}B_x$

$$(1-x)\,T_A(E) + xT_B(E) = 0 \tag{3.92}$$

For the T-matrices $T_A(E)$ and $T_B(E)$ we use (3.36). Thus, to find the coherent potential $V_{ch}(E)$, one has to solve the equation:

$$\frac{(1-x)V_{0A}}{1 - V_{0A}\,F(E)} + \frac{xV_{0B}}{1 - V_{0B}F(E)} = 0, \tag{3.93}$$

with

$$F(E) = \int \frac{d^3\mathbf{k}}{(2\pi)^3} G_{\mathbf{k}}(E) = \int \frac{\rho_0(t)\,dt}{E - t - V_{ch}(E) + i\delta\, sign(E - \mu)}. \tag{3.94}$$

Real and imaginary parts of $V_{ch}(E)$ follow from (3.93) and (3.94). To avoid divergence in integral (3.94) with the parabolic energy spectrum, one can use the tight-binding spectrum of a simple rectangular lattice (a_0 is the lattice constant):

$$\varepsilon_{\mathbf{k}} = \frac{\hbar^2}{ma_0^2}\left[3 - \cos(k_xa_0) - \cos(k_ya_0) - \cos(k_za_0)\right] \tag{3.95}$$

which gives the parabolic limit at small $ka_0 \ll 1$.

3.12 Electron in a Smooth Disorder Potential

Gaussian correlator (3.71) describes the short-ranged disorder. If the random field is smoothly varying in space, the correlation length a of disorder potential $V(\mathbf{r})$ could be much larger than the electron wavelength, $ka \gg 1$. In this limit, the smooth disorder potential model is justified, and so semiclassical approximation is applicable [5,6]. In what follows, we deal with the problem using Green's function formalism.

The Hamiltonian of electrons with parabolic energy spectrum in disorder field $V(\mathbf{r})$ is the operator,

$$H = -\frac{\hbar^2\Delta}{2m} + V(\mathbf{r}) \equiv H_0 + V(\mathbf{r}), \tag{3.96}$$

where the potential $V(\mathbf{r})$ is a random function with the correlator,

$$\langle V(\mathbf{r})\rangle = 0, \quad \langle V(\mathbf{r})V(\mathbf{r}')\rangle = \gamma_0\, g(\mathbf{r}-\mathbf{r}'), \tag{3.97}$$

where $\langle\ldots\rangle$ means averaging over randomness, $g(\mathbf{r}-\mathbf{r}')$ has the characteristic length a, so that at $|\mathbf{r}-\mathbf{r}'| > a$ the correlator exponentially decays to zero. For example, we can assume that $g(\mathbf{r}-\mathbf{r}') \sim \exp\left(-(\mathbf{r}-\mathbf{r}')^2/a^2\right)$.

Green's function of electrons, described by Hamiltonian (3.96) with a given configuration of random field $V(\mathbf{r})$, reads (see also Eqs. (3.29),(3.30))

$$\hat{G}(E) = (E-H)^{-1}. \tag{3.98}$$

The equation can be presented as a series (for brevity, we omit E-dependence)

$$\hat{G} = \hat{G}_0 + \hat{G}_0 V(\mathbf{r})\hat{G}_0 + \hat{G}_0 V(\mathbf{r})\hat{G}_0 V(\mathbf{r})\hat{G}_0 + \ldots \tag{3.99}$$

or in the matrix form

$$\begin{aligned} G(\mathbf{r},\mathbf{r}') &= G_0(\mathbf{r},\mathbf{r}') + \sum_{r_1} G_0(\mathbf{r},\mathbf{r}_1)\, V(\mathbf{r}_1) G_0(\mathbf{r}_1,\mathbf{r}') \\ &+ \sum_{r_1,r_2} G_0(\mathbf{r},\mathbf{r}_1)\, V(\mathbf{r}_1) G_0(\mathbf{r}_1,\mathbf{r}_2) V(\mathbf{r}_2) G_0(\mathbf{r}_2,\mathbf{r}') + \ldots \end{aligned} \tag{3.100}$$

After averaging, only the terms of even order in $V(\mathbf{r})$ survive and we get

$$\begin{aligned} \langle G(\mathbf{r},\mathbf{r}')\rangle &= G_0(\mathbf{r},\mathbf{r}') + \gamma_0 \sum_{r_1,r_2} g(\mathbf{r}_1-\mathbf{r}_2) G_0(\mathbf{r},\mathbf{r}_1) G_0(\mathbf{r}_1,\mathbf{r}_2) G_0(\mathbf{r}_2,\mathbf{r}) \\ &+ \gamma_0^2 \sum_{r_1,r_2,r_3,r_4} \left[g(\mathbf{r}_1-\mathbf{r}_2) g(\mathbf{r}_3-\mathbf{r}_4) + g(\mathbf{r}_1-\mathbf{r}_3) g(\mathbf{r}_2-\mathbf{r}_4) + g(\mathbf{r}_1-\mathbf{r}_4) g(\mathbf{r}_2-\mathbf{r}_3) \right] \\ &\times G_0(\mathbf{r},\mathbf{r}_1) G_0(\mathbf{r}_1,\mathbf{r}_2) G_0(\mathbf{r}_2,\mathbf{r}_3) G_0(\mathbf{r}_3,\mathbf{r}_4) G_0(\mathbf{r}_4,\mathbf{r}') + \ldots \end{aligned} \tag{3.101}$$

Graphically, this equation is equivalent to Feynman diagrams in Figure 3.8, where the thick line is $G(\mathbf{r},\mathbf{r}')$, the thin line is $G_0(\mathbf{r},\mathbf{r}')$, and the dashed line corresponds to the disorder correlator $g(\mathbf{r}-\mathbf{r}')$.

Analyzing the topology of diagrams, we conclude that the infinite series in the right-hand side is the equation for the averaged Green's function $G(\mathbf{r},\mathbf{r}')$ (last row in Figure 3.8):

$$\langle G(\mathbf{r},\mathbf{r}')\rangle = G_0(\mathbf{r},\mathbf{r}') + G_0(\mathbf{r},\mathbf{r}_1)\Sigma(\mathbf{r}_1,\mathbf{r}_2) G(\mathbf{r}_2,\mathbf{r}'), \tag{3.102}$$

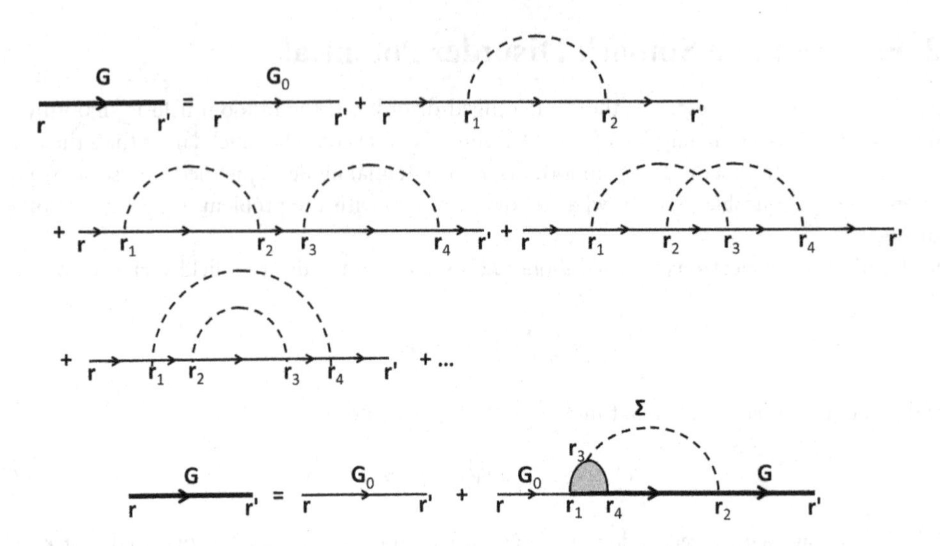

FIGURE 3.8 Average Green's function for electron scattered by the random field (upper rows) and the same series written as an integral equation (lower row).

where

$$\Sigma(r_1,r_2)=\gamma_0\sum_{r_3,r_4} g(r_3-r_2)\Gamma(r_1,r_3,r_4)\langle G(r_4,r_2)\rangle \tag{3.103}$$

is *the self-energy part* and $\Gamma(r_1,r_2)$ is the *vertex function*. Bare Green's function G_0 and the disorder-averaged full function $\langle G\rangle$ are spatially homogeneous, meaning that they depend only on the coordinate difference: $G_0(\boldsymbol{r},\boldsymbol{r}')=G_0(\boldsymbol{r}-\boldsymbol{r}')$ and $\langle G(\boldsymbol{r},\boldsymbol{r}')\rangle=\langle G(\boldsymbol{r}-\boldsymbol{r}')\rangle$. Therefore, one can use the Fourier transformation over $(\boldsymbol{r}-\boldsymbol{r}')$ and present Eq. (3.102) as follows (from now on, we omit the angle brackets for averaged Green's function)

$$G(\boldsymbol{k})=G_0(\boldsymbol{k})+G_0(\boldsymbol{k})\Sigma(\boldsymbol{k})G(\boldsymbol{k}) \tag{3.104}$$

and

$$\Sigma(\boldsymbol{k})=\gamma_0\int\frac{d^3q}{(2\pi)^3}\Gamma(\boldsymbol{k},\boldsymbol{q},\boldsymbol{k}-\boldsymbol{q})g(\boldsymbol{q})G(\boldsymbol{k}-\boldsymbol{q}), \tag{3.105}$$

where $G_0(\boldsymbol{k})=(E-\varepsilon_k\pm i\delta)^{-1}$. The sign plus or minus refers to retarded (R) or advanced (A) Green's function, respectively.

Assuming that the function $g(\boldsymbol{r}-\boldsymbol{r}')$ slowly depends on $(\boldsymbol{r}-\boldsymbol{r}')$ as compared to $G_0(\boldsymbol{r}-\boldsymbol{r}')$ and $\langle G(\boldsymbol{r},\boldsymbol{r}')\rangle$, we conclude that the main contribution to (3.105) stems from $q \ll k$ so we can substitute $g(\boldsymbol{q})$ by the delta function, $g(\boldsymbol{q})\simeq\delta(\boldsymbol{q})$. Our assumption corresponds to the *semiclassical approximation* to quantum mechanics (electron wavelength is much smaller than the characteristic length a of disorder variation). Then we come to the equation,

$$G=G_0+\gamma_0 G_0\Gamma_0 G^2, \tag{3.106}$$

where we denote $\Gamma_0\equiv\Gamma(\boldsymbol{k},0,\boldsymbol{k})$ and omit $\boldsymbol{k}$-dependence.

As a next step, we express the vertex function at zero momentum transfer $\boldsymbol{q}$ using the *Ward identity*:

$$\Gamma_0 = \frac{dG^{-1}}{dE}. \tag{3.107}$$

Indeed, differentiating the Green's function presented by an arbitrary number of diagrams with dashed lines, we are taking derivative of every function $G_0 = \left(E - \varepsilon_k \pm i\delta\right)^{-1}$ in the series of diagrams. After differentiating, one obtains the series,

$$\frac{dG}{dE} = -G\Gamma_0 G, \tag{3.108}$$

and then comparing (3.108) with $\partial\left(GG^{-1}\right)/\partial E = 0$, we obtain (3.107). Substituting Ward identity (3.107) in (3.106), one obtains

$$\gamma_0 \frac{dG}{dx} + Gx - 1 = 0, \tag{3.109}$$

where $x = E - \varepsilon_k \pm i\delta$. Corresponding homogeneous equation

$$\gamma_0 \frac{dy}{dx} + xy = 0 \tag{3.110}$$

has the solution $y(x) = \exp\left(-x^2/2\gamma_0\right)$. Hence, we are looking for a solution to (3.109) in the form

$$G(x) = c(x)\exp\left(-\frac{x^2}{2\gamma_0}\right). \tag{3.111}$$

Substituting (3.111) in (3.109) we find the equation for function $c(x)$:

$$\gamma_0 \frac{dc}{dx} = \exp\left(\frac{x^2}{2\gamma_0}\right), \tag{3.112}$$

which has the general solution

$$c(x) = \frac{1}{\gamma_0}\int_0^x dt \exp\left(\frac{t^2}{2\gamma_0}\right) + C_1, \tag{3.113}$$

where C_1 is a constant. One can modify this expression multiplying the integrand by

$$1 = \frac{1}{\sqrt{2\pi\gamma_0}}\int_{-\infty}^{\infty} dV \exp\left(-\frac{V^2}{2\gamma_0}\right) = \frac{1}{\sqrt{2\pi\gamma_0}}\int_{-\infty}^{\infty} dV \exp\left(-\frac{(V-t)^2}{2\gamma_0}\right) \tag{3.114}$$

and then integrating over t:

$$\begin{aligned} c(x) &= \frac{1}{\sqrt{2\pi\gamma_0}}\int_{-\infty}^{\infty} \frac{dV}{V} \exp\left(-\frac{V^2}{2\gamma_0}\right)\left[\exp\left(-\frac{Vx}{\gamma_0}\right) - 1\right] + C_1 \\ &= \frac{1}{\sqrt{2\pi\gamma_0}} \exp\left(\frac{x^2}{2\gamma_0}\right)\int_{-\infty}^{\infty} \frac{dV}{x+V} \exp\left(-\frac{V^2}{2\gamma_0}\right) - \frac{1}{\sqrt{2\pi\gamma_0}}\int_{-\infty}^{\infty} \frac{dV}{V} \exp\left(-\frac{V^2}{2\gamma_0}\right) + C_1. \end{aligned} \tag{3.115}$$

Choosing constant C_1 to cancel the second integral in (3.115), we finally obtain the solution for the Gauss-averaged Green's function in fluctuating field V:

$$G(\boldsymbol{k})=\frac{1}{\sqrt{2\pi\gamma_0}}\int_{-\infty}^{\infty}\frac{\exp\left(-\frac{V^2}{2\gamma_0}\right)dV}{E-\varepsilon_k-V\pm i\delta}. \tag{3.116}$$

From Eq. (3.116), one can calculate the density of states:

$$\rho(E)=-\frac{2}{\pi}\int\frac{d^3k}{(2\pi)^3}G^R(\boldsymbol{k})=\frac{1}{\sqrt{2\pi\gamma_0}}\int_{-\infty}^{\varepsilon}\exp\left(-\frac{V^2}{2\gamma_0}\right)\rho_0(E-V)dV$$

$$=\frac{m^{3/2}}{\sqrt{\gamma_0\pi^5}\hbar^3}\int_{-\infty}^{\varepsilon}\exp\left(-\frac{V^2}{2\gamma_0}\right)\sqrt{E-V}\,dV, \tag{3.117}$$

where $\rho_0(E)=(2m)^{3/2}\sqrt{E}/2\pi^2\hbar^3$. If $E<0$ and $|E|\gg\sqrt{2\gamma_0}$ (i.e., deep in the tail of the density of states), the main contribution follows from large V, for which the probability is small. Introducing notation $u=E-V/\sqrt{2\gamma_0}$, for $E<0$, we obtain

$$\rho(\tilde{\varepsilon})=\frac{2^{3/4}\gamma_0^{1/4}m^{3/2}\exp\left(-\tilde{\varepsilon}^2\right)}{\pi^{5/2}\hbar^3}\int_0^{\infty}\exp\left(-u^2-2|\tilde{\varepsilon}|u\right)\sqrt{u}\,du, \tag{3.118}$$

where $\tilde{\varepsilon}=E/\sqrt{2\gamma_0}$. For $|\tilde{\varepsilon}|\gg 1$, we find

$$\int_0^{\infty}\exp\left(-u^2-2|\tilde{\varepsilon}|u\right)\sqrt{u}\,du\simeq\int_0^{\infty}\exp\left(-2|\tilde{\varepsilon}|u\right)\sqrt{u}\,du=\left(2|\tilde{\varepsilon}|\right)^{-3/2}\Gamma\left(\frac{3}{2}\right)=\frac{\sqrt{\pi}}{\sqrt{32}\,|\tilde{\varepsilon}|^{3/2}}, \tag{3.119}$$

where $\Gamma(z)$ is the Gamma function. Substituting (3.119) in (3.118), we finally find the asymptotic formula for the electron density of state deep in the tail

$$\rho(\tilde{\varepsilon})\simeq\frac{\gamma_0^{1/4}m^{3/2}\exp\left(-\tilde{\varepsilon}^2\right)}{2^{7/4}\pi^2\hbar^3|\tilde{\varepsilon}|^{3/2}}. \tag{3.120}$$

The model that includes the smoothly fluctuating-in-space random potential describes the optical properties of heavily doped compensated semiconductors. Compensation means high impurity density at a relatively small carrier concentration, which is possible when the concentration of donors and acceptors are comparable. In this case, the density-of-state tail provides for light absorption for photon energies below the bandgap. The main reason for the smooth potential to be a good approximation is the large screening length at low carrier density, $L>>n_i^{-1/3}$.

References

1. R.D. Mattuck, *Guide to Feynman Diagrams in the Many-Body Problem* (2nd ed., McGraw- Hill 1967, Dover New York, 2012).
2. A.A. Abrikosov, L.P. Gorkov, and I.E. Dzyaloshinskii, *Methods of Quantum Field Theory in Statistical Physics* (Dover, New York, 1963).
3. G.D. Mahan, *Many-Particle Physics*, Chapter 2 (3rd ed., Kluwer, New York, 2000).
4. B. Shklovskii and A. Efros, *Electronic Properties of Doped Semiconductors* (Springer-Verlag, Berlin, 1984).
5. L.V. Keldysh and G.P. Proshko, "Infrared absorption in heavily doped germanium", *Sov. Phys. - Solid State*. **5**, 2481 (1964).
6. A.L. Efros, "Theory of electron states in heavily doped semiconductors", *Sov. Phys. JETP*. **32**, 479 (1971).

4

Statistics of Electrons in Semiconductors

A variety of quantum states of an individual electron in a solid state is a set of eigenstates of the Schrödinger equation. An electron, being injected into the system, could fill any of these states and make transitions between them when subjected to external fields. In a crystal, we deal with a large number of electrons in contact with a reservoir. The system under consideration exchanges particles and energy with the reservoir in the course of evolution toward thermodynamic equilibrium. The Schrödinger equation, as a single-particle problem, says nothing about how a large number of non-interacting electrons would fill the energy levels and also how many electrons would comprise the system in thermodynamic equilibrium. That is quantum statistical physics, which gives answers to these questions.

4.1 Statistical Physics: Gibbs Distribution

The base of statistical theory is the Gibbs distribution function that is the probability for the N-particle system be in the state with energy E_{nN} [1]:

$$w_{nN} = \exp\left(\frac{\Omega + \mu N - E_{nN}}{k_B T}\right), \quad \Omega = F - \mu N, \tag{4.1}$$

where Ω is the grand thermodynamic potential, F is the Helmholtz free energy, μ is the chemical potential, T is the temperature in Kelvins (K), and $k_B = 1.38\times10^{-23}$ Joule/K is the Boltzmann constant. The normalization condition reads as

$$\sum_{n,N} w_{nN} = \exp\left(\frac{\Omega}{k_B T}\right)\sum_{n,N}\exp\left(\frac{\mu N - E_{nN}}{k_B T}\right) = 1, \tag{4.2}$$

where summation goes over all quantum states and all particle numbers. The thermodynamic potential follows from (4.2):

$$\Omega = -k_B T\ log\ Z, \quad Z = \sum_{N,n} exp\left(\frac{\mu N - E_{nN}}{k_B T}\right). \tag{4.3}$$

The grand partition function Z represents a sum of the Gibbs factors over the statistical ensemble – a collection of independent replicas of the system. Replicas enumerate by the number of particles and accessible energy states for a given particle number.

Measurements upon a system determine the average value of any physical property over the finite time interval, the characteristic instrumentation time, which is much larger than the time during which

DOI: 10.1201/9780429285929-4

the system exchanges energy and particles with the reservoir. As the observable physical property in equilibrium is a time average, it is convenient to apply the ergodic hypothesis, which equates the time average to the ensemble average. So, the average value of physical property $A(E_{nN}, N)$ we calculate using the Gibbs distribution as follows:

$$\langle A(E_{nN}, N)\rangle = \sum_{N,n} A(E_{nN}, N) w_{nN} = \sum_{N,n} A(E_{nN}, N) \exp\left(\frac{\Omega + \mu N - E_{nN}}{k_B T}\right)$$
$$= \frac{1}{Z} \sum_{N,n} A(E_{nN}, N) \exp\left(\frac{\mu N - E_{nN}}{k_B T}\right). \tag{4.4}$$

Using Eq. (4.4), one may express the thermodynamic properties through the grand partition function. The average particles number and energy expressed through the partition function are

$$\langle N\rangle = \frac{1}{Z} \sum_{N,n} N \exp\left(\frac{\mu N - E_{nN}}{k_B T}\right) = -\frac{\partial \Omega}{\partial \mu} = k_B T \frac{\partial \log Z}{\partial \mu},$$
$$\langle E\rangle = \frac{1}{Z} \sum_{N,n} E_{nN} \exp\left(\frac{\mu N - E_{nN}}{k_B T}\right) = k_B \left(\mu T \frac{\partial}{\partial \mu} + T^2 \frac{\partial}{\partial T}\right) \log Z. \tag{4.5}$$

Below we apply the general approach to the electron statistics in semiconductors along with a discussion of the difference in statistics for band electrons and electrons localized on impurities.

In free electron gas with energy spectrum $\varepsilon_{\mathbf{k}}$, the total energy and number of electrons are

$$E = \sum_{\mathbf{k}} \varepsilon_{\mathbf{k}} \left(n_{\mathbf{k}\uparrow} + n_{\mathbf{k}\downarrow}\right), \; N = \sum_{\mathbf{k}} \left(n_{\mathbf{k}\uparrow} + n_{\mathbf{k}\downarrow}\right), \tag{4.6}$$

where $n_{\mathbf{k},\uparrow,\downarrow}$ is the number of electrons of spin $\uparrow,\downarrow$ in state $\mathbf{k}$. Two values of the spin variable correspond to electron spin $S = 1/2$. By the Pauli principle, the number of particles in a quantum state is 0 or 1. Partition function (4.3) contains the sum over $n_{\mathbf{k}\uparrow\downarrow} = 0,1$ and has the form:

$$Z = \sum_{n_{\mathbf{k}\uparrow}, n_{\mathbf{k}\downarrow}} \exp\left[\frac{1}{k_B T} \sum_{\mathbf{k}} (\mu - \varepsilon_{\mathbf{k}})\left(n_{\mathbf{k}\uparrow} + n_{\mathbf{k}\downarrow}\right)\right] = \prod_{\mathbf{k}} Z_{\mathbf{k}},$$
$$Z_{\mathbf{k}} = \sum_{n_{\mathbf{k}\uparrow}, n_{\mathbf{k}\downarrow}} \exp\left[\frac{1}{k_B T} (\mu - \varepsilon_{\mathbf{k}})\left(n_{\mathbf{k}\uparrow} + n_{\mathbf{k}\downarrow}\right)\right] = \left[1 + \exp\left(\frac{\mu - \varepsilon_{\mathbf{k}}}{k_B T}\right)\right]^2. \tag{4.7}$$

The state $n_{\mathbf{k}\uparrow} + n_{\mathbf{k}\downarrow} = 2$ does not violate the Pauli principle and contributes to Z. The total number of electrons in thermal equilibrium follows from Eq. (4.5) and (4.7):

$$\langle N\rangle = k_B T \frac{\partial \log Z}{\partial \mu} = 2 \sum_{\mathbf{k}} f_{\mathbf{k}}, \quad f_{\mathbf{k}} = \frac{1}{\exp\left(\dfrac{\varepsilon_{\mathbf{k}} - \mu}{k_B T}\right) + 1}, \tag{4.8}$$

where $f_{\mathbf{k}}$ is the Fermi-Dirac distribution function (occupation number of state $\mathbf{k}$). Equation (4.8) relates μ and $\langle N\rangle$. The multiplying factor 2 in (4.8) comes from two spin states of the electron and should be replaced with $2S + 1$ if one considers fermions of higher spin.

For Bose particles (phonons, photons), there are no restrictions on how many particles might be in state $\mathbf{k}$, then the sum in Z_k runs from zero to infinity and defines the Bose statistics:

$$Z_k = \sum_{n_k=0}^{\infty} \exp\left(\frac{(\mu-\varepsilon_k)n_k}{k_B T}\right) = \sum_{n=0}^{\infty}\left[\exp\left(\frac{\mu-\varepsilon_k}{k_B T}\right)\right]^n = \left[1-\exp\left(\frac{\mu-\varepsilon_k}{k_B T}\right)\right]^{-1},$$

$$\langle n_k \rangle = k_B T \frac{\partial \log Z_k}{\partial \mu} = \frac{1}{\exp\left(\frac{\varepsilon_k-\mu}{k_B T}\right)-1}. \tag{4.9}$$

The geometrical progression (4.9) converges if $\exp\left[(\mu-\varepsilon_k)/k_B T\right] < 1$. As the condition must be satisfied for all ε_k, including $\varepsilon_k = 0$, one may conclude that the chemical potential must not be a positive number. For photons in a box, the number of photons is not fixed and determined by the thermal equilibrium between radiation and the wall, so the chemical potential equals zero. For further details on Bose statistics, see Ref's [1–3].

If $\exp\left((\varepsilon_k-\mu)/k_B T\right) \gg 1$, both Fermi and Bose distributions coincide and belong to the classical Maxwell–Boltzmann statistics:

$$f(\varepsilon_k) \approx \exp\left(\frac{\mu-\varepsilon_k}{k_B T}\right) \tag{4.10}$$

The approximation describes the electron statistics in non-degenerate semiconductors discussed below.

4.2 Metals and Semiconductors

In metals, the chemical potential lies in the energy band, and the corresponding Fermi distribution (4.9) is illustrated in Figure 4.1.

The chemical potential referenced to the band edge is called the Fermi energy, $\mu = \varepsilon_F$. At $T = 0$ ε_F represents the upper limit of the energy region filled with electrons: states $\varepsilon_k < \mu$ are filled and states $\varepsilon_k > \mu$ are vacant. If $\varepsilon_F \gg k_B T$, the electron gas is called degenerate. That is what metal is from a statistical point of view. At $T = 0$ the relation between the electron density and the Fermi energy can be explicitly obtained from Eq. (4.8), providing we consider a single-band metal with parabolic spectrum $\varepsilon_k = \hbar^2 k^2/2m$, m is the effective electron mass. The average electron number in equilibrium has the form,

$$\langle N \rangle = 2\sum_k f(\boldsymbol{k}) = \frac{2V}{(2\pi)^3}\int f(\varepsilon_k)\, d^3k, \tag{4.11}$$

where V is the crystal volume, and the discrete quantum number $\boldsymbol{k}$ runs over the first Brillouin zone. The factor $(2\pi)^3/V$ in Eq. (4.11) is the volume in the Brillouin zone per one quantum state $\boldsymbol{k}$. In the thermodynamic limit ($V \to \infty, N/V \to \text{const}$), we consider $\boldsymbol{k}$ a continuous variable and replace the sum with an integral. An explicit calculation is possible as the Fermi distribution is the step function at $T = 0$,

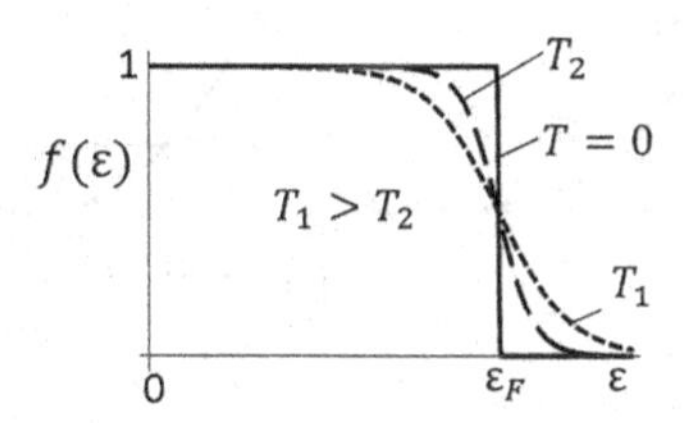

FIGURE 4.1 Fermi-Dirac distribution function in metals.

$$f(\varepsilon_k)=\begin{cases} 1, \varepsilon_k \le \varepsilon_F \\ 0, \ \varepsilon_k > \varepsilon_F \end{cases}, \tag{4.12}$$

and the electron density is related to ε_F as

$$n=\frac{\langle N\rangle}{V}=\frac{2}{(2\pi)^3}\int_0^{\pi}\sin\theta\, d\theta\int_0^{2\pi}d\varphi\int_0^{k_F}k^2dk=\frac{k_F^3}{3\pi^2},\ k_F=\frac{\sqrt{2m\varepsilon_F}}{\hbar},$$
$$\varepsilon_F=\frac{\hbar^2}{2m}\left(3\pi^2 n\right)^{2/3}. \tag{4.13}$$

where $\hbar k_F$ is the Fermi momentum. Average electron energy:

$$\langle\varepsilon\rangle=\frac{2}{N}\sum_k \varepsilon_k f(\boldsymbol{k})=\frac{3\hbar^2\pi^2}{k_F^3 m(2\pi)^3}\int k^2 f(\varepsilon_k)dk=\frac{3}{5}\varepsilon_F. \tag{4.14}$$

In semiconductors, the chemical potential could fall into the conduction or valence band, as well as in the bandgap. At $T=0$, the Fermi-Dirac distribution function requires all energy levels below the chemical potential to be occupied, while all states above vacant. In a nondegenerate semiconductor, μ lies in the bandgap, so at $T=0$ (the ground state called "vacuum"), the valence band is fully occupied by electrons while the conduction band is empty. Conduction electrons and holes (empty states in the valence band) may appear as a result of injection through the contact with a reservoir or, if $T\neq 0$, due to the thermal excitation of valence electrons or (and) occupied impurity states located in the bandgap. Carriers determine the kinetic, thermodynamic, and optical properties of a semiconductor.

In an n-type semiconductor, μ is close to the edge of the conduction band E_c, and the degree of electron degeneracy determined by the parameter $(\mu-E_c)/k_BT$ may vary depending on the relative positions of μ and E_c as illustrated in Figure 4.2.

As illustrated in Figure 4.2 for $\mu=\mu_2$ and $\mu=\mu_3$, the condition of degeneracy $(\mu-E_c)/k_BT\gg 1$ cannot hold, and the electron gas is nondegenerate. At $E_c-\mu_{2,3}>4k_BT$ the Fermi function approximately equals Maxwell–Boltzmann distribution.

In the limit $T\to 0$, the filling factors in the conduction band behave differently depending on the position of the chemical potential: the conduction band becomes empty if $\mu=\mu_{2,3}$ and degenerate if $\mu=\mu_1$.

We obtained the electron density (4.13) under the condition of strong degeneracy typical for metals. So, Eq. (4.13) is not valid if μ is either in the bandgap, $\mu-E_c<0$, or in the band, $\mu-E_c\ll k_BT$. To obtain the relation between μ and n, one should use the exact Fermi distribution. Below we calculate $n(\mu)$ in a

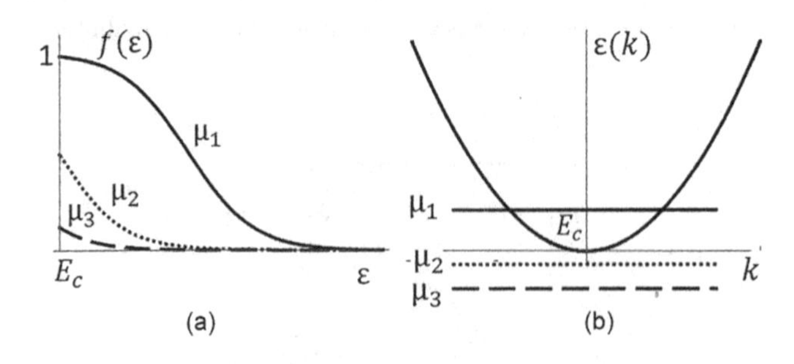

FIGURE 4.2 Variable degree of electron degeneracy in the conduction band. (a) Occupation number of electron states. (b) Electron spectrum in the conduction band and possible positions of the chemical potential.

parabolic conduction band with any degree of degeneracy-the arbitrary relative positions of μ and E_c. Assuming the spectrum is given by $\varepsilon = E_c + \hbar^2 k^2 / 2m_c$, we use Eq. (4.11) to obtain

$$n = \frac{1}{\pi^2} \int_0^\infty \frac{k^2 dk}{\exp\left(\frac{\varepsilon - \mu}{k_B T}\right) + 1} = \int_{E_c}^\infty f(\varepsilon) N(\varepsilon)\, d\varepsilon,$$

$$N(\varepsilon) = \frac{\sqrt{2} m_c^{3/2}}{\pi^2 \hbar^3} \sqrt{\varepsilon - E_c}. \tag{4.15}$$

Change of variables in (4.15) from $\boldsymbol{k}$- to ε-integration generates the *density of states* $N(\varepsilon)$.

The density of states (4.13) is specific for three-dimensional solids. If one deals with quantum wells, a particle at each size-quantized energy level has the wavevector with two in-plane components: $\boldsymbol{k} = (k_x, k_x)$. In a quantum well of width d, the volume density of electrons belonging to sub-band $\varepsilon(\boldsymbol{k})$:

$$n = \frac{\langle N \rangle}{V} = \frac{2}{V} \sum_{\boldsymbol{k}} f(\boldsymbol{k}) = \frac{2S}{V(2\pi)^2} \int f(\boldsymbol{k})\, d^2\boldsymbol{k} = \frac{1}{2\pi^2 d} \int_0^{2\pi} d\varphi \int_0^\infty \frac{k dk}{\exp\left(\frac{\varepsilon - \mu}{k_B T}\right) + 1}$$

$$= \frac{m k_B T}{\pi \hbar^2 d} \int_0^\infty f(\varepsilon) d\varepsilon = \frac{m k_B T}{\pi \hbar^2 d} \log\left[1 + \exp\left(\frac{E_F}{k_B T}\right)\right], \tag{4.16}$$

where S is the area of quantum well, coefficient $m k_B T / \pi \hbar^2$ represents the energy-independent $2D$ density of states, $E_F = \mu - E_C$ is the Fermi energy-chemical potential measured from the sub-band edge.

Strictly speaking, wavenumber $\boldsymbol{k}$ in Eqs. (4.15) and (4.16) runs through the finite region - entire Brillouin zone, whereas the parabolic approximation for the spectrum works well only at small $\boldsymbol{k}$ (close to the band minimum). As the Fermi distribution function is negligible at large $\boldsymbol{k}$, one may extend the upper limit of integration to infinity and use the approximate spectrum, not violating the accuracy of calculations.

In the parabolic model assumed here, the effective mass does not depend on energy. In a non-parabolic spectrum, m_c in Eq. (4.15) implies the *density-of-states effective mass*, in other words, the mass at the band minimum. In anisotropic semiconductors, masses are different in three directions $k_{x,y,z}$, which results in the density-of-states effective mass $m_c = \sqrt[3]{m_x m_y m_z}$. If a semiconductor has g equivalent minima (minima of the same energy), this gives an additional multiplying factor g in the density of states.

Replacing variables, $x = (\varepsilon - E_c)/k_B T$, $\tilde{\mu} = (\mu - E_c)/k_B T$, we express Eq. (4.15) as

$$n = \frac{\sqrt{2}(m_c k_B T)^{3/2}}{\pi^2 \hbar^3} \int_0^\infty \frac{\sqrt{x} dx}{\exp(x - \tilde{\mu}) + 1} \equiv N_c\, F_{1/2}(\tilde{\mu}),$$

$$F_{1/2}(\tilde{\mu}) = \frac{2}{\sqrt{\pi}} \int_0^\infty \frac{\sqrt{x} dx}{\exp(x - \tilde{\mu}) + 1}, \quad N_c = 2\left(\frac{m_c k_B T}{2\pi \hbar^2}\right)^{3/2}, \tag{4.17}$$

where $F_{1/2}(\tilde{\mu})$ is the Fermi-Dirac integral, N_c is the *effective conduction band density of states*. In the Maxwell–Boltzmann regime, $E_c - \mu > 4 k_B T$, integration in Eq. (4.17) gives

$$n = N_c \exp\left(\frac{\mu - E_c}{k_B T}\right). \tag{4.18}$$

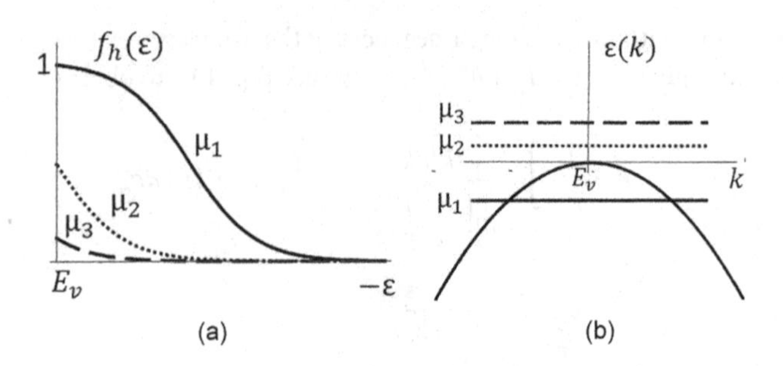

FIGURE 4.3 Variable degree of degeneracy in the valence band. (a) Occupation number of hole states. (b) Electron spectrum in the valence band and possible positions of the chemical potential.

Expression (4.18) gives physical meaning to N_c: in an n -type non-degenerate semiconductor, μ increases with n and crosses the conduction band edge when $n > N_c$. Upon further increase in n, a semiconductor becomes degenerate if condition $\mu - E_c > 4k_BT$ holds.

In p-type semiconductors, the chemical potential is close to the valence band edge E_v. Similarly to the n -type semiconductor discussed above, we can calculate the occupation of the valence band. The difference is that thermally excited particles are holes – empty states in the valence band predominantly filled with electrons. That is why the hole occupation number expressed through the electron occupation number reads as $f_h = 1 - f$. For details, let us consider the valence band with electron spectrum $\varepsilon = E_v - \hbar^2k^2 / 2m_v$. The electron occupation number is the Fermi-Dirac function $f(\varepsilon)$. The hole occupation number is

$$f_h(\varepsilon) = 1 - f = 1 - \frac{1}{\exp\left(\dfrac{\varepsilon - \mu}{k_BT}\right) + 1} = \frac{1}{\exp\left(\dfrac{\mu - \varepsilon}{k_BT}\right) + 1}. \tag{4.19}$$

Valence band filling versus position of the chemical potential is shown in Figure 4.3.

Similarly to (4.17) and (4.18), we get the hole density in a p-type semiconductor for an arbitrary degree of carrier degeneracy,

$$p = N_v\ F_{1/2}\left(\frac{E_v - \mu}{k_BT}\right),\ N_v = 2\left(\frac{m_v k_B T}{2\pi\hbar^2}\right)^{3/2}, \tag{4.20}$$

and for nondegenerate holes, $\mu - E_v > 4k_BT$,

$$p = N_v\ \exp\left(\frac{E_v - \mu}{k_BT}\right). \tag{4.21}$$

Formally, the "vacuum" state in a nondegenerate semiconductor follows from Eqs. (4.18) and (4.21), $T \to 0, n, p \to 0$.

4.3 Intrinsic Semiconductors

In Section 4.2, we related the equilibrium electron and hole densities as a function of the position of chemical potential. If the system were in contact with a capacious reservoir, that reservoir would determine the chemical potential. Otherwise – in an isolated system – one needs an additional condition to find μ. Electrical charges of electrons and holes are opposite, and we cannot change the number of conduction electrons and valence holes arbitrarily, not violating the electrical neutrality of a crystal.

Deviation from neutrality longs shortly during the Maxwellian relaxation time, while in thermal equilibrium, a crystal is neutral. The additional condition – neutrality equation – serves to find μ. In intrinsic semiconductors (no impurity states in the bandgap), the charge neutrality equates numbers of thermally excited electrons and holes, $n = p$, and, in the Maxwell–Boltzmann regime, the condition has the form:

$$N_c \exp\left(\frac{\mu - E_c}{k_B T}\right) = N_v \exp\left(\frac{E_v - \mu}{k_B T}\right). \tag{4.22}$$

Solving Eq. (4.22), we obtain

$$\mu = \frac{E_c + E_v}{2} + \frac{k_B T}{2}\log\frac{N_v}{N_c} = E_M + \frac{3k_B T}{4}\log\frac{m_v}{m_c}, \tag{4.23}$$

where E_M is the mid-gap energy. At $T = 0$, μ is in the middle of the bandgap. In the most simple case of mirror-like conduction and valence bands $(m_c = m_v)$, μ does not depend on temperature.

The product np does not depend on μ and relates concentrations directly to temperature and material parameters:

$$np = N_c N_v \exp\left(-\frac{E_g}{k_B T}\right). \tag{4.24}$$

Expression (4.24) is called a *mass-action law in semiconductors.* In an intrinsic semiconductor, $n = p \equiv n_i$ and $np = n_i^2$. This law is valid in nondegenerate semiconductors that obey the Maxwell–Boltzmann statistics. At an arbitrary degree of degeneracy, the product of n and p from Eqs. (4.17) and (4.20), respectively, depends on μ.

4.4 Electron Distribution in Doped Semiconductors

Intrinsic (undoped) semiconductors with bandgap $E_g >> k_B T$ exhibit high electrical resistance and find their use as substrates for various electronic devices and templates for multiple device integration. However, most often, the functionality of the devices depends on the doped n- and p-type regions brought into contact. Doping with donors and acceptors is not a trivial task for some semiconductors. However, materials for practical use should be able to be doped with both donors and acceptors during growth or postgrowth treatment. Moreover, in the search for new semiconductors, the technological ability to manufacture and process both n - and p-type materials is a must to guarantee their use in device applications.

4.4.1 Simple Donors and Acceptors

Impurity atoms and other defects in semiconductors violate the translation symmetry that may result in localized electron states with energy in the bandgap. Atoms of dopant may lose a weakly bound atomic valence electron by donating it to the lattice (donor) or accept an electron from the lattice to a weakly bound atomic valence state (acceptor). For example, in silicon, Al impurity substituting atom Si produces an acceptor state. Because Al has one less valence electron than Si and Al $3p$-state has higher energy than that of Si $3p$-state from which the valence band originates, an unoccupied impurity state appears above the top of valence band maximum.

A donor (acceptor) creates the energy level $E_D\,(E_A)$ in the bandgap. The energy level of a shallow donor (acceptor) lies near the conduction (valence) band. For deep donors and acceptors, the levels sit in the bandgap deep enough to prevent their thermal ionization at ambient temperature. The band diagram of a doped semiconductor is illustrated in Figure 4.4.

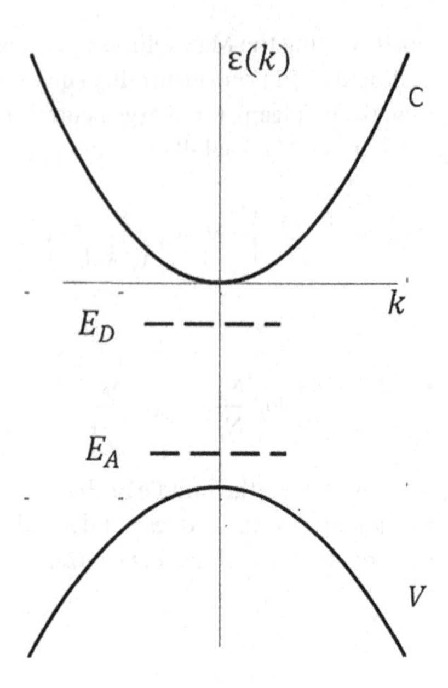

FIGURE 4.4 Electron bands in a doped semiconductor. $E_D\,(E_A)$ is the donor (acceptor) energy level.

Quantum states E_D, E_A could be degenerate and occupied with several electrons. First, we consider monovalent donors and acceptors, which can bind no more than one electron. The charge state of a level depends on occupation: a donor with a bound electron and an acceptor stripped of an electron (empty) are both neutral. At finite temperature, some donors (acceptors) are thermally ionized and positively (negatively) charged. In thermal equilibrium, the electron filling and the ratio between the number of neutral and charged impurities are regulated by the Fermi-Dirac distribution function, as discussed below.

To study the electron filling of the impurity levels, we consider a semiconductor doped with donors only and use the partition function similar to (4.7):

$$E = E_D\left(n_\uparrow + n_\downarrow\right),\ N = n_\uparrow + n_\downarrow,$$

$$Z = \sum_{n_\uparrow, n_\downarrow} \exp\left[\frac{1}{k_BT}(\mu - E_D)\left(n_\uparrow + n_\downarrow\right)\right] = 1 + 2\exp\left(\frac{\mu - E_D}{k_BT}\right), \tag{4.25}$$

$$f_d = k_BT\frac{\partial \log Z}{\partial \mu} = \frac{1}{1 + \frac{1}{2}\exp\left(\frac{E_D - \mu}{k_BT}\right)},$$

where f_d is the electron distribution function (occupation number of the donor level). Summation over all accessible states in Eq. (4.25) means $n_{\uparrow,\downarrow} = 0,1$ with the exclusion of state $n_\uparrow + n_\downarrow = 2$. The physical meaning of this exclusion is related to the Coulomb interaction and will be discussed later in more detail.

The occupation number of the acceptor level f_a is Eq. (4.25), with E_D replaced by E_A.

The density of occupied (neutral) donors is $n_d^0 = n_d f_d$, where n_d is the total donor concentration. The concentration of empty (positively charged) donors is

$$n_d^+ = n_d(1 - f_d) = \frac{n_d}{1 + 2\exp\left(\frac{\mu - E_D}{k_BT}\right)}. \tag{4.26}$$

The difference in statistics for band electrons (4.8) and electrons at the impurity level (4.25) comes from different numbers of available states. The band state ε_k can be filled-up with two electrons with opposite spins (see comment after Eq. (4.7)). The two-electron state, bound to impurity, has energy much higher than E_D due to the strong on-site Coulomb interaction. That is why we exclude the two-electron state from Eq. (4.25) and obtain the modified Fermi-Dirac distribution function.

If the bandgap is large enough $E_g >> k_BT$, one may neglect thermal excitation of the valence band, meaning that all electrons in the conduction band come from thermally excited donors. The neutrality condition reads as follows: density of conduction electrons (4.17) equals the density of empty (ionized) donors:

$$\frac{n_d}{2\exp\left(\dfrac{\mu - E_D}{k_BT}\right)+1} = N_c F_{1/2}\left(\frac{\mu - E_c}{k_BT}\right) \tag{4.27}$$

The chemical potential that follows from Eq. (4.27) can be calculated explicitly in the Maxwell Boltzmann regime, $E_c - \mu > 4k_BT$,

$$\frac{n_d}{2\exp\left(\dfrac{\mu - E_D}{k_BT}\right)+1} = N_c \exp\left(\frac{\mu - E_c}{k_BT}\right), \tag{4.28}$$

and we thus obtain

$$\mu = E_D + k_BT \log\frac{1}{4}\left(\sqrt{1+\frac{8n_d}{N_c}\exp\left(\frac{E_c - E_D}{k_BT}\right)}-1\right), \tag{4.29}$$

where $E_c - E_D$ is the donor ionization energy. At low temperature, $\exp\left[(E_c - E_D)/k_BT\right] \gg 1$ and

$$\mu \approx \frac{1}{2}\left(E_c + E_D + k_BT \log\frac{n_d}{N_c}\right). \tag{4.30}$$

At $T = 0$, μ is in the middle between the donor level and the conduction band edge. In typical nondegenerate semiconductors, $n_d < N_c$, μ decreases with temperature and becomes lower than E_D. Further increase in temperature up to $k_BT \approx E_g$ makes (4.28) invalid as the valence band comes into play. Thermal excitations across the bandgap make a semiconductor almost intrinsic, and μ tends to the value determined by Eq. (4.23).

Similarly, in *p*-type semiconductors doped with acceptors, the neutrality requires the equality of hole density in the valence band and electron density on acceptor levels:

$$N_v \exp\left(\frac{E_v - \mu}{k_BT}\right) = \frac{n_a}{\dfrac{1}{2}\exp\left(\dfrac{E_A - \mu}{k_BT}\right)+1}, \tag{4.31}$$

where n_a is the total density of acceptors. Thus, at $T = 0$ μ lies between the acceptor level and the valence band edge.

Compensated semiconductors are doped with both donors and acceptors, so the full neutrality equation has the form,

$$\frac{n_d}{2\exp\left(\dfrac{\mu - E_D}{k_BT}\right)+1} + N_v F_{1/2}\left(\frac{E_v - \mu}{k_BT}\right) = N_c F_{1/2}\left(\frac{\mu - E_c}{k_BT}\right) + \frac{n_a}{\dfrac{1}{2}\exp\left(\dfrac{E_A - \mu}{k_BT}\right)+1}. \tag{4.32}$$

Condition (4.32) equates the density of positive charges (valence holes and empty donors) and negative charges (conduction electrons and filled acceptors). At an arbitrary level of doping, the equation has a numerical solution. In a nondegenerate semiconductor, one calculates the chemical potential analytically in the Maxwell–Boltzmann approximation. If μ is located deep in the bandgap, so that $E_D - \mu >> k_B T$ and $\mu - E_A >> k_B T$, Eq. (4.32) simplifies to the condition $n - p = n_d - n_a$. So, when $n_d > n_a$ ($n_d < n_a$) a semiconductor is of n - (p-) type. For Si and Ge doped with group-V donors and group-III acceptors, this condition holds in the whole range from solid nitrogen to room temperature.

4.4.2 Degenerate Impurities

In the previous section, we dealt with monovalent donor and acceptor levels, which are doubly spin-degenerate. In a more general situation, additional degeneracy of impurity levels may originate from the degenerate conduction or valence bands like those in Si and zinc-blende crystals. As the additional degeneracy increases the number of available states, the arrangement of electrons among the states changes and affects the statistics. We assume that the donor level E_D can be occupied with no more than one electron and has the degree of degeneracy $g(N)$, which may depend on the level occupancy: $g(0) = g_0$, $g(1) = g_1$. We express partition function (4.3) as

$$Z = \sum_{N=0,1} g(N)\exp\left[\frac{1}{k_B T}(\mu - E_D)N\right] = g_0 + g_1 \exp\left(\frac{\mu - E_D}{k_B T}\right)$$
$$f_d = k_B T \frac{\partial \log Z}{\partial \mu} = \frac{1}{1 + \frac{g_0}{g_1}\exp\left(\frac{E_D - \mu}{k_B T}\right)}. \tag{4.33}$$

Factor $g(N)$ in the first line of Eq. (4.33) appears as a result of the implicitly assumed summation over degeneracy of each state N. Providing the additional degeneracy is absent, and level E_D has spin degeneracy only ($g_0 = 1, g_1 = 2$) the distribution function f_d coincides with (4.25). In a most general case, the neutrality equation (4.32) would contain the density of neutral and positively charged donors,

$$n_d^0 = \frac{n_d}{1 + \frac{g_0}{g_1}\exp\left(\frac{E_D - \mu}{k_B T}\right)}, \quad n_d^+ = n_d(1 - f_d) = \frac{n_d}{1 + \frac{g_1}{g_0}\exp\left(\frac{\mu - E_D}{k_B T}\right)}, \tag{4.34}$$

as well as the density of negatively charged and neutral acceptors,

$$n_a^- = \frac{n_a}{1 + \frac{g_0'}{g_1'}\exp\left(\frac{E_A - \mu}{k_B T}\right)}, \quad n_a^0 = \frac{n_a}{1 + \frac{g_1'}{g_0'}\exp\left(\frac{\mu - E_A}{k_B T}\right)}, \tag{4.35}$$

where $g_{1,0}'$ are degeneracy factors of the acceptor level.

4.4.3 Multivalent Impurities and Electron Interaction

Multivalent impurities could bind or donate several electrons. Calculating the electron distribution function (4.25), we excluded the two-electron state from the number of accessible states. By doing this, we justified the trick by referring to strong Coulomb interaction. However, so that the verbal justification did not look like a spell, we have to reconsider the problem by explicitly including the two-electron state and the Coulomb interaction. Including on-site Coulomb repulsion U in the partition function (4.25) and taking the sum over all $n_\uparrow = 0,1$ and $n_\downarrow = 0,1$ we obtain:

$$E = E_D(n_\uparrow + n_\downarrow) + U\, n_\uparrow n_\downarrow, \quad N = n_\uparrow + n_\downarrow,$$

$$Z = \sum_{n_\uparrow, n_\downarrow} \exp\left[\frac{(\mu - E_D)(n_\uparrow + n_\downarrow) - U\, n_\uparrow n_\downarrow}{k_B T}\right]$$

$$= 1 + 2\exp\left(\frac{\mu - E_D}{k_B T}\right) + \exp\left[\frac{2(\mu - E_D) - U}{k_B T}\right],$$

$$f_d = k_B T \frac{\partial \log Z}{\partial \mu} = 2\frac{1 + \exp\left[\dfrac{\mu - E_D - U}{k_B T}\right]}{2 + \exp\left[\dfrac{E_D - \mu}{k_B T}\right] + \exp\left[\dfrac{\mu - E_D - U}{k_B T}\right]}. \tag{4.36}$$

By taking a limit $U \to \infty$, we exclude the two-electron state from the partition function, so the distribution function (4.36) tends to that obtained in (4.25). If repulsion is weak, we have to use f_d from Eq. (4.36), which accounts for two levels in the bandgap: one-electron level $E_1 = E_D$ and two-electron level $E_2 = E_D + U$. Note that level E_2 does not exist until one electron occupies E_1. In a semiconductor doped with multivalent centers, the picture, described above for a divalent impurity, can be extended to many E_j which appear during sequential filling by j electrons. Partition function (4.33) includes summation on all charge states: $j = 0,1,\ldots,M$ each of those $g(j)$-fold degenerate:

$$Z = \sum_{j=0}^{M} g(j)\exp\left[\frac{(\mu j - E_j)}{k_B T}\right] = g(0) + \sum_{j=1}^{M} g(j)\exp\left[\frac{(\mu j - E_j)}{k_B T}\right], \tag{4.37}$$

The total number of electrons bound to an M-valent state:

$$f_M = \frac{\partial \log Z}{\partial \mu} = \sum_{j=1}^{M} j f_j,$$

$$f_j = \frac{1}{Z}\, g(j)\exp\left[\frac{(\mu j - E_j)}{k_B T}\right]. \tag{4.38}$$

If an impurity is monovalent, $M = 1$, the distribution (4.38) coincides with that obtained in (4.33).

4.4.4 Amphoteric Impurities

The amphoteric behavior of the impurity center is the ability to change its charge state, creating positive and negative ions in a semiconductor host. An amphoteric center is neutral at some Fermi level position μ_0. When μ moves within the bandgap, the center acquires charges of opposite signs on both sides of μ_0. Figure 4.5 shows a classic example of the Au dopant in germanium.

The scheme in Figure 4.5 illustrates how the charge state of Au varies when the chemical potential is moving within the bandgap. The diagram renders levels starting with the deep donor $E1$, which at low temperature and $\mu = \mu_0$ is occupied and neutral.

Variation of μ can be implemented by compensation with additional shallow donors and acceptors. Starting from the valence band maximum upward $\mu < E1$, the donor is empty and positively charged, so the charge state of Au is $z = +1$. In the range $E1 < \mu < E2$, the donor becomes occupied ($z = 0$), and further movement of μ leads to sequential filling of $E2, E3, E4$, which gives $z = -1, -2, -3$, respectively. Negatively charged states render acceptors (A) in Figure 4.5.

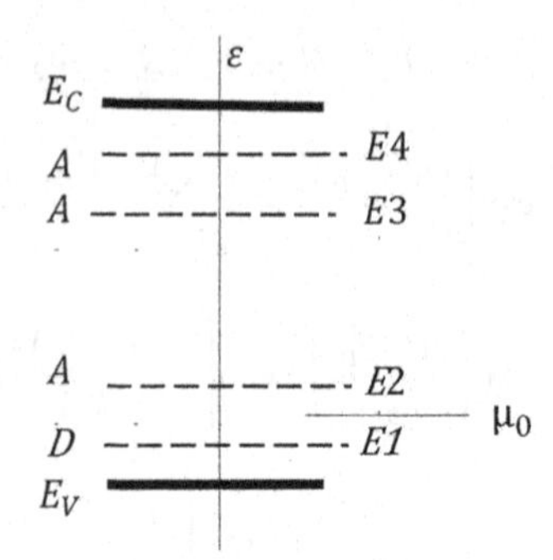

FIGURE 4.5 Energy levels of Au impurity in germanium.

Another type of amphoteric impurity, cation-anion amphoteric centers, exists in binary semiconductor compounds. An impurity atom could be either a donor or an acceptor. It depends on the sublattice in which it resides as a substitutional impurity. An example is the group IV dopants in III-V semiconductors: the dopant has an intermediate valence relative to the valence of the host components.

Real semiconductors are often doped with several donors (acceptors) and compensated with several acceptors (donors). Besides, excited states of impurity often come into play if they happen to be in the bandgap. For more details on compensated semiconductors doped with particular donors and acceptors, see Ref. [4].

References

1. L.D. Landau and E.M. Lifshitz, *Statistical Physics, Part 1* (3rd ed., Elsevier, Amsterdam, 1980).
2. G. Cook and R.H. Dickerson, "Understanding the chemical potential", *Am. J. Phys.* **63**, 737 (1995).
3. R.K. Pathria, *Statistical Mechanics* (2nd ed., Butterworth-Heinemann, Oxford, 1996).
4. J.S. Blakemore, *Semiconductor Statistics* (Pergamon Press, Oxford, 1962).

5

Electrons in a Magnetic Field

5.1 Lorentz Force

The behavior of carriers (electrons and holes) in external fields is at the origin of solid-state device applications. Electric and magnetic fields in conducting and dielectric media obey four Maxwell equations. One more equation controls the dynamics of a charged particle in the fields. That is the Lorentz force equation we derive below.

In Hamilton formulation of mechanics, instead of six variables, electric field $\boldsymbol{E}$, and magnetic field $\boldsymbol{B}$, we deal with four auxiliary potentials: electric potential V and vector potential $\boldsymbol{A}$:

$$\boldsymbol{B} = \nabla \times \boldsymbol{A}, \quad \boldsymbol{E} = -\frac{\partial \boldsymbol{A}}{\partial t} - \nabla V. \tag{5.1}$$

The price we pay for a reduced number of variables is the ambiguity in potentials, which is now to be fixed by choice of gauge. The Hamiltonian of a nonrelativistic electron subjected potentials V and $\boldsymbol{A}$ has the form,

$$H = \frac{1}{2m}\left(\boldsymbol{p} - e\boldsymbol{A}\right)^2 + eV. \tag{5.2}$$

In Chapter 12, one finds more details on the gauge choice, gauge transformations, and justification for the Hamiltonian (5.2). We choose the Hamiltonian approach as it works for both classical and quantum descriptions of physical phenomena.

For now, we do not include the electron spin and consider the classical orbital motion of a charged particle using the Hamilton function (5.2). Equations of motion are

$$\dot{\boldsymbol{p}} = -\frac{\partial H}{\partial \boldsymbol{r}}, \quad \dot{\boldsymbol{r}} = \frac{\partial H}{\partial \boldsymbol{p}}, \tag{5.3}$$

where dot accent means the full time-derivative. It is convenient to write (5.3) in Cartesian components:

$$\begin{aligned} \dot{r}_x &= \frac{1}{m}\left(p_x - eA_x\right) \equiv \mathrm{v}_x, \\ \dot{p}_x &= -e\frac{\partial V}{\partial x} + \frac{e}{m}\left(\boldsymbol{p} - e\boldsymbol{A}\right)\cdot\frac{\partial \boldsymbol{A}}{\partial x} = eE_x + e\mathbf{v}\cdot\frac{\partial \boldsymbol{A}}{\partial x}, \end{aligned} \tag{5.4}$$

where $\boldsymbol{E} = -\partial V / \partial \boldsymbol{r}$ is the electric field. The force acting on a particle follows from Newton's law:

DOI: 10.1201/9780429285929-5

$$F_x = m\ddot{r}_x = \dot{p}_x - e\frac{dA_x}{dt} = eE_x + e\mathbf{v}\cdot\frac{\partial \boldsymbol{A}}{\partial x} - e\left(\frac{\partial A_x}{\partial t} + \frac{\partial A_x}{\partial \boldsymbol{r}}\cdot\mathbf{v}\right)$$

$$= eE_x + ev_y\left(\frac{\partial A_y}{\partial x} - \frac{\partial A_x}{\partial y}\right) + ev_z\left(\frac{\partial A_z}{\partial x} - \frac{\partial A_x}{\partial z}\right)$$

$$= eE_x + ev_y\left[\nabla\times\boldsymbol{A}\right]_z - ev_z\left[\nabla\times\boldsymbol{A}\right]_y = eE_x + e\left[\mathbf{v}\times\boldsymbol{B}\right]_x. \tag{5.5}$$

Assuming fields time-independent, we neglect the time-derivative of vector-potential in the second line (5.5). The same procedure for other components gives the force $\boldsymbol{F} = e\boldsymbol{E} + e\left[\mathbf{v}\times\boldsymbol{B}\right]$, where the second term is the Lorentz force that follows from Hamiltonian (5.2).

5.2 Circular Motion in a Magnetic Field

Classical equation of motion of a free electron in a uniform magnetic field follows from (5.5) providing the electric field is zero,

$$\frac{d\boldsymbol{p}}{dt} = e\left[\mathbf{v}\times\boldsymbol{B}\right]. \tag{5.6}$$

Since the force is perpendicular to the velocity, an electron rotates about the direction of the magnetic field while uniformly moving in the magnetic field direction that makes the motion helicoidal. Circular motion in the uniform magnetic field is illustrated in Figure 5.1.

Let's dwell on a circular motion. Energy conservation follows from Eq. (5.6):

$$\frac{d\varepsilon}{dt} = \frac{d\varepsilon}{d\boldsymbol{p}}\cdot\frac{d\boldsymbol{p}}{dt} = \mathbf{v}\cdot\boldsymbol{F} = 0\ ,\ \varepsilon = \text{const}. \tag{5.7}$$

If the momentum and magnetic field are collinear, the force in the r-h side of (5.6) is zero, so an electron moves uniformly and rectilinearly. If the magnetic field points in the z-direction, $p_z = \text{const}$. In components, (5.6) has the form,

$$\frac{dp_x}{dt} = ev_yB,$$

$$\frac{dp_y}{dt} = -ev_xB, \tag{5.8}$$

or

$$\frac{dl}{dt} = ev_\perp B,$$

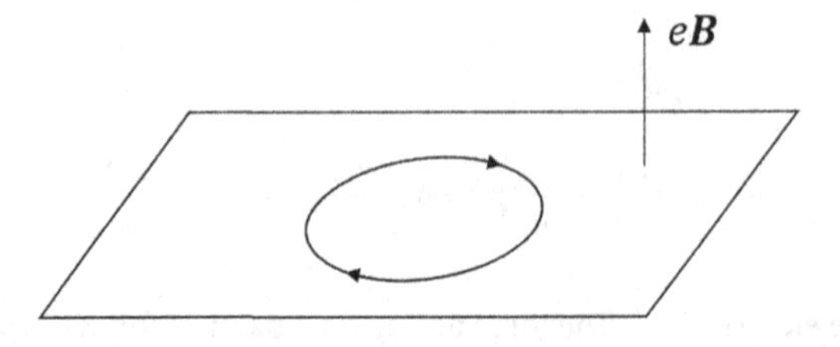

FIGURE 5.1 Electron motion under Lorentz force (5.6) in the perpendicular to the magnetic field plane.

where $v_\perp = \sqrt{v_x^2 + v_y^2}$, $dl = \sqrt{(dp_x)^2 + (dp_y)^2}$ is the infinitesimal length of electron trajectory in a momentum space. Solving (5.8) one finds

$$dt = \frac{dl}{ev_\perp B} \tag{5.9}$$

The trajectory could be closed or not depending on the energy spectrum $\varepsilon(\boldsymbol{p})$. Providing the trajectory is closed, we find the period of circular motion as an integral over the closed trajectory,

$$T = \int_0^T dt = \frac{1}{ev_\perp B}\oint dl. \tag{5.10}$$

Integral in Eq. (5.10) goes over contour C in the momentum space, as illustrated in Figure 5.2a.

To calculate T in Eq. (5.10), we consider the area of the dashed region in Figure 5.2a:

$$S(\varepsilon, p_z) = \int dp_x dp_y. \tag{5.11}$$

Instead of integration over $p_{x,y}$, one can integrate along path C and in the perpendicular direction $\boldsymbol{p}_\perp$. The width of the ring (see Figure 5.2b) along $\boldsymbol{p}_\perp$ is $|d\boldsymbol{p}_\perp / d\varepsilon|\, d\varepsilon$, so the total area is

$$S(\varepsilon, p_z) = \int d\boldsymbol{p}_\perp \oint dl = \oint dl \int \left| \frac{d\boldsymbol{p}_\perp}{d\varepsilon} \right| d\varepsilon = \frac{1}{v_\perp}\int d\varepsilon \oint dl. \tag{5.12}$$

Combining (5.12) and (5.10), one obtains

$$\begin{aligned} T &= \frac{1}{eB}\frac{\partial S}{\partial \varepsilon} = \frac{2\pi m_c}{eB}, \\ m_c &\equiv \frac{1}{2\pi}\frac{\partial S}{\partial \varepsilon}, \end{aligned} \tag{5.13}$$

In Eq. (5.13), we have introduced the cyclotron effective mass m_c, which depends on the electron energy and p_z. Angular frequency of electron rotation $\omega_c = eB / m_c$ is called *cyclotron frequency.*

Cyclic motion along contour C is the motion in a momentum space that obeys Eq. (5.6). After one introduces displacements in coordinate space, Eq. (5.6) takes the form $d\boldsymbol{p} = e\,[d\boldsymbol{r} \times \boldsymbol{B}]$, and if projected onto the plane perpendicular to the magnetic field, it takes the form

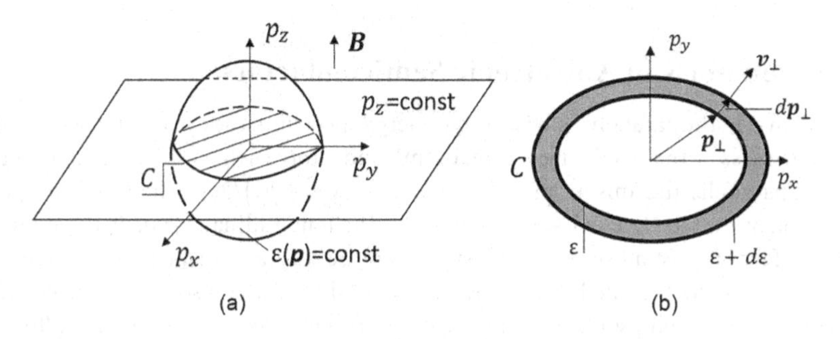

FIGURE 5.2 (a) Contour C is the electron trajectory in a momentum space $p_z = \text{const}$, $\varepsilon(\boldsymbol{p}) = \text{const}$. (b) The shaded ring is an infinitesimal element of the area bounded by contour C (dashed area on panel (a)).

$$dp_{\perp} = e\left[dr_{\perp} \times B\right], \tag{5.14}$$

which implies that vectors $dp_{\perp}, dr_{\perp}$, and B are mutually perpendicular to each other. Integrating both sides in (5.14), one finds

$$|p_{\perp}| = |e| B |r_{\perp}| \tag{5.15}$$

So, the electron trajectory in real space in the plane perpendicular to the field repeats the position in the momentum space up to the scaling factor $|e|B$. In other words, an electron rotates with frequency ω_c in real space in the plane normal to the field and makes uniform rectilinear motion along the field ($p_z = \text{const}$). So, the trajectory is a spiral around the direction of the magnetic field.

The energy spectrum of low energy electrons in metals and semiconductors is a parabolic dispersion law, $\varepsilon(p) = p^2 / 2m^*, m^*$ is the effective mass. For $p_z = \text{const}$, $S = \pi p_{\perp}^2 = \pi\left(p^2 - p_z^2\right) = \pi\left(2m^*\varepsilon - p_z^2\right)$ and then (5.13) gives $m_c = m^*$. Cyclotron mass coincides with the effective mass near the band edge only if the electron spectrum is parabolic and isotropic. In real metals and semiconductors, experimentally observed oscillations may involve high-energy electrons. In such experimental settings, the observable cyclotron mass differs from that at the band edge when non-parabolic corrections to spectrum come to effect, thus making electrons of different energies rotate with different frequencies. Besides, rotation direction depends on the sign of the cyclotron mass, which could be positive or negative depending on the details of the Fermi surface. On cyclotron motion in crystals with complex Fermi surface, see Refs. [1,2].

In real space, the size of the spiral in the plane perpendicular to the field (Larmor radius) follows from Eq. (5.15), $R = |p_{\perp}| / |e| B$ and depends on the electron energy and the magnetic field. The classical approach to the problem is valid if R is much larger than the electron de Broglie wavelength: $R >> \lambda$. For electrons at the Fermi level in metals, $\lambda \approx \hbar / p_F$ is of the order of the lattice constant a, thus classical description works if

$$B << \frac{\hbar}{|e| a^2}. \tag{5.16}$$

In most metals, experimental settings satisfy the condition (5.16), and classical theory holds.

In nondegenerate semiconductors, the electron wavelength relates to the average thermal energy $3k_BT/2$: $\lambda_T = h/\sqrt{3m^*k_BT}$. At low temperatures, condition $R \gg \lambda_T$ breaks in experimentally achievable fields, so at $k_BT < h^2/3m^*R^2$, the finite motion in perpendicular to the magnetic field plane becomes quantized. Quantum description and corresponding Landau levels are discussed in the next section.

Cyclic motion affects electrical and magnetic properties in clean enough samples where electrons could make at least one rotation before they scatter on impurities and phonons: $l > R$, l is the electron mean free path. The condition defines a strong field and, if expressed through time between collisions τ, reads as $\omega_c \tau > 1$.

5.2.1 Cyclotron Mass in Anisotropic Semiconductors

As mentioned above, in the parabolic and isotropic energy spectrum, the cyclotron mass (5.13) coincides with the effective mass at the band edge. In real semiconductors such as silicon and germanium, the dispersion law is parabolic and anisotropic: $\varepsilon(p) = p_z^2/2m_l + \left(p_x^2 + p_y^2\right)/2m_t$, where m_l and m_t are the longitudinal and transverse effective masses, respectively. The longitudinal z-axis is the main axis of an ellipsoid – the surface of constant energy in p-space. There are several equivalent minima in the p-space, and thus, the magnetic field is tilted differently to axes of different ellipsoids. One needs to solve the problem separately for each ellipsoid assuming the magnetic field has the direction $B = (B\alpha_1, B\alpha_2, B\alpha_3)$, where $\alpha_1, \alpha_2, \alpha_3$ are direction cosines between the magnetic field and the axes of the ellipsoid, $\alpha_1^2 + \alpha_2^2 + \alpha_3^2 = 1$. For ellipsoid of revolution, the solution to Eq. (5.6) gives the cyclotron frequency as

$$\omega_c^2 = \omega_1^2 \alpha_1^2 + \omega_2^2 \alpha_2^2 + \omega_3^2 \alpha_3^2,$$
$$\omega_1 = \omega_2 = \frac{eB}{\sqrt{m_l m_t}}, \ \omega_3 = \frac{eB}{m_t}. \tag{5.17}$$

If the magnetic field makes angle θ with z-direction, $\alpha_3 = \cos\theta$, $\alpha_1^2 + \alpha_2^2 = \sin^2\theta$, and

$$\omega_c^2 = e^2 B^2 \left(\frac{\sin^2\theta}{m_l m_t} + \frac{\sin^2\theta}{m_t^2} \right) \equiv \frac{e^2 B^2}{m_c^2}, \tag{5.18}$$

where the cyclotron mass

$$m_c = \left(\frac{\sin^2\theta}{m_l m_t} + \frac{\sin^2\theta}{m_t^2} \right)^{-2}. \tag{5.19}$$

The cyclotron mass depends on the effective mass tensor as well as on the direction of the magnetic field. Experimental measurements of the absorption of microwave radiation in the magnetic field (cyclotron resonance) allow finding m_l and m_t from data on ω_c at different θ.

5.3 Landau Quantization

The classical circular motion of a free electron in the external magnetic field presents the circular electric current, which, in turn, creates a magnetic field directed against the external field. The process is called the *diamagnetic response*. However, calculations based on classical statistical physics give the diamagnetic susceptibility zero. Diamagnetism of free electrons in solids, that is, nonzero diamagnetic susceptibility, is essentially a quantum effect arising from Landau quantization. For details, see Ref. [3].

First, we consider a simplistic model of the conduction band in metal or semiconductor where spin-orbit interaction is negligible. Once an electron is a quantum particle, the Hamiltonian in a magnetic field accounts for both the orbital motion (Eq. 5.2) and spin orientation (Pauli term):

$$H = \frac{1}{2m^*}(\boldsymbol{p} - e\boldsymbol{A})^2 \pm \frac{1}{2} g\mu_B B, \tag{5.20}$$

where g is the electron g-factor, $\mu_B = e\hbar / 2m_0$ is the Bohr magneton. Momentum in (5.20) is the differential operator $\boldsymbol{p} = -i\hbar\nabla$, m_0 is the free electron mass.

For now, we omit Zeeman spin splitting (last term in Eq. (5.20)) and dwell on the orbital part of the Hamiltonian by choosing a magnetic field in the z-direction and Landau gauge for the vector potential: $A_x = A_z = 0, \ A_y = Bx$. One can write the Schrödinger equation as

$$-\frac{\hbar^2}{2m^*}\frac{\partial^2\psi}{\partial x^2} + \frac{1}{2m^*}\left(-i\hbar\frac{\partial\psi}{\partial y} - eBx \right)^2 - \frac{\hbar^2}{2m^*}\frac{\partial^2\psi}{\partial z^2} = E\psi. \tag{5.21}$$

Since coefficients in Eq. (5.21) do not contain variables y, z, we look for a solution in the form

$$\psi = \varphi(x)\exp\left(ik_y y + ik_z z\right). \tag{5.22}$$

Substituting (5.22) in (5.21), one obtains

$$-\frac{\hbar^2}{2m^*}\frac{\partial^2\varphi}{\partial x^2}+\frac{1}{2}m^*\omega_c^2(x-x_0)^2\varphi=E'\varphi,$$
$$x_0=\frac{\hbar}{eB}k_y,\ E'=E-\frac{p_z^2}{2m^*},\ \omega_c=\frac{|e|B}{m^*}. \tag{5.23}$$

Since we consider the parabolic electron spectrum, the cyclotron frequency contains the effective mass, which in this case, coincides with the cyclotron mass. Equation (5.23) is the Schrödinger equation for the one-dimensional harmonic quantum oscillator of mass m^* and spring constant $m^*\omega_c^2/2$ oscillating about equilibrium position x_0. The eigenfunctions correspond to quantum levels enumerated by number $n=0,1,2.....$, the *Landau subbands*. Normalized to unity eigenfunctions are expressed through Hermite polynomials of the order n, $H_n(z)$,

$$\varphi_n=\frac{1}{\left(\sqrt{\pi}\ n!2^n\ l_B\right)^{1/2}}H_n\left(\frac{x-x_0}{l_B}\right)\exp\left(-\frac{(x-x_0)^2}{2l_B^2}\right), \tag{5.24}$$

where $l_B=\sqrt{\hbar/|eB|}$ is the *magnetic length*. Eigenvalues are

$$E'=\hbar\omega_c\left(n+\frac{1}{2}\right). \tag{5.25}$$

Using notations in Eq. (5.23), one obtains the electron energy in Landau subbands: the free motion along $\boldsymbol{B}$ and discrete levels for perpendicular to field motion: $E(n,p_z)=\hbar\omega_c(n+1/2)+p_z^2/2m^*$. So, the magnetic field quantizes the area of closed electron trajectory transverse to the magnetic field and leaves untouched the longitudinal motion,

$$p_z=\text{const.}$$
$$p_\perp^2=eB\hbar(2n+1). \tag{5.26}$$

Zeeman term in Eq. (5.20) splits each Landau energy level into spin-up and down states:

$$E(n,p_z,s)=\frac{p_z^2}{2m^*}+\hbar\omega_c\left[\left(n+\frac{1}{2}\right)\right]+\frac{1}{2}sg\mu_B B,\ s=\pm 1. \tag{5.27}$$

Note that for free electrons, $m^*=m_0$, $g=2$, and then $\hbar\omega_c=g\mu_B B$ – the energy distance between neighboring Landau levels is equal to the spin splitting. In real metals and semiconductors, m^* and g deviate from their free-electron values, and then the orbital and spin splittings are different. Often to describe the deviation, one introduces "*spin mass*" m_s: $g=2m_0/m_s$ and presents the spin splitting as $e\hbar B/m_s\equiv 2\mu_B^*B$, μ_B^* is the "effective" magneton. If a semiconductor were anisotropic, spectrum (5.27) would contain m_z^* in the z-direction, and ω_c being the cyclotron mass.

Landau conduction subbands in a degenerate semiconductor or simple metal are illustrated in Figure 5.3.

Degenerate bands in III-V semiconductors and almost degenerated conduction and valence bands in narrow-gap II-VI (or IV-VI) materials have a more complicated structure, so Landau quantization is worth special consideration briefly discussed in Section 5.9.

When the magnetic field increases, Landau levels move up in energy, cross the Fermi level, and thus cause oscillations in the density of states at the Fermi level. The density of states determines the kinetic and thermodynamic properties of solids making both longitudinal electrical conductivity (Shubnikov-de Haas effect) and magnetic susceptibility (de Haas-van Alphen effect) oscillating

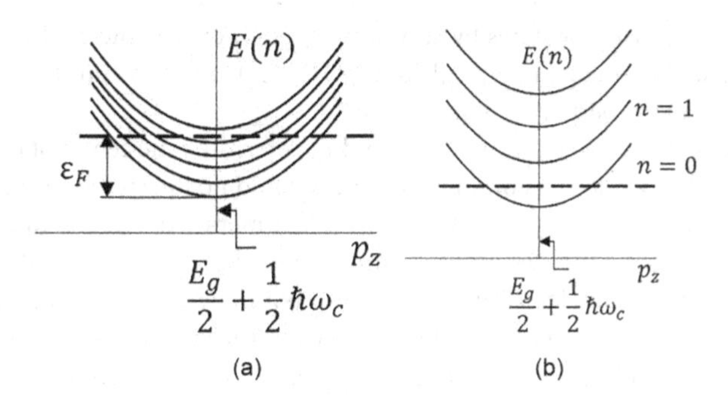

FIGURE 5.3 Landau levels in the conduction band of a degenerate semiconductor. (a) A weak magnetic field, (b) ultra-quantum limit, ε_F is the Fermi energy.

functions of the magnetic field. On the axis $1/B$, the period of oscillation equals $2\pi e/\hbar S$, S being the extremal cross-sectional area of the Fermi surface in the plane perpendicular to the magnetic field. Based on experimental information obtained at various directions of the magnetic field, one can reconstruct the Fermi surface and determine effective masses.

In the *ultra-quantum limit*, $\hbar\omega_c > \varepsilon_F$, only $n = 0$ Landau level is occupied by electrons, as shown in Figure 5.3b. The regime is hardly achievable in metals due to a very high magnetic field required. In degenerate semiconductors, relatively small Fermi energy makes the limit accessible in experiments.

5.3.1 Landau Levels: Degeneracy and Density of States

Three quantum numbers that define spectrum (5.27): n, p_z, and s. Inspecting the wave function (5.24), we find one more quantum number x_0 – the guiding center position (position of the axis of the spiral in the x–y plane), which does not enter the energy spectrum. That means the energy level (n, p_z, s) is degenerate. The degree of degeneracy equals the number of possible $x_0 = \hbar k_y / eB$ in the range of $0 \le x_0 \le L_x$, L_x is the crystal size in the x-direction. Thus, the interval for k_y is $0 \le k_y \le k_{y\,max}$, $k_{ymax} = |e| BL_x/\hbar$. In solids, wavenumber k_y is a discrete variable with a step of $q = 2\pi/L_y$. The number of states per given n, p_z, s is

$$N = \frac{k_{y\max}}{q} = \frac{|e| BL_x L_y}{2\pi\hbar} = \frac{L_x L_y}{2\pi l_B^2}. \tag{5.28}$$

So, four quantum numbers describe the quantum state of an electron in a magnetic field: n, p_z, s, x_0. As follows from Eq. (5.28), the degeneracy of level (n, p_z, s) is proportional to the magnetic field.

Translational motion along the z-direction forms the Landau band. The density of states in that band per spin is

$$\rho_{n,s}(\varepsilon) = N\sum_{k_z} \delta(\varepsilon - E(n, k_z, s)), \tag{5.29}$$

where N is the number of states for a particular (n, k_z, s) (5.28). Making the transition to integration $\sum_{k_z}(\ldots) = (L_z/2\pi)\int dk_z$, one obtains the number of states per unit energy per spin in the Landau subband n:

$$\rho_{n,s}(\varepsilon) = \frac{L_x L_y L_z \sqrt{2m_z^*}}{4\hbar\pi^2 l_B^2} \frac{1}{\sqrt{\varepsilon - \hbar\omega_c\left[\left(n + \frac{1}{2}\right)\right] + \frac{s}{2} g\mu_B B}}. \tag{5.30}$$

One can obtain the total density of states by summation (5.30) over n and s. The singular behavior of $\rho_{n,s}(\varepsilon)$ at the Landau subband edge is typical for the 1D electron spectrum that is a consequence of quantized motion in the plane perpendicular to the field.

Within the classical picture, two coordinates, X and Y, determine the center of rotation. In the quantum approach, the coordinates do not commute and thus could not be specified simultaneously, meaning that only one quantum number x_0 $(\sim k_y)$ remains to enumerate the degenerate states. Why is k_y of particular importance, whereas the system has symmetry in the $x-y$ plane, and no preference seems justified? It is the gauge choice that determines the quantum number. Choice $\boldsymbol{A}=(0,Bx)$ preserves translational invariance of the Hamiltonian in the y-direction and thus makes x_0 special. If we chose Landau gauge as $\boldsymbol{A}=(-By,0)$, it would preserve translations in the x-direction, making y_0 $(\sim k_x)$ a quantum number. One more option is the symmetric gauge $A=(-By/2, Bx/2)$, which preserves rotational invariance. Note that $\boldsymbol{A}$ is defined up to gauge transformation (see Chapter 12): for example, one can go from $\boldsymbol{A}=(-By/2,\ Bx/2)$ to $\tilde{\boldsymbol{A}}=(-By,0)$ by gauge transformation $\tilde{\boldsymbol{A}}=\boldsymbol{A}-\nabla(Bxy/2)$.

An alternative approach to degeneracy relates to the magnetic flux threading the area bound by electron trajectory. It follows from (5.13) that in the momentum space, the change in energy betwen Landau states n and $n\pm1$ correspond to the discrete change in the area bound by the trajectory perpendicular to the magnetic field (contour C in Figure 5.2):

$$\Delta S(p_\perp)=2\pi m^*\Delta E=2\pi m^*\hbar\omega_c, \tag{5.31}$$

In real space, we apply a scaling factor (5.15) to find

$$\Delta S(r_\perp)=\frac{2\pi m^*\Delta E}{(eB)^2}=\frac{2\pi\hbar}{|e|B}. \tag{5.32}$$

By calculating corresponding magnetic flux change

$$\Delta\Phi=B\Delta S(r_\perp)=\frac{2\pi\hbar}{|e|}=2\pi l_B^2\ B, \tag{5.33}$$

we find that on n and $n\pm1$ Landau levels, the magnetic flux through area bound by the trajectory differs by one flux quantum Φ_0 that is the universal constant, $\Phi_0=2\pi\hbar/|e|=4.136\times10^{-15}$ Wb – the magnetic flux through the area $2\pi l_B^2$.

The degree of degeneracy of the Landau level (5.28) is equal to the number of flux quanta threading the space area transverse to the magnetic field:

$$N=|e|BL_xL_y/2\pi\hbar=\Phi/\Phi_0. \tag{5.34}$$

Since N does not depend on n, all Landau levels are equally degenerate.

5.4 Landau Levels in Symmetric Gauge

The symmetric form of Schrödinger equation (5.21) reads as

$$\frac{1}{2m^*}\left(\boldsymbol{p}-\frac{|e|}{2}(\boldsymbol{B}\times\boldsymbol{r})\right)^2\Psi=E\Psi. \tag{5.35}$$

If the magnetic field points in the z-direction, it is convenient to use cylindrical coordinates and look for solution $\Psi=\psi(z)\ \phi(\varrho,\varphi)$, ϱ,φ are polar coordinates in the x - y plane. Electron motion in the z-direction is not affected by the magnetic field, while for transverse motion, the solution reads as [4,5]:

$$\phi(\varrho,\varphi)=\phi_{n,m}(\varrho,\varphi)=\frac{e^{im\varphi}}{l_B\sqrt{2\pi}}\left(\frac{n!}{(n+|m|)!}\right)^{1/2}\exp\left(-\frac{\varrho^2}{4l_B^2}\right)\left(\frac{\varrho}{\sqrt{2}l_B}\right)^{|m|}L_n^{|m|}\left(\frac{\varrho^2}{2l_B^2}\right),$$
$$E-\frac{p_z^2}{2m^*}=\varepsilon_{n,m}=\hbar\omega_c\left(n+\frac{m+|m|+1}{2}\right), \tag{5.36}$$

where n numerates Landau levels, $0\le n<\infty$, m is the magnetic quantum number (eigenvalues of the angular momentum), $-\infty<m<\infty$, and $L_n^k(x)=\frac{e^x x^{-k}}{n!}\frac{d^n}{dx^n}\left(e^{-x}x^{n+k}\right)$ are the associated Laguerre polynomials. All solutions $m\le 0$ describe degenerate states that correspond to a particular Landau level n.

5.5 Ladder Operators

The electron spectrum in a magnetic field can be calculated directly by diagonalization of the Hamiltonian in the second quantization representation (see Appendix 1), https://www.routledge.com/Modern-Semiconductor-Physics-and-Device-Applications/Dugaev-Litvinov/p/book/9780367250829#. Below we apply the method to $2D$ electrons.

If a quantum well thickness is of the order of the electron wavelength, confinement modifies the spectrum in the z-direction (growth direction perpendicular to the QW plane). Instead of a continuous band, we deal with a set of discrete levels. If electrons have low enough energy to be in the ground state (first size-quantized band in the $x-y$ plane), they can be considered two-dimensional.

For a magnetic field in the z-direction, we choose the vector potential in a symmetric gauge: $\mathbf{A}=(-By/2, Bx/2)$. Hamiltonian (5.2) for a quantum particle becomes

$$H=\frac{1}{2m^*}(\mathbf{p}-e\mathbf{A})^2=\frac{1}{2m^*}\left[(p_x-eA_x)^2+(p_y-eA_y)^2\right]$$
$$=\frac{1}{2m^*}\left(P_x^2+P_y^2\right), \tag{5.37}$$

where $\mathbf{p}=-i\hbar\nabla$, so that kinetic momenta $P_x=p_x-eA_x$ and $P_y=p_y-eA_y$ do not commute:

$$\left[P_x,P_y\right]=P_xP_y-P_yP_x=ie\hbar[\nabla\times\mathbf{A}]_z+e\left(A_yp_x-A_xp_y\right)$$
$$=ie\hbar B+\frac{eB}{2}\left(xp_x+yp_y\right)=i\hbar eB. \tag{5.38}$$

Term $\left(xp_x+yp_y\right)$ is equal to zero in virtue of commutation rules for canonical coordinate and momentum, $\left[x,p_x\right]=\left[y,p_y\right]=i\hbar$: $xp_x=i\hbar+p_xx=i\hbar-i\hbar=0$.

Assuming $eB>0$, we introduce *ladder* operators a^+ and a as:

$$a=\frac{1}{\sqrt{2e\hbar B}}\left(P_x+iP_y\right),$$
$$a^+=\frac{1}{\sqrt{2e\hbar B}}\left(P_x-iP_y\right), \tag{5.39}$$
$$\left[a,a^+\right]=1,$$

It should be noted that to preserve commutation relation (5.39), for $eB < 0$, the ladder operators look as follows:

$$a = \frac{1}{\sqrt{2\hbar|eB|}}\left(P_x - iP_y\right),$$
$$a^+ = \frac{1}{\sqrt{2\hbar|eB|}}\left(P_x + iP_y\right). \tag{5.40}$$

For $eB > 0$, by calculating product

$$\left(P_x + iP_y\right)\left(P_x - iP_y\right) = P_x^2 + P_y^2 + i\left(P_yP_x - P_xP_y\right) = P_x^2 + P_y^2 + e\hbar B, \tag{5.41}$$

we express Hamiltonian (5.37) as

$$H = \frac{e\hbar B}{m^*}\left(a^+a + \frac{1}{2}\right). \tag{5.42}$$

Eigenvalues of the Hamiltonian $E = \hbar\omega_c(n+1/2)$ is the $2D$ discrete Landau spectrum, where n is the eigenvalue of the number operator a^+a : if $|n\rangle$ is the oscillator wavefunction of order n,

$$a^+a|n\rangle = n|n\rangle. \tag{5.43}$$

To satisfy (5.43), ladder operators a^+, a should increase and decrease the number of oscillator quanta as

$$a^+|n\rangle = \sqrt{n+1}|n+1\rangle,$$
$$a|n\rangle = \sqrt{n}|n-1\rangle. \tag{5.44}$$

The ladder operator technique is a convenient tool for finding the Landau spectrum in semiconductors of arbitrary dimensions and complex band structure. The recipe is to use (5.40) to replace transverse to magnetic field components of momentum and then find eigenvalues of the resulting Hamiltonian. Below we discuss several examples in 2D and 3D semiconductors.

5.6 Localized States and Extended Chiral Modes

The diagram shown in Figure 5.4 presents Landau levels in the interior of the 2D-electron gas subjected perpendicular magnetic field. The electron energy increases on the edge of the sample. The example shows three Landau levels filled with electrons and, correspondingly, three conductive edge channels.

If the Fermi energy lies in the gap between Landau levels at $T = 0$, electrons fill up ν levels (ν is an integer number). In the example shown in Figure 5.4, $\nu = 3$.

If the Fermi energy falls into the energy gap between Landau levels, as shown in Figure 5.4, all states in the interior of the sample cannot carry electric current. Electron motion in these states is rendered in Figure 5.5 by circular trajectories. Only edge states lie on the Fermi level (hollow circles in Figure 5.4) and thus can take part in transport. States at the edges are delocalized and represent chiral modes in which electrons can move in one direction along the edge of the sample, as shown in Figure 5.5.

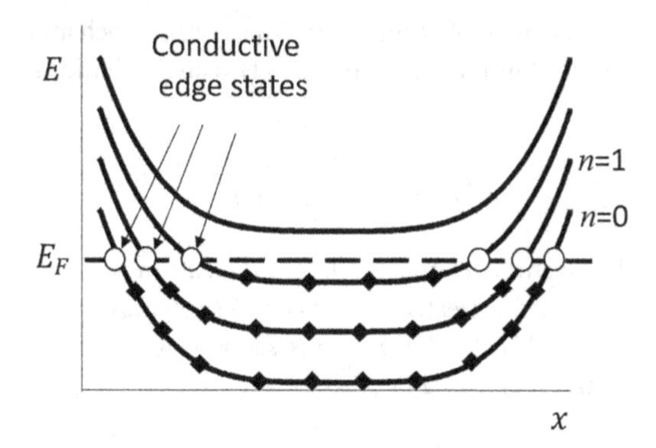

FIGURE 5.4 Landau levels in a 2D sample as functions of coordinate. Diamonds mark localized states, delocalized edge states are hollow circles.

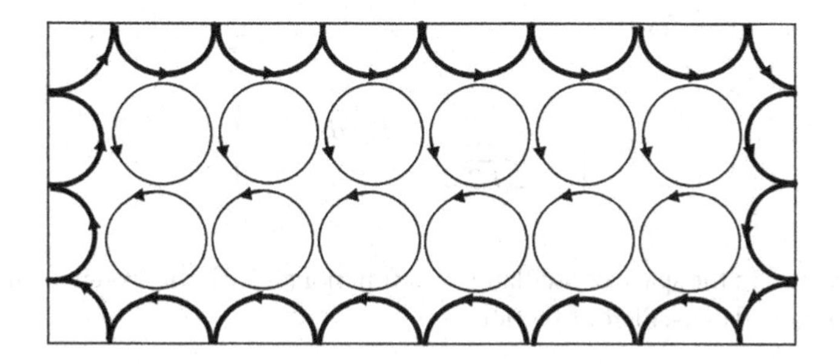

FIGURE 5.5 Semiclassical electron trajectories. Chiral extended edge state (bold line) and localized states inside the bulk (circles).

Given N_e electrons are on N-fold degenerate Landau level, the Landau level filling factor follows from (5.34):

$$\nu = \frac{N_e}{N} = \frac{\Phi_0 n_e}{B}, \tag{5.45}$$

where n_e is the electron sheet density. If the Fermi energy lies in the gap between Landau levels, levels below the gap are fully occupied by electrons at $T = 0$, so ν is an integer – the number of occupied levels. In the example shown in Figure 5.4, $\nu = 3$.

The presence of localized electrons in the interior of the 2D sample and chiral electrons at edges are at the origin of the quantum Hall effect (Chapter 7).

5.7 Dirac Electrons in a Magnetic Field

Low-energy excitations in narrow-gap semiconductors, interface states in band-inverted contacts, and surface states in topological insulators are described by the Dirac-type Hamiltonian (Chapter 2, Eq. (2.65)). In what follows, we consider orbital quantization only and neglect spin, so we deal with one block in Eq. (2.65):

$$H = \begin{pmatrix} \Delta & v p_- \\ v p_+ & -\Delta \end{pmatrix}, \quad p_\pm = p_x \pm i p_y, \tag{5.46}$$

where v is the interband matrix element of momentum operator on Bloch amplitudes at the band minimum, [m/s]. The spectrum of the Hamiltonian comprises two energy bands $E_{1,2}$ separated by the bandgap 2Δ:

$$E_{1,2}(\boldsymbol{p}) = \pm\sqrt{\Delta^2 + \mathrm{v}^2 p^2}. \tag{5.47}$$

Two bands formally resemble the Dirac relativistic electron–positron spectrum where, however, parameters originate from the semiconductor electron velocity and the bandgap.

Following the procedure described in the previous section, we replace canonical momenta with kinetic ones, $\boldsymbol{p} \to \boldsymbol{P} = \boldsymbol{p} - e\boldsymbol{A}$, and then using Eq. (5.39)

$$\begin{aligned} P_x + iP_y &\to \sqrt{2e\hbar B}a, \\ P_x - iP_y &\to \sqrt{2e\hbar B}a^+ \end{aligned} \tag{5.48}$$

obtain the Hamiltonian in the form:

$$H = \begin{pmatrix} \Delta & \sqrt{2e\hbar B}\ a^+ \\ \sqrt{2e\hbar B}\ a & -\Delta \end{pmatrix}, \tag{5.49}$$

Hamiltonian (5.49) acts in the space of oscillator wavefunctions $\varphi_n(\boldsymbol{r}) = |n\rangle$. By choosing the wave function as a spinor composed of oscillator functions,

$$\Psi = \begin{pmatrix} f_1\varphi_n \\ f_2\varphi_{n-1} \end{pmatrix}, \quad n = 1, 2, \ldots, \tag{5.50}$$

we write down Schrödinger equation $H\Psi = E\Psi$ and, using Eq. (5.44), obtain

$$\begin{cases} (\Delta - E)\ \varphi_n\ f_1 + \sqrt{2e\hbar Bn}\ \varphi_n\ f_2 = 0, \\ \sqrt{2e\hbar Bn}\ \varphi_{n-1} f_1 - (\Delta + E)\ \varphi_{n-1}\ f_2 = 0. \end{cases} \tag{5.51}$$

Multiplying the first line in Eq. (5.51) by φ_n^*, the second by φ_{n-1}^*, taking into account orthogonality of oscillator functions $\langle \varphi_i | \varphi_j^* \rangle = \delta_{ij}$, one gets a new representation of the Hamiltonian:

$$\begin{aligned} \tilde{H}\begin{pmatrix} f_1 \\ f_2 \end{pmatrix} &= E\begin{pmatrix} f_1 \\ f_2 \end{pmatrix} \\ \tilde{H} &= \begin{pmatrix} \Delta & \sqrt{2e\hbar Bn} \\ \sqrt{2e\hbar Bn} & -\Delta \end{pmatrix}. \end{aligned} \tag{5.52}$$

The eigenvalues of (5.52) are the Landau spectrum of Dirac electrons:

$$E_n = \pm\sqrt{\Delta^2 + 2e\hbar Bn}, \quad n = 1, 2, \ldots \tag{5.53}$$

If $n=0$, $\Psi=\begin{pmatrix} f_1\,\varphi_0 \\ 0 \end{pmatrix}$, and in Eq. (5.51) remains the first equation only with the eigenvalue of $E_0=\Delta$. Full Landau spectrum has the form:

$$E_n=\begin{cases} \pm\sqrt{\Delta^2+2e\hbar Bn}, & n=1,2,\ldots \\ \Delta, & n=0. \end{cases} \tag{5.54}$$

When $eB<0$, we use (5.40), and instead of (5.49), one gets

$$H=\begin{pmatrix} \Delta & \sqrt{2|eB|\hbar}\;a \\ \sqrt{2|eB|\hbar}\;a^{+} & -\Delta \end{pmatrix}. \tag{5.55}$$

With eigenspinor

$$\Psi=\begin{pmatrix} f_1\varphi_{n-1} \\ f_2\varphi_n \end{pmatrix},\quad n=1,2,\ldots, \tag{5.56}$$

the Schrödinger equation reads as

$$\begin{cases} (\Delta-E)\,\varphi_{n-1}\,f_1+\sqrt{2|eB|\hbar n}\;\varphi_{n-1}\,f_2=0, \\ \sqrt{2|eB|\hbar n}\;\varphi_n f_1-(\Delta+E)\,\varphi_n\,f_2=0. \end{cases} \tag{5.57}$$

Again, the case $n=0$ needs to be considered separately with $\Psi=\begin{pmatrix} 0 \\ f_2\varphi_0 \end{pmatrix}$, and then in (5.57), the only second equation remains with the eigenvalue $E=-\Delta$. The whole spectrum becomes

$$E_n=\begin{cases} \pm\sqrt{\Delta^2+2|eB|\hbar n}, & n=1,2,\ldots \\ -\Delta, & n=0. \end{cases} \tag{5.58}$$

Inspecting (5.54) and (5.58), one may conclude on the general properties of the Landau levels in a Dirac semiconductor. For $n\geq 1$, the spectrum is symmetric to the magnetic field inversion and presents electron-like and hole-like branches denoted by plus and minus signs, respectively. However, for level $n=0$, the symmetry breaks, and at $\Delta>0$, the level belongs to either of two bands: conduction if $eB>0$, or valence if $eB<0$ (see Figure 5.6a), or another way around if $\Delta<0$, as illustrated in Figure 5.6c. In other words, the electron (hole)-like zero-mode level does not have a partner in the valence (conduction) band. That is the manifestation of the parity anomaly in a magnetic field [6].

In the gapless Dirac spectrum (for instance, at the band-inverted interface), the $n=0$ level has zero energy and belongs to both conduction and valence bands, as shown in Figure 5.6b.

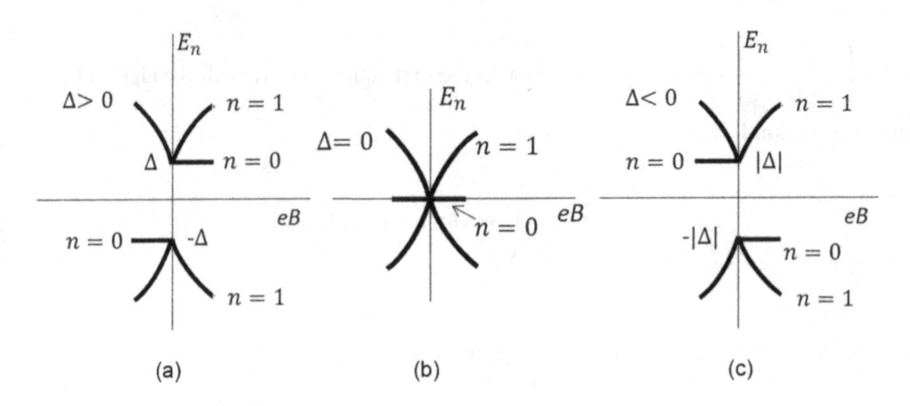

FIGURE 5.6 Zero-mode Landau level in Dirac spectrum. Parity anomaly. Panels (a–c) correspond to the gap parameter positive, zero, and negative, respectively.

5.8 Chiral Anomaly

Energy bands in three-dimensional semiconductors with no time or spatial inversion symmetry may have accidental (not symmetry-related) degeneracy at general points in the Brillouin zone. The points in the BZ, in which bands touch each other (*nodes*), come in pairs residing at momenta $\pm p_{z0}$, as illustrated in Figure 5.7.

The energy spectrum near a node p_{z0} is described by the Weyl Hamiltonian, which looks similar to that in (5.46):

$$H = \begin{pmatrix} v_z p_z & v p_- \\ v p_+ & -v_z p_z \end{pmatrix}. \tag{5.59}$$

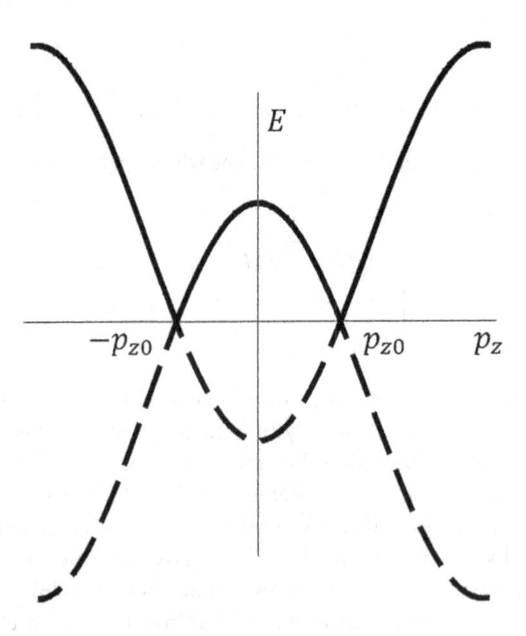

FIGURE 5.7 Accidental degeneracy in nodes residing at $\pm p_{z0}$.

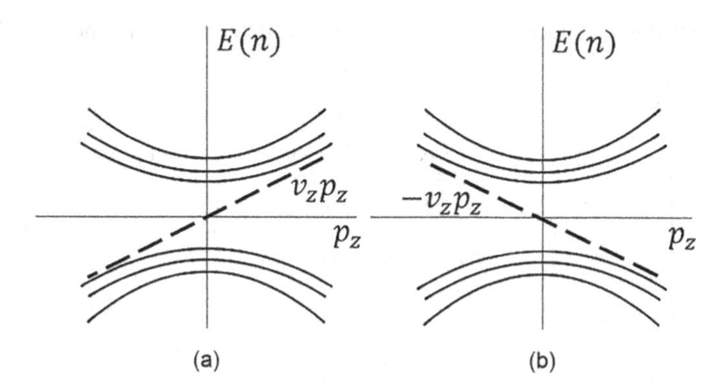

FIGURE 5.8 Chiral Landau zero-modes in Weyl semimetals. Momenta in nodes (a) and (b) are measured relative to corresponding node locations.

The second member of the pair has the Hamiltonian (5.59) with $v_z p_z$ replace to $-v_z p_z$. In the vicinity of nodes $\pm p_{z0}$, there is no energy gap in the spectrum, so if the Fermi energy lies close to $E=0$, a semiconductor is called *Weyl semimetal* that is the matter with approximately equal numbers of electrons and holes. An example semiconductor is TaAs.

In a magnetic field $B \| z$, one can use all the arguments of the previous section to get the Landau spectrum similar to (5.54) for one Weyl node,

$$E_n = \begin{cases} \pm\sqrt{v_z^2 p_z^2 + 2e\hbar Bn}, & n = 1,2,\ldots \\ v_z p_z, & n = 0. \end{cases} \tag{5.60}$$

and similar to (5.58) for the other:

$$E_n = \begin{cases} \pm\sqrt{v_z^2 p_z^2 + 2|eB|\hbar n}, & n = 1,2,\ldots \\ -v_z p_z, & n = 0. \end{cases} \tag{5.61}$$

Due to the parity anomaly, the zero Landau level does not have a partner in the same node, so only $v_z p_z$ branch appears in the spectrum (5.60) (see Figure 5.8a). The other Weyl node (5.61) has branch $-v_z p_z$, as shown in Figure 5.8b.

In each of the two modes in Figure 5.8, the p-linear branch corresponds to *chiral particles* moving along the z-axis, the direction of the magnetic field. In chiral modes, backscattering is suppressed as no branch with opposite momentum exists near the same node, while internode transition requires large and thus hardly probable momentum transfer.

Electric field $\boldsymbol{E} \| \boldsymbol{B}$ breaks the balance between chiral branches, meaning that numbers of particles moving in opposite directions become different, and thus the current flows along $\boldsymbol{B}$. That is the *chiral anomaly*, which, in experiments, reveals itself in longitudinal magnetoresistance.

5.9 Landau Spectrum in a Narrow-Gap Semiconductor

Within the k-p perturbation theory, the Hamiltonian of narrow-gap semiconductors is a matrix which dimensions equals the number of almost degenerate bands divided by a small bandgap. Remote energy bands bring corrections to the parameters of matrix Hamiltonian. Several examples of such Hamiltonians we discussed in Chapter 1. Below, for illustration purposes, we consider IV-IV materials where 4×4 k-p Hamiltonian describes conduction and valence bands in L-point in BZ (Chapter 1, (1.100)):

$$H=\begin{pmatrix} \Delta & 0 & v_z p_z & v_t(p_x-ip_y) \\ 0 & \Delta & v_t(p_x+ip_y) & -v_z p_z \\ v_z p_z & v_t(p_x-ip_y) & -\Delta & 0 \\ v_t(p_x+ip_y) & -v_z p_z & 0 & -\Delta \end{pmatrix} \tag{5.62}$$

$$\varepsilon_{1,2}(\boldsymbol{p})=\pm\sqrt{\Delta^2+v_t^2\left(p_x^2+p_y^2\right)+v_z^2 p_z^2}\,.$$

If the magnetic field is parallel to <111> direction (z-axis), the transition to number representation requires replacement (5.48) in the Hamiltonian:

$$p_x+ip_y=\frac{\hbar\sqrt{2}}{l_B}a,\ p_x-ip_y=\frac{\hbar\sqrt{2}}{l_B}a^+,$$

$$H=\begin{pmatrix} \Delta-g\mu_B B/2 & 0 & v_z p_z & \hbar\omega_c a^+ \\ 0 & \Delta+g\mu_B B/2 & \hbar\omega_c a & -v_z p_z \\ v_z p_z & \hbar\omega_c a^+ & -\Delta-g\mu_B B/2 & 0 \\ \hbar\omega_c a & -v_z p_z & 0 & -\Delta+g\mu_B B/2 \end{pmatrix} \tag{5.63}$$

$$\hbar\omega_c\equiv v_t\sqrt{2eB\hbar},$$

where μ_B is the Bohr magneton, and we assume electron g-factors equal in conduction and valence bands. Using ansatz (5.50),

$$\Psi=\begin{pmatrix} f_1\varphi_n \\ f_2\varphi_{n-1} \\ f_3\varphi_n \\ f_4\varphi_{n-1} \end{pmatrix},\ n=1,2,\ldots, \tag{5.64}$$

and following steps described after Eq. (5.50), we obtain the Hamiltonian acting on coefficients f_i:

$$\tilde{H}\begin{pmatrix} f_1 \\ f_2 \\ f_3 \\ f_4 \end{pmatrix}=E\begin{pmatrix} f_1 \\ f_2 \\ f_3 \\ f_4 \end{pmatrix},$$

$$\tilde{H}=\begin{pmatrix} \Delta-g\mu_B B/2 & 0 & \mathrm{v}_z p_z & \hbar\omega_c\sqrt{n} \\ 0 & \Delta+g\mu_B B/2 & \hbar\omega_c\sqrt{n} & -\mathrm{v}_z p_z \\ \mathrm{v}_z p_z & \hbar\omega_c\sqrt{n} & -\Delta-g\mu_B B/2 & 0 \\ \hbar\omega_c\sqrt{n} & -\mathrm{v}_z p_z & 0 & -\Delta+g\mu_B B/2 \end{pmatrix} \tag{5.65}$$

Landau and Zeeman splitting in conduction and valence bands follow as eigenvalues of $\tilde{H}$:

$$\varepsilon_{c,v}(n,\,p_z,s)=\pm\sqrt{\Delta^2+\left(\frac{1}{2}g\mu_B B\right)^2+n\hbar^2\omega_c^2+\mathrm{v}_z^2 p_z^2+g\mu_B Bs\sqrt{\Delta^2+\mathrm{v}_z^2 p_z^2}}\,, \tag{5.66}$$

$s=\pm 1,\ n=1,2,\ldots$

Again, similarly to (5.54), zero Landau mode we consider separately. Corresponding eigenvalues at $n=0$:

$$\varepsilon_{c,v}(0,p_z)=-\frac{1}{2}g\mu_B B\pm\sqrt{\Delta^2+\mathrm{v}_z^2 p_z^2},\quad n=0. \tag{5.67}$$

Both conduction and valence bands levels $n=0$ have Zeeman shift of the same sign, meaning that these states in conduction and valence bands correspond to the same spin state. If the Fermi level is in the gap, the valence $n=0$ Landau level is spin-polarized.

Since there are four non-equivalent L-points in the BZ, the spectrum in the other three extrema instead of $\boldsymbol{B}$ in Eq. (5.66) would contain projections of the magnetic field on the axis to a corresponding L-point.

The approach illustrated here allows finding Landau levels in degenerate or almost degenerate bands in semiconductors described by matrix k-p Hamiltonian of any dimension.

5.10 Landau Levels in Rashba Electron Gas

The energy spectrum (5.62) accounts for the spin–orbit interaction (parameter v_t) and, nevertheless, is spin degenerate because of the spatial inversion symmetry in IV-VI materials.

In noncentrosymmetric materials, for example, wurtzite III-V and II-VI semiconductors as well as asymmetric quantum wells, the spin–orbit interaction requires the electron spectrum to obey the Kramers relation: $\varepsilon_\uparrow(\boldsymbol{k})=\varepsilon_\downarrow(-\boldsymbol{k})$, which implies spin splitting at $k\neq 0$. In Chapter 2, we discussed the 2D electrons in inversion-asymmetric heterostructure that can be described by Rashba Hamiltonian,

$$H=\frac{p^2}{2m^*}+\alpha\left(\sigma_x p_y-\sigma_y p_x\right)=\begin{pmatrix} \varepsilon_p & i\alpha p_- \\ -i\alpha p_+ & \varepsilon_p \end{pmatrix}, \tag{5.68}$$

where α is the constant of Rashba spin–orbit interaction, $p_\pm=p_x\pm ip_y$, and $\varepsilon_p=p^2/2m^*$. Spin-polarized bands are eigenvalues of (5.68): $\varepsilon_{\uparrow\downarrow}(p)=\varepsilon_p\pm\alpha p$.

Following the recipe (5.63), to get the Landau spectrum in the z-directed magnetic field, one needs to make a replacement $p_+=(\hbar\sqrt{2}/l_B)a,\ p_-=(\hbar\sqrt{2}/l_B)a^+$ in the Hamiltonian (5.68):

$$H=\begin{pmatrix} \hbar\omega_c\left(a^+a+\frac{1}{2}\right)-\frac{1}{2}g\mu_B B & i\alpha\sqrt{2m^*\hbar\omega_c}\ a^+ \\ -i\alpha\sqrt{2m^*\hbar\omega_c}\ a & \hbar\omega_c\left(a^+a+\frac{1}{2}\right)+\frac{1}{2}g\mu_B B \end{pmatrix},\ \omega_c=\frac{eB}{m^*}. \tag{5.69}$$

Following steps described after Eq. (5.49), we use

$$\Psi = \begin{pmatrix} f_{\uparrow}\ \varphi_n \\ f_{\downarrow}\ \varphi_{n-1} \end{pmatrix}, \quad n = 1, 2, \ldots \tag{5.70}$$

and obtain

$$\tilde{H} = \begin{pmatrix} \hbar\omega_c\left(n + \frac{1}{2}\right) - \frac{1}{2}g\mu_B B & i\alpha\sqrt{2m^* n\hbar\omega_c} \\ -i\alpha\sqrt{2m^* n\hbar\omega_c} & \hbar\omega_c\left(n + \frac{1}{2} - 1\right) + \frac{1}{2}g\mu_B B \end{pmatrix}, \tag{5.71}$$

Eigenvalues are given below:

$$\begin{aligned} \varepsilon_{\uparrow\downarrow} &= n\hbar\omega_c \mp \frac{1}{2}\sqrt{\left(g\mu_B B - \hbar\omega_c\right)^2 + 8\alpha^2 m^* n\hbar\omega_c}, \ n = 1, 2, \ldots \\ \varepsilon_{\uparrow} &= \frac{1}{2}\hbar\omega_c - \frac{1}{2}g\mu_B B, \ n = 0. \end{aligned} \tag{5.72}$$

Zero Landau level is spin-polarized and does not experience Zeeman spin-splitting.

5.11 Aharonov–Bohm Effect

In classical electrodynamics, scalar and vector potentials V and $\boldsymbol{A}$ are auxiliary variables in the sense that they are convenient, but not necessary, to use. Maxwell equations operate with observable fields $\boldsymbol{E}$ and $\boldsymbol{B}$. The Lorentz force (5.5) also contains the fields. In quantum physics, however, the role of potentials is more essential. Potentials enter the Schrödinger equation, and as a consequence of this fact, a quantum particle feels potentials V and $\boldsymbol{A}$ even though it is moving in the region where fields $\boldsymbol{E}$ and $\boldsymbol{B}$ are absent. The Aharonov–Bohm effects demonstrate the sensitivity of the electron wavefunction to potentials $\boldsymbol{V}$ and $\boldsymbol{A}$ rather than to fields $\boldsymbol{E}$ and $\boldsymbol{B}$.

Below we consider the Aharonov–Bohm effect related to the magnetic field. We are going to follow the spatial motion of the quantum particle which, according to the quantum mechanics recipe, is described by a wave packet $\Psi(\boldsymbol{r},t)$, $\boldsymbol{r}$ is the center of the packet. Particle evolution in space and time follows from the Schrödinger equation:

$$i\hbar\frac{\partial\Psi(\boldsymbol{r},t)}{\partial t} = \frac{1}{2m^*}\left(-i\hbar\nabla - e\boldsymbol{A}(\boldsymbol{r})\right)^2 \Psi(\boldsymbol{r},t). \tag{5.73}$$

Let us consider the region where the magnetic field is absent, $\boldsymbol{B} = \nabla\times\boldsymbol{A} = 0$. The vector-potential, generally, is not zero and can be expressed as a gauge field $\boldsymbol{A} = \nabla\Lambda(\boldsymbol{r},t)$, Λ is any real function of coordinates and time. Due to the gauge freedom, one may choose the point $\boldsymbol{R}_0$ where the vector-potential is zero $\boldsymbol{A}(\boldsymbol{R}_0) = 0$. If $\boldsymbol{A} = 0$, the ground state solution to (5.73) we denote as $\Psi(\boldsymbol{r} - \boldsymbol{R}_0)$. If slowly move the particle from $\boldsymbol{R}_0$ to arbitrary point $\boldsymbol{R}$, the state acquires the phase factor

$$\Psi(\boldsymbol{r} - \boldsymbol{R}) = \exp\left(-\frac{ie}{\hbar}\int_{\boldsymbol{R}_0}^{\boldsymbol{R}}\boldsymbol{A}(\boldsymbol{r})\cdot d\boldsymbol{r}\right)\Psi(\boldsymbol{r} - \boldsymbol{R}_0). \tag{5.74}$$

Substituting (5.74) to (5.73), one obtains the derivative of (5.74) over the upper limit $\left(\text{term } \nabla_R\right)$, which cancels the $e\boldsymbol{A}$ -term in the Hamiltonian. That proves $\Psi(\boldsymbol{r}-\boldsymbol{R})$ is the solution to the Schrödinger equation with $\boldsymbol{A} \neq 0$ if $\Psi(\boldsymbol{r}-\boldsymbol{R}_0)$ is the solution at $\boldsymbol{A}=0$. Slow (adiabatic) motion preserves the same quantum state on the path from $\boldsymbol{R}_0$ to $\boldsymbol{R}$, while otherwise, the wavefunction would transform in a more complicated way than just changing the phase. If the path is a closed contour ending in initial point $\boldsymbol{R} \to \boldsymbol{R}_0$,

$$\Psi(\boldsymbol{r}-\boldsymbol{R}_0) \to \exp(-i\gamma)\,\Psi(\boldsymbol{r}-\boldsymbol{R}_0),\ \gamma = \frac{e}{\hbar}\oint \boldsymbol{A}\cdot d\boldsymbol{l}. \tag{5.75}$$

According to Stokes theorem,

$$\oint \boldsymbol{A}\cdot d\boldsymbol{l} = \int d\boldsymbol{S}\cdot(\nabla\times\boldsymbol{A}) \equiv \int d\boldsymbol{S}\cdot\boldsymbol{B} = \Phi. \tag{5.76}$$

Surface integration runs over the area enclosed by the trajectory, and Φ is the magnetic flux threading the area. The phase $\gamma = e\Phi/\hbar$ is the *Aharonov-Bohm phase*, which is one of the realizations of the more general Berry phase (see Chapter 1).

To observe how the vector potential affects the electron motion, we use a setting typical for an interference experiment. The double-slit experiment setting is shown in Figure 5.9.

Interference pattern depends on the phase difference between waves coming from through two slits:

$$\exp\left[\frac{ie}{\hbar}\left(\int_{L1}\boldsymbol{A}\cdot d\boldsymbol{l} - \int_{L2}\boldsymbol{A}\cdot d\boldsymbol{l}\right)\right] = \exp\left[\frac{ie}{\hbar}\oint \boldsymbol{A}\cdot d\boldsymbol{l}\right] = \exp\left(\frac{ie}{\hbar}\Phi\right), \tag{5.77}$$

where we changed the integration to a backward direction in the second term, obtained the loop integral, and used (5.76). So, the interference pattern is sensitive to the magnetic flux inside the tube despite electrons move along paths L_1 and L_2 where $\boldsymbol{B}=0$. That is the *Aharonov–Bohm effect*, which underlines the particular role of vector-potential in the interference of quantum particles.

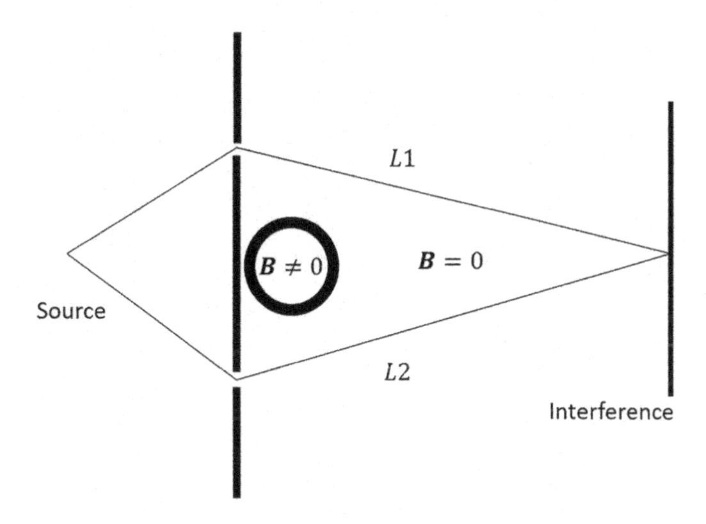

FIGURE 5.9 Double-slit interference setting. The magnetic field inside the cylinder can be created by rotating the charged cylinder or by placing a long solenoid with an electric current.

References

1. C. Kittel, *Quantum Theory of Solids* (John Wiley & Sons Inc., New York-London, 1963).
2. I.M. Lifshits, M.Y. Azbel and M.I. Kaganov, *Electron Theory of Metals* (Consultants Bureau, New York, 1973).
3. A.A. Abrikosov, *Fundamentals of the Theory of Metals* (North Holland, Amsterdam, 1988; Dover, New York, 2017).
4. V. Fock, "Bemerkung zur Quantelung des harmonischen Oszillators im Magnetfeld", *Z. Phys.*, **47**, 446 (1928).
5. L.D. Landau and E.M. Lifshits, *Quantum Mechanics, Non-Relativistic Theory* (3rd ed., Pergamon Press, New York, 1997).
6. F.D.M. Haldane, "Model for a quantum Hall effect without landau levels: Condensed-matter realization of the "Parity Anomaly", *Phys. Rev. Lett.* **61**, 2015 (1988).

6

Phonons and Electron–Phonon Interaction

A solid-state is a complex system of interacting electrons and nuclei. Electron-ion, electron-electron, and ion-ion Coulomb interactions determine bonding forces in molecules and solids. If bonding exists, the potential profile $U(R)$ as a function of interatomic distance R always has the form shown in Figure 6.1. Generally, function $U(R)$ consists of two terms, one repulsive at short distances and the other attractive forming the minimum at R_0-equilibrium interatomic distance.

The repulsive part of interatomic interaction comes to effect when atoms are close enough, so their valence electron states strongly overlap. Since low-energy valence levels are filled, overlap brings the Pauli exclusion principle into effect by forcing electrons to fill up high-energy states. An increase in the overall energy means repulsion at short distances. The profile $U(R)$ in solids depends on the type of chemical bonding.

6.1 Types of Chemical Bonds

There are five types of chemical bonding in molecules and solid crystals.

Ionic bonding is typical for I–VII compounds such as NaCl and KBr and presents a considerable contribution to bonding in II–VI semiconductors. When two constituent atoms come close, it becomes energetically favorable to ionize Na, then to have transferred an electron to the valence shell of Cl, thus forming an electrostatic bond between two ions of opposite electric charge. At short distances, the valence shells strongly overlap, resulting in repulsion, so the energy of ion interaction looks like in Figure 6.1.

Metallic bonding. Metallic bonding is similar to ionic bonding in that the positively charged metal ions attract delocalized electrons. Due to a large number of delocalized electrons, the average distance between attracting each other electrons and ions is much less than the distance between ions

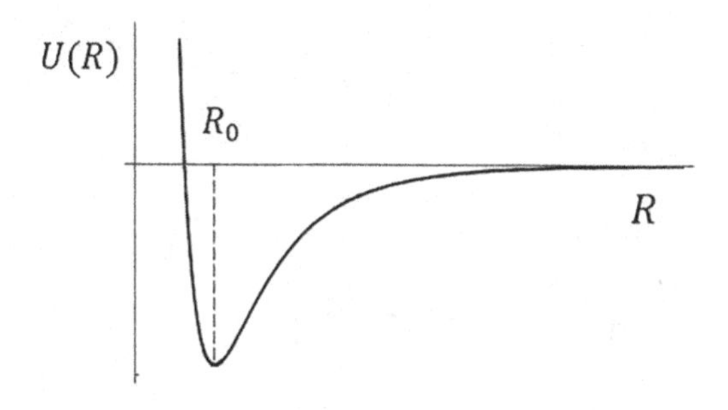

FIGURE 6.1 Potential for ions in molecules and crystals. R_0 is the ion equilibrium position.

DOI: 10.1201/9780429285929-6

contributing to repulsion. It creates an overall binding force. At the same time, the energy of degenerate electron gas contributes to repulsion at short distances, thus forming the minimum in Figure 6.1.

It is instructive to look at metal like a gigantic molecule with a large number of electrons common for all ions. In such a system, the bonding force weakly depends on the distance between ions explaining the stability of metal structure to plastic deformation.

Covalent bonding prevails in group-IV semiconductors such as Si, Ge, C, as well as molecules comprising identical atoms like hydrogen, H_2. While ionic bonding is about electron transfer from one atom to another, the covalent bond appears when two atoms share valence electrons, in other words, when electrons from both atoms are in the same quantum state provided by a Coulomb field of two nuclei. Due to the Pauli exclusion principle, the process of filling of the same state is sensitive to the spins of electrons, thus cannot be understood within classical considerations. The classical approach to Coulomb forces between valence electrons and nuclei always results in repulsion between two identical atoms. Within the quantum mechanical description, one can divide Coulomb interaction between atoms into two parts: always repulsive direct interaction and exchange interaction, which could be of either sign depending on the total spin of the pair. The state with antiparallel spins of valence electrons (singlet state) is spatially symmetric, providing for finite electron density between atoms and thus creating chemical bonding. In other words, in a singlet state, the exchange part of Coulomb interaction between two ions turns attractive and, together with the repulsive direct part, forms the potential profile of Figure 6.1. Quantum state with parallel spins-triplet state has an antisymmetric wave function, which prevents valence electrons from being in interatomic space. Depicted in Figure 6.2 is the valence electron density distribution in singlet and triplet states in molecule H_2.

In the singlet state shown in Figure 6.2b, the electron density distribution presents the electron cloud common for both nuclei that manifests the formation of a stable molecule. In the triplet state (Figure 6.2a), electron clouds repulse each other, making the electron density between nuclei absent and bonding ceases to exist.

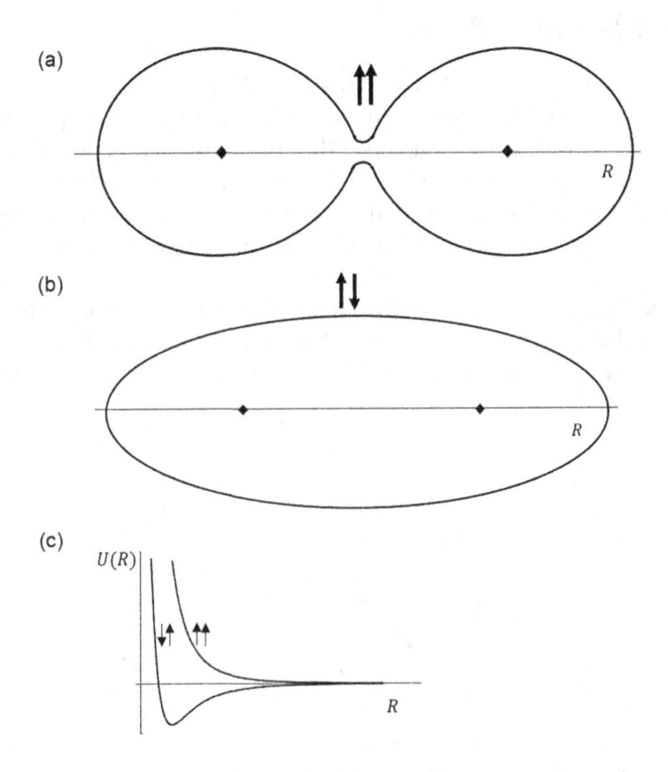

FIGURE 6.2 Electron density in triplet (a) and singlet (b) states. (c) Interionic potential energy.

The exchange part of interaction exists as long as the valence states of two atoms overlap. As valence shells are not necessarily spherically symmetric, the bonds have spatial orientation and thus determine the point symmetry of the crystal. Chemical bonding in III–V semiconductors is mostly covalent with some admixture of ionic bonding. Intermediate between ionic and covalent types of bonding is typical for II–VI semiconductors such as CdS and ZnO.

Van-der-Waals bonding. Ionic, metallic, and covalent bonds belong to the class of chemical bonds in the sense that they modify atomic valence shells. Two atoms with filled valence shells cannot form chemical bonds. It applies to atoms of noble (inert) gases and solid molecular crystals such as H_2 and N_2, and various organic compounds. In molecular crystals, the distance between neutral molecules is much larger than the distance between atoms in the molecule, so the overlap of wave functions of neighboring molecules is negligible that makes covalent bonding impossible. Molecules are neutral, but fluctuations in spatial electron distribution generate time-dependent dipole moment d_1 and corresponding electric field $\sim d_1/R^3$. This field induces the dipole moment in a neighboring molecule, $d_2 \sim \kappa d_1/R^3$, κ is the polarizability. Dipole-dipole interaction: $V(R) = -\kappa d_1 d_2/R^3 = -\kappa d_1^2/R^6$. The time-averaged moment is zero, $\langle d_1 \rangle = 0$ but fluctuation is not, $\langle d_1^2 \rangle \neq 0$, so $\langle V(R) \rangle \neq 0$ and explains the origin of the Van der Waals bonding. So, the Van der Waals force $F = -\partial V / \partial R \sim 1/R^7$ attracts neutral inert atoms or molecules and causes their condensation into a solid-state. The process may take place if a low enough temperature does not prevent condensation. The Van der Waals binding energy is of the order of 10^{-2} eV, so in semiconductors, it is much weaker than other types of bonding and may be considered irrelevant.

Hydrogen bonding. To explain the origin of hydrogen bond it is instructive to look at the molecule of water shown in Figure 6.3. Hydrogen gives up its electron to oxygen, creating a covalent bond. As a result, the proton electrostatically attracts another electronegative oxygen atom forming the hydrogen bond.

Hydrogen bonding exists in ferroelectrics and organic compounds. Its strength is one order of magnitude larger than that of Van der Waals bond, but it is still small as compared to ionic and covalent bonds.

Bonding between atoms determines various static elastic properties of the solid-state, while lattice vibrations and their coupling to electrons are responsible for the thermal characteristics of crystals [1–3]. Elastic constants for most common semiconductors can be found in Ref. [4].

It is virtually impossible to study the many-particle system by analyzing equations of motion of each particle. The progress in solid-state physics became possible due to two fundamental approximations: Born-Oppenheimer approximation, which allows separating dynamics of electrons and nuclei, and the concept of quasiparticles, which reduces the collective behavior of particles to motion of independent quasiparticles.

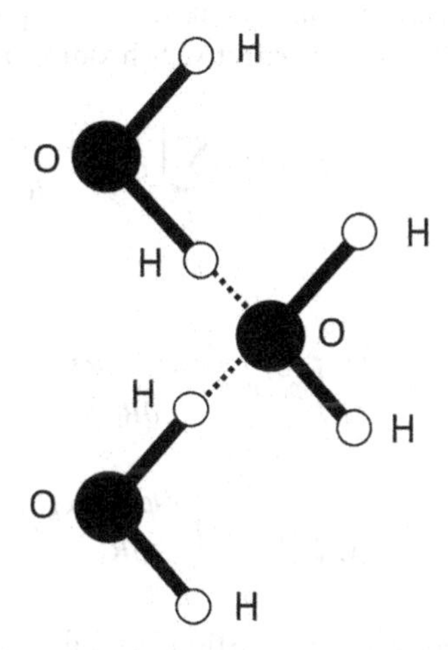

FIGURE 6.3 A molecule of water. Solid lines – covalent bonds; broken lines – hydrogen bonds.

6.2 Born-Oppenheimer Approximation

Molecules and solid crystals present systems of interacting electrons and nuclei for which exact quantum mechanical solution becomes virtually impossible because of a large number of constituent particles. Total Hamiltonian of the system is a sum of kinetic and potential energy written as

$$H = H_e + T_i,$$

$$H_e = T_e + V_{ee} + V_{ii} + V_{ie}, \tag{6.1}$$

where T_e (T_i) is the kinetic energy of electrons (ions). Terms V_{ii}, V_{ee}, and V_{ie} are the interion, interelectron, and electron-ion Coulomb interactions, respectively. As ion mass M is much larger than electron mass m, the kinetic energy $T_i = (2M)^{-1}\sum_k P_k^2$, $P_k = -i\hbar\nabla_k$, (index k numerates ions), may be considered as a weak perturbation, and in zero-order approximation (frozen ions), one has to solve the Schrödinger equation for electrons only,

$$H_e\varphi_n(\boldsymbol{r},\boldsymbol{R}) = U_n(\boldsymbol{R})\varphi_n(\boldsymbol{r},\boldsymbol{R}), \tag{6.2}$$

where $\boldsymbol{r}$ is the coordinate of an electron and $\boldsymbol{R}$ is the configuration of ions. The main point of zero-order approximation is that H_e does not contain derivatives on ion coordinates, so $\varphi_n(\boldsymbol{r},\boldsymbol{R})$ and $U_n(\boldsymbol{R})$ include $\boldsymbol{R}$ as a parameter. This means (6.2) determines the electron states and energy when ions are frozen. The total wave function, expressed as a superposition of zero-order electron eigenfunctions, has the form:

$$\psi(\boldsymbol{r},\boldsymbol{R},t) = \sum_n \Phi_n(\boldsymbol{R},t)\varphi_n(\boldsymbol{r},\boldsymbol{R}), \tag{6.3}$$

where $\Phi_n(\boldsymbol{R},t)$ are the coefficients of expansion that describe the ion motion. The next step is to use representation (6.3) in full Schrödinger equation,

$$i\hbar\frac{\partial\psi(\boldsymbol{r},\boldsymbol{R},t)}{\partial t} = (H_e + T_i)\psi(\boldsymbol{r},\boldsymbol{R},t). \tag{6.4}$$

In (6.4), operator H_e acts on electron coordinates only, while T_i does on ion coordinates. As $\varphi_n(\boldsymbol{r},\boldsymbol{R})$ are eigenfunctions of equation (6.2) they are orthonormalized, so multiplying (6.4) by $\varphi_l^*(\boldsymbol{r},\boldsymbol{R})$ and integrating over electron coordinates $\boldsymbol{r}$, one obtains the effective Schrödinger equation for ion motion:

$$i\hbar\frac{\partial\Phi_l(\boldsymbol{R},t)}{\partial t} = \left(T_i + U_l(\boldsymbol{R})\right)\Phi_l(\boldsymbol{R},t) + \sum_{n\mathrm{k}}\left[A_{ln}^{\mathrm{k}}\frac{\partial\Phi_n(\boldsymbol{R},t)}{\partial \boldsymbol{R}_{\mathrm{k}}} + B_{ln}^{\mathrm{k}}\Phi_n(\boldsymbol{R},t)\right], \tag{6.5}$$

where

$$A_{ln}^{\mathrm{k}} = -\frac{\hbar^2}{M}\int \varphi_l^*(\boldsymbol{r},\boldsymbol{R})\frac{\partial\varphi_n(\boldsymbol{r},\boldsymbol{R})}{\partial \boldsymbol{R}_{\mathrm{k}}}d^3r,$$

$$B_{ln}^{\mathrm{k}} = -\frac{\hbar^2}{2M}\int \varphi_l^*(\boldsymbol{r},\boldsymbol{R})\frac{\partial^2\varphi_n(\boldsymbol{r},\boldsymbol{R})}{\partial \boldsymbol{R}_{\mathrm{k}}^2}d^3r. \tag{6.6}$$

The first two terms on the right-hand side (6.5) constitute the adiabatic (Born–Oppenheimer) approximation. The adiabatic approximation implies that electrons do not change their state when ions change configuration $\boldsymbol{R}$. The expectation value of electron energy at frozen ions positions, $U_n(\boldsymbol{R})$, plays the role of the potential energy for ions. The typical shape of the potential has the minimum at ions equilibrium positions.

A key point of adiabatic approximation reads: electrons instantaneously adapt their quantum state to a particular ion configuration while remaining in the same energy level $U_n(\boldsymbol{R})$. The terms in sum (6.5) are nonadiabatic corrections which involve electron transitions caused by ions motion. Born and Oppenheimer have shown that the smallness parameter that allows neglecting non-adiabatic terms is $\delta R / R_0$, where δR is the amplitude of ion oscillations and R_0 is the average interionic distance. For bound electrons in the ground state, the characteristic energy is Rydberg, I, and the spatial dimension is Bohr radius, a_0. Bohr radius determines the average distance between minima in electron energy $U_0(R)$, and thus one may estimate R_0 as $R_0 \approx a_0$. Ions oscillate around their equilibrium positions. Oscillation energy is $E_{osc} \approx K(\delta R)^2/2$, where the stiffness constant has a pure electronic origin, $K = U''(R_0) \approx I/a_0^2$. By applying the virial theorem - the average kinetic and potential energy are equal - we express the oscillation energy as $K(\delta R)^2 \approx P^2/M$, where ion momentum estimate follows from the uncertainty principle: $P \approx \hbar / \delta R$. As a result, the oscillation amplitude is $\delta R \approx \sqrt[4]{\hbar^2 a_0^2/MI}$. Using atomic constants $I = me^4/\hbar^2$ and $a_0 = \hbar^2/me^2$, one gets $Ia_0^2 = \hbar^2/m$, and finally, the smallness parameter becomes

$$\frac{\delta R}{a_0} \approx \sqrt[4]{\frac{\hbar^2}{MIa_0^2}} = \sqrt[4]{\frac{m}{M}}, \tag{6.7}$$

where m is the free electron mass. For a detailed analysis of nonadiabatic corrections, see Refs. [1,5].

6.3 Lattice Dynamics

In the adiabatic approximation, one separates motions of electrons and ions so that each ion experiences an effective field–Coulomb interaction from electrons and other nuclei. In crystalline solids, nuclei (ions) undergo oscillating movement near their equilibrium positions corresponding to minimum $U(\boldsymbol{R})$. Adiabatic approximation allows calculating electron states in crystals assuming ions frozen in their equilibrium positions, which form the Bravais lattice $\boldsymbol{R}_n$ and if the unit cell contains more than one atom, the basis $\boldsymbol{\rho}_s$ (see Chapter 1). Equilibrium positions are

$$\boldsymbol{R}_{ns0} = \boldsymbol{R}_n + \boldsymbol{\rho}_s,$$

$$\boldsymbol{R}_n = n_1\boldsymbol{a} + n_2\boldsymbol{b} + n_3\boldsymbol{c}, \tag{6.8}$$

where $\boldsymbol{n} = (n_1, n_2, n_3)$ are integers, $\boldsymbol{a}, \boldsymbol{b}, \boldsymbol{c}$ are basic lattice vectors, $s = 1, 2 \ldots r$ numerates the basis-atoms in the unit cell. Actual positions of ions include vibrating displacement $\boldsymbol{u}_s$: $\boldsymbol{R}_{ns} = \boldsymbol{R}_{ns0} + \boldsymbol{u}_s$, $|\boldsymbol{u}_s| \ll a$, a is the interatomic distance. A crystal starts melting at $|\boldsymbol{u}_s| \approx 0.1a$.

Potential energy, U, can be approximated by expansion in powers of small oscillation amplitude:

$$U = U(\boldsymbol{R}_{ns0}) + \frac{1}{2}\sum_{nn'}\sum_{ss'}\sum_{\alpha\alpha'} D_{nn'}^{\substack{\alpha\alpha' \\ ss'}} u_{ns\alpha}\, u_{n's'\alpha'} + \ldots,$$

$$D_{nn'}^{\substack{\alpha\alpha' \\ ss'}} = \left.\frac{\partial^2 U}{\partial u_{ns\alpha}\, \partial u_{n's'\alpha'}}\right|_{u=0}, \tag{6.9}$$

where dynamical matrix $D_{nn'}^{\substack{\alpha\alpha' \\ ss'}}$ carries information on the type of bonding forces between ions, $U(\boldsymbol{R}_{ns0}) = \min[U\{r\}]$ is the potential energy of the lattice with ions in equilibrium positions, $\boldsymbol{n}$ runs over unit cells, and $\alpha=(x,y,z)$ is the displacement polarization vector. The term linear in $u_{ns\alpha}$ is absent in expansion (6.9) as in equilibrium $\partial U/\partial u_{ns\alpha}=0$. The total energy of the lattice comprises kinetic and potential energy terms:

$$H=\sum_{ns\alpha}\frac{P_{ns\alpha}^2}{2M_s}+\frac{1}{2}\sum_{nn'}\sum_{ss'}\sum_{\alpha\alpha'}D_{nn'}^{\substack{\alpha\alpha'\\ ss'}}u_{ns\alpha}u_{n's'\alpha'}, \tag{6.10}$$

where $P_{ns\alpha}=M_s\dot{u}_{ns\alpha}$ is the ion momentum.

The spatial periodicity of the Bravais lattice requires translational invariance of matrix D. Operation of translation on Bravais vector $\boldsymbol{m}$ acting on a function of coordinates shifts the spatial coordinate as $T_m\varphi(\boldsymbol{R}_n+\boldsymbol{r})=\varphi(\boldsymbol{R}_n+\boldsymbol{R}_m+\boldsymbol{r})$, thus in the matrix representation, it is expressed as

$$\sum_{n'}(T_m)_{nn'}\ \varphi(\boldsymbol{R}_{n'})=\varphi(\boldsymbol{R}_n+\boldsymbol{R}_m), \tag{6.11}$$

As follows from (6.11),

$$(T_m)_{nn'}=\delta_{n+m,n'}\ ;\ \left(T_m^{-1}\right)_{nn'}=\delta_{n,n'+m}\ . \tag{6.12}$$

Invariance under translation $D=T_m^{-1}DT_m$ written in the matrix form takes the form:

$$D_{nn'}=\sum_{pp'}\left(T_m^{-1}\right)_{np}D_{pp'}(T_m)_{p'n'}=D_{n-m,n'-m} \tag{6.13}$$

The left- and right-hand sides in (6.13) are equal only if $D_{nn'}$ is a function of difference $\boldsymbol{n}-\boldsymbol{n}'$. It is consistent with the physical argument that the interaction between atoms depends on the distance between them rather than their absolute positions in space. Renormalizing displacements, momenta, and dynamical matrix as

$$\tilde{u}_{ns\alpha}=u_{ns\alpha}\sqrt{\frac{M_s}{M}},\quad \tilde{P}_{ns\alpha}=P_{ns\alpha}\sqrt{\frac{M}{M_s}},\quad M=\sum_s M_s,\quad \tilde{D}_{n-n'}^{\substack{\alpha\alpha'\\ ss'}}=\frac{M}{\sqrt{M_sM_{s'}}}D_{n-n'}^{\substack{\alpha\alpha'\\ ss'}}, \tag{6.14}$$

one comes to a more convenient form of the Hamilton function:

$$H=\frac{1}{2M}\sum_{ns\alpha}\tilde{P}_{ns\alpha}^2+\frac{1}{2}\sum_{nn'}\sum_{ss'}\sum_{\alpha\alpha'}\tilde{D}_{n-n'}^{\substack{\alpha\alpha'\\ ss'}}\tilde{u}_{ns\alpha}\tilde{u}_{n's'\alpha'}, \tag{6.15}$$

Hamilton function (6.15) contains terms up to the second-order displacements and corresponds to the harmonic approximation. Hamilton equations of motion

$$\frac{\partial u_{ns\alpha}}{\partial t}=\frac{\partial H}{\partial P_{ns\alpha}},\ \frac{\partial P_{ns\alpha}}{\partial t}=-\frac{\partial H}{\partial u_{ns\alpha}}, \tag{6.16}$$

can be written as

$$M\frac{\partial^2\tilde{u}_{ns\alpha}}{\partial t^2}+\sum_{n's'\alpha'}\tilde{D}_{n-n'}^{\substack{\alpha\alpha'\\ ss'}}\tilde{u}_{n's'\alpha'}=0. \tag{6.17}$$

As follows from (6.17), if all atoms experience equal displacements $u_{ns\alpha} = c_\alpha$,

$$\sum_{n's'\alpha'} D_{n-n'}^{\substack{\alpha\alpha' \\ ss'}} c_{\alpha'} = 0. \tag{6.18}$$

As $c_{\alpha'} \neq 0$ are independent, (6.18) results in

$$\sum_{n's'} D_{n-n'}^{\substack{\alpha\alpha' \\ ss'}} = 0. \tag{6.19}$$

We are looking for a monochromatic time-oscillating solution to (6.17) in the form,

$$\tilde{u}_{ns\alpha} = \tilde{u}_{s\alpha} \exp\left(-i\omega t + i\boldsymbol{q}\cdot\boldsymbol{R}_n\right), \tag{6.20}$$

where $\tilde{\boldsymbol{u}}_s$ is the displacement of the s-atom in the unit cell. The eigenvalue of the operator of translation-$\exp\left(i\boldsymbol{q}\cdot\boldsymbol{R}_n\right)$- is the factor that enters any solution related to an excitation propagating in a lattice. Note that we look for a solution in a complex-valued form, meaning that displacement is $\mathrm{Re}\left(\tilde{u}_{ns\alpha}\left(\boldsymbol{q}\right)\right)$.

Wavevector of lattice vibrations is introducing similarly to electron quasiparticles discussed in Chapter 1. The lattice periodicity and cyclic boundary conditions make $\boldsymbol{q}$ a discrete variable defined modulo the reciprocal lattice vector, $\boldsymbol{K}_g = g_1\boldsymbol{b}_1 + g_1\boldsymbol{b}_1 + g_1\boldsymbol{b}_1$, where g_i are integers and $\boldsymbol{b}_i$ are basis vectors of the reciprocal lattice. The number of discrete $\boldsymbol{q}$-modes is N – the number of unit cells in a crystal. Displacements propagating with wavevectors $\boldsymbol{q}$ and $\boldsymbol{q}+\boldsymbol{K}_g$ are physically indistinguishable. Formally, it follows from relation $\exp\left(\boldsymbol{R}_n\cdot\boldsymbol{K}_g\right)=1$, which is the definition of the reciprocal lattice. Using (6.20) in (6.17), one gets the system of homogeneous equations for displacements:

$$-M\omega^2\tilde{u}_{s\alpha} + \sum_{s'\alpha'} \tilde{D}^{\substack{\alpha\alpha' \\ ss'}}\left(\boldsymbol{q}\right)\tilde{u}_{s'\alpha'} = 0, \tag{6.21}$$

where

$$\tilde{D}^{\substack{\alpha\alpha' \\ ss'}}\left(\boldsymbol{q}\right) \equiv \sum_m \tilde{D}_m^{\substack{\alpha\alpha' \\ ss'}} \exp\left(-i\boldsymbol{q}\cdot\boldsymbol{R}_m\right) \tag{6.22}$$

is the Hermitian matrix, $\tilde{D}^{\substack{\alpha\alpha' \\ ss'}}\left(\boldsymbol{q}\right) = \left[\tilde{D}^{\substack{\alpha'\alpha \\ s's}}\left(\boldsymbol{q}\right)\right]^* = \left[\tilde{D}^{\substack{\alpha\alpha' \\ ss'}}\left(-\boldsymbol{q}\right)\right]^*$.

System (6.21) comprises $3r$ homogeneous equations for polarization vectors. The nontrivial solution requires that the determinant of $3r \times 3r$ matrix be zero:

$$\det\left(\tilde{D}^{\substack{\alpha\alpha' \\ ss'}}\left(\boldsymbol{q}\right) - M\omega^2\delta_{ss'}\delta_{\alpha\alpha'}\right) = 0. \tag{6.23}$$

The solution to (6.23) determines $\boldsymbol{q}$-dependent eigenvalues $\omega_j^2\left(\boldsymbol{q}\right)$ of the Hermitian dynamical matrix, $j = 1,2\ldots3r$, that is, $3r$ branches of the vibrational spectrum. As a function of $\boldsymbol{q}$, $\omega_j\left(\boldsymbol{q}\right)$ is called the *dispersion law* of branch j.

The potential energy in (6.15) reaches the minimum (zero) in equilibrium ($u=0$), meaning it is positive for any $u \neq 0$ so D is the definite-positive matrix, which has positive eigenvalues, $\omega_j^2(\boldsymbol{q})>0$ that guarantees finite displacements in (6.20), which otherwise would infinitely grow with time.

Matrix $\tilde{D}_{ss'}^{\alpha\alpha'}(\boldsymbol{q})$ is Hermitian, which implies $\omega_j(\boldsymbol{q})=\omega_j(-\boldsymbol{q})$: the same eigenvalue corresponds to propagation in opposite directions. That is the consequence of the invariance of mechanical equations of motion to time reversal. The crystal point symmetry may cause additional constraints on $\omega_j(\boldsymbol{q})$. For example, if the lattice has a reflection symmetry plane $x=0$, the spectrum is symmetric as $\omega_j(q_x)=\omega_j(-q_x)$.

For each eigenvalue $\omega_j^2(\boldsymbol{q})$ there exists the eigenvector of (6.23) – dimensionless polarization vectors $\tilde{\boldsymbol{e}}_s^j(\boldsymbol{q})$. As eigenvectors, those $\tilde{\boldsymbol{e}}_s^j(\boldsymbol{q})$ that belong to different branches are orthogonal, and those belonging to the same branch are normalized to unity:

$$\sum_s \left[\tilde{\boldsymbol{e}}_s^j(\boldsymbol{q})\right]^* \cdot \tilde{\boldsymbol{e}}_s^{j'}(\boldsymbol{q})=\delta_{jj'},$$

and

$$\tilde{\boldsymbol{e}}_s^j(\boldsymbol{q})=\left[\tilde{\boldsymbol{e}}_s^j(-\boldsymbol{q})\right]^*. \tag{6.24}$$

It follows from (6.24) that $\tilde{\boldsymbol{e}}_s^j(\boldsymbol{q})$ becomes a real-valued vector if $\boldsymbol{q}\to 0$. As a solution to dynamical equations (6.21), complex-valued displacements (6.20) have the form:

$$\tilde{\boldsymbol{u}}_{ns}^j(\boldsymbol{q})=A_{qj}\tilde{\boldsymbol{e}}_s^j(\boldsymbol{q})\exp\left(-i\omega_j(\boldsymbol{q})t+i\boldsymbol{q}\cdot\boldsymbol{R}_n\right), \tag{6.25}$$

that is displacements of atom s in unit cell $\boldsymbol{n}$, which corresponds to mode $\boldsymbol{q}$ in the branch j, A_{qj} is the displacement amplitude of dimension $[m]$.

To analyze the long-wavelength limit of $\omega_j(\boldsymbol{q}\to 0))$, one can apply general arguments. It follows from (6.18) and (6.22) that

$$\sum_{s'\alpha'} \tilde{D}_{ss'}^{\alpha\alpha'}(\boldsymbol{q}=0)=0. \tag{6.26}$$

Equation (6.26) states that the sum of columns in matrix $\tilde{D}_{ss'}^{\alpha\alpha'}(\boldsymbol{q}=0)$ is zero that makes the determinant zero as well:

$$\det \tilde{D}_{ss'}^{\alpha\alpha'}(\boldsymbol{q}=0)=0. \tag{6.27}$$

Comparing to (6.23) one concludes that (6.27) is the dispersion equation with $\omega^2(\boldsymbol{q}=0)=0$. The above formal exercise proves the existence of *acoustic* branches and physically means that due to space homogeneity, no energy to be spent to shift the lattice as a whole. Frequency $\omega(\boldsymbol{q}=0)=0$ is three-fold degenerate as it corresponds to displacements in three possible directions x, y, z. These are three acoustic branches in which all atoms in the unit cell oscillate in phase. These oscillations propagate as sound waves in the medium that justifies the name acoustic. If the wavelength is much larger than interatomic distance, the atomic structure is irrelevant, and vibrations propagate the same way as in a homogeneous elastic medium.

In an isotropic lattice, among the three, there are two transverse (*TA*) and one longitudinal (*LA*) branches. In two *TA* branches, the displacement polarization is perpendicular to propagation vector $\boldsymbol{q}$, while in the *LA* branch, the displacements and $\boldsymbol{q}$ are collinear. In the limit $\boldsymbol{q} \to 0$, the lattice experiences parallel shift as a whole. Other $3r-3$ branches $\omega_j(0) \neq 0$ are called *optical*. Optical branches exist in the lattice with a basis (more than one atom in the unit cell). Optical-mode atomic displacements can be studied using the basic equation (6.21) which can be written in terms of displacements $u_{s\alpha}$ as follows:

$$M_s\omega^2(\boldsymbol{q})u_{s\alpha} = \sum_{s'\alpha'} D_{ss'}^{\alpha\alpha'}(\boldsymbol{q})\, u_{s'\alpha'}. \tag{6.28}$$

Using definition (6.22) and assuming $\omega^2(0) \neq 0$, we present (6.28) in the form,

$$M_s\omega^2(0)u_{s\alpha} = \sum_{m,s'\alpha'} D_{\boldsymbol{m}}^{\alpha\alpha'}{}_{ss'}\, u_{s'\alpha'} \tag{6.29}$$

By summing both sides over s (over the basis), one obtains

$$\omega^2(0)\sum_s M_s u_{s\alpha} = \sum_{s'\alpha'} u_{s'\alpha'} \sum_{m,s} D_{\boldsymbol{m}}^{\alpha\alpha'}{}_{ss'}, \tag{6.30}$$

which in virtue of (6.19) becomes

$$\sum_s M_s u_{s\alpha} = 0. \tag{6.31}$$

Equation (6.31) states that atoms in each unit cell are moving relative to each other, leaving the center of mass at rest. In ionic crystals where two ions in the unit cell have opposite electric charges, the relative motion can be induced by an electromagnetic field in the infrared optical spectral range, thus justifying the name "optical" for atomic vibrations. Similarly to acoustical branches, in an isotropic lattice, the optical modes are distinguished as transverse (*TO*) and longitudinal (*LO*) depending on the direction of displacement vectors (polarization) relative to the propagation direction. In anisotropic crystals, acoustic and optical branches are not purely longitudinal or transverse.

Vibrational spectra are periodic in the reciprocal space, $\omega_j(\boldsymbol{q}) = \omega_j(\boldsymbol{q} + \boldsymbol{K}_g)$, so it is enough to define them in the first Brillouin zone (BZ).

Shown in Figure 6.4 is the spectrum along a particular crystallographic direction. In an anisotropic crystal, one identifies the longitudinal acoustic branch as the one with maximum propagation velocity $\boldsymbol{v} = \partial\omega / \partial\boldsymbol{q}$. In an isotropic crystal, two lower transverse acoustic branches are degenerate. Analytical analysis of phonon branches in model systems follows below in Section 6.3.2.

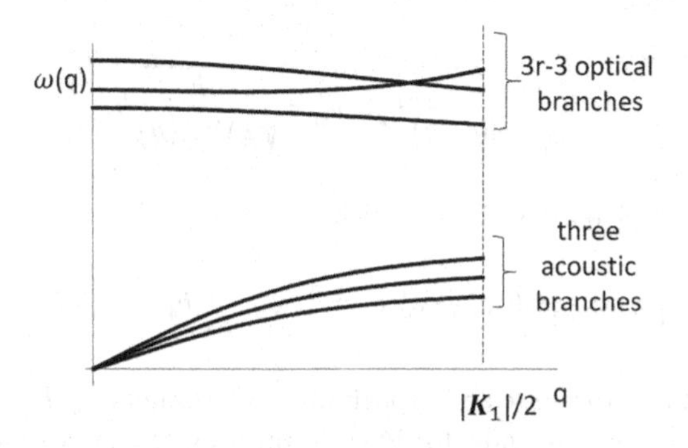

FIGURE 6.4 Vibrational spectrum. $\boldsymbol{K}_1$ is the reciprocal lattice vector.

6.3.1 Phonons

Arbitrary real-valued atomic displacement can be composed by summing oscillations (6.25) over all branches and all modes as follows:

$$\tilde{u}_{ns\alpha} = \frac{1}{\sqrt{N}} \sum_{jq} \left(\tilde{e}^{j}_{s\alpha}(\boldsymbol{q}) a_{qj} \exp(i\boldsymbol{q}\cdot\boldsymbol{R}_n) + \left[\tilde{e}^{j}_{s\alpha}(\boldsymbol{q})\right]^* a^*_{qj} \exp(-i\boldsymbol{q}\cdot\boldsymbol{R}_n) \right),$$

$$a_{qj} = A_{qj} \exp\left(-i\omega_j(\boldsymbol{q})t\right). \quad (6.32)$$

To calculate the total energy of lattice oscillations (6.15), one needs displacements (6.32) and momenta

$$\tilde{P}_{ns\alpha} = M\frac{\partial \tilde{u}_{ns\alpha}}{\partial t} = \frac{iM}{\sqrt{N}} \sum_{jq} \omega_j(\boldsymbol{q}) \left(-\tilde{e}^{j}_{s\alpha}(\boldsymbol{q}) a_{qj} \exp(i\boldsymbol{q}\cdot\boldsymbol{R}_n) + \left[\tilde{e}^{j}_{s\alpha}(\boldsymbol{q})\right]^* a^*_{qj} \exp(-i\boldsymbol{q}\cdot\boldsymbol{R}_n) \right). \quad (6.33)$$

Substituting (6.32) and (6.33) into (6.15), one gets

$$\frac{1}{2M} \sum_{ns\alpha} \tilde{P}^2_{ns\alpha} = \frac{M}{2} \sum_{qj} \omega^2_j(\boldsymbol{q}) \left(-a_{qj}a_{-qj} + a^*_{qj}a_{qj} + a_{qj}a^*_{qj} - a^*_{qj}a^*_{-qj} \right),$$

$$\frac{1}{2} \sum_{nn'} \sum_{ss'} \sum_{\alpha\alpha'} \tilde{D}^{\alpha\alpha'}_{n-n'\,ss'}\; \tilde{u}_{ns\alpha}\tilde{u}_{n's'\alpha'} = \frac{M}{2} \sum_{qj} \omega^2_j(\boldsymbol{q}) \left(a_{qj}a_{-qj} + a^*_{qj}a_{qj} + a_{qj}a^*_{qj} + a^*_{qj}a^*_{-qj} \right),$$

$$H = M \sum_{qj} \omega^2_j(\boldsymbol{q}) \left(a^*_{qj}a_{qj} + a_{qj}a^*_{qj} \right). \quad (6.34)$$

On the way of getting (6.34), we used (6.21), (6.22), (6.24), and $\sum_n \exp(i\boldsymbol{q}\cdot\boldsymbol{R}_n + i\boldsymbol{q}'\cdot\boldsymbol{R}_n) = N\delta_{q,-q'}$.

What has been done above is we moved from Hamiltonian (6.15) nondiagonal in lattice displacements due to the coupling between them to Hamiltonian (6.34) diagonal in new variables a^*_{qj}, a_{qj}. These variables are called normal coordinates. In normal coordinates, the energy of lattice vibrations (6.34) is a sum of independent contributions, each corresponding to particular mode $\boldsymbol{q}$ and branch j.

The total number of oscillating states is $3rN$. The notion of phonons, the quasiparticles carrying lattice excitations, naturally arises once, instead of normal coordinates a^*_{qj}, a_{qj}, we describe lattice vibrations in terms of operators b^+_{qj}, b_{qj}:

$$a_{qj} = \sqrt{\frac{\hbar}{2M\omega_j(\boldsymbol{q})}}\, b_{qj}, \quad a^*_{qj} = \sqrt{\frac{\hbar}{2M\omega_j(\boldsymbol{q})}}\, b^+_{qj}, \quad (6.35)$$

which satisfy commutation relations for Bose particles:

$$\left[b_{qj}, b^+_{q'j'} \right] \equiv b_{qj}b^+_{q'j'} - b^+_{q'j'}b_{qj} = \delta_{qq'}\delta_{jj'}, \quad \left[b_{qj}, b_{q'j'} \right] = 0. \quad (6.36)$$

For the second quantization procedure and explicit form of operators b_{qj}, b^+_{qj}, see Appendix 1, https://www.routledge.com/Modern-Semiconductor-Physics-and-Device-Applications/Dugaev-Litvinov/p/book/9780367250829#. Substituting (6.35) and (6.36) in (6.34), we express the total Hamiltonian as

$$H = M\sum_{qj}\omega_j^2(q)\left(a_{qj}^* a_{qj} + a_{qj}a_{qj}^*\right) = \sum_{qj}\hbar\omega_j(q)\left(n_{qj} + \frac{1}{2}\right), \tag{6.37}$$

where $n_{qj} = b_{qj}^+ b_{qj}$ is the number operator. The expectation value of the Hamiltonian is the energy of lattice vibrations,

$$E = \sum_{qj}\hbar\omega_j(q)m_{qj} + E_0, \tag{6.38}$$

where $m_{qj} = 0,1,2$ is the eigenvalue of number operator n_{qj}, that is, the number of phonons in mode $\boldsymbol{q}j$. *Zero-point energy* E_0 follows from ½-term in (6.37) and often is called *vacuum energy* if one understands "vacuum" as $|0\rangle$, the state with no phonons, $m_{qj} = 0$, for all $\boldsymbol{q}, j$.

Hamiltonian (6.37) acts on wave functions in the occupation number representation $|m_{qj}\rangle$ – a quantum state with m phonons in mode $\boldsymbol{q}j$. Operators b_{qj}^+ (b_{qj}) increase (decrease) the occupation number by one: $b_{qj}^+|m_{qj}\rangle = \sqrt{m_{qj}+1}\,|m_{qj}+1\rangle$; $b_{qj}|m_{qj}\rangle = \sqrt{m_{qj}}\,|m_{qj}-1\rangle$. The state with an arbitrary number of phonons one can create from vacuum state $|0\rangle$: $|m_{qj}\rangle = \left(b_{qj}^+\right)^m|0\rangle$, $b_{qj}|0\rangle = 0$. So, the total energy of the vibrating lattice is the sum of terms from independent oscillators. Zero-point energy is the finite number as the sum in (6.37) has $3rN$ terms. So, from the theory standpoint, this term can easily be swept under the rug choosing $E_0 = 0$ as an energy reference point. In a theory that deals with observable properties of solids, this trick is always implied.

Zero-point oscillations manifest quantum mechanical vibrations related to Heisenberg's uncertainty in simultaneous atomic positions and momenta, $\Delta p\,\Delta x \geq \hbar$. Localization within Δx increases the kinetic energy by $(\Delta p)^2/2M$. Minimum Δx in each phonon mode $\boldsymbol{q}j$ – the amplitude of zero-point oscillations follows from $(\Delta p)^2/2M = \hbar\omega_{qj}/2$: $\Delta x = \sqrt{\hbar/M\omega_{qj}}$. If Δx is close to interatomic distance, the medium is called quantum crystal, the object where atoms are not strictly localized in lattice sites but can hop from site to site by tunneling. If Δx is larger than the lattice constant, the object becomes a quantum liquid. Examples are liquid He^3 and He^4.

The process of quantization of lattice vibrations is quite similar to that of the electromagnetic field (see Appendix 3), https://www.routledge.com/Modern-Semiconductor-Physics-and-Device-Applications/Dugaev-Litvinov/p/book/9780367250829#. So, in some respects, phonon quasiparticles are similar to photons. However, phonons exist in a lattice only that makes them different from real particles – photons and electrons. What distinguishes them is that in thermal equilibrium, the total momentum of phonons is zero as it follows from (6.24) and (6.33):

$$\sum_n \boldsymbol{P}_{ns} = 0. \tag{6.39}$$

Phonon momentum $\hbar\boldsymbol{q}$ is not related to mass or energy transfer, as is the case with real particles. Nevertheless, in a somewhat restricted sense, we may use the notion of phonon momenta when studying phonon–phonon and electron–phonon scattering processes. More details follow.

In thermal equilibrium, the thermal average phonon number in state $\boldsymbol{q}j$ equals the Planck distribution function:

$$\langle n_{qj}\rangle = \frac{1}{\exp\left[\hbar\omega_j(q)/k_BT\right] - 1}, \tag{6.40}$$

where T is the absolute temperature. The distribution function (6.40) is the Bose–Einstein filling factor with zero chemical potential, $\mu = 0$ (see Chapter 4). In thermal equilibrium, characteristics of the ideal Bose gas are independent macroscopic parameters: volume V, temperature T, and the number of particles N. As mentioned in Chapter 4, Bose gas has nonpositive chemical potential. Phonon number N is not fixed and determined by minimum free energy, $(\partial F/\partial N)_{T,V}=0$. Since by definition, $\mu=(\partial F/\partial N)_{T,V}$, the chemical potential in (6.40) equals zero.

At high temperature, $T\gg\hbar\omega_j(\boldsymbol{q})$, the number of phonons increases as $\langle n_{qj}\rangle\approx k_BT/\hbar\omega_j(\boldsymbol{q})$. If temperature tends to absolute zero, the number of equilibrium phonons decreases so that at achievable low temperatures, there exist only long-wavelength acoustical phonons.

In three-dimensional lattices with more than one atom in the unit cell, dynamic equations (6.21) have a numerical solution. However, it is instructive to consider model lattices that allow analytical solutions and thus provide a simple demonstration of the general properties of a vibrating lattice.

6.3.2 One-Dimensional Chain: Acoustical Vibrations

The particular shape of the dynamical matrix D depends on the system under consideration. Below we consider the Born–von Karman chain that is the one-dimensional atomic lattice with cyclic boundary conditions. Periodic boundary conditions provide for displacements in the form (6.20), where q runs over N discrete numbers, N being the number of atoms in the crystal: $q=(2\pi/L)n, n=0,1,2\ldots N$, where $L=Na$ is the length of the crystal and a is the lattice constant. Figure 6.5 illustrates atom displacements in a simple lattice (one atom in the unit cell). There are two types of displacements: longitudinal (along the chain) and transverse (perpendicular to the chain).

As we consider simple lattice (no basis, r=1), there exist acoustic vibrations only. First, we will consider a pure one-dimensional model, which implies displacement polarization along the chain. In this case, (6.28) reduces to

$$M\omega^2(q)u=D(q)u,$$

$$\omega^2(q)=\frac{1}{M}D(q)=\frac{1}{M}\sum_m D_m\exp(-iqR_m), \tag{6.41}$$

where $D_{n-n'}=\partial^2U/\partial u_n\partial u_{n'}\big|_{u=0}$, $D_0>0$. In what follows, we choose site $m=0$ and account for elastic interaction with neighbors only (nearest-neighbor approximation), that means summation in (6.41) over $R_m=-a,0,a$:

$$\omega^2(q)=\frac{1}{M}\left[D_0+D_1\exp(-aq)+D_1\exp(aq)\right]. \tag{6.42}$$

Dynamical constants that describe elastic interaction between pairs of neighbors m=0,1 and m=−1,0 are equal: $D_1=D_{-1}\equiv$ D. Constants D_i obey condition (6.19). In a simple one-dimensional simple lattice, it reads as

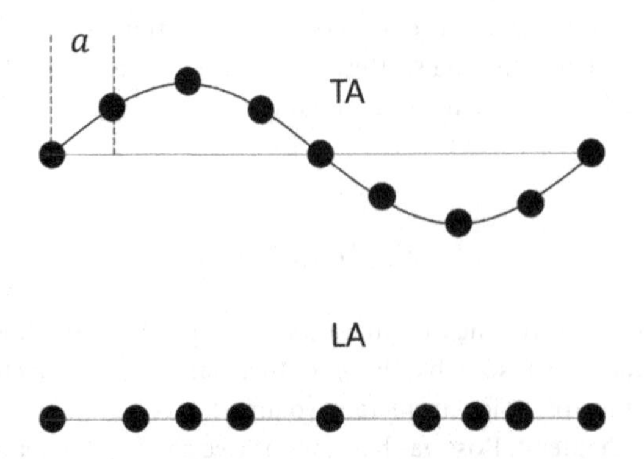

FIGURE 6.5 One dimensional monoatomic lattice. Transverse and longitudinal acoustic vibrations.

$$D_0 + 2D = 0. \tag{6.43}$$

Using (6.43), we calculate (6.42) as

$$\omega^2(q) = \frac{1}{M}\left[D_0 + D\exp(-iqa) + D\exp(iqa)\right],$$

$$\omega(q) = \omega_{\max}\left|\sin\left(\frac{qa}{2}\right)\right|, \quad \omega_{\max} = 2\sqrt{\frac{|D|}{M}}. \tag{6.44}$$

The dispersion of longitudinal acoustical branch $\omega(q)$ is shown in Figure 6.6

As shown in Figure 6.6, the phonon dispersion is a periodic function of wavevector with a period equal to the reciprocal lattice vector, $K = 2\pi/a : \omega(q) = \omega(q+K)$, that is why it is enough to define the spectrum only within the first BZ ($-\pi/a \le q \le \pi/a$). Phonon wavelength is $\lambda \equiv 2\pi/q$, and in the long-wavelength limit, $q \ll \pi/a$, the spectrum is linear,

$$\omega(q) = \mathrm{v}q, \quad \mathrm{v} = a\sqrt{\frac{|D|}{M}} = \frac{a\,\omega_{\max}}{2}, \tag{6.45}$$

where v is the sound velocity in a crystal. In the short-wavelength limit $q \to \pi/a$, the dispersion law is not of the sound type, so that group velocity is zero on the boundary of the BZ: $\mathrm{v} = \partial\omega(q)/\partial q = 0$ at $q = \pm\pi/a$. At this wavevector, Bragg's reflection turns the wave to the standing one with displacements $u_n = u_0\exp[-i\omega t + in\pi] = (-1)^n u_0 \exp[-i\omega t]$, which means neighbors oscillate in antiphase.

Analysis of the *TA* phonon spectrum is similar to that for the *LA* branch, with one exception. Elastic force arising from longitudinal displacements is larger than that corresponding to transverse motion, so dynamical constant $|D|$ for *LA* modes is the largest, and thus out of three acoustic branches, the *LA* branch has the highest frequency. Along arbitrary direction in three-dimensional anisotropic crystals, the phonon branches are neither longitudinal nor transverse. Usually, one identifies as *LA*, the phonon branch of the highest frequency. Dispersion relation (6.44) is obtained from the written in Fourier components equation (6.41). The same result follows from the coordinate equation of motion:

$$M\frac{\partial^2 u_n}{\partial t^2} = -\frac{\partial U}{\partial u_n}, \quad U = \frac{1}{2}\sum_{nn'} D_{nn'}\, u_n u_{n'}, \tag{6.46}$$

where U is the potential energy (6.15) adapted to one-dimensional simple lattice. In a nearest-neighbor approximation,

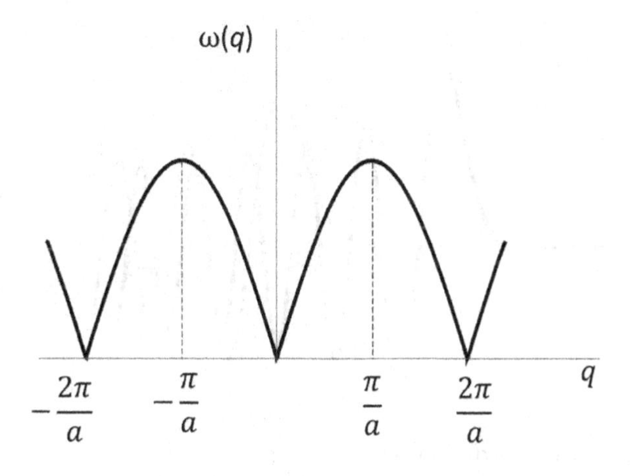

FIGURE 6.6 Dispersion law for acoustic waves.

$$U = \frac{1}{2}\sum_{n} D_0 (u_n)^2 + D_{-1}\, u_n u_{n+1} + D_1\, u_n u_{n-1}. \tag{6.47}$$

Using (6.43), one gets

$$-\frac{\partial U}{\partial u_k} = |D|(-2\, u_k + u_{k+1} + u_{k-1}) = |D|(u_{k-1} - u_k) + |D|(u_{k+1} - u_k),$$

and

$$\frac{\partial^2 u_n}{\partial t^2} = \frac{\omega_{\max}^2}{4}(-2\, u_n + u_{n+1} + u_{n-1}). \tag{6.48}$$

With plane-wave ansatz, $u_n = u\exp(-i\omega t + iqR_n)$, one gets (6.44) – stationary monochromatic oscillations propagating with group velocity $\partial\omega(q)/\partial q$.

Equation of motion (6.48), though, is more general than (6.41) as it admits nonmonochromatic solutions. One may guess a solution, using the identity for Bessel functions of the first kind:

$$\frac{\partial^2 J_{2n}(bx)}{\partial x^2} = \frac{b^2}{4}\left[-2J_{2n}(bx) + J_{2(n-1)}(bx) + J_{2(n+1)}(bx)\right]. \tag{6.49}$$

Comparing (6.48) and (6.49), we find a solution in the form,

$$u_n = u_0\, J_{2n}(\omega_{\max} t), \tag{6.50}$$

where u_0 is the local displacement of atom at site $n = 0$ and $t = 0$. Initially, local displacement propagates through the crystal according to (6.50) with gradually decaying amplitude, as illustrated in Figure 6.7.

The perturbation does not have a single front because dispersion in phase velocity $\omega(q)/q$ spreads the signal. The first maximum in u_n is moving at distance na going from $n = 0$ to $n \gg 1$ during the time $t \approx 2n/\omega_{\max}$. So the perturbation propagates with speed close to $a\omega_{\max}/2$ – the sound velocity. If one takes into account the cubic anharmonicity, the signal may propagate faster than sound with a single front called soliton [6].

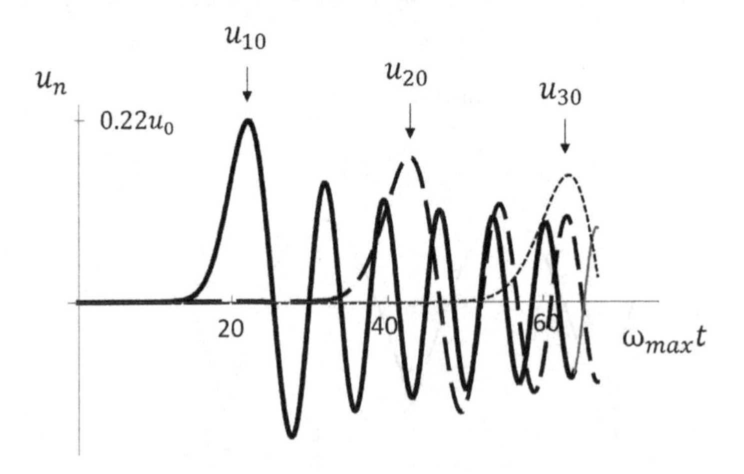

FIGURE 6.7 Propagation of local perturbation $u_n(t)$.

6.3.3 One-Dimensional Chain: Optical Vibrations

For illustration purposes, we restrict our consideration by a pure one-dimensional model that is the longitudinal modes in the 1D chain with the basis $s = 1, 2$, as shown in Figure 6.8. Longitudinal modes imply along-the-chain displacements of two atoms in the unit cell: $u_n^{(1)}$ and $u_n^{(2)}$. Using force constants similar to (6.47), we express equations of motion as

$$M_1 \frac{\partial^2 u_n^{(1)}}{\partial t^2} = \beta_1 \left(u_{n-1}^{(2)} - u_n^{(1)} \right) + \beta_2 \left(u_n^{(2)} - u_n^{(1)} \right),$$

$$M_2 \frac{\partial^2 u_n^{(2)}}{\partial t^2} = \beta_2 \left(u_n^{(1)} - u_n^{(2)} \right) + \beta_1 \left(u_{n+1}^{(1)} - u_n^{(2)} \right) \tag{6.51}$$

where $M_{1,2}$ and $\beta_{1,2}$ are masses and elastic constants, respectively.

We look for space- and time-periodic solution in the form:

$$u_n^{(1,2)} = u_0^{(1,2)} \exp\left(-i\omega t + iqan\right), \tag{6.52}$$

where a is the lattice constant shown in Figure 6.8. Substituting (6.51) into (6.52) we obtain the system of homogeneous equations for amplitudes $u_0^{(1,2)}$:

$$\begin{pmatrix} -\beta_1 - \beta_2 + \omega^2 M_1 & \beta_1 e^{-iqa} + \beta_2 \\ \beta_2 + \beta_1 e^{iqa} & \omega^2 M_2 - \beta_2 - \beta_1 \end{pmatrix} \begin{pmatrix} u_0^{(1)} \\ u_0^{(2)} \end{pmatrix} = 0. \tag{6.53}$$

The nontrivial solution to (6.53) exists if

$$\det \begin{pmatrix} -\beta_1 - \beta_2 + \omega^2 M_1 & \beta_1 e^{-iqa} + \beta_2 \\ \beta_2 + \beta_1 e^{iqa} & \omega^2 M_2 - \beta_2 - \beta_1 \end{pmatrix} = 0. \tag{6.54}$$

Solving (6.54), one finds the spectrum of lattice vibrations:

$$\omega^2(q) = \frac{\omega_0^2}{2} \left(1 \pm \sqrt{1 - \gamma^2 \sin^2\left(\frac{qa}{2}\right)} \right),$$

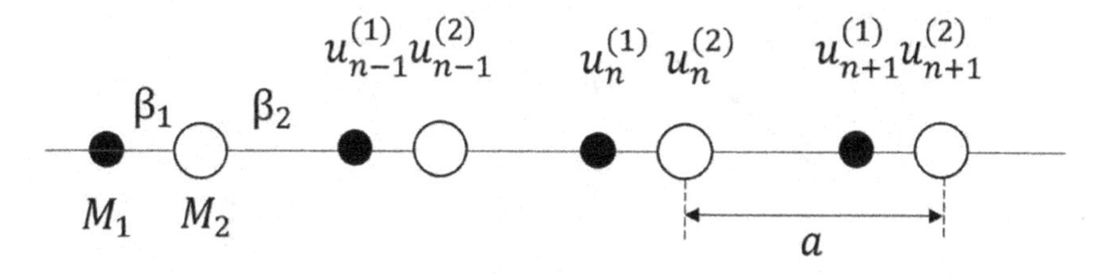

FIGURE 6.8 One-dimensional lattice with the basis of two atoms.

$$\omega_0 = \sqrt{\frac{\beta}{\mu}}, \quad \beta = \beta_1 + \beta_2, \quad \mu = \frac{M_1 M_2}{M_1 + M_2},$$

$$\gamma^2 \equiv \frac{16\beta_1\beta_2\mu}{\beta^2(M_1 + M_2)}. \tag{6.55}$$

The spectrum consists of two branches, acoustic $\omega_{ac}^2(q \to 0) \to 0$, and optical $\omega_{op}^2(q = 0) \neq 0$. In the long-wavelength limit, $qa \ll 1$:

$$\omega_{ac}^2(q) \approx \frac{\omega_0^2}{16}\gamma^2(aq)^2,$$

$$\omega_{op}^2(q) \approx \omega_0^2\left[1 - \frac{1}{16}\gamma^2(aq)^2\right]. \tag{6.56}$$

Branches (6.56) are illustrated in Figure 6.9.

Solving (6.53) for $u_0^{(1)}/u_0^{(2)}$, one gets

$$\frac{u_0^{(1)}}{u_0^{(2)}} = \left.\frac{\beta - M_2\omega^2(q)}{\beta_2 + \beta_1 \exp(iqa)}\right|_{q \to 0} = \begin{cases} 1, & ac \\ -\dfrac{M_2}{M_1}, & opt \end{cases}. \tag{6.57}$$

In the long-wavelength limit (6.57), two atoms in the unit cell move in-phase in acoustic mode and out-of-phase in optical one. Amplitudes in optical mode (lower line in (6.57)) satisfy condition (6.31), which guarantees the center of mass is at rest.

In a three-dimensional lattice with a basis of two atoms, there are two additional optical phonon branches, which in isotropic crystals are transverse to the direction of propagation. Limiting frequency ω_0 for longitudinal and transverse branches is determined by corresponding elastic constants and, if the unit cell contains two ions of opposite electric charges, by an internal electric field. The coupling between mechanical vibrations and electric fields will be discussed in the next section.

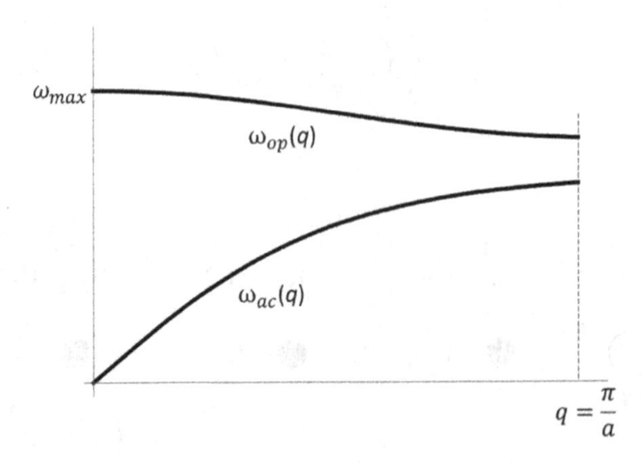

FIGURE 6.9 Longitudinal phonon branches in a diatomic chain.

6.3.4 Optical Vibrations in Ionic Dielectric Crystals: Lyddane–Sachs–Teller Relation

In acoustic modes, the unit cell oscillates as a whole, so no oscillating electric field takes part in lattice vibrations. The relative motion of ions in optical modes induces an electric field, which affects lattice dynamics and modifies equations (6.51). In long-wavelength optical modes, in-phase oscillations of neighboring unit cells are electrically inactive, while intra-unit-cell vibrations induce an oscillating electric field. That is the ionic motion in a single unit cell that determines optical frequencies at $\boldsymbol{q} = 0$. Equations (6.51) for a single unit cell ($\beta_1 = 0$) modified due to the presence of an internal electric field have the form:

$$M_+ \frac{\partial^2 \boldsymbol{u}_+}{\partial t^2} = \beta_2 (\boldsymbol{u}_- - \boldsymbol{u}_+) + e^* \boldsymbol{E}_i,$$

$$M_- \frac{\partial^2 \boldsymbol{u}_-}{\partial t^2} = \beta_2 (\boldsymbol{u}_+ - \boldsymbol{u}_-) - e^* \boldsymbol{E}_i, \tag{6.58}$$

where $\boldsymbol{u}_+ (\boldsymbol{u}_-)$ is the displacement of positively (negatively) charged ion of mass $M_+ (M_-)$ in the unit cell, e^* is the effective ion charge, $\boldsymbol{E}_i$ is the internal electric field which is the sum of external field and field of other ions in the crystal. Equations (6.58) can be generalized to three-dimensional lattice and rewritten as the equation of motion for the relative displacement, which is proportional to the dipole moment:

$$\frac{\partial^2 (\boldsymbol{u}_+ - \boldsymbol{u}_-)}{\partial t^2} = -\frac{\beta_2}{\mu} (\boldsymbol{u}_+ - \boldsymbol{u}_-) + \frac{e^*}{\mu} \boldsymbol{E}_i, \tag{6.59}$$

where μ is the reduced mass defined in (6.55).

Effective local field $\boldsymbol{E}_i$ is a sum of an external field and Lorentz field, which is induced by polarization. In cubic crystals, the local field at a cubic site [7],

$$\boldsymbol{E}_i = \boldsymbol{E} + \frac{\boldsymbol{P}}{3\varepsilon_0}, \tag{6.60}$$

where $\boldsymbol{P}$ is the polarization - the dipole moment per unit volume and ε_0 is the vacuum permittivity. Total polarization comprises two contributions. One is related to the relative displacement of ions, and another stems from the polarization of each ion due to the electric field-induced shift of electronic shells relative to nuclei:

$$\boldsymbol{P} = \frac{N}{\Omega} \left[e^* (\boldsymbol{u}_+ - \boldsymbol{u}_-) + \alpha \boldsymbol{E}_i \right], \tag{6.61}$$

where N is the number of unit cells (dipoles), Ω is the crystal volume, $\alpha = \alpha_+ + \alpha_-$ is the total polarizability of the unit cell. Excluding $\boldsymbol{E}_i$ from (6.60) and (6.61), one gets

$$\boldsymbol{P} = \frac{N/\Omega}{1 - \dfrac{\alpha N}{3\varepsilon_0 \Omega}} \left(e^* (\boldsymbol{u}_+ - \boldsymbol{u}_-) + \alpha \boldsymbol{E} \right). \tag{6.62}$$

Polarizability is not directly measurable, so it is convenient to express it through the dielectric constant. The dielectric constant $\varepsilon(\omega)$ is defined through the electrical induction in the medium (see Chapter 12): $\boldsymbol{D}(\omega) = \varepsilon_0 \boldsymbol{E}(\omega) + \boldsymbol{P}(\omega) = \varepsilon(\omega)\, \varepsilon_0 \boldsymbol{E}$, or $\boldsymbol{P}(\omega) = \varepsilon_0 \left(\varepsilon(\omega) - 1 \right) \boldsymbol{E}(\omega)$.

Polarizability in (6.62) can be expressed through the dielectric constant at the frequency much higher than the maximum frequency of lattice vibrations, $\varepsilon(\infty)$. As heavy ions unable to follow the alternating electric field, the polarization contains the only term related to the polarizability of atoms. Infinity in $\varepsilon(\infty)$ is conditional in the sense that the frequency is lower than the inverse response time of bound electrons in atoms. Further increase in frequency would imply $\varepsilon \to 1$. So, $\varepsilon(\infty)$ describes an electronic contribution to the dielectric properties of media.

Then, polarization(6.62) takes the form,

$$\boldsymbol{P} = \varepsilon_0\left(\varepsilon(\infty)-1\right)\boldsymbol{E} = \frac{N\alpha \boldsymbol{E}/\Omega}{1-\alpha N/3\Omega\varepsilon_0}, \quad \alpha = \frac{3\Omega\varepsilon_0\left(\varepsilon(\infty)-1\right)}{N\left(\varepsilon(\infty)+2\right)}. \tag{6.63}$$

Using α in (6.60) and (6.62), one obtains

$$\boldsymbol{P} = \frac{N}{3\Omega}\left(\varepsilon(\infty)+2\right)e^*(\boldsymbol{u}_+ - \boldsymbol{u}_-) + \varepsilon_0\left(\varepsilon(\infty)-1\right)\boldsymbol{E},$$

$$\boldsymbol{E}_i = \frac{N}{9\varepsilon_0\Omega}\left(\varepsilon(\infty)+2\right)e^*(\boldsymbol{u}_+ - \boldsymbol{u}_-) + \frac{1}{3}\left(\varepsilon(\infty)+2\right)\boldsymbol{E},$$

$$\frac{\partial^2(\boldsymbol{u}_+ - \boldsymbol{u}_-)}{\partial t^2} = -\omega_0^2(\boldsymbol{u}_+ - \boldsymbol{u}_-) + \frac{e^*}{3\mu}\left(\varepsilon(\infty)+2\right)\boldsymbol{E},$$

$$\omega_0^2 \equiv \frac{\beta_2}{\mu}\left(1 - e^{*2}\frac{N\left(\varepsilon(\infty)+2\right)}{9\varepsilon_0\beta_2\Omega}\right). \tag{6.64}$$

Expressions (6.64) do not include polarizability; however, they still contain effective charge, which is also not a directly measurable parameter. In order to exclude e^*, we consider the static limit, $\omega = 0$. Static displacement and polarization follow from (6.64):

$$(\boldsymbol{u}_+ - \boldsymbol{u}_-) = \frac{e^*}{3\mu\omega_0^2}\left(\varepsilon(\infty)+2\right)\boldsymbol{E},$$

$$\boldsymbol{P} = \frac{N}{9\Omega}\left(\varepsilon(\infty)+2\right)^2\frac{e^{*2}}{\mu\omega_0^2}\boldsymbol{E} + \varepsilon_0\left(\varepsilon(\infty)-1\right)\boldsymbol{E} \tag{6.65}$$

Making use of static polarization $\boldsymbol{P} = \varepsilon_0\left(\varepsilon(0)-1\right)\boldsymbol{E}$, one finds

$$\frac{1}{3}\left(\varepsilon(\infty)+2\right)e^* = \omega_0\sqrt{\frac{\mu\varepsilon_0\Omega\left(\varepsilon(0)-\varepsilon(\infty)\right)}{N}}. \tag{6.66}$$

Substituting (6.66) into (6.64), we obtain the equation of motion and polarization expressed in terms of measurable parameters:

$$\frac{\partial^2 \boldsymbol{x}}{\partial t^2} = -\omega_0^2\boldsymbol{x} + \omega_0\sqrt{\varepsilon_0\left(\varepsilon(0)-\varepsilon(\infty)\right)}\,\boldsymbol{E},$$

$$P = \omega_0 \sqrt{\varepsilon_0 \left(\varepsilon(0) - \varepsilon(\infty)\right)}\, x + \varepsilon_0 \left(\varepsilon(\infty) - 1\right) E,$$

$$x \equiv \sqrt{\frac{\mu N}{\Omega}} \left(u_+ - u_-\right). \tag{6.67}$$

Any vector can be divided into longitudinal (curl-free) and transverse (divergence-free) parts, $A = A_l + A_t$, so that $\nabla \cdot A_t = 0$, $\nabla \times A_l = 0$. Dividing displacement, one gets the equation (6.67) in the form:

$$\frac{\partial^2}{\partial t^2}\left(x_l + x_t\right) = -\omega_0^2 \left(x_l + x_t\right) + \omega_0 E \sqrt{\varepsilon_0 \left(\varepsilon(0) - \varepsilon(\infty)\right)}\,. \tag{6.68}$$

If there are no free charges in the system, $\nabla \cdot D = 0$, or

$$\nabla \cdot \left(\varepsilon_0 E + P\right) = 0. \tag{6.69}$$

Substituting (6.67) in (6.69), we solve the equation for E_l:

$$E_l = -\frac{\omega_0}{\varepsilon(\infty)\, \varepsilon_0} \sqrt{\varepsilon_0 \left(\varepsilon(0) - \varepsilon(\infty)\right)}\, x_l. \tag{6.70}$$

Using (6.70) in (6.68), one obtains

$$\frac{\partial^2}{\partial t^2}\left(x_l + x_t\right) = -\omega_0^2\, x_t - \frac{\omega_0{}^2 \varepsilon(0)}{\varepsilon(\infty)} x_l + \omega_0 \sqrt{\varepsilon_0 \left(\varepsilon(0) - \varepsilon(\infty)\right)}\, E_t, \tag{6.71}$$

Taking operation curl or divergence from both parts of (6.71), one picks out the transverse or longitudinal equation, respectively:

$$\frac{\partial^2 x_t}{\partial t^2} = -\omega_0^2 x_t + \omega_0 \sqrt{\varepsilon_0 \left(\varepsilon(0) - \varepsilon(\infty)\right)}\, E_t,$$

$$\frac{\partial^2 x_l}{\partial t^2} = -\frac{\omega_0{}^2 \varepsilon(0)}{\varepsilon(\infty)} x_l \tag{6.72}$$

Assuming the transverse external field is zero, $E_t = 0$, and using the time-periodic ansatz $x_{t,l} = x_0 \exp(-i\omega t)$ in (6.72), we get $q \to 0$ limit of *TO*- and *LO*-mode frequencies: $\omega_{TO} = \omega_0$, $\omega_{LO} = \omega_0 \sqrt{\varepsilon(0)/\varepsilon(\infty)}$, and thus *Lyddane–Sachs–Teller relation*:

$$\frac{\omega_{LO}^2 (q = 0)}{\omega_{TO}^2 (q = 0)} = \frac{\varepsilon(0)}{\varepsilon(\infty)}. \tag{6.73}$$

The relation allows finding the *LO* phonon energy once the *TO* one is determined from an experiment. As follows from $\varepsilon(0) > \varepsilon(\infty)$, in the long-wavelength limit, longitudinal phonons carry higher energy than transverse ones.

The actual frequency of long-wavelength optical modes in ionic dielectrics such as NaCl, KCl, and semiconductors such as ZnS and GaAs is of the order of $10^{13}\ \text{s}^{-1}$, so it falls in the far-infrared spectral range.

6.4 Thermodynamics of Lattice Vibrations. Heat Capacity

In the quantum description, the total energy of lattice vibrations is the energy sum of $3rN$ independent oscillators. The vibrational energy is the expectation value of the Hamiltonian (6.37):

$$H = \sum_{qj} \hbar\omega_j(\mathbf{q})\left(n_{qj} + \frac{1}{2}\right), \tag{6.74}$$

where expectation values of phonon number operators n_{qj} may take arbitrary number $0, 1, 2, \ldots, \infty$.

The partition function determines all thermodynamic properties of the system (see Chapter 4):

$$Z = \sum_{n_{qj}} \exp\left(-\frac{\mu N - H}{k_B T}\right) = \prod_{qj} Z_{qj},$$

$$Z_{qj} = \exp\left(-\frac{\hbar\omega_j(\mathbf{q})}{2k_B T}\right)\sum_{n_{qj}=0}^{\infty} \exp\left(\frac{n_{qj}\left(\mu - \hbar\omega_j(\mathbf{q})\right)}{k_B T}\right)$$

$$= \exp\left(-\frac{\hbar\omega_j(\mathbf{q})}{2k_B T}\right)\left[1 - \exp\left(\frac{\mu - \hbar\omega_j(\mathbf{q})}{k_B T}\right)\right]^{-1}. \tag{6.75}$$

Differentiating Z_{qj} (see Eq. (4.9)), and equating μ to zero, we get Planck distribution function (6.40):

$$\langle n_{qj}\rangle = k_B T \left.\frac{\partial \log Z_{qj}}{\partial \mu}\right|_{\mu=0} = \frac{1}{\exp\left[\frac{\hbar\omega_j(\mathbf{q})}{k_B T}\right] - 1}. \tag{6.76}$$

The free energy calculated form partition function (6.75) has the form:

$$F = -k_B T \log Z = E_0 + k_B T \sum_{qj} \log\left[1 - \exp\left(-\frac{\hbar\omega_j(\mathbf{q})}{k_B T}\right)\right],$$

$$E_0 = \frac{1}{2}\sum_{qj} \hbar\omega_j(\mathbf{q}). \tag{6.77}$$

Parameter E_0 denotes the energy of lattice at $T = 0$ when atoms are in quantum zero oscillations state.

Summation in (6.77) depends on the types of phonons contributing to the free energy. One may calculate the optical phonons contribution within the Einstein model, which neglects momentum dependence in optical branches ω_j:

$$F_{opt} = E_0 + k_B T N \sum_{j=1}^{m} \log\left[1 - \exp\left(-\frac{\hbar\omega_j}{T}\right)\right], \tag{6.78}$$

where N is the number of $\mathbf{q}$-modes equal to the number of unit cells, $m = 3r - 3$ is the number of optical branches, and r being the basis.

For acoustic modes, the momentum dependence is essential, so we perform $\boldsymbol{q}$-summation in (6.77) by replacing it with integration,

$$\sum_{qj}(\ldots)=\frac{\Omega}{(2\pi)^3}\sum_j\int(\ldots)\,d^3q, \tag{6.79}$$

where Ω is the crystal volume. The upper limit of integration is the boundary of the BZ that requires detailed knowledge $\omega_j(\boldsymbol{q})$ for large $\boldsymbol{q}$. If one knew phonon spectrum in the whole BZ, it would be possible to calculate the free energy numerically. Analytical calculation is possible for the model acoustic spectrum, which is the *Debye model*. The model implies (a) the acoustic phonon spectrum is linear and isotropic in the whole range of integration $0<q<q_D$, (b) two transverse and one longitudinal acoustic branches have the same sound velocity, which equals the average value: $\omega_j(\boldsymbol{q})=\bar{\mathrm{v}}q$, $\bar{\mathrm{v}}=(2\mathrm{v}_t+\mathrm{v}_l)/3$; (c) one determines maximum momentum q_D from the condition that the total number of acoustic modes equals $3N$:

$$\sum_{qj}1=\frac{\Omega}{2\pi^2}\sum_{j=1}^{3}\int_0^{q_D}q^2\,dq=3N,$$

$$q_D=\left(\frac{6\pi^2N}{\Omega}\right)^{1/3}. \tag{6.80}$$

Acoustic phonons contribute to free energy as

$$F_{ac}=E_0+\frac{\Omega k_BT}{2\pi^2}\sum_{j=1}^{3}\int_0^{q_D}\log\left[1-\exp\left(-\frac{\hbar\omega_j(q)}{k_BT}\right)\right]q^2\,dq. \tag{6.81}$$

Using integration variable $\omega=\bar{\mathrm{v}}q$, one obtains

$$F_{ac}=E_0+\frac{9Nk_BT}{\omega_D^3}\int_0^{\omega_D}\log\left[1-\exp\left(-\frac{\hbar\omega}{k_BT}\right)\right]\omega^2\,d\omega,$$

$$\omega_D=\bar{\mathrm{v}}q_D. \tag{6.82}$$

After integration by parts, one gets

$$F_{ac}=E_0+Nk_BT\left[3\log\left[1-\exp\left(-\frac{T_D}{T}\right)\right]-D\left(\frac{T_D}{T}\right)\right].$$

$$D(z)\equiv\frac{3}{z^3}\int_0^z\frac{x^3dx}{\exp(x)-1}\approx\begin{cases}1-\frac{3}{8}z+\frac{1}{20}z^2, & z\ll1\\ \frac{\pi^4}{5z^3}, & z\gg1\end{cases}, \tag{6.83}$$

where $D(z)$ is the Debye function, $T_D\equiv\hbar\omega_D/k_B$ is the Debye temperature. Once free energy is known, average phonon energy and specific heat at constant volume have the form,

$$\langle E\rangle = F - T\frac{\partial F}{\partial T} = E_0 + 3Nk_BT\,D\left(\frac{T_D}{T}\right),$$

$$C_V = \frac{\partial\langle E\rangle}{\partial T} = 3Nk_B\left[D\left(\frac{T_D}{T}\right) - \frac{T_D}{T}D'\left(\frac{T_D}{T}\right)\right] \approx \begin{cases} 3Nk_B\left[1 - \dfrac{1}{20}\left(\dfrac{T_D}{T}\right)^2\right], & T \gg T_D \\ \dfrac{12\pi^4 Nk_B}{5}\left(\dfrac{T}{T_D}\right)^3, & T \ll T_D, \end{cases} \tag{6.84}$$

where $D'(x) \equiv \partial D(x)/\partial x$. At high temperature, the main contribution to the specific heat equals $3Nk_B$ and thus proportional to the number of degrees of freedom in acoustic vibrations.

6.4.1 The Density of States

Replacing q-integration in (6.81) to variable ω in (6.82), we have implicitly introduced the density of states $g(\omega)$ as follows:

$$\frac{\Omega}{(2\pi)^3}\sum_j\int(\ldots)\,d^3q = \int(\ldots)g(\omega)d\omega,$$

$$g(\omega) = \begin{cases} \dfrac{3\Omega\omega^2}{2\pi^2\bar{v}^3}, & \omega \le \omega_D \\ 0, & \omega \ge \omega_D. \end{cases} \tag{6.85}$$

Function $g(\omega)$ is the spectral density of phonon states, which is the number of states per unit frequency interval. It is easy to check

$$\int g(\omega)\,d\omega = 3N, \tag{6.86}$$

that is, the total number of states in three acoustic branches.

The shape of $g(\omega)$ in (6.85) is specific to the Debye model for acoustic branches. Once the actual phonon spectrum is known, the free energy reads as

$$F = E_0 + Nk_BT\sum_{j=1}^{3r}\int\log\left[1 - \exp\left(-\frac{\hbar\omega}{k_BT}\right)\right]g_j(\omega)d\omega, \tag{6.87}$$

where $g_j(\omega)$ is the partial density of states for branch j. For arbitrary phonon spectrum,

$$g_j(\omega) = \frac{\Omega}{(2\pi)^3}\int d^3q\,\delta\big(\omega - \omega_j(\mathbf{q})\big), \tag{6.88}$$

and normalization is similar to (6.86):

$$\sum_{j=1}^{3r}\int g_j(\omega)d\omega = 3rN, \tag{6.89}$$

For more details on singularities of the density of states, see Ref. [6].

Expression (6.87) represents the lattice part of total energy. Besides that, there is an electronic part shortly discussed in Chapter 4.

6.5 Phonon–Phonon Interaction

Picture of lattice vibrations as an ideal gas of phonons with the Hamiltonian (6.37) and distribution function (6.40) is valid as far as we restrict ourselves by second-order terms in the expansion (6.9) (harmonic approximation). Harmonic approximation implies linear dynamic equations (6.21). The whole picture is valid for small oscillation amplitudes so that one can approximate potential in Figure 6.1 by a parabolic one in the vicinity of the minimum. Once the amplitude increases, the potential becomes asymmetric, meaning that the force acting on atoms is nonlinear in displacements. In a simple one-dimensional model, asymmetry appears as an additional term in the expansion of potential energy:

$$U = \frac{\beta}{2}u^2 - g\frac{u^3}{3}, \tag{6.90}$$

where coefficient g accounts for deviation from harmonic potential.

Similarly, anharmonic contributions to the general Hamiltonian (6.9) are presented by higher-order terms in displacements. Third- and fourth-order terms result in nonlinear effects such as phonon–phonon interaction: interacting waves can exchange energy by converting one phonon into two or more depending on a type of nonlinearity, as well as by merging phonons. Cubic terms correspond to phonon's decay into two phonons of lower energy or to the reverse process of the merger of two phonons into one, as illustrated in Figure 6.10a. Fourth-order terms describe phonon–phonon scattering (see Figure 6.10b). In the course of phonon–phonon scattering, conservation laws hold energy conservation as well as momentum conservation within the first BZ.

Energy conservation for the process shown in Figure 6.10a looks as follows: $\omega_{j''}(\boldsymbol{q}'') = \omega_j(\boldsymbol{q}) + \omega_{j'}(\boldsymbol{q}')$. Conservation of quasi-momentum reads as $\boldsymbol{q}'' = \boldsymbol{q} + \boldsymbol{q}'$, where all three momenta are within the first BZ. The process is called normal (N-process). Normal processes do not change the total momentum of phonons in a crystal, which is zero in virtue (6.39). If $\boldsymbol{q}''$ goes beyond the first BZ, the momentum conservation law fails and reads instead as $\boldsymbol{q} + \boldsymbol{q}' = \boldsymbol{q}'' - \boldsymbol{K}_1$, where $\boldsymbol{K}_1$ is the minimum reciprocal vector, which brings $\boldsymbol{q}'' - \boldsymbol{K}_1$ into the first BZ. The process is called Umklapp (U) process. In the course of the U-process, phonon experiences reflection from the lattice as a whole so that the whole lattice absorbs the momentum change. In thermal equilibrium, U-processes with $\pm\boldsymbol{K}_1$ are equally probable, so the total momentum is still zero. For nonequilibrium phonons, U-processes are responsible for their thermalization on the way to equilibrium.

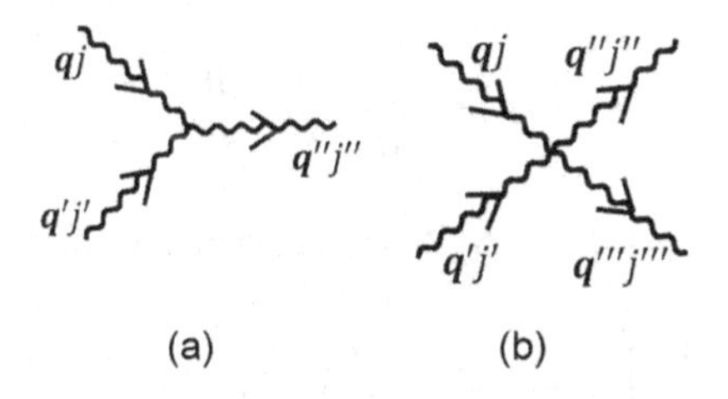

FIGURE 6.10 Diagrams illustrated cubic and fourth-order anharmonicity: (a) phonon-merging event; (b) phonon–phonon scattering. A wavy line with an arrow depicts the phonon mode $\boldsymbol{q}j$.

6.5.1 Thermal Expansion

One can explain linear lattice expansion using a one-dimensional model (6.90). In the harmonic potential, the equilibrium distance between atoms R_0 (see Figure 6.1) does not depend on oscillation amplitude that means that the thermal expansion is absent. The cubic term induces an asymmetric force, which depends on the displacement direction and thus shifts the equilibrium positions of atoms. That is thermal expansion. We calculate the thermal average of the displacement using classical Boltzmann distribution function,

$$f(u) = A\exp\left(-\frac{U}{k_B T}\right) \approx A\left(1+\frac{gu^3}{3k_B T}\right)\exp\left(-\frac{\beta u^2}{2k_B T}\right), \tag{6.91}$$

Normalization of distribution function gives

$$\int_{-\infty}^{\infty} f(u)du = 1, \quad A = \sqrt{\frac{\beta}{2\pi k_B T}}\exp\left(-\frac{\beta u^2}{2k_B T}\right). \tag{6.92}$$

Average displacement and coefficient of linear expansion (relative elongation per kelvin) are

$$\langle u \rangle = \int_{-\infty}^{\infty} u f(u)du = \frac{gk_B T}{\beta^2},$$

$$\alpha \equiv \frac{\langle u \rangle / R_0}{T} = \frac{gk_B}{\beta^2 R_0}, \quad \left[\text{kelvin}^{-1}\right]. \tag{6.93}$$

One can obtain (6.93) not explicitly integrating but using physical arguments. In thermal equilibrium, the force following from (6.90) equals zero:

$$\langle F \rangle \equiv -\left\langle \frac{\partial U}{\partial u} \right\rangle = -\beta\langle u \rangle + g\langle u^2 \rangle = 0,$$

$$\langle u \rangle = \frac{g}{\beta}\langle u^2 \rangle. \tag{6.94}$$

The average potential energy of the oscillator equals the number of degrees of freedom times $k_B T/2$. In one-dimensional lattice, $\beta\langle u^2 \rangle/2 = k_B T/2$, so from (6.94), we obtain $u = gk_B T/\beta^2$, which coincides with (6.93).

Parameters β and g introduced above are phenomenological constants. To give them microscopic meaning, below we estimate the expansion coefficient in a monovalent ionic crystal where the force acting on two neighboring ions of opposite signs comprises attractive Coulomb part and repulsive force at a short distance:

$$F(R) = -\frac{e^2}{4\pi\varepsilon_0 R^2} + \frac{B}{R^{10}}. \tag{6.95}$$

In equilibrium, $F(R_0) = 0$, thus $B = e^2 R_0^8 / 4\pi\varepsilon_0$. A small deviation from equilibrium distance gives

$$F(u) = -\frac{e^2}{4\pi\varepsilon_0}\left(\frac{1}{(R_0+u)^2} - \frac{R_0^8}{(R_0+u)^{10}}\right) \approx \frac{e^2}{4\pi\varepsilon_0}\left(-\frac{8u}{R_0^3} + \frac{52u^2}{R_0^4}\right) \tag{6.96}$$

Comparing (6.96) to the force following from (6.90), $F = -\partial U / \partial u$, one obtains

$$\beta = \frac{8e^2}{4\pi\varepsilon_0 R_0^3}, \quad g = \frac{52e^2}{4\pi\varepsilon_0 R_0^4}, \quad \alpha = \frac{13k_B\pi\varepsilon_0 R_0}{4e^2}. \tag{6.97}$$

Assuming the lattice constant to be $R_0 = 3\dot{A}$, one gets $\alpha \approx 1.5\times10^{-5}\,\mathrm{K}^{-1}$-the order of magnitude observable in solid crystals.

6.5.2 Thermal Conductivity and Resistance

As atoms in solids interact with each other, atomic vibrations excited in a heated part of a crystal transfer to the other parts, thus providing heat conduction. Here we consider thermal conductivity caused by lattice vibrations only, putting aside an electronic part. The energy flux in isotropic media, $\boldsymbol{J}$ [W / m^2], obeys equation $\boldsymbol{J} = -\kappa\,\nabla T$. Here, $\kappa\left(\kappa^{-1}\right)$ is the thermal conductivity (thermal resistivity) and T_K is the absolute temperature. The minus sign means that the heat flux direction is opposite to the temperature gradient.

As it follows from (6.40), the number of equilibrium phonons increases with temperature. If there is a temperature gradient, phonons diffuse into the low-temperature region that is the phonon picture of thermal conductivity. Diffusion implies phonon–phonon scattering, which occurs if anharmonicity is taking into account. Moreover, normal phonon–phonon collisions conserve energy and total phonons momentum, making thermal resistance zero. So, to conduct heat with nonzero thermal resistance, phonons should take part in U-processes if they scatter against phonons. Scattering on other than phonons agents such as defects and electrons also maintains finite thermal resistance. The mean free path between collisions is $\lambda = 1/n_{ph}\sigma$, where n_{ph} is the phonon density and σ is the cross-section of U-processes. At high temperature, $T \gg T_D$, $n_{ph} \sim T$, the number of high-energetic phonons is enough to make probabilities of N- and U-processes almost equal. One may consider phonons as a nonideal classical gas for which thermal conductivity equals

$$\kappa = \frac{1}{3\Omega} C_V \bar{v}\lambda \sim \frac{1}{T}, \quad \left[\frac{\mathrm{W}}{\mathrm{kelvin.m}}\right], \tag{6.98}$$

where $C_V \approx 3Nk_B$ is given in (6.84).

At low temperatures $(T \ll T_D)$, the heat capacity (6.84) behaves as T^3. In (6.98), λ carries more strong temperature dependence. The point is that only phonons of high enough energy, let say exceeding E_0, may take part in U-collisions. At low temperatures ($k_BT \ll E_0$), the density of energetic phonons $n_{ph} \sim \exp\left(-E_0/k_BT\right)$. The mean free path $\lambda \sim \exp\left(E_0/k_BT\right)$ exponentially increases when the temperature goes down, meaning an increase in thermal conductivity $\kappa \sim \exp\left(E_0/k_BT\right)$. However, the rise of the mean free path is restricted by sample dimensions or by phonon scattering against impurities, making $\lambda = const$. So, when the temperature tends to absolute zero, the thermal conductivity behaves again as T^3. The typical plot $\kappa(T)$ is shown in Figure 6.11.

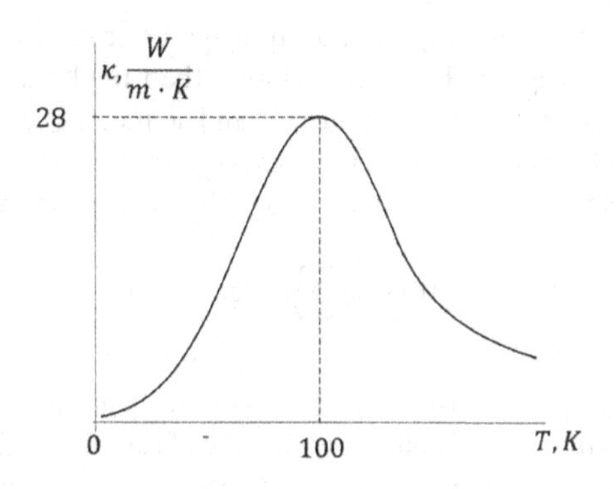

FIGURE 6.11 Thermal conductivity in a diamond.

The maximum value depends on the lattice structure. One more example is sapphire: thermal conductivity reaches the maximum of 62 W/mK at $T \approx 40$ K.

6.6 Electron–Phonon Interaction

Adiabatic approximation treats ions as frozen in a lattice, which forms ideally periodic potential for electrons $V(\boldsymbol{r})$. An electron, moving in a periodic potential, acquires the effective mass, and its wave function, being the plane wave in free space, becomes the Bloch wave instead (see Chapter 1). Periodic potential does not scatter electrons, so it does not contribute to electrical resistance, or said differently, a free electron in a vacuum turns to a free quasiparticle in an ideal crystal. Terms "quasiparticle" and "quasi-momentum" characterize an electron in a lattice and reflect specific kinematics, in which an electron momentum is a discrete number defined within the first BZ.

Atom displacements from equilibrium positions violate lattice periodicity and create an additional potential $\delta V(\boldsymbol{r})$ that is at the origin of *electron–phonon interaction*. The interaction scatters electrons and contributes to electrical conductivity.

Potential $\delta V(\boldsymbol{r})$ affects electrons in different ways, depending on the type of crystal and phonon branch under consideration. There are two main mechanisms the electrons would interact with phonons. (a) Long-wavelength phonons induce lattice deformation, which locally changes the electron energy spectrum. That creates the *deformation potential* that scatters electrons, and the mechanism works in all materials. In covalent and molecular crystals, it is the only mechanism of electron–phonon interaction. (b) Phonons propagating through the crystal may give rise to polarization and macroscopic electric fields. That is what happens in ionic and piezoelectric materials. Below we discuss typical examples.

6.6.1 Deformation Potentials

Acoustic phonons. In the limit $\boldsymbol{q} \to 0$, phonons represent the parallel shift of the lattice as a whole and thus cannot affect the electron energy. So, energy correction should not depend on atom displacements $\boldsymbol{u}(\boldsymbol{r})$, but rather on their derivatives, which turn zero when the lattice shifts as a whole. Electrons feel lattice deformation formally expressed by symmetric strain tensor:

$$\xi_{\alpha\beta} = \frac{1}{2}\left(\frac{\partial u_\alpha}{\partial x_\beta} + \frac{\partial u_\beta}{\partial x_\alpha}\right), \quad \alpha,\beta = 1,2,3 \leftrightarrow (x,y,z). \tag{6.99}$$

Deformation-induced volume dilation expresses through diagonal matrix elements,

$$\frac{\delta\Omega}{\Omega} = \sum_\alpha \xi_{\alpha\alpha} = \operatorname{div} \boldsymbol{u}(\boldsymbol{r}), \tag{6.100}$$

where Ω is the crystal volume. At finite $\boldsymbol{q}$, phonon propagation causes deformation in macroscopic regions of the linear size of $\sim 1/|\boldsymbol{q}|$ which are much larger than the lattice constant. In each of these regions, the adiabatic approach (Section 6.2) is applicable and defines the electron energy at frozen atoms positions. As a result, the electron energy varies from region to region, affecting the electron dynamics, so we deal with an acoustic electron–phonon interaction. The interaction shifts electron energy as

$$W(r) = \sum_{\alpha\beta} a_{\alpha\beta}\, \xi_{\alpha\beta}, \tag{6.101}$$

where coefficients $a_{\alpha\beta}$ have energy dimension and are called the *deformation potentials*. The number of independent matrix elements $a_{\alpha\beta}$ depends on the symmetry of the electronic state at a particular point in the BZ. Energy minimum (maximum) in conduction (valence) band lies in a high symmetry point (for example, cubic crystals with minimum electron energy located in point Γ), all nondiagonal components in $a_{\alpha\beta}$ equal zero, and deformation potential becomes a scalar:

$$a_{11} = a_{22} = a_{33} = \sigma,$$

$$W(\boldsymbol{r}) = \sigma \operatorname{div} \boldsymbol{u}(\boldsymbol{r}), \tag{6.102}$$

where σ weakly depends on $\boldsymbol{k}$ at small $|\boldsymbol{k}|$. Under deformation, edges of conduction and valence bands shift differently, so deformation potentials carry the band index $\sigma_{c,v}$.

In uniaxial crystals with a band minimum at $\boldsymbol{k}=0$, the electron energy spectrum is the ellipsoid elongated in, let say, z-direction. In this case, $a_{11} = a_{22} \neq a_{33}$, and thus two deformation potentials describe the shift of the energy spectrum (6.101):

$$W(\boldsymbol{r}) = \Xi_d \operatorname{div} \boldsymbol{u}(\boldsymbol{r}) + \Xi_u \xi_{33}(\boldsymbol{r}), \tag{6.103}$$

where $a_{11} = a_{22} \equiv \Xi_d$, $a_{33} - (a_{11} + a_{22})/2 \equiv \Xi_u$. Similarly, one needs two deformation potentials in cubic n-Ge and n-Si, where the conduction band minima are not in the center of BZ [8].

In metals, the deformation potential varies with point $\boldsymbol{k}_F$ on the Fermi surface. One can estimate σ_m assuming parabolic spectrum and using the volume dependence of the Fermi energy (Chapter 4, Eq. (4.1)):

$$W(\boldsymbol{r}) = \delta\varepsilon_F = -\frac{2}{3}\frac{\hbar^2}{2m}\left(3\pi^2 \frac{N}{\Omega}\right)^{2/3} \frac{\delta\Omega}{\Omega} = \sigma_m \operatorname{div} \boldsymbol{u}(\boldsymbol{r}),$$

$$\sigma_m = -\frac{2}{3}\varepsilon_F. \tag{6.104}$$

The sign means a decrease in energy with lattice expansion. The order of magnitude $\sigma_m \approx (1 \div 10)\,\text{eV}$ is also applicable to semiconductors. Usually, deformation potentials are found to be from experiments. In the long-wavelength limit, one can use continuous medium approximation, so displacements (6.32), expressed through phonon creation and annihilation operators (6.35), take the form,

$$\boldsymbol{u}(\boldsymbol{r}) = \sqrt{\frac{\hbar}{2MN}} \sum_{jq} \frac{1}{\sqrt{\omega_j(\boldsymbol{q})}} \left(\boldsymbol{e}^j(\boldsymbol{q}) b_{qj} \exp(i\boldsymbol{q}\cdot\boldsymbol{r}) + \left[\boldsymbol{e}^j(\boldsymbol{q})\right]^* b_{qj}^+ \exp(-i\boldsymbol{q}\cdot\boldsymbol{r}) \right), \tag{6.105}$$

where $\boldsymbol{e}^j$ are polarization vectors which obey conditions (6.24) and (6.31):

$$\sum_s M_s \left[\boldsymbol{e}_s^j(\boldsymbol{q})\right]^* \cdot \boldsymbol{e}_s^{j'}(\boldsymbol{q}) = M\delta_{jj'},$$

$$\sum_s M_s e_{s\alpha}(\boldsymbol{q}) = 0. \tag{6.106}$$

Since all atoms in the unit cell displace in the same direction, $\boldsymbol{e}_s^j(\boldsymbol{q}) = e^j(\boldsymbol{q})$, $|e^j(\boldsymbol{q})| = 1$. Displacements (6.105) substituted in (6.99) give

$$\xi_{\alpha\beta} = \sqrt{\frac{\hbar}{2MN}} \sum_{jq} \frac{iq}{2\sqrt{\omega_j(\boldsymbol{q})}} \left[\left(e_\alpha^j(\boldsymbol{q}) n_\beta + e_\beta^j(\boldsymbol{q}) n_\alpha \right) b_{qj} \exp(i\boldsymbol{q}\cdot\boldsymbol{r}) - \left(e_\alpha^{j*}(\boldsymbol{q}) n_\beta + e_\beta^{j*}(\boldsymbol{q}) n_\alpha \right) b_{qj}^+ \exp(-i\boldsymbol{q}\cdot\boldsymbol{r}) \right]. \tag{6.107}$$

where $\boldsymbol{n} \equiv \boldsymbol{q}/q$ is the unit vector in the direction of phonon propagation. Using (6.107) in (6.101), one obtains

$$W(r)=\sum_{\alpha\beta} a_{\alpha\beta}\xi_{\alpha\beta}=\sqrt{\frac{\hbar}{2MN}}\sum_{jq}\frac{iq}{\sqrt{\omega_j(\boldsymbol{q})}}\left[D_q^j b_{qj}\exp(i\boldsymbol{q}\cdot\boldsymbol{r})-\left(D_q^j\right)^*(\boldsymbol{q})b_{qj}^+\exp(-i\boldsymbol{q}\cdot\boldsymbol{r})\right], \quad (6.108)$$

where effective acoustic deformation potentials are

$$D_q^j=\frac{1}{2}\sum_{\alpha\beta} a_{\alpha\beta}\left(e_\alpha^j(\boldsymbol{q})n_\beta+e_\beta^j(\boldsymbol{q})n_\alpha\right). \quad (6.109)$$

Effective potentials D_q^j depend on phonon polarization and propagation directions. In high-symmetry crystals from (6.102), one gets

$$D_q^j=\sigma\boldsymbol{n}\cdot\boldsymbol{e}^j(\boldsymbol{q})=\begin{cases}\sigma, & LA\\ 0, & TA\end{cases} \quad (6.110)$$

So, in cubic crystals with electron maximum in the Γ-point of BZ, the interaction of electrons with acoustic phonons involves longitudinal phonons only. As it follows from (6.109), in crystals of lower symmetry, nondiagonal terms $a_{\alpha\beta}$ call for shear components of deformation, then electrons interact also with transverse acoustic phonons.

Optical phonons. In covalent crystals, optical phonons are not electrically active, so the mechanism of electron–phonon interaction relies on a deformation-induced shift in electron energy. For optical phonons, atom displacements do not shift the center of mass in the unit cell, so the interaction not necessarily depends on derivatives, but, contrary to acoustic phonons, it is proportional to displacement itself (see text preceding Eq. (6.99)). As the electron energy shift is a scalar, one may write it as a dot product:

$$W(\boldsymbol{r})=\sum_{j\geq 4}\boldsymbol{A}_j\cdot\boldsymbol{u}_j(\boldsymbol{r}), \quad (6.111)$$

where summation goes over optical branches ($j=1,2,3$ are acoustic branches), $\boldsymbol{u}$ is the optical displacement (in one-dimensional model (6.51), $u=u_n^{(2)}-u_n^{(1)}$), $\boldsymbol{A}_j$ is the vector of *optical deformation potential*, which is defined by the symmetry of electron spectrum. The necessary but insufficient condition for the existence of vector $\boldsymbol{A}_j$ is the symmetry-defined direction in the BZ like that in uniaxial crystals or crystals with electron extrema not in the center of BZ. If unit vector $\boldsymbol{l}$ points from $\boldsymbol{k}=0$ to the band minimum, $\boldsymbol{A}_j=E_{0j}\boldsymbol{l}$. More general expression reads as $A_{\alpha j}=E_{0j}^{\alpha\beta}l_\beta$, where we imply summation over repeating indexes, $E_0^{\alpha\beta}$ comprises phenomenological constants [J/m]. The symmetry of a crystal dictates the form of tensor $E_{0j}^{\alpha\beta}$. In cubic crystals with electron minimum in $\boldsymbol{k}=0$, there is nowhere to direct vector $\boldsymbol{A}$, thus $\boldsymbol{A}=0$. Detailed symmetry analysis prescribes $\boldsymbol{A}=0$ in Si and $\boldsymbol{A}\neq 0$ in Ge [9].

Substituting $A_{\alpha j}$ and (6.105) in (6.111), we obtain the electron energy shift due to electron-optical phonon interaction in covalent crystals:

$$W(\boldsymbol{r})=\sqrt{\frac{\hbar}{2MN}}\sum_{j\geq 4}\sum_{q}\frac{E_{0j}^{\alpha\beta}l_\beta}{\sqrt{\omega_j(\boldsymbol{q})}}\left\{e_\alpha^j(\boldsymbol{q})b_{qj}\exp(i\boldsymbol{q}\cdot\boldsymbol{r})+e_\alpha^{j*}(\boldsymbol{q})b_{qj}^+\exp(-i\boldsymbol{q}\cdot\boldsymbol{r})\right\}. \quad (6.112)$$

Interaction (6.112) depends on the relative orientation between the phonon polarization vector and the axis to the electron minimum. Sometimes $E_0^{\alpha\beta}$ is written as $E_0^{\alpha\beta}=\Xi_0^{\alpha\beta}K$, K being the absolute value of the reciprocal lattice vector. Constants $\Xi_0^{\alpha\beta}$ have energy dimension and are called *optical deformation potentials*. Note that, strictly speaking, "optical deformation" does not mean lattice deformation (tensor) but rather a relative displacement in the unit cell (vector).

6.6.2 Electron–Phonon Interaction in Ionic Crystals

In ionic crystals, optical lattice vibrations generate oscillating electrical polarization in each unit cell. The resulting electric field shifts electron energy. The electric field created by optical phonons does not involve lattice deformation and relies on relative ions displacements within the unit cell. In noncentrosymmetric ionic crystals, lattice deformation in acoustic vibrations generates electric fields through the piezoelectric effect. Both mechanisms are to discuss in the next sections.

6.6.2.1 Fröhlich Hamiltonian

In ionic crystals, optical lattice vibrations generate oscillating electrical polarization in each unit cell. The resulting electric field shifts electron energy, so this mechanism of electron–phonon interaction has an electrostatic origin and does not rely on lattice deformation. Since we consider optical phonons, the simplest object is the crystal with two ions in the unit cell. We still can use long-wavelength displacements (6.105) written for two ions, $s=1,2$:

$$\boldsymbol{u}_s(\boldsymbol{r})=\sqrt{\frac{\hbar}{2MN}}\sum_{j\geq 4}\sum_{q}\frac{1}{\sqrt{\omega_j(\boldsymbol{q})}}\left(\boldsymbol{e}_s^j(\boldsymbol{q})b_{qj}\exp(i\boldsymbol{q}\cdot\boldsymbol{r})+\left[\boldsymbol{e}_s^j(\boldsymbol{q})\right]^*b_{qj}^+\exp(-i\boldsymbol{q}\cdot\boldsymbol{r})\right), \tag{6.113}$$

Polarization-induced electric field is given in (6.70):

$$\boldsymbol{E}_l=-\frac{\omega_0}{\varepsilon(\infty)\varepsilon_0}\sqrt{\varepsilon_0\left(\varepsilon(0)-\varepsilon(\infty)\right)}\,\boldsymbol{x}_l=-\omega_{LO}\sqrt{\frac{\mu N}{\Omega\varepsilon_0}\left(\frac{1}{\varepsilon(\infty)}-\frac{1}{\varepsilon(0)}\right)}(\boldsymbol{u}_1-\boldsymbol{u}_2), \tag{6.114}$$

where $\boldsymbol{u}_{1,2}$ are displacements (6.113) for $s=1,2$, respectively; ω_{LO} appears in (6.114) from Lyddane-Sachs-Teller relation (6.73).

The electrons energy correction due to electric field $\boldsymbol{E}_l$ is:

$$W(\boldsymbol{r})=-e\varphi(\boldsymbol{r})=e\int E_l\,d^3r,\quad e>0. \tag{6.115}$$

Substituting (6.114) in (6.115), one gets

$$W(\boldsymbol{r})=ie\omega_{LO}\sqrt{\frac{\mu\hbar}{2M\Omega\varepsilon_0}\left(\frac{1}{\varepsilon(\infty)}-\frac{1}{\varepsilon(0)}\right)}\sum_{q}\frac{1}{q\sqrt{\omega_j(\boldsymbol{q})}}\left[(e_1-e_2)b_{qj}\exp(i\boldsymbol{q}\cdot\boldsymbol{r})-(e_1-e_2)^*b_{qj}^+\exp(-i\boldsymbol{q}\cdot\boldsymbol{r})\right]. \tag{6.116}$$

Transverse optical phonons do not contribute to the integral (6.115), so equation (6.116) presents interaction with the longitudinal phonon branch, in which we neglect the dispersion: $\omega_j(\boldsymbol{q})=\omega_{LO}$. Longitudinal components of polarization vectors $e_{1,2}$ at small $\boldsymbol{q}$ are real-valued numbers. From condition (6.31) and normalization (6.106), we find

$$M_1e_1+M_2e_2=0,$$

$$M_1e_1^2+M_2e_2^2=M,$$

or

$$e_1 - e_2 = \frac{M}{\sqrt{M_1 M_2}}. \tag{6.117}$$

Using (6.117) in (6.116), we find the Fröhlich electron–longitudinal phonon interaction in the form,

$$W(\mathbf{r}) = i\sqrt{\frac{e^2\hbar\omega_{LO}}{2\Omega\varepsilon_0}\left(\frac{1}{\varepsilon(\infty)} - \frac{1}{\varepsilon(0)}\right)}\sum_q \frac{1}{q}\left[b_{qj}\exp(i\mathbf{q}\cdot\mathbf{r}) - b_{qj}^+\exp(-i\mathbf{q}\cdot\mathbf{r})\right]. \tag{6.118}$$

The interaction is divergent at $q = 0$. Momentum and energy conservation forbid intraband interaction at exact zero phonon momentum. Besides, it is an unscreened electric field that causes an $1/q$ anomaly. Were the screening by free carriers taken into account, the divergence would disappear. The total Hamiltonian of the electron–phonon system comprises three terms: the kinetic energy of free electron quasiparticles, the energy of free phonons (harmonic approximation), and electron–phonon interaction: $H = H_e + H_p + H_{ep}$. Following the procedure described in Appendix 1, https://www.routledge.com/Modern-Semiconductor-Physics-and-Device-Applications/Dugaev-Litvinov/p/book/9780367250829# we present the Hamiltonian in the second quantization representation. Taking wavefunctions $\psi_k = e^{i\mathbf{k}\cdot\mathbf{r}}/\sqrt{\Omega}$ as a basis, we expand electron field operators:

$$\Psi(\mathbf{r}) = \sum_k c_k\psi_k(\mathbf{r}),\quad \Psi^+(\mathbf{r}) = \sum_k c_k^+\psi_k^*(\mathbf{r}), \tag{6.119}$$

where $c_k\left(c_k^+\right)$ is the destruction (creation) operator acting on occupation number states.

From (A1.16) and (6.37):

$$H_e + H_p = \sum_{ks} E_{ks}c_{ks}^+c_{ks} + \hbar\omega_{LO}\sum_q\left(b_{qj}^+b_{qj} + \frac{1}{2}\right). \tag{6.120}$$

Second quantization representation for interaction $W(\mathbf{r})$ takes the form,

$$H_{ep} = \int_\Omega \Psi^+(\mathbf{r})W(\mathbf{r})\Psi(\mathbf{r})d^3r = \sum_{kk'} W_{kk'}c_k^+c_{k'},$$

$$W_{kk'} = \int_\Omega \psi_k^*(\mathbf{r})W(\mathbf{r})\,\psi_{k'}(\mathbf{r})d^3\mathbf{r} \tag{6.121}$$

After substitution (6.118) in (6.121) and integration over the crystal volume

$$\int e^{-i\mathbf{k}\cdot\mathbf{r}}e^{i\mathbf{q}\cdot\mathbf{r}}e^{i\mathbf{k}'\cdot\mathbf{r}}d^3\mathbf{r} = \Omega\delta_{q,k-k'},$$

one obtains the Fröhlich Hamiltonian

$$H_{ep} = i\sqrt{\frac{e^2\hbar\omega_{LO}}{2\Omega\varepsilon_0}\left(\frac{1}{\varepsilon(\infty)} - \frac{1}{\varepsilon(0)}\right)}\sum_{kq}\frac{1}{q}\left(b_{qj}c_{k+q}^+c_k - b_{qj}^+c_{k-q}^+c_k\right). \tag{6.122}$$

Diagram representation of two scattering terms in (6.122) is shown in Figure 6.12

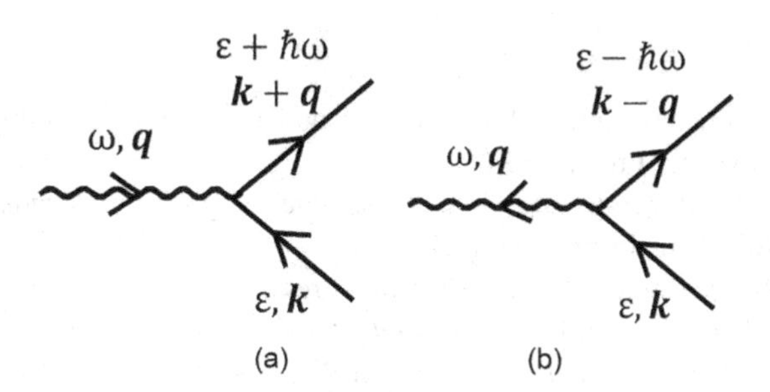

FIGURE 6.12 Electron–phonon interaction. A straight-arrow depicts an electron, a wavy arrow - a phonon. (a) The first term in (6.122). (b) The second term.

In deriving (6.122), we imply the continuous approximation for electrons, which allows using the free-electron wavefunction, effective mass for electrons, and the long-wavelength limit for optical phonons. In crystals, it comes to the same Hamiltonian with a more complicated interaction constant, which includes Bloch amplitudes. Besides, $\boldsymbol{k}$ and $\boldsymbol{q}$ become quasi-momenta for which Umklapp scattering processes are possible. Processes shown in Figure 6.12 occur with energy conservation. There are two types of processes: normal (N) with strict momentum conservation and U-processes, which conserve momentum to the accuracy of the reciprocal lattice vector. If slow electrons interact with long-wavelength phonons, their momenta before and after scattering do not go beyond the first BZ, and exact momentum conservation holds.

6.6.2.2 Piezoelectric Interaction

In inversion-asymmetric ionic materials, homogeneous acoustic deformation (6.99) induces macroscopic polarization:

$$P_\alpha = \beta_{\alpha,\beta\gamma}\,\xi_{\beta\gamma}, \tag{6.123}$$

where $(\alpha\beta\gamma)=(x,y,z)$, $\beta_{\alpha,\beta\gamma}$ piezoelectric moduli and $\xi_{\beta\gamma}$ is the deformation tensor (6.107). Electrostatic potential induced by polarization (6.123) is given by the Poisson equation, which under the assumption of local electrical neutrality, has the form $\nabla\cdot\boldsymbol{D}=0$. Using induction $\boldsymbol{D}=\varepsilon_0\boldsymbol{E}+\boldsymbol{P}$, and $\boldsymbol{E}=-\nabla\varphi$, one obtains the Poisson equation for potential and the interaction Hamiltonian,

$$\Delta\varphi=\nabla\cdot\boldsymbol{P},\quad W(\boldsymbol{r})=-e\varphi. \tag{6.124}$$

Substituting (6.123) into (6.124), one finds the Hamiltonian that describes the interaction between electrons and piezoelectrically active phonons:

$$W(\boldsymbol{r})=-\sqrt{\frac{\hbar}{2MN}}\sum_{jq}\frac{e\beta_{\gamma,\alpha\beta}n_\gamma}{2\varepsilon_0\sqrt{\omega_j(q)}}\Big[\big(e_\alpha^j(\boldsymbol{q})n_\beta+e_\beta^j(\boldsymbol{q})n_\alpha\big)b_{qj}\exp(i\boldsymbol{q}\cdot\boldsymbol{r})+h.c.\Big], \tag{6.125}$$

where $h.c.$ is for Hermitian conjugate, $\omega_j(\boldsymbol{q})=\mathrm{v}_j q$. The potential φ enters (6.125) as a solution to the Poisson equation that can be verified by direct substitution in (6.124).

Unlike the interaction with polar optical phonons (6.116) which is active for longitudinal mode only, the piezoelectric mechanism (6.125) includes acoustic phonons of all polarizations.

In noncentrosymmetric crystals, optical phonons also interact with electrons via the piezoelectric mechanism described above for acoustic phonons.

References

1. M. Born and K. Huang, Dynamical Theory of Crystal Lattices (Claredon Press, Oxford, UK, 1954; reprint Oxford University Press, 1988).
2. A. Maradudin, E.W. Montroll, and G.H. Weiss, *Theory of Lattice Dynamics in the Harmonic Approximation* (Academic Press, New York, 1963).
3. J.F. Nye, *Physical Properties of Crystals* (Claredon Press, Oxford, UK, 1985).
4. S. Adachi, *Physical Properties of III-V Semiconductor Compounds* (John Wiley, New York, 1992).
5. Yu. A. Ill'inskii and L.V. Keldysh, *Electromagnetic Response of Material Media* (Plenum Press, New York, 1994)
6. A.M. Kosevich, *The Crystal Lattice, Phonons, Solitons, Dislocations, Superlattices* (2nd ed., Wiley-VCH Verlag GmbH, Weinheim, 2005).
7. C. Kittel, *Introduction to Solid State Physics* (7th ed., Wiley, New York, 1996).
8. P.Y. Yu and M. Cardona, *Fundamentals of Semiconductors*, (3rd ed., Springer, Berlin, 1994).
9. G.L, Bir and G.E. Pikus, *Symmetry and Strain Effects in Semiconductors* (Wiley, New York, 1974).

7 Transport Properties

7.1 Electrons in Electric and Magnetic Fields

As was discussed in Chapter 5, in the framework of classical electrodynamics, an electron in electric and magnetic fields experiences the Lorenz force (we use the notation, in which the charge of electron $e < 0$)

$$F = eE + e\mathbf{v} \times B, \tag{7.1}$$

where E and B are the electric field and magnetic induction, respectively, $\mathbf{v}$ is the electron velocity. In the medium, B is the microscopic field averaged over a macroscopic volume.

In addition to the Lorentz force acting on any charged classical particle, an electron feels the magnetic field through its quantum state – spin: corresponding term in the Hamiltonian is the Zeeman interaction

$$V_Z = \frac{1}{2} g\mu_B\, \sigma \cdot B, \tag{7.2}$$

where g is the Landé factor, μ_B is the Bohr magneton, $\sigma = \left(\sigma_x, \sigma_y, \sigma_z\right)$ are the spin variables (Pauli matrices).

The interaction of an electron with the electromagnetic field is described by coupling to field potentials φ (scalar potential) and A (vector potential). To introduce the electromagnetic field and account for the gauge invariance, one has to make a substitution in the wave equation:

$$\frac{\partial}{\partial t} \to \frac{\partial}{\partial t} + \frac{ie\varphi}{\hbar}, \quad \nabla \to \nabla - \frac{ieA}{\hbar} \tag{7.3}$$

We discuss gauge invariance in more detail in Chapter 12.

7.2 Nonequilibrium State under Electric Field or Temperature Gradient

In equilibrium, an electron fills the state with energy ε_n with a probability determined by the Fermi–Dirac distribution function (see Chapter 4),

$$f_0\left(\varepsilon_n\right) = \frac{1}{\exp\left(\frac{\varepsilon_n - \mu}{k_B T}\right) + 1}, \tag{7.4}$$

DOI: 10.1201/9780429285929-7

where μ is the electron chemical potential and k_B is the Boltzmann constant.

In equilibrium, there is no transport of neither charge nor energy. The external magnetic field affects the electron energy states but does not violate equilibrium in the system. It means that the distribution function of the electron system in the magnetic field has the form of (7.4) with the electron energy spectrum depending on the magnetic field.

Nonzero electric field $\boldsymbol{E}$ and temperature gradient ∇T are the driving forces that cause flows of electric charges and energy. The flows mean the electron system is out of equilibrium and thus to be described by nonequilibrium distribution function $f(\varepsilon_n)$. The regime of small deviation from equilibrium is called a *linear response*. In this regime, one can present the distribution function as $f(\varepsilon_n) = f_0(\varepsilon_n) + \delta f(\varepsilon_n)$, where δf is a small correction proportional to $\boldsymbol{E}$ or ∇T. In linear approximation, the electric current is proportional to δf, and thus to $\boldsymbol{E}$ or ∇T.

7.3 Electric Current: Conductivity Tensor

The linear dependence of current density on electric field $\boldsymbol{E}$ has the form,

$$j_i = \sigma_{ij} E_j, \tag{7.5}$$

where j_i and E_i are the components of current density and electric field, respectively, and σ_{ij} is the *conductivity tensor*.[1] Ohm's law (7.5) is local in nature, so in an inhomogeneous semiconductor, the current in point $\boldsymbol{r}$ is proportional to electric field $\boldsymbol{E}(\boldsymbol{r})$ in the same point.[2]

Mathematically, this is a general form of the linear relation between two vectors. The number of independent components in σ_{ij} depends on the symmetry of the crystal lattice. In an isotropic homogeneous medium, this tensor is diagonal, $\sigma_{ij} = \sigma_0 \delta_{ij}$. In crystals of cubic symmetry, the conductivity tensor has a diagonal form, as long as we choose the coordinate system along the symmetry axes of the lattice. As polar vector changes sign under inversion $\boldsymbol{r} \to -\boldsymbol{r}$, linear relation (7.5) between two polar vectors $\boldsymbol{E}$ and $\boldsymbol{j}$ preserves invariance to spatial inversion. Time-inversion symmetry, however, is broken explicitly as under time inversion, $t \to -t$, vector $\boldsymbol{j}$ changes sign but vector $\boldsymbol{E}$ does not. Lack of time-inversion invariance means that the kinetic coefficient, relating $\boldsymbol{j}$ and $\boldsymbol{E}$, describes nonreversible electron transport accompanied by entropy production.

In the absence of a magnetic field, the thermodynamics imposes condition $\sigma_{ij} = \sigma_{ji}$, which follows from *Onsager's principle* of the symmetry of kinetic coefficients discussed in Section 7.9. In the magnetic field, the conductivity tensor acquires nondiagonal components. One can present the tensor of conductivity as a sum of symmetric $\sigma_{ij}^{(s)}$ and antisymmetric $\sigma_{ij}^{(a)}$ parts. If $\boldsymbol{B} = 0$, the antisymmetric part is zero. From Onsager's principle, it follows that $\sigma_{ij}(\boldsymbol{B}) = \sigma_{ji}(-\boldsymbol{B})$ (see Section 7.9 for details). Then for a small magnetic field, one can present the antisymmetric part as $\sigma_{ij}^{(a)} = \beta \epsilon_{ijk} B_k$, where β is a constant, and ϵ_{ijk} is the unit antisymmetric tensor (the *Levi-Civita symbol*). Tensor $\sigma_{ij}^{(a)}$ is antisymmetric, linear in $\boldsymbol{B}$, and satisfies Onsager's principle. Besides, after multiplying it by $\boldsymbol{E}$, we get the polar vector as it should be for current.[3]

In a homogeneous medium, we get the general expression,

$$\boldsymbol{j} = \boldsymbol{j}_0 + \boldsymbol{j}_B = \sigma_0 \boldsymbol{E} + \beta \boldsymbol{B} \times \boldsymbol{E}. \tag{7.6}$$

---

[1] We use the rule of summation over repeated indices $a_i b_i \equiv \sum_i a_i b_i$.

[2] As we discuss below, in the inhomogeneous case, there is one more contribution to the current caused by the diffusion of electrons and holes.

[3] Antisymmetric tensor $\sigma_{ij}^{(a)}$ is equivalent to an axial vector $a_i = \epsilon_{ijk} \sigma_{jk}^{(a)}$ (this is called *duality*). Correspondingly, the product of vectors $\boldsymbol{a}$ and $\boldsymbol{B}$ gives the polar vector.

Here the first term presents the Ohm's law with dissipative longitudinal current $j_0 = \sigma_0 E$ determined by usual conductivity σ_0. The second term is the transverse current j_B, dissipationless as it flows in the direction perpendicular to the electric field (*Hall current*).

7.4 Drude Theory

The Drude theory relates the transport properties of semiconductors with the motion of a single electron under electric and magnetic fields in a continuous medium. The classical equation of motion reads

$$m^* \frac{d\mathbf{v}}{dt} = e\mathbf{E} - \frac{m^*\mathbf{v}}{\tau}, \tag{7.7}$$

where $\mathbf{v}$ is the electron velocity and τ is the *electron relaxation time* – the time in which an electron loses its velocity due to scattering against defects, impurities, and phonons after $\mathbf{E}$ turns zero. The second term mimics the resistive force due to scattering.

The stationary state is the motion with constant velocity $\mathbf{v} = e\tau\mathbf{E}/m^*$. Quantity $\mu_n = e\tau/m^*$ is called the *electron mobility* as it relates the drift electron velocity with an applied electric field: $\mathbf{v} = \mu_n \mathbf{E}$. It is convenient to introduce the *mean free path* $\ell = v\tau$, which is the mean distance at which the electron does not substantially change the direction of motion. If one assumes that drift velocity $\boldsymbol{v}$ is the same for all electrons, we obtain the expression for current density,

$$j = en\mathbf{v} = \frac{ne^2\tau}{m^*}\mathbf{E}, \tag{7.8}$$

where n is the electron density. Correspondingly, the static electric conductivity is

$$\sigma_0 = \frac{ne^2\tau}{m^*} = en\mu_n. \tag{7.9}$$

Conductivity (7.9) depends on the density of all electrons. In degenerate semiconductors at $T \to 0$, only electrons near the Fermi level contribute to the conductivity. Nevertheless, as will be shown below by more rigorous calculations, expression (7.9) is exact.

Let us consider the periodic in time electric field $\mathbf{E}(t) = \mathbf{E}_0 \exp(-i\omega t)$ where E_0 is the field amplitude and ω is the frequency. Then solving Eq. (7.7) with $\mathbf{v}(t) = \mathbf{v}(\omega)\exp(-i\omega t)$, we find

$$\sigma(\omega) = \frac{\sigma_0}{1 - i\omega\tau}, \tag{7.10}$$

which is the conductivity in the *Drude model*.

Macroscopic electrodynamics relates conductivity to dielectric function as

$$\varepsilon(\omega) = \varepsilon + \frac{i\sigma(\omega)}{\omega\varepsilon_0}, \tag{7.11}$$

so that Im $\sigma(\omega)$ contributes to the frequency-dependent dielectric constant. For details, see Chapter 12.

7.5 Hall Effect

Simultaneous action of electric and magnetic fields results in the *Hall effect*. One can describe the effect using the Drude model by adding the Lorentz force (7.1) to Eq.(7.7):

$$m^* \frac{d\mathbf{v}}{dt} = e\mathbf{E} + e\mathbf{v} \times \mathbf{B} - \frac{m^*\mathbf{v}}{\tau}, \tag{7.12}$$

Hall setting implies that the electric field is in the (x, y)-plane, and the magnetic field is along axis z. In components, Eq. (7.12) has the form

$$m^* \frac{dv_x}{dt} + \frac{m^* v_x}{\tau} = eE_x + ev_y B, \tag{7.13}$$

$$m^* \frac{dv_y}{dt} + \frac{m^* v_y}{\tau} = eE_y - ev_x B. \tag{7.14}$$

In the stationary regime, $dv_x/dt = dv_y/dt = 0$, we find

$$v_x = \frac{e\tau}{m^*} \frac{E_x + \omega_c \tau E_y}{1 + \omega_c^2 \tau^2} \tag{7.15}$$

$$v_y = \frac{e\tau}{m^*} \frac{E_y - \omega_c \tau E_x}{1 + \omega_c^2 \tau^2}, \tag{7.16}$$

where $\omega_c = eB/m^*$ is the cyclotron frequency.

In a homogeneous semiconductor, driving electric field in the x-direction induces the longitudinal ohmic current in the same direction. Since the drift in the y-direction is absent, we have to put $v_y = 0$ in Eqs. (7.15) and (7.16). The electric field $E_y \neq 0$ (the Hall field) results from accumulated charges at the sample side boundaries. Previously discussed dissipationless Hall current flows in the transverse to driving current direction until field E_y stops the process. Lorenz force is at the origin of the effect.

The solution of Eqs. (7.15) and (7.16) with $v_y = 0$ gives

$$v_x = \frac{e\tau E_x}{m^*}, \tag{7.17}$$

$$E_y = \omega_c \tau E_x. \tag{7.18}$$

Linear velocity-field relation (7.17) can be rewritten as $v_x = \mu_n E_x$, where $\mu_n = e\tau/m^*$ is the electron mobility. For the transverse electric field, we obtain

$$E_y = RBj_x, \tag{7.19}$$

where $j_x = \sigma_0 E_x$, $R = 1/en$ is the *Hall constant*. By measuring j_x and E_y, one can find the Hall constant and, correspondingly, the electron density n. The geometry of the Hall effect is illustrated in Figure 7.1.

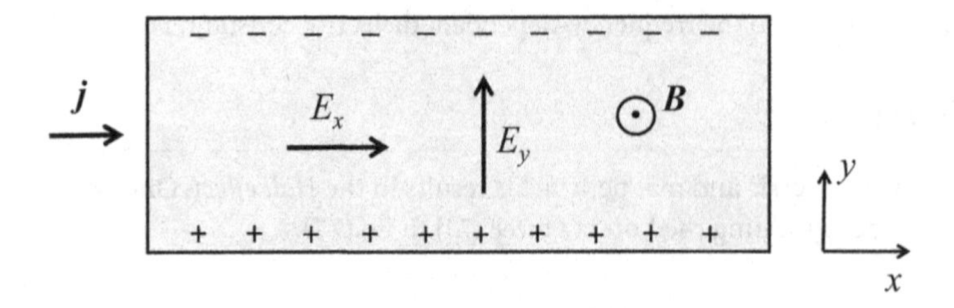

FIGURE 7.1 The geometry of the Hall effect.

Hall effect is the widely used method of semiconductor sample characterization. Indeed, if we know the resistivity and Hall constant in a semiconductor with predominantly one type of carrier, we can determine the density and mobility of electrons or holes.

In other geometry, the electric field is along the x-direction, and current may flow in both x- and y-directions. In this setting, we should put $E_y = 0$ in Eqs. (7.15) and (7.16). It results in longitudinal and transverse currents,

$$j_x = \sigma_0 E \frac{1}{1+\omega_c^2\tau^2}, \tag{7.20}$$

$$j_y = -\sigma_0 E \frac{\omega_c\tau}{1+\omega_c^2\tau^2}. \tag{7.21}$$

Thus, the longitudinal conductivity becomes

$$\sigma = \frac{\sigma_0}{1+\omega_c^2\tau^2}. \tag{7.22}$$

As follows from Eq. (7.22), the longitudinal conductivity falls with an increasing magnetic field, which is called the *magnetoresistance effect.*

So far, we assumed all electrons are of the same velocity because the Drude theory does not account for the energy distribution of carriers and thus for a relative contribution of electrons with different energies to transport characteristics. Therefore, the results obtained above do not always describe correctly observable transport properties. Nevertheless, the Drude theory helps enormously to find a simple explanation of various electron transport phenomena.

7.6 Thermoelectric and Thermo-Electromagnetic Effects

In this section, we will give a phenomenological description of thermoelectric effects in semiconductor samples with the temperature gradient.

If the temperature difference between sample ends is small, $|\Delta T| \ll \varepsilon/k_B$, one can use the linear response approach to determine the temperature gradient-induced charge (and energy) transport properties. Here ε is the characteristic electron energy, $\varepsilon \simeq k_B T$ for nondegenerate or $\varepsilon \simeq \mu$ for degenerate carriers.

Seebeck effect. As follows from semiconductor statistics, the higher temperature, the larger carrier density. Thus, in a homogeneous semiconductor, the temperature gradient induces diffusion of carriers in the opposite to the gradient direction. For a small perturbation, the diffusion flux is proportional to the gradient. Diffusion fluxes of charged carriers produce an electric field (*Seebeck effect*):

$$\boldsymbol{E} = S\nabla T, \tag{7.23}$$

where S is the *Seebeck coefficient* $[V/T]$.

Thermal conductivity. The transport of particles under the temperature gradient is accompanied by the transport of heat from hot to cold regions. Heat transfer occurs by various mechanisms and different particles (electrons, phonons, etc.). Generally, the heat flux, heat amount per unit square per second, can be written as $\boldsymbol{j}_Q = -\kappa\ \nabla T, \kappa$ is the thermal conductivity [W/Km]. In Chapter 6, we discussed the phonon part of thermal conductivity.

If flux $\boldsymbol{j}_Q$ depends on coordinates, the in-coming and out-coming powers are not equal, and the power generated per unit volume is $dQ/dt = \nabla\cdot(\kappa\nabla T)$.

Thomson effect. If the electric current is parallel to the temperature gradient, the power produced per unit volume is $dQ/dt = K_T\boldsymbol{j}\cdot\nabla T$, K_T is the *Thomson constant*, $[V/T]$. Linear in $\boldsymbol{j}$ dependence means that heating or cooling depends on mutual directions of current and gradient. So, in the simultaneous presence of electric current and temperature gradient, the power dissipates due to Joule heat, thermal conductivity, and *Thomson effect*:

$$\frac{dQ}{dt} = \frac{j^2}{\sigma} + \nabla \cdot (\kappa \nabla T) + K_T \ \boldsymbol{j} \cdot \nabla T. \tag{7.24}$$

Peltier effect. Not only the temperature gradient, but the electric current also causes heat transfer. Heat flux, induced by the electric current, is referred to as the *Peltier effect*:

$$\boldsymbol{j}_Q = \Pi \boldsymbol{j}, \tag{7.25}$$

where Π is the *Peltier coefficient*, $[V]$. If we have a thermocouple, the junction of two materials with Peltier coefficients Π_A and Π_B, the difference in heat transport through each of them generates power per unit area of the contact, $dQ / dt = (\Pi_A - \Pi_B) j$.

One more class of effects – the *transverse* thermo-electromagnetic effects – includes an external magnetic field in combination with temperature gradient and electric field.

Nernst effect. For example, the temperature gradient along axis $\boldsymbol{x}$ combined with magnetic field along z gives the electric field along axis y

$$E_y = NB_z \ \nabla_x T, \tag{7.26}$$

where N is the *Nernst constant*. This effect can be considered as a Hall effect, in which temperature gradient generates the driving current.

Righi-Leduc effect. The x-component of the temperature gradient and the magnetic field in the z-direction generate the temperature gradient in the y-direction:

$$\frac{dT}{dy} = A_{RL} B_z \nabla_x T \tag{7.27}$$

where A_{RL} is the *Righi-Leduc coefficient*.

Thermoelectric coefficients introduced above are related to each other as they all originate from kinetic coefficients which obey Onsager relations. The relations follow from thermodynamic considerations and thus have general nature irrespective of microscopic mechanisms. More details will follow.

7.7 Kinetic Equation

The method of the kinetic equation allows us to determine the nonequilibrium distribution function of electrons $f(\boldsymbol{r},\boldsymbol{k},t)$.[4] This function describes the probability of finding an electron in state $\boldsymbol{k}$ in point $\boldsymbol{r}$ in the time moment t. In its general form, the kinetic equation looks as follows:

$$\frac{df}{dt} = St \ f, \tag{7.28}$$

where the left-hand side is the total derivative, which takes into account the dependence of both $\boldsymbol{r}$ and $\boldsymbol{k}$ on time. The right-hand side, the collision integral (from German word *Stoß*, in English – *Collision*), accounts for all processes of electron scattering that change the electron state: impurities and defects,

[4] In this section, we omit the electron spin variable. If we accounted for the electron spin, the distribution function would be a matrix in spin space.

phonon absorption and emission, electron–electron interactions, and so on. For definiteness, in this section, we consider the scattering of electrons from impurities.

Calculating the total derivative in Eq. (7.28), then introducing quantum-mechanical probability per second $w(\boldsymbol{k},\boldsymbol{k}')$ of impurity scattering between $\boldsymbol{k}$- and $\boldsymbol{k}'$-states, we obtain

$$\frac{\partial f(\boldsymbol{k})}{\partial t}+\mathbf{v}\cdot\nabla f(\boldsymbol{k})+\frac{e}{\hbar}(\boldsymbol{E}+\mathbf{v}\times\boldsymbol{B})\nabla_k f(\boldsymbol{k})=-\sum_{k'}\{w(\boldsymbol{k},\boldsymbol{k}')f(\boldsymbol{k})[1-f(\boldsymbol{k}')] \\ -[w(\boldsymbol{k}',\boldsymbol{k})f(\boldsymbol{k}')][1-f(\boldsymbol{k})]\}, \tag{7.29}$$

where $\mathbf{v}=\partial \boldsymbol{r}/\partial t$ is the electron velocity, and the Lorentz force $\boldsymbol{F}=e\boldsymbol{E}+e\mathbf{v}\times\boldsymbol{B}$ comes from the equation of motion $\hbar(\partial\boldsymbol{k}/\partial t)=\boldsymbol{F}$. The first term in the right-hand is scattering $\boldsymbol{k}\to\boldsymbol{k}'$. The collision integral is proportional to the number of filled states $\boldsymbol{k}$ and the number of unoccupied states $\boldsymbol{k}'$. The second term on the right-hand side is scattering $\boldsymbol{k}'\to\boldsymbol{k}$ that increases $f(\boldsymbol{k})$ due to scattering from all $\boldsymbol{k}'$ states.

As $\boldsymbol{r}$ and $\boldsymbol{k}$ are assumed independent variables, collision-induced change of state $\boldsymbol{k}\to\boldsymbol{k}'$ does not alter the particle position. Equation (7.29) holds in a particular spatial point $\boldsymbol{r}$. Strictly speaking, this contradicts quantum mechanics (uncertainty principle) and may be accepted only as an approximation – the semiclassical approximation to electron transport. Still, the kinetic equation accounts for the quantum nature of electrons through both band structure and quantum-mechanical transition probability in the collision integral. The approximation is valid if external fields change negligibly on a length scale of the mean distance between collisions, which, in turn, is greater than the size of the electron wave packet (thermal or Fermi wavelength). The semiclassical description does not apply to quantum wells and other nanostructures with size bound to the scale of the wavelength of the electron.

By the *principle of detailed balance* related to the time inversion symmetry [1], the probability of scattering with $\boldsymbol{k}\to\boldsymbol{k}'$ is equal to the probability of the inverse process, $w(\boldsymbol{k},\boldsymbol{k}')=w(\boldsymbol{k}',\boldsymbol{k})$. Then the kinetic Eq. (7.29) acquires the form,

$$\frac{\partial f(\boldsymbol{k})}{\partial t}+\mathbf{v}\cdot\nabla f(\boldsymbol{k})+\frac{e}{\hbar}(\boldsymbol{E}+\mathbf{v}\times\boldsymbol{B})\cdot\nabla_k f(\boldsymbol{k})=-\sum_{k'}w(\boldsymbol{k},\boldsymbol{k}')[f(\boldsymbol{k})-f(\boldsymbol{k}')]. \tag{7.30}$$

To calculate function $w(\boldsymbol{k},\boldsymbol{k}')$, one needs the scattering theory of quantum mechanics. In the Born approximation, we have

$$w(\boldsymbol{k},\boldsymbol{k}')=\frac{2\pi n_i}{\hbar\Omega}|V_{k-k'}|^2\,\delta(\varepsilon_k-\varepsilon_{k'}), \tag{7.31}$$

where n_i is the impurity density, and $V_{k-k'}$ is the Fourier transform of single-impurity potential $V(\boldsymbol{r})$ (see Chapter 3).

In the absence of external fields, the solution of the kinetic equation is the equilibrium Fermi-Dirac distribution function:

$$f_0(\boldsymbol{k})=\frac{1}{\exp\left(\frac{\varepsilon_k-\mu}{k_BT}\right)+1}, \tag{7.32}$$

where $\varepsilon_k=\hbar^2k^2/2m^*$. If the perturbation related to the electric and magnetic fields is small, one can present the distribution function as $f(\boldsymbol{k})=f_0(\boldsymbol{k})+\delta f(\boldsymbol{k})$, $|\delta f(\boldsymbol{k})|\ll f_0(\boldsymbol{k})$.

Let us consider the steady-state solution to Eq. (7.30) as a response to weak external fields. Also, we apply a small temperature gradient, which makes the distribution function depending on $\boldsymbol{r}$. One can account for inhomogeneity by assuming T in Eq. (7.32) to be a smooth function of $\boldsymbol{r}$. It gives us

$$\nabla f(\boldsymbol{k}) \simeq -\frac{(\varepsilon_k - \mu)\,\nabla T}{T}\frac{\partial f_0(k)}{\partial \varepsilon_k} \tag{7.33}$$

The third term in Eq. (7.30) presents a weak perturbation caused by electric and magnetic fields. After substituting $f_0(\boldsymbol{k}) + \delta f(\boldsymbol{k})$ into (7.30), the first nonzero term proportional to electric field contains equilibrium distribution function,

$$\nabla_{\boldsymbol{k}} f(\boldsymbol{k}) \to \frac{\hbar^2 \boldsymbol{k}}{m^*}\frac{\partial f_0(\boldsymbol{k})}{\partial \varepsilon_k}, \tag{7.34}$$

Proportional to $\boldsymbol{B}$ term in Eq. (7.30) turns zero with $f(\boldsymbol{k}) = f_0(\boldsymbol{k})$, so we have to keep the first-order correction to the distribution function. As a result, we obtain the kinetic equation for steady state,

$$\begin{aligned} &-\frac{\hbar(\varepsilon_k - \mu)\,\boldsymbol{k}\cdot\nabla T}{m^* T}\frac{\partial f_0(\boldsymbol{k})}{\partial \varepsilon_k} + \frac{e\hbar\boldsymbol{k}\cdot\boldsymbol{E}}{m^*}\frac{\partial f_0(\boldsymbol{k})}{\partial \varepsilon_k} + \frac{e}{m^*}(\boldsymbol{k}\times\boldsymbol{B})\cdot\nabla_{\boldsymbol{k}}\delta f(\boldsymbol{k}) \\ &= -\sum_{\boldsymbol{k}'} w(\boldsymbol{k},\boldsymbol{k}')\left[f(\boldsymbol{k}) - f(\boldsymbol{k}')\right]. \end{aligned} \tag{7.35}$$

We will be looking for a solution of Eq. (7.35) in the form $\delta f(\mathbf{k}) = \mathbf{k}\cdot\mathbf{g}(\mathbf{k})$ with a vector function $\mathbf{g}(\mathbf{k})$ depending on $\mathbf{k} = |\mathbf{k}|$. Then the right-hand side of (7.35) becomes

$$\begin{aligned} St\,f &= -\sum_{k'} w(\mathbf{k},\mathbf{k}')\,(\mathbf{k}-\mathbf{k}')\cdot\mathbf{g}(\mathbf{k}) \\ &= -\mathbf{k}\cdot\mathbf{g}(\mathbf{k})\sum_{k'} w(\mathbf{k},\mathbf{k}')\left(1 - \frac{\mathbf{k}\cdot\mathbf{k}'}{\mathbf{k}^2}\right) = -\frac{\delta f(\mathbf{k})}{\tau(\mathbf{k})}, \end{aligned} \tag{7.36}$$

where we introduced notation,

$$\frac{1}{\tau(\boldsymbol{k})} = \sum_{k'} w(\boldsymbol{k},\boldsymbol{k}')\,(1-\cos\theta), \tag{7.37}$$

and θ is the angle between $\boldsymbol{k}$ and $\boldsymbol{k}'$. Parameter $\tau(\boldsymbol{k})$ determines various transport properties and called *transport relaxation time* or *momentum relaxation time*. Without factor $(1-\cos\theta)$ in Eq. (7.37), value $\sum_{k'} w(\boldsymbol{k},\boldsymbol{k}')$ is the probability of electron transition from the $\boldsymbol{k}$-state to any other state with the same energy. Factor $(1-\cos\theta)$ makes scattering to small angles less important for conductivity.

Note that in the general case (e.g., when multiple mechanisms of electron relaxation take place), the solution to the kinetic equation can present a difficult problem. Nevertheless, expressing the collision integral in the *τ-approximation*, $St\,f = -\delta f/\tau$, one substantially simplifies the problem. Within the semiclassical approximation, one can neglect k-derivatives of $\boldsymbol{g}(k)$ (i.e., $kg' \ll g$) in Eq. (7.36). Then we have $\nabla_{\boldsymbol{k}}\delta f(\boldsymbol{k}) = \boldsymbol{g}(k)$, and the equation for $\boldsymbol{g}(k)$ reads (here we can omit the arguments in the functions)

$$-\frac{\hbar\,(\varepsilon_k - \mu)\,\nabla T}{m^* T} f_0' + \frac{e\hbar E}{m^*} f' + \frac{e}{m^*}(\boldsymbol{B}\times\boldsymbol{g}) = -\frac{\boldsymbol{g}}{\tau}. \tag{7.38}$$

Components of vector function $\boldsymbol{g}$ depend on directions of ∇T, $\boldsymbol{E}$, and $\boldsymbol{B}$. In the geometry $\nabla T, \boldsymbol{E} \| \boldsymbol{x}$, and $\boldsymbol{B} \| \boldsymbol{z}$, we obtain

$$-\frac{\hbar\left(\varepsilon_k-\mu\right)\nabla T}{m^* T} f_0'+\frac{e\hbar E}{m^*} f_0'-\frac{eB}{m^*} g_y=-\frac{g_x}{\tau}, \tag{7.39}$$

$$\frac{eB}{m^*} g_x=-\frac{g_y}{\tau}, \tag{7.40}$$

and $g_z=0$. Solving (7.39) and (7.40), we get

$$g_x=\frac{\hbar\left(\varepsilon_k-\mu\right)\tau\nabla T}{m^* T\left(1+\omega_c^2\tau^2\right)} f_0'-\frac{e\hbar E\tau}{m^*\left(1+\omega_c^2\tau^2\right)} f_0', \tag{7.41}$$

$$g_y=-\frac{\hbar\omega_c\tau^2\left(\varepsilon_k-\mu\right)\nabla T}{m^* T\left(1+\omega_c^2\tau^2\right)} f_0'+\frac{e\hbar\omega_c\tau^2 E}{m^*\left(1+\omega_c^2\tau^2\right)} f_0', \tag{7.42}$$

where $\omega_c=eB/m^*$ is the cyclotron frequency. Finally, we obtained the nonequilibrium correction to the distribution function,

$$\delta f(\boldsymbol{k})=k_x g_x+k_y g_y, \tag{7.43}$$

where $g_{x,y}(k)$ are determined by Eqs. (7.41) and (7.42).

7.8 Kinetic Coefficients

Using the solution of the kinetic equation, one can calculate the electric-field- or temperature-gradient-induced transport of the electric charge and heat, as well as the effect of magnetic field on transport characteristics. For example, to calculate the electric current, we use the following expression:

$$\boldsymbol{j}=\frac{2e}{\Omega}\sum_{\boldsymbol{k}}\mathbf{v}(\boldsymbol{k})f(\boldsymbol{k})=\frac{2e\hbar}{\Omega m^*}\sum_{\boldsymbol{k}}\boldsymbol{k}\delta f(\boldsymbol{k}) \tag{7.44}$$

where Ω is the crystal volume. Equation (7.44) states that an electron in the state with wavevector $\boldsymbol{k}$ is transmitting current density $e\mathbf{v}(\boldsymbol{k})/\Omega$, where $\mathbf{v}(\boldsymbol{k})=\hbar\boldsymbol{k}/m^*$ is the electron velocity. Factor 2 is due to the spin degeneration.

In the absence of the magnetic field and temperature gradient, the electron current flows along the electric field. Using (7.43) and (7.41), we get

$$j_x=\frac{2e^2\hbar^2 E}{m^{*2}}\int\frac{d^3k}{(2\pi)^3}k_x^2\tau(k)\left(-\frac{\partial f_0}{\partial\varepsilon_k}\right)=\frac{2e^2E}{3m^*}\int d\varepsilon\,\rho(\varepsilon)\,\varepsilon\tau(\varepsilon)\left(-\frac{\partial f_0}{\partial\varepsilon}\right), \tag{7.45}$$

where $\rho(\varepsilon)$ is the density of states (DOS), which takes into account spin degeneration (see Eq. (3.12), Chapter 3). With (7.45), one can calculate conductivity $\sigma=j_x/E$ once the scattering problem is solved, and thus, the relaxation time as a function of the electron energy is known.

If the magnetic field along axis z is not zero and $\nabla T=0$, then using (7.41)–(7.43), we find

$$j_x=\frac{2e^2\hbar^2 E}{m^{*2}}\int\frac{d^3k}{(2\pi)^3}\frac{k_x^2\,\tau}{1+\omega_c^2\tau^2}\left(-\frac{\partial f_0}{\partial\varepsilon_k}\right), \tag{7.46}$$

$$j_y = -\frac{2e^2\hbar^2 E\omega_c}{m^{*2}} \int \frac{d^3k}{(2\pi)^3} \frac{k_y^2 \tau^2}{1+\omega_c^2\tau^2} \left(-\frac{\partial f_0}{\partial \varepsilon_k}\right). \tag{7.47}$$

Similarly, one can calculate temperature-gradient-induced heat flux:

$$j_Q = \frac{2}{\Omega} \sum_k \varepsilon_k \mathbf{v}(\boldsymbol{k}) \delta f(\boldsymbol{k}) \tag{7.48}$$

Assuming $\boldsymbol{E} = \boldsymbol{B} = 0$ and using Eq. (7.41), we find from Eq. (7.48) the heat flux along axis x:

$$j_Q = \frac{2\hbar^2 \nabla T}{m^{*2} T} \int \frac{d^3k}{(2\pi)^3} k_x^2 \tau(k)\, \varepsilon_k\, (\varepsilon_k - \mu) \left(-\frac{\partial f_0}{\partial \varepsilon_k}\right)$$

$$= \frac{2\, \nabla T}{3m^* T} \int d\varepsilon\, \rho(\varepsilon)\, \varepsilon^2\, (\varepsilon - \mu)\, \tau(\varepsilon) \left(-\frac{\partial f_0}{\partial \varepsilon}\right). \tag{7.49}$$

In arbitrarily oriented $\boldsymbol{E}$, $\boldsymbol{B}$, and ∇T, it is convenient to use the so-called *generalized kinetic coefficients* defined as

$$K_{ij}^{rs} = \int d\varepsilon\, \rho(\varepsilon) \frac{\mathrm{v}_i \mathrm{v}_j\, \varepsilon^{r-1}\, \tau^s}{1+\omega_c^2\tau^2} \left(-\frac{\partial f_0}{\partial \varepsilon}\right). \tag{7.50}$$

Tensor K_{ij}^{rs} is symmetric on i, j-components, r, s are integers. Using the kinetic coefficients, after rather cumbersome calculations, one can express current and heat flux as follows:

$$j_i = eK_{ij}^{11}\left(eE_j + \frac{\mu\, \nabla_j T}{T}\right) - \frac{e}{T} K_{ij}^{21} \nabla_j T + \frac{e^2}{m^*} K_{ij}^{12} \left[\left(e\boldsymbol{E} + \frac{\mu\, \nabla T}{T}\right) \times \boldsymbol{B}\right]_j$$

$$- \frac{e^2}{m^* T} K_{ij}^{22}\, (\nabla T \times \boldsymbol{B})_j + \frac{e^3}{m^{*2}} \left[\left(e\boldsymbol{E} + \frac{\mu \nabla T}{T}\right) \cdot \boldsymbol{B}\right] K_{ij}^{13} B_j - \frac{e^3 (\nabla T \cdot \boldsymbol{B})}{m^{*2} T} K_{ij}^{23} B_j, \tag{7.51}$$

$$j_{Qi} = K_{ij}^{21}\left(eE_j + \frac{\mu\, \nabla_j T}{T}\right) - \frac{1}{T} K_{ij}^{31} \nabla_j T + \frac{e}{m^*} K_{ij}^{22} \left[\left(e\boldsymbol{E} + \frac{\mu\, \nabla T}{T}\right) \times \boldsymbol{B}\right]_j$$

$$- \frac{e}{m^* T} K_{ij}^{32}\, (\nabla T \times \boldsymbol{B})_j + \frac{e^2}{m^{*2}} \left[\left(e\boldsymbol{E} + \frac{\mu \nabla T}{T}\right) \cdot \boldsymbol{B}\right] K_{ij}^{23} B_j - \frac{e^2 (\nabla T \cdot \boldsymbol{B})}{m^{*2} T} K_{ij}^{33} B_j. \tag{7.52}$$

These equations contain various thermoelectric and thermo-magnetoelectric effects in which electric and thermal currents appear as a result of electric field and temperature gradient in the presence of an external magnetic field.

To calculate the electrical conductivity from Eq. (7.51), we put $\boldsymbol{B} = 0$ and $\nabla T = 0$, which gives us $j = \sigma E$ with

$$\sigma = \frac{2e^2}{3m^*} \int d\varepsilon \rho(\varepsilon)\, \varepsilon \tau(\varepsilon) \left(-\frac{\partial f_0}{\partial \varepsilon}\right). \tag{7.53}$$

Let us assume $\tau = \text{const}$. In a nondegenerate semiconductor, $\varepsilon - \mu \gg k_B T$, we can use the classical distribution $f_0(\varepsilon) \simeq \exp\left[-(\varepsilon - \mu)/k_B T\right]$ and obtain

$$\sigma = \frac{2e^2\tau \exp\left(\mu/k_BT\right)}{3m^* k_BT} \int d\varepsilon \rho(\varepsilon)\, \varepsilon \exp\left(-\frac{\varepsilon}{k_BT}\right)$$

$$= \frac{2e^2\tau \left(k_BT\right)^{3/2} \rho_0 \exp\left(\mu / k_BT\right)}{3m^*} \int_0^\infty dx\, x^{3/2} \exp(-x) = \frac{2e^2 n\tau}{3m^*} \frac{\Gamma\left(\frac{5}{2}\right)}{\Gamma\left(\frac{3}{2}\right)} = \frac{e^2 n\tau}{m^*} \tag{7.54}$$

where $\Gamma(x)$ is the Gamma-function.[5] Here we denoted $\rho(\varepsilon) = \rho_0 \varepsilon^{1/2}$ and used the electron density

$$n = \int d\varepsilon \rho(\varepsilon) \exp\left(-\frac{\varepsilon - \mu}{k_BT}\right) = \exp\left(\frac{\mu}{k_BT}\right) \rho_0 \int_0^\infty d\varepsilon\, \varepsilon^{1/2} \exp\left(-\frac{\varepsilon}{k_BT}\right). \tag{7.55}$$

Conductivity (7.54) coincides with the result of the Drude theory (7.9).

7.9 Symmetry of Kinetic Coefficients: Onsager's Principle

The electric current and heat flux have general forms:

$$j_i = \sigma_{ij} E_j - \alpha_{ij} \nabla_j T, \tag{7.56}$$

$$\tilde{j}_{Qi} = \beta_{ij} E_j - \kappa_{ij} \nabla_j T, \tag{7.57}$$

where we denote $\tilde{j}_Q = j_Q - \varphi \boldsymbol{j}$ and $\varphi = \mu/e$. This way, we eliminate from the heat flux the part related to the energy transport by electric current $\boldsymbol{j}$. Phenomenological coefficients σ_{ij}, β_{ij}, κ_{ij}, and α_{ij} denote tensors of conductivity, Seebeck coefficients, heat conductivity, and thermoelectric tensor, respectively. The explicit forms of these tensors depend on material parameters and an external magnetic field. In Eqs. (7.56) and (7.57), both the electric current and heat flux are of the first order in $\boldsymbol{E}$ and ∇T as prescribed by the linear response theory.

Making use of (7.51) and (7.52), one can present the tensors through combinations of kinetic coefficients K_{ij}^{rs} and the magnetic field components. Note that Eq. (7.52) describes the heat flux transferred by electrons, but there is also heat transport due to the phonons. The heat conductivity includes both electron and phonon contributions, $\kappa_{ij} = \kappa_{ij}^{(e)} + \kappa_{ij}^{(ph)}$.

Onsager's principle establishes some symmetry properties and relations between the tensor coefficients in Eqs. (7.56) and (7.57). This principle grounds on the thermodynamics of fluctuations at a small deviation from equilibrium. First, we introduce a set of macroscopic parameters x_i, which characterize a nonequilibrium state and equal zero in equilibrium. On the way to equilibrium, the relaxation rate undergoes the linear response equation,

$$\dot{x}_i = -\gamma_{ij} X_j, \tag{7.58}$$

where γ_{ij} are the *Onsager kinetic coefficients* and X_i are thermodynamically conjugate quantities (generalized forces),

$$X_i = -\frac{\partial S}{\partial x_i}, \tag{7.59}$$

[5] $\Gamma(z) \equiv \int_0^\infty dx\, x^{z-1} e^{-x}$. $\Gamma(z+1) = z\Gamma(z)$.

where S is the entropy. Onsager's principle states that if the quantities x_i and x_j behave in the same way under time reversal, then tensor γ_{ij} should be symmetric, $\gamma_{ij}=\gamma_{ji}$. If one of these quantities changes sign and the other remains unchanged under time reversal, then $\gamma_{ij}=-\gamma_{ji}$. If these quantities depend on magnetic field $\boldsymbol{B}$, then $\gamma_{ij}(\boldsymbol{B})=\gamma_{ji}(-\boldsymbol{B})$.

To apply this principle to kinetic coefficients in Eqs. (7.56) and (7.57) we have to identify the parameters in these equations as x_i and X_i in Onsager's theory. Identification occurs with the help of an equation for entropy production, which in terms of Onsager's parameters has the form

$$\frac{\partial S}{\partial t}=-\int_{\Omega} d^3 r X_i \frac{\partial x_i}{\partial t}, \tag{7.60}$$

where Ω is the crystal volume.

Now, turning to the problem of heat transport, we express the heat dissipated per unit volume per unit time as

$$\delta Q=-\,div\, \boldsymbol{j}_Q=-\,div\left(\boldsymbol{j}_Q-\varphi \boldsymbol{j}\right)-div\left(\varphi \boldsymbol{j}\right)=-div\, \tilde{\boldsymbol{j}}_Q+\boldsymbol{E}\cdot \boldsymbol{j}, \tag{7.61}$$

where we used $div\, \boldsymbol{j}=0$ and $\boldsymbol{E}=-\nabla\varphi$. Correspondingly, the entropy production reads as

$$\frac{\partial S}{\partial t}=\int_{\Omega} d^3 r\left(-\frac{div\, \tilde{\boldsymbol{j}}_Q}{T}+\frac{\boldsymbol{j}\cdot \boldsymbol{E}}{T}\right)=\int_{\Omega} d^3 r\left(-\frac{\tilde{\boldsymbol{j}}_Q\cdot \nabla T}{T^2}+\frac{\boldsymbol{j}\cdot \boldsymbol{E}}{T}\right), \tag{7.62}$$

where we integrated by parts the first term and assumed the heat flux disappears at the sample surface. Comparing (7.62) with (7.60), one concludes that for the electric current part, the choice is $\partial x_i/\partial t=j_i$ and $X_i=-E_i/T$, while for the heat flux part, $\partial x_i/\partial t=\tilde{j}_{Qi}$ and $X_i=\nabla_i T/T^2$. Then, Eqs. (7.56) and (7.57) can be unified as follows:

$$\begin{pmatrix} j_x \\ j_y \\ j_z \\ \tilde{j}_{Qx} \\ \tilde{j}_{Qy} \\ \tilde{j}_{Qz} \end{pmatrix}=\begin{pmatrix} T\sigma_{xx} & T\sigma_{xy} & T\sigma_{xz} & T^2\alpha_{xx} & T^2\alpha_{xy} & T^2\alpha_{xz} \\ T\sigma_{yx} & T\sigma_{yy} & T\sigma_{yz} & T^2\alpha_{yx} & T^2\alpha_{yy} & T^2\alpha_{yz} \\ T\sigma_{xx} & T\sigma_{xy} & T\sigma_{xz} & T^2\alpha_{zx} & T^2\alpha_{zy} & T^2\alpha_{zz} \\ T\beta_{xx} & T\beta_{xy} & T\beta_{xz} & T^2\kappa_{xx} & T^2\kappa_{xy} & T^2\kappa_{xz} \\ T\beta_{yx} & T\beta_{yy} & T\beta_{yz} & T^2\kappa_{yx} & T^2\kappa_{yy} & T^2\kappa_{yz} \\ T\beta_{zx} & T\beta_{zy} & T\beta_{zz} & T^2\kappa_{zx} & T^2\kappa_{zy} & T^2\kappa_{zz} \end{pmatrix}\begin{pmatrix} -E_x/T \\ -E_y/T \\ -E_z/T \\ \nabla_x T/T^2 \\ \nabla_y T/T^2 \\ \nabla_z T/T^2 \end{pmatrix} \tag{7.63}$$

and due to Onsager's principle, we obtain

$$\sigma_{ij}(\boldsymbol{B})=\sigma_{ji}(-\boldsymbol{B}),\quad \kappa_{ij}(\boldsymbol{B})=\kappa_{ji}(-\boldsymbol{B}),\quad \beta_{ij}(\boldsymbol{B})=T\alpha_{ij}(-\boldsymbol{B}). \tag{7.64}$$

Thus, Onsager's principle establishes reciprocity relations in conductivity and heat conductivity tensors and also relates the tensor of Seebeck coefficients to the thermoelectric tensor.

7.10 Macroscopic Equations

Violation of thermal equilibrium in semiconductors creates an inhomogeneous distribution of macroscopic material parameters such as electron and hole density n and p, mobility of electrons and holes μ_n and μ_p, electric conductivity σ, electrostatic potential φ, and chemical potential μ. Coordinate-dependent material parameters also take place in equilibrium semiconductor structures consisting of semiconductors with different properties such as *p-n* junctions. As far as spatial variations are smooth, meaning that characteristic length of variation a is larger than the electron wavelength, the semiclassical description applies. By assuming spatially variable chemical potential, we include in consideration nonequilibrium structures subjected to an applied voltage. Moreover, we assume that the mean free paths of electrons ℓ_n and holes ℓ_p are much less than a. All these assumptions allow us to use the macroscopic equations to calculate carrier concentrations and currents and thus to study the operation of semiconductor devices.

In the absence of magnetic field, after summation kinetic equation (7.30) over $\boldsymbol{k}$, we obtain

$$\frac{\partial}{\partial t}\sum_k f(\boldsymbol{k}) + div \sum_k \mathbf{v}\, f(\boldsymbol{k}) + e\boldsymbol{E}\sum_k \mathbf{v}\,\frac{\partial f_0}{\partial \varepsilon_k} = 0, \tag{7.65}$$

where f_0 depends on the electrostatic potential $\varphi(\boldsymbol{r})$, $f_0(\boldsymbol{k},\boldsymbol{r}) = \left\{\exp\left[(\varepsilon_k - e\varphi)/k_BT\right]+1\right\}^{-1}$. In the third term, we replaced f by f_0, keeping the first order in electric field $\boldsymbol{E}(\boldsymbol{r})$, which we assumed to be weak. After replacing the sum with integral and integrating over angles (directions of vector $\mathbf{v} = \hbar\boldsymbol{k}/m^*$), the third term disappears. The right-hand side of (7.65) is zero since $w(\boldsymbol{k},\boldsymbol{k}') = w(\boldsymbol{k}',\boldsymbol{k})$. As a result, Eq.(7.65) becomes the charge conservation equation

$$\frac{\partial \rho(\mathbf{r})}{\partial t} + \mathrm{div}\,\mathbf{j}(\mathbf{r}) = 0, \tag{7.66}$$

where

$$\rho(\mathbf{r}) = e\, n(\mathbf{r}) = \frac{2e}{\Omega}\sum_k f(\mathbf{k},\mathbf{r}) \tag{7.67}$$

is the charge density and

$$\mathbf{j}(\mathbf{r}) = \frac{2e}{\Omega}\sum_k \mathbf{v}\, f(\mathbf{k},\mathbf{r}) \tag{7.68}$$

is the current density.

To calculate $\boldsymbol{j}(\boldsymbol{r})$ we use the kinetic equation in τ-approximation:

$$\frac{\partial f}{\partial t} + div\,(\mathbf{v}f) + e\,(\mathbf{E}\cdot\mathbf{v})\, f_0' = -\frac{\delta f}{\tau}. \tag{7.69}$$

In the stationary state, one finds

$$\delta f \simeq -\tau\, div(\mathbf{v}f_0) - e\tau(\boldsymbol{E}\cdot\mathbf{v})\, f_0'. \tag{7.70}$$

Substituting δf into (7.68) and integrating over angles, we obtain[6]

$$\mathbf{j} = -\frac{2e\tau}{3\Omega}\sum_k \mathbf{v}^2\nabla f_0 - \frac{2e^2\tau\boldsymbol{E}}{3\Omega}\sum_k \mathbf{v}^2 f_0' \simeq -eD_n\,\nabla n + \sigma\boldsymbol{E}, \tag{7.71}$$

where $D_n = \overline{\mathbf{v}^2}\tau/3$ is the *diffusion coefficient of electrons* and $\overline{\mathbf{v}^2}$ is the mean square of electron velocity. Note that in the kinetic theory of gases, $m^*\overline{\mathbf{v}^2}/2 = 3k_BT/2$, which leads to *Einstein's relation* $eD_n = \mu_n k_BT$,

[6] We use $\mathbf{v}_i = \hbar k_i/m^*$ and the rule for averaging over spherical angles $\int k_i k_j\, d\boldsymbol{n}_k = 4\pi k^2\delta_{ij}/3$, $\boldsymbol{n}_k \equiv \boldsymbol{k}/k$.

where $\mu_n = e\tau / m^*$ is the electron mobility. So, in inhomogeneous semiconductors, the current density consists of two terms: *diffusion current*, $j_{dif} = -D_n \nabla \rho$, which is due to the inhomogeneity of charge distribution, and *drift current* $j_E = \sigma E$ caused by the driving electric field. In ambipolar semiconductors, the total electric current includes contributions from electrons and holes

$$\mathbf{j} = -eD_n \nabla n + eD_p \nabla p + \sigma \mathbf{E}, \tag{7.72}$$

where D_p is the hole diffusion coefficient, and the total conductivity also comprises contributions of electrons and holes: $\sigma = \sigma_n + \sigma_p$.

Within the macroscopic description, carrier distribution functions and thus the concentration of electrons and holes contain $\boldsymbol{r}$-dependent potential energy $e\varphi(\boldsymbol{r})$:

$$n(\mathrm{r}) = \frac{2}{\Omega} \sum_k f_0(\varepsilon_k + e\varphi) \tag{7.73}$$

$$p(\mathbf{r}) = \frac{2}{\Omega} \sum_k \left[1 - f_0\left(\varepsilon_k^h + e\varphi\right)\right], \tag{7.74}$$

where $\varepsilon_k^h = -E_g - \hbar^2 k^2 / 2m_p^*$ is the energy spectrum of electrons in the valence band, E_g is the bandgap, and m_p^* is the effective mass of holes. The chemical potential also depends on r. The electrostatic potential to be determined from the Poisson equation:

$$\Delta\varphi(\mathbf{r}) = -\frac{e}{\varepsilon_r \varepsilon_0}\left[n(\mathbf{r}) - p(\mathbf{r}) + n_a^- - n_d^+\right], \tag{7.75}$$

where n_a^- and n_d^+ are the densities of ionized acceptors and donors, respectively.

In the nonequilibrium state, the variation of chemical potential μ is related to electric current. Assuming semiconductor structure inhomogeneous along $\boldsymbol{x}$ and the voltage applied along the same axis, one writes the current,

$$j_x = \frac{\sigma(x)}{e} \nabla_x \mu(x) \tag{7.76}$$

where j_x is constant and $\sigma(x) = |e|\left[n(x)\mu_n + p(x)\mu_p\right]$. From Eq. (7.76) follows

$$\mu(x) = \mu(x_1) + ej_x \int_{x_1}^{x} \frac{dx}{\sigma(x)}, \tag{7.77}$$

and $\mu(x_2) - \mu(x_1) = eU$, where U is the voltage between points x_2 and x_1.

The system of macroscopic equations discussed in this section creates a background in the theory of semiconductor devices [2].

7.11 Electron Wave Packet in Electric and Magnetic Fields

Transport properties – the response of electrons in solids to external forces – deal with concepts of scattering, mean free path, cyclotron orbit, etc. All these concepts imply that an electron has a trajectory in space, which can be introduced only in a semiclassical theory. The semiclassical approach to electron dynamics does not contradict quantum mechanics if we replace the wave function with a wave packet, which also satisfies the Schrödinger equation. That is the essence of the correspondence principle in quantum mechanics. The center of the electron wave packet has a trajectory, which we call an electron trajectory. An electron as a wave packet has coordinate and momentum – coordinate of the center and the average momentum of the partial waves that make up the packet, respectively. This approach is instrumental in the study of electron dynamics and allows us to relate the nontrivial topology of the Brillouin zone with anomalous and spin Hall effects.

The wave packet localized in spatial point $\boldsymbol{R}$ consists of Bloch functions with $\boldsymbol{k}$ close to the packet momentum $\boldsymbol{Q}$ [3,4]:

$$\psi_{QR}(\boldsymbol{r}) = \sum_{k} C_{QR}(\boldsymbol{k})\, e^{i\boldsymbol{k}\cdot(\boldsymbol{r}-\boldsymbol{R})}\, u_k(\boldsymbol{r}), \quad \int d^3r\, \psi^*_{QR}(\boldsymbol{r})\, \psi_{QR}(\boldsymbol{r}) = 1, \tag{7.78}$$

where $u_k(\boldsymbol{r})$ is the Bloch amplitude in a single band, coefficient $C_{QR}(\boldsymbol{k})$ is nonzero when $\boldsymbol{k}$ is in close vicinity of $\boldsymbol{Q}$. That is the semiclassical description of the state of an electron in point $\boldsymbol{R}$ and wavevector $\boldsymbol{Q}$.

Let us calculate the expectation value of position $\boldsymbol{r}$ using Eq. (7.78):

$$\begin{aligned}
\langle \boldsymbol{r} \rangle &= \int d^3r \sum_{kk'} C^*_{QR}(\boldsymbol{k})\, C_{QR}(\boldsymbol{k}')\, e^{-i\boldsymbol{k}\cdot(\boldsymbol{r}-\boldsymbol{R})} u^*_k(\boldsymbol{r})\; \boldsymbol{r}\; e^{i\boldsymbol{k}'\cdot(\boldsymbol{r}-\boldsymbol{R})} u_{k'}(\boldsymbol{r}) \\
&= \int d^3r \sum_{kk'} C^*_{QR}(\boldsymbol{k})\, C_{QR}(\boldsymbol{k}')\, e^{-i\boldsymbol{k}\cdot(\boldsymbol{r}-\boldsymbol{R})} u^*_k(\boldsymbol{r})\; u_{k'}(\boldsymbol{r}) \left(-i\frac{\partial}{\partial \boldsymbol{k}'} + \boldsymbol{R} \right) e^{i\boldsymbol{k}'\cdot(\boldsymbol{r}-\boldsymbol{R})} \\
&= i \int d^3r \sum_{kk'} C^*_{QR}(\boldsymbol{k}) \left(\frac{\partial}{\partial \boldsymbol{k}'} C_{QR}(\boldsymbol{k}') \right) e^{-i\boldsymbol{k}\cdot(\boldsymbol{r}-\boldsymbol{R})} u^*_k(\boldsymbol{r})\, u_{k'}(\boldsymbol{r})\; e^{i\boldsymbol{k}'\cdot(\boldsymbol{r}-\boldsymbol{R})} \\
&\quad + i \int d^3r \sum_{kk'} C^*_{QR}(\boldsymbol{k})\, C_{QR}(\boldsymbol{k}')\, e^{-i\boldsymbol{k}\cdot(\boldsymbol{r}-\boldsymbol{R})} u^*_k(\boldsymbol{r}) \left(\frac{\partial}{\partial \boldsymbol{k}'} u_{k'}(\boldsymbol{r}) \right) e^{i\boldsymbol{k}'\cdot(\boldsymbol{r}-\boldsymbol{R})} + \boldsymbol{R} \\
&= i \sum_{k} C^*_{QR}(\boldsymbol{k}) \frac{\partial}{\partial \boldsymbol{k}} C_{QR}(\boldsymbol{k}) + i \sum_{k} |C_{QR}(\boldsymbol{k})|^2 \int d^3r\; u^*_k(\boldsymbol{r}) \frac{\partial}{\partial \boldsymbol{k}} u_k(\boldsymbol{r}) + \boldsymbol{R} \\
&= \frac{d\xi_{QR}}{d\boldsymbol{Q}} + i \int d^3r\; u^*_Q(\boldsymbol{r})\, \frac{\partial}{\partial \boldsymbol{Q}} u_Q(\boldsymbol{r}) + \boldsymbol{R},
\end{aligned} \tag{7.79}$$

where $C_{QR}(\boldsymbol{k}) \equiv |C_{QR}(\boldsymbol{k})| \exp[-i\xi_{QR}]$ and we take into account that $|C_{QR}(\boldsymbol{k})|$ has a maximum at $\boldsymbol{k} = \boldsymbol{Q}$.

The second term in Eq. (7.79) defines the *Berry connection*,

$$\mathcal{A}(\boldsymbol{Q}) = i \int d^3r\; u^*_Q(\boldsymbol{r}) \frac{\partial}{\partial \boldsymbol{Q}} u_Q(\boldsymbol{r}) \tag{7.80}$$

characterizing the variation of Bloch amplitude in the Brillouin zone ($\boldsymbol{k}$-space).

To be consistent, we have to require the expectation value of coordinate to be $\langle \boldsymbol{r} \rangle = \boldsymbol{R}$, so (7.81) takes the form,

$$\frac{d\xi_Q}{d\boldsymbol{Q}} = -\,\mathcal{A}(\boldsymbol{Q}), \tag{7.81}$$

which shows how the Berry connection is related to the phase change when the wave packet moves in the Brillouin zone.

The equation of motion of the wave packet in electric and magnetic fields one can find using the Lagrangian mechanics. After averaging the Lagrangian of electrons in the state (7.78), we come to the Lagrangian describing the wave packet,

$$L(\mathbf{Q},\mathbf{R}) = \int d^3r\ \psi^*_{QR}(\mathbf{r})\left(i\hbar\frac{\partial}{\partial t} - H\right)\psi_{QR}(\mathbf{r}), \tag{7.82}$$

where Hamiltonian H includes the crystal potential and interaction with electric and magnetic fields. Calculation of the integral gives us:

$$L(\mathbf{Q},\mathbf{R}) = \int d^3r \sum_{kk'} C^*_{QR}(\mathbf{k})\ C_{QR}(\mathbf{k}')\ e^{-i\mathbf{k}\cdot(\mathbf{r}-\mathbf{R})}\ u^*_k(\mathbf{r})$$
$$\times\left[-\hbar\dot{\mathbf{Q}}\cdot\mathcal{A}_Q + \hbar\mathbf{k}'\cdot\dot{\mathbf{R}} - \varepsilon_{k'} + \frac{e\hbar\mathbf{k}'\cdot\mathbf{A}(\mathbf{r})}{m} - e\varphi(\mathbf{r})\right] e^{i\mathbf{k}'\cdot(\mathbf{r}-\mathbf{R})}\ u_{k'}(\mathbf{r})$$
$$= -\hbar\dot{\mathbf{Q}}\cdot\mathcal{A}_Q + \hbar\mathbf{Q}\cdot\dot{\mathbf{R}} - \varepsilon_Q + \frac{e\hbar\mathbf{Q}\cdot\mathbf{A}(\mathbf{R})}{m} - e\varphi(\mathbf{R}), \tag{7.83}$$

where $\varepsilon_k = \hbar^2k^2/2m$, $\mathbf{A}(\mathbf{r})$ and $\varphi(\mathbf{r})$ are the vector and scalar potentials related to the magnetic and electric fields as $\mathbf{B} = rot\,\mathbf{A}$, and $\mathbf{E} = -\nabla\varphi$, respectively. From the Lagrangian (7.83) and equations of motion

$$\frac{\partial L}{\partial \mathbf{R}} - \frac{d}{dt}\frac{\partial L}{\partial \dot{\mathbf{R}}} = 0, \qquad \frac{\partial L}{\partial \mathbf{Q}} - \frac{d}{dt}\frac{\partial L}{\partial \dot{\mathbf{Q}}} = 0, \tag{7.84}$$

we obtain

$$\hbar\dot{\mathbf{Q}} = e\mathbf{E} + \frac{e\hbar}{m}\ \mathbf{Q}\times\mathbf{B}, \tag{7.85}$$

$$\dot{\mathbf{R}} = \frac{1}{m}(\hbar\mathbf{Q} - e\mathbf{A}) - \dot{\mathbf{Q}}\times\mathcal{B}(\mathbf{Q}), \tag{7.86}$$

where

$$\mathcal{B}(\mathbf{k}) = \nabla_k\times\mathcal{A}\,(\mathbf{k}) \tag{7.87}$$

is the Berry curvature. Using Eqs. (7.80) and (7.87), one can present the Berry curvature as

$$\epsilon_i(\mathbf{k}) = i\,\epsilon_{ijl}\int d^3r\ \frac{\partial u^*_k}{\partial k_j}\frac{\partial u_k}{\partial k_l}, \tag{7.88}$$

where ϵ_{ijl} is the 3D unit antisymmetric tensor.

As we see, Eq. (7.85) is the usual equation of motion with the Lorentz force on the right-hand side. Equation (7.86) is the relation between velocity $\dot{\mathbf{R}}$ and momentum. It includes the Berry curvature, which turns out to be nonzero for the nontrivial topology of the energy band in a crystal lattice. This term is called *anomalous velocity.*

7.12 Quantum Transport: Green's Functions and Feynman Diagrams

The discussed above approach to the calculation of transport properties of solids is based mostly on the semiclassical approximation. In some cases, this approximation cannot describe quantum transport effects such as quantum Hall or anomalous Hall effects. Besides, electron tunneling in quantum wells and dots, as well as point contacts, require more rigorous methods of quantum transport theory.

Let us consider electrons in semiconductors subjected to a weak external field with vector potential $A(t)$. The Hamiltonian of the system is $H = H_0 + H_{int}$, where

$$H_0 = -\frac{\hbar^2 \Delta}{2m^*} + W(\mathbf{r}), \tag{7.89}$$

$$H_{int} = -\frac{ie\hbar}{m^*} \mathbf{A} \cdot \nabla. \tag{7.90}$$

Here $W(\mathbf{r}) = \sum_i V(\mathbf{r} - \boldsymbol{R}_i)$ is the potential of impurities randomly distributed in points $\boldsymbol{R}_i$, which we assume to be weak. The electric field is related to the vector potential $\boldsymbol{A}$ by

$$\boldsymbol{E} = -\frac{\partial \boldsymbol{A}}{\partial t}. \tag{7.91}$$

We assume $\boldsymbol{A}(t) = \boldsymbol{A}_\omega e^{-i\omega t}$ and following (7.91), $\boldsymbol{E}_\omega = i\omega \boldsymbol{A}_\omega$.

7.12.1 Green's Function Technique at $T = 0$

The quantum-mechanical current density at temperature $T = 0$ has the form (factor 2 is due to spin)

$$\mathbf{j}(\mathbf{r},t) = 2e \sum_n \psi_n^*(\mathbf{r},t)\ \mathbf{v}\ \psi_n(\mathbf{r},t)\ \theta(\mu - \varepsilon_n), \tag{7.92}$$

where $\theta(x)$ is the Heaviside's step function, and μ is the chemical potential. Here

$$\mathbf{v} = -\frac{i\hbar\nabla}{m^*} - \frac{eA}{m^*} \tag{7.93}$$

is the operator of electron velocity, $\psi_n(\mathbf{r},t)$ and ε_n are the eigenfunctions and eigenvalues of H_0, respectively.

In Green's function formalism, the electric current density (7.89) has the form,

$$\mathbf{j}(\mathbf{r},t) = -2ie\hbar\ \mathbf{v}\ G(\mathbf{r},t;\mathbf{r}',t+\delta)|_{r=r'}, \tag{7.94}$$

where causal Green's function

$$G(\mathbf{r},t;\mathbf{r}',t') = -i\left\langle Ta(\mathbf{r},t)\, a^\dagger(\mathbf{r}',t')\right\rangle \tag{7.95}$$

contains the time-ordered product averaged over the ground state. Operators $a^\dagger(\mathbf{r},t)$ and $a(\mathbf{r},t)$ (for details, see Chapter 3, Appendix 1, https://www.routledge.com/Modern-Semiconductor-Physics-and-Device-Applications/Dugaev-Litvinov/p/book/9780367250829# and Ref. [5]) create and destroy an electron in point $\boldsymbol{r}$ at the moment t. The time-ordering indicated by T in (7.95) means

$$G(\mathbf{r},t;\mathbf{r}',t') = \begin{cases} -i\,\langle a(\mathbf{r},t)\, a^\dagger(\mathbf{r}',t')\rangle, & t > t', \\ i\,\langle a^\dagger(\mathbf{r}',t')\, a(\mathbf{r},t)\rangle, & t < t'. \end{cases} \tag{7.96}$$

Using the one-particle eigenfunctions $\psi_n(\mathbf{r},t)$ as a basis, we expand operators as

$$a(\mathbf{r},t) = \sum_n c_n\, \psi_n(\mathbf{r},t), \quad a^\dagger(\mathbf{r},t) = \sum_n c_n^\dagger\, \psi_n^*(\mathbf{r},t), \tag{7.97}$$

where $c_n^\dagger$, c_n are the creation and annihilation operators of electron in state n. Substituting (7.96) into (7.94) and using Eq. (7.97), we obtain Eq. (7.92), in which the sum goes over the states occupied by electrons.

The Green's function obeys the equation

$$\left(i\hbar\frac{\partial}{\partial t}-H\right)G(\mathbf{r},t;\mathbf{r}',t')=\delta(\mathbf{r}-\mathbf{r}')\,\delta(t-t'). \tag{7.98}$$

Assuming H_{int} to be a small perturbation, we find from (7.98) (we already used perturbation theory in Chapter 3)

$$\begin{aligned}G(\mathbf{r},t;\mathbf{r}',t')&\simeq G_0(\mathbf{r},t;\mathbf{r}',t')+\int d^3\mathrm{r}_1\,dt_1\,G_0(\mathbf{r},t;\mathbf{r}_1,t_1)\,H_{int}(t_1)\,G_0(\mathbf{r}_1,t_1;\mathbf{r}',t')\\&=G_0(\mathbf{r},t;\mathbf{r},t')-\frac{ie\hbar}{m^*}\int d^3\mathrm{r}_1\,dt_1\,G_0(\mathbf{r},t;\mathbf{r}_1,t_1)\,A(t_1)\cdot\nabla_{r_1}G_0(\mathbf{r}_1,t_1;\mathbf{r}',t'),\end{aligned} \tag{7.99}$$

where $G_0(\mathbf{r},t;\mathbf{r}',t')\equiv G_0(\mathbf{r},\mathbf{r}';t-t')$ is the solution of Eq. (7.98) with unperturbed Hamiltonian H_0. Note that G_0 is not translationally invariant due to impurity potential $W(\mathbf{r})$ included in H_0.

Substituting (7.93) and (7.99) into (7.94), after Fourier transformation on t, we find in the linear approximation in $\boldsymbol{A}$.

$$\mathbf{j}(\mathbf{r},\omega)=-\frac{2ie^2\hbar^2}{m^{*2}}\int\frac{d\varepsilon}{2\pi}\,d^3\mathrm{r}_1\,\nabla_r\,G_0(\mathbf{r},\mathbf{r}_1;\varepsilon+\hbar\omega)\,A_\omega\cdot\nabla_{r_1}\,G_0(\mathbf{r}_1,\mathbf{r}';\varepsilon)|_{r=r'}-\frac{e^2n(\mathbf{r})\boldsymbol{A}_\omega}{m^*}. \tag{7.100}$$

where $n(\mathbf{r})$ is the electron density. The second term in Eq. (7.100) describes *the diamagnetic response* and has nothing to do with electrical resistivity. The first term is the dissipative current, and using the relation $\boldsymbol{A}_\omega=-i\boldsymbol{E}_\omega/\omega$ determines conductivity,

$$\sigma_{ij}(\mathbf{r},\omega)=-\frac{2e^2\hbar^2}{m^{*2}\omega}\int\frac{d\varepsilon}{2\pi}\,d^3\mathrm{r}_1\nabla_{r_i}G_0(\mathbf{r},\mathbf{r}_1;\varepsilon+\hbar\omega)\cdot\nabla_{r_{1j}}\,G_0(\mathbf{r}_1,\mathbf{r}';\varepsilon)|_{r=r'}\,. \tag{7.101}$$

The next step is to average (7.101) over random impurity potential $W(\mathbf{r})$. For this, it is convenient to use the Feynman diagrams. In graphs, averaging over impurity potential is rendered by dashed impurity lines, which connect two points on Green's function lines, as shown in Figure 7.2. The procedure is discussed already in Chapter 3.

First, by this method, we present the conductivity in the absence of impurities. In this case, due to translational invariance $G_0(\mathbf{r},\mathbf{r}',\varepsilon)\equiv G_0(\mathbf{r}-\mathbf{r}',\varepsilon)$, and after Fourier transformation on $\boldsymbol{r}$ we obtain from (7.101):

$$\begin{aligned}\sigma_{ij}(\omega)&=\frac{2e^2\hbar^2}{m^{*2}\omega\Omega}\int\frac{d\varepsilon}{2\pi}\sum_k k_ik_j\,G_0(\boldsymbol{k},\varepsilon+\hbar\omega)\,G_0(\boldsymbol{k},\varepsilon)\\&=\frac{2e^2\hbar^2\delta_{ij}}{3m^{*2}\omega\Omega}\int\frac{d\varepsilon}{2\pi}\sum_k k^2\,G_0(\boldsymbol{k},\varepsilon+\hbar\omega)\,G_0(\boldsymbol{k},\varepsilon),\quad W(\mathbf{r})=0,\end{aligned} \tag{7.102}$$

where

$$G_0(\boldsymbol{k},\varepsilon)=\frac{1}{\varepsilon-\varepsilon_k+\mu+i\delta\ sign\ \varepsilon} \tag{7.103}$$

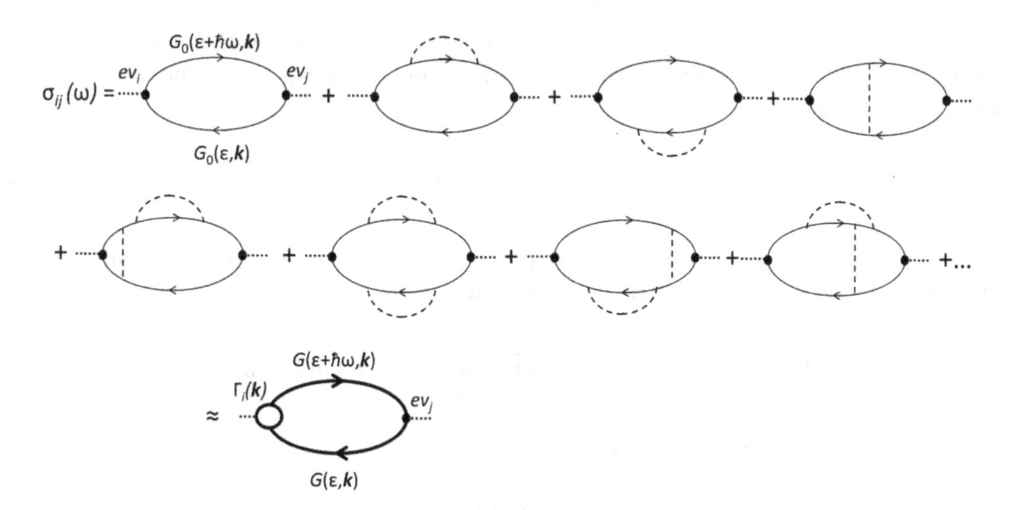

FIGURE 7.2 Feynman diagrams for the conductivity tensor.

being casual Green's function of free electron, $\varepsilon_k = \hbar^2 k^2 / 2m^*$, energy parameter ε counts from μ.

Note that for $\omega \to 0$, the impurity-free conductivity is divergent as it should be in the absence of scattering.

Loop diagram (first diagram in Figure 7.2) represents conductivity (7.102). The solid line is the Green's function G_0, and each vertex corresponds to the current operator $e\mathbf{v} = e\hbar\mathbf{k} / m^*$.

Dashed lines denote the interaction with impurities. Assuming that the impurity potential is weak, we restrict ourselves by the Born approximation for scattering (second-order perturbation theory interaction to the impurity potential). The analytic expression for each impurity line is $n_i V^2(\boldsymbol{q})$, where n_i is the impurity density, and $V(\boldsymbol{q})$ is the Fourier transform of a single impurity potential. Among diagrams, shown in Figure 7.2, some renormalize Green's functions, and others renormalize the vertex part. Renormalization of Green's function occurs with *self-energy* in Figure 7.3b:

$$\Sigma(\boldsymbol{k},\varepsilon) = n_i \int \frac{d^3k'}{(2\pi)^3} V^2(\boldsymbol{k}-\boldsymbol{k}')\, G_0(\boldsymbol{k}',\varepsilon). \tag{7.104}$$

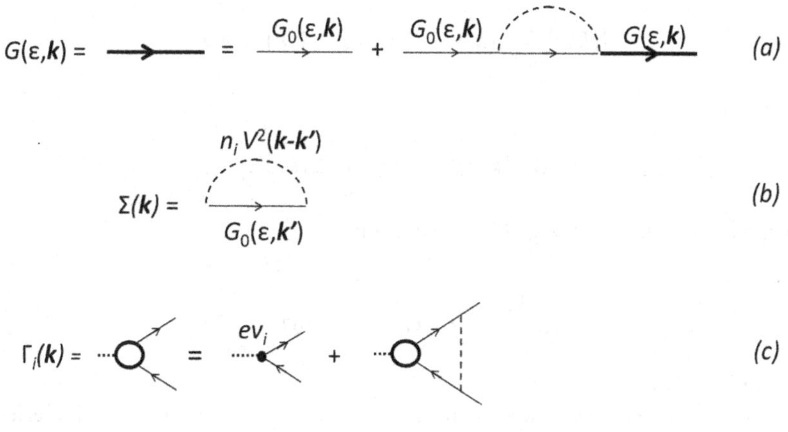

$G(\varepsilon,\boldsymbol{k})$ = ——→—— = $G_0(\varepsilon,\boldsymbol{k})$ + $G_0(\varepsilon,\boldsymbol{k})$ ⌒ $G(\varepsilon,\boldsymbol{k})$ (a)

$n_i V^2(\boldsymbol{k}-\boldsymbol{k}')$

$\Sigma(\boldsymbol{k})$ = $G_0(\varepsilon,\boldsymbol{k}')$ (b)

$\Gamma_i(\boldsymbol{k})$ = = ev_i + (c)

FIGURE 7.3 (a) Feynman diagrams for the Green's function, (b) self-energy part, (c) ladder approximation for vertex diagram.

The self-energy has real and imaginary parts. We neglect the real part since weak impurity interaction gives a small correction to the energy spectrum. Substituting (7.103) into (7.104) and calculating the imaginary part, we get

$$\text{Im } \Sigma(\boldsymbol{k},\varepsilon) = -\pi n_i \text{ } sign \text{ } \varepsilon \int \frac{d^3k'}{(2\pi)^3} V^2(\boldsymbol{k}-\boldsymbol{k}') \delta(\varepsilon_{k'} - \varepsilon - \mu). \tag{7.105}$$

Energy parameter ε in diagrams varies from zero to infinity and corresponds to so-called virtual particles. For real electrons with spectrum ε_k, we have to keep $\varepsilon = \varepsilon_k - \mu$:

$$\text{Im } \Sigma(\boldsymbol{k}) = -\pi n_i \text{ } sign \left(\varepsilon_k - \mu\right) \int \frac{d^3k'}{(2\pi)^3} V^2(\boldsymbol{k}-\boldsymbol{k}') \delta(\varepsilon_{k'} - \varepsilon_k), \tag{7.106}$$

or by introducing the relaxation time $\tau_0(\boldsymbol{k})$,

$$\text{Im } \Sigma(\boldsymbol{k}) = -\frac{\hbar \text{ } sign \left(\varepsilon_k - \mu\right)}{2\tau_0(\boldsymbol{k})}. \tag{7.107}$$

Then the expression for τ_0 reads

$$\frac{1}{\tau_0(\boldsymbol{k})} = \frac{2\pi n_i}{\hbar} \int \frac{d^3k'}{(2\pi)^3} V^2(\boldsymbol{k}-\boldsymbol{k}') \text{ } \delta(\varepsilon_{k'} - \varepsilon_k), \tag{7.108}$$

which corresponds to our calculation of the collision integral in the kinetic equation but does not contain $(1-\cos\theta)$. The relaxation time τ_0 characterizes the lifetime of an electron in state $\boldsymbol{k}$. The value $\hbar / \tau_0$ is usually called the *level width*.

If scattering occurs on short-range impurities, we take $V(\boldsymbol{k}-\boldsymbol{k}') = \text{const} = V_0$. Then from (7.108), follows:

$$\frac{1}{\tau_0} = \frac{\pi n_i V_0^2 \rho(\varepsilon_k)}{\hbar}, \tag{7.109}$$

where $\rho(\varepsilon)$ is the DOS. In this case, the transport relaxation time τ from Eq. (7.37) and τ_0 are the same.

The Dyson equation for the Green function follows from the summation of diagrams rendered in Figure 7.3a,

$$G(\boldsymbol{k},\varepsilon) = G_0(\boldsymbol{k},\varepsilon) + G_0(\boldsymbol{k},\varepsilon) \text{ } \Sigma(\boldsymbol{k},\varepsilon) \text{ } G(\boldsymbol{k},\varepsilon), \tag{7.110}$$

which gives us

$$G^{-1}(\boldsymbol{k},\varepsilon) = G_0^{-1}(\boldsymbol{k},\varepsilon) - \Sigma(\boldsymbol{k},\varepsilon). \tag{7.111}$$

Using the imaginary part of self-energy (7.107), we obtain

$$G(\boldsymbol{k},\varepsilon) = \frac{1}{\varepsilon - \varepsilon_k + \mu + i\gamma \text{ } sign \text{ } \varepsilon}, \tag{7.112}$$

where we denote $\gamma = \hbar / 2\tau_0$.

The main contribution to the vertex part comes from diagrams in Figure 7.3c. The series is the sum of diagrams, which include an infinite number of impurity (dashed) lines. There are two types of vertex diagrams: diagrams in which dashed lines cross each other (not shown in the picture) and ladder-like graphs with parallel lines. Upon neglecting crossed-impurity-line diagrams, the series presents the equation for the vertex in so-called *ladder approximation*.[7] Thus, the equation for the vertex has the form:

$$\Gamma(\boldsymbol{k}) = \frac{e\hbar \boldsymbol{k}}{m^*} + \frac{n_i}{\Omega}\sum_{k'} \Gamma(\boldsymbol{k}')\, G(\boldsymbol{k}',\varepsilon)\, G(\boldsymbol{k}',\varepsilon+\hbar\omega)\, V^2(\boldsymbol{k}-\boldsymbol{k}') \tag{7.113}$$

We will be looking for the solution of this equation in the form

$$\Gamma(\mathrm{k}) = \frac{e\hbar \mathrm{k}}{m^*}\, Q(\mathrm{k}), \tag{7.114}$$

where $Q(\mathrm{k})$ is the scalar function. Substituting (7.114) into (7.113), we get the equation for $Q(\mathrm{k})$

$$Q(k) = 1 + \frac{1}{\Omega}\sum_{k'} \frac{n_i\,(\boldsymbol{k}\cdot\boldsymbol{k}')\, Q(k')\, V^2(\boldsymbol{k}-\boldsymbol{k}')}{k^2\left[\varepsilon+\hbar\omega-\varepsilon_{k'}+\mu+i\gamma\ sign\,(\varepsilon+\hbar\omega)\right]\left[\varepsilon-\varepsilon_{k'}+\mu+i\gamma\ sign\ \varepsilon\right]} \tag{7.115}$$

Replacing sums with integrals, we obtain,

$$Q(k) = 1 + \frac{n_i}{2\pi k}\int_0^{\pi} d\theta \int \frac{\rho(\varepsilon_{k'})d\varepsilon_{k'}\ k'\cos\theta\ Q(k')V^2(\boldsymbol{k}-\boldsymbol{k}')}{\left[\varepsilon_{k'}-\hbar\omega-\varepsilon-\mu-i\gamma\ sign\,(\varepsilon+\hbar\omega)\right]\left[\varepsilon_{k'}-\varepsilon-\mu-i\gamma\ sign\ \varepsilon\right]} \tag{7.116}$$

where θ is the angle between $\boldsymbol{k}$ and $\boldsymbol{k}'$.

Calculating the second integral over $\varepsilon_{k'}$ in complex ε_k-plane, we find the integral is nonzero if $\varepsilon+\hbar\omega$ and ε have different signs (otherwise the poles are at the same half-plane of complex $\varepsilon_{k'}$). Since we assume $\omega > 0$, the corresponding condition is $-\hbar\omega < \varepsilon < 0$. Thus, we obtain

$$Q(k) = 1 + \frac{n_i\tau_0\ \rho(\varepsilon+\mu)\ k_0\ Q(k_0)}{\hbar k\,(1-i\omega\tau)}\int_0^{\pi} d\theta\ \cos\theta V^2(\boldsymbol{k}-\boldsymbol{k}_0), \tag{7.117}$$

where k_0 follows from $\hbar^2 k_0^2/2m^* = \varepsilon+\mu$.

Taking $k = k_0$, we find the solution for $Q(k_0)$:

$$Q(k_0) = \left[1 - \frac{n_i\tau_0\ \rho(\varepsilon+\mu)}{\hbar\,(1-i\omega\tau)}\int_0^{\pi} d\theta\ \cos\theta\ V^2(\theta)\right]^{-1}. \tag{7.118}$$

In the limit $\omega \to 0$,

$$Q(k_F) = \left[1 - \frac{n_i\tau_0\ \rho(\mu)}{\hbar}\int_0^{\pi} d\theta\ \cos\theta\ V^2(\theta)\right]^{-1}, \tag{7.119}$$

where $k_F = (2m^*\mu)^{1/2}/\hbar$ is the Fermi wavevector. Using Eq. (7.108), $Q(k_F)$ becomes

$$Q(k_F) = \tau/\tau_0, \tag{7.120}$$

[7] Detailed analysis shows that contribution of crossed-line diagrams is small if $\varepsilon\tau/\hbar \gg 1$, where ε is the characteristic energy of electrons: $\varepsilon \sim k_B T$ in nondegenerate and $\varepsilon \sim \mu$ in a degenerate semiconductor.

where τ is the transport relaxation time at the Fermi level,

$$\frac{1}{\tau}=\frac{n_i\,\rho(\mu)}{\hbar}\int_0^{\pi}d\theta\,(1-\cos\theta)\;V^2(\theta). \tag{7.121}$$

Correspondingly, the vertex function at the Fermi level is

$$\Gamma(\boldsymbol{k}_F)=\frac{e\hbar\boldsymbol{k}_F}{m^*}\,\frac{\tau}{\tau_0}. \tag{7.122}$$

Note that for short-range scatterers, $V(\theta)\to V_0$, the transport relaxation time τ coincides with τ_0.

7.12.2 Green's Function Technique at Finite Temperatures

To apply Green's functions technique at finite temperatures, $T\neq 0$, we will be using Matsubara *formalism*. Now instead of continuous energy parameter ε in electron Green's functions, one uses discrete imaginary energies $i\varepsilon_n=i(2n+1)\pi k_BT$, n is an integer. Hence, the free-electron Matsubara function has the form,

$$G_0(\boldsymbol{k},i\varepsilon_n)=\frac{1}{i\varepsilon_n-\varepsilon_k+\mu}. \tag{7.123}$$

Correspondingly, the electron self-energy in Born approximation at $T\neq 0$ acquires the form similar to (7.104)

$$\Sigma(\boldsymbol{k},i\varepsilon_n)=n_i\int\frac{d^3k'}{(2\pi)^3}\,\frac{V^2(\boldsymbol{k}-\boldsymbol{k}')}{i\varepsilon_n-\varepsilon_{k'}+\mu}, \tag{7.124}$$

and the renormalized by impurity interactions Matsubara's function is now

$$G(\boldsymbol{k},i\varepsilon_n)=\frac{1}{i\varepsilon_n-\varepsilon_k+\mu-\Sigma(\boldsymbol{k},i\varepsilon_n)}. \tag{7.125}$$

To calculate Feynman diagrams at $T\neq 0$ ε-integration is substituted by a sum over integers n,

$$i\int\frac{d\varepsilon}{2\pi}\ldots\to -k_BT\sum_n\ldots \tag{7.126}$$

As a result, the $T\neq 0$ analog of Eq. (7.102) reads as

$$\begin{aligned}\sigma_{ij}(i\omega_m)&=-\frac{2e^2\hbar^2\delta_{ij}k_BT}{m^{*2}i\omega_m\Omega}\sum_n\sum_k k_ik_j\;G_0(\boldsymbol{k},i\varepsilon_n+i\hbar\omega_m)\,G_0(\boldsymbol{k},i\varepsilon_n)\\&=-\frac{2e^2\hbar^2\delta_{ij}k_BT}{3m^{*2}i\omega_m\Omega}\sum_n\sum_k k^2\;G_0(\boldsymbol{k},i\varepsilon_n+i\hbar\omega_m)\,G_0(\boldsymbol{k},i\varepsilon_n),\quad W(\boldsymbol{r})=0,\end{aligned} \tag{7.127}$$

where instead of continuous $\hbar\omega$, the discrete values $i\hbar\omega_m=2mi\pi k_BT$ used (m is an integer). In the next section, we will use Eqs. (7.126) and (7.127) to calculate the conductivity. More details on Green's function method and Feynman diagrams one finds in Refs. [6,7].

7.13 Quantum Transport Approach to Conductivity

In the ladder approximation, the conductivity, shown in Figure 7.1, can be expressed with Green's functions (7.112) and vertex correction (7.122). In a semiconductor with degenerate electron gas (it is also valid for $T \neq 0$ as long as $k_B T \ll \mu$):

$$\sigma_{ij}(\omega)=\frac{2e^2\hbar^2\delta_{ij}}{3m^{*2}\omega\Omega}\int\frac{d\varepsilon}{2\pi}\sum_k\frac{k^2\tau}{\tau_0}G(\mathbf{k},\varepsilon+\hbar\omega)\,G(\mathbf{k},\varepsilon)$$

$$=\frac{2e^2\hbar^2\delta_{ij}}{3m^{*2}\omega\Omega}\int\frac{d\varepsilon}{2\pi}\sum_k\frac{k^2\tau}{\tau_0\left[\varepsilon+\hbar\omega-\varepsilon_k+\mu+i\gamma\ sign\,(\varepsilon+\hbar\omega)\right]\left[\varepsilon-\varepsilon_k+\mu+i\gamma\ sign\ \varepsilon\right]}$$

$$=\frac{2e^2\hbar^2\delta_{ij}}{3m^{*2}\omega\Omega}\int_{-\hbar\omega}^{0}\frac{d\varepsilon}{2\pi}\sum_k\frac{k^2\tau}{\tau_0\left[\varepsilon+\hbar\omega-\varepsilon_k+\mu+i\gamma\right]\left[\varepsilon-\varepsilon_k+\mu-i\gamma\right]}. \tag{7.128}$$

Then for $\omega\to 0$, we obtain the conductivity (a diagonal component of the conductivity tensor)

$$\sigma=\frac{e^2\hbar^3}{3\pi m^{*2}\Omega}\sum_k{}'\frac{k^2\tau}{\tau_0\left(\varepsilon_k-\mu-i\gamma\right)\left(\varepsilon_k-\mu+i\gamma\right)}=\frac{e^2\hbar^2\rho(\mu)\,k_F^2\tau}{3m^{*2}}=\frac{ne^2\tau}{m^*}, \tag{7.129}$$

which coincides with the results of Drude theory and kinetic equation. In Eq. (7.129) we used $n=2\mu\ \rho(\mu)/3$ that follows from electron density in the degenerate semiconductor, $n=k_F^2/3\pi^2$ (see Chapter 4).

At $T\neq 0$, we start from

$$\sigma_{ij}(i\omega_m)=-\frac{2e^2\hbar^2\delta_{ij}k_BT}{3m^{*2}i\omega_m\Omega}\sum_n\sum_k\frac{k^2\tau}{\tau_0}G(\mathbf{k},i\varepsilon_n+i\hbar\omega_m)\,G(\mathbf{k},i\varepsilon_n). \tag{7.130}$$

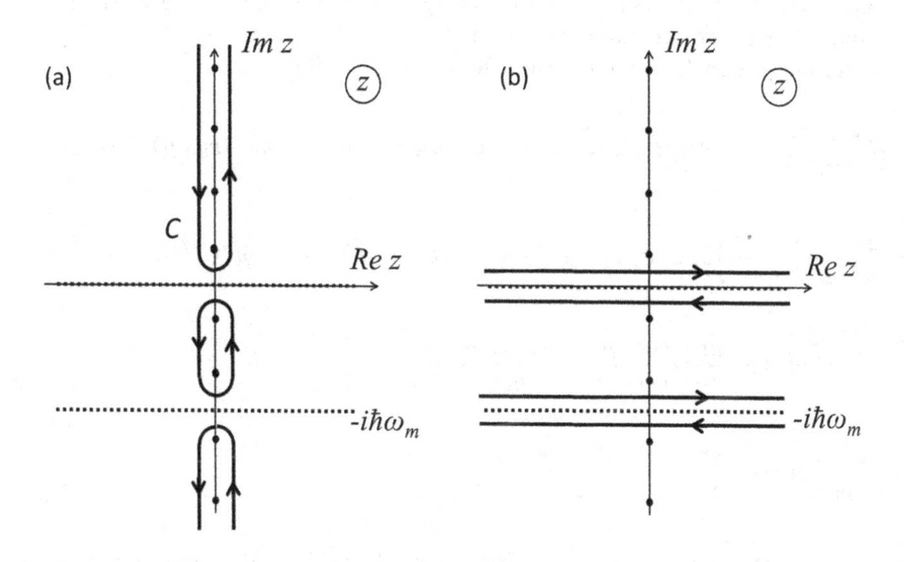

FIGURE 7.4 (a) The contour of integration C consists of three parts; (b) transformed contour C.

The standard method for the calculation of sum over n is to express it as a contour integral in a complex z-plane:

$$k_BT\sum_n \ldots = \frac{i}{2\pi}\int_C f(z)\,dz\ldots, \tag{7.131}$$

where $f(z)=\left[\exp(z/k_BT)+1\right]^{-1}$ and the contour C is shown in Figure 7.4a. The reason behind this representation is that the function $f(z)$ has the poles at $z=i\varepsilon_n$, and the integration around these poles leads exactly to the sum over n. So, one can represent (7.130) as

$$\sigma(i\omega_m)=-\frac{2e^2\hbar^2}{3\pi m^{*2}i\omega_m\Omega}\int_C f(z)\,dz\sum_k \frac{k^2\tau}{\tau_0}\,G(\mathbf{k},z+i\hbar\omega_m)\,G(\mathbf{k},z). \tag{7.132}$$

Since there are no other singularities in the complex z-plane but the poles of both $f(z)$ and Green's functions, the contour C can be transformed to the contours along real axes $\text{Im}\,z=0$ and $\text{Im}\,z=-\hbar\omega_m$ as shown in Figure 7.4b. The Green's functions have the poles on these lines. Integration over transformed contour becomes an integral over branches:

$$\sigma(i\omega_m)=-\frac{2e^2\hbar^2}{3\pi m^{*2}i\omega_m\Omega}\sum_k\frac{k^2\,\tau}{\tau_0}\left\{\int_{-\infty}^{\infty}f(z)dz\,\left[G^R(\mathbf{k},z)-G^A(\mathbf{k},z)\right]G^A(\mathbf{k},z+i\hbar\omega_m)\right.$$

$$\left.+\int_{-\infty-i\hbar\omega_m}^{\infty-i\hbar\omega_m} f(z)\,dz\left[G^R(\mathbf{k},z+i\hbar\omega_m)-G^A(\mathbf{k},z+i\hbar\omega_m)\right]G^R(\mathbf{k},z)\right\}, \tag{7.133}$$

where we used the definition of *retarded* G^R and *advanced* G^A functions

$$G^{R,A}(\mathbf{k},z)=\frac{1}{z-\varepsilon_k+\mu\pm i\gamma}. \tag{7.134}$$

Here $i\gamma$ comes from the imaginary part of the self-energy (7.124), and $\pm$ signs are related to the upper and lower contours along the branch-cuts in Figure 7.4b.[8]

After performing the analytical continuation to real ω, we obtain

$$\begin{aligned}\sigma(\omega)&=\frac{2ie^2\hbar^2}{3m^{*2}\omega\Omega}\sum_k\frac{k^2\tau}{\tau_0}\int_{-\infty}^{\infty}f(\varepsilon)d\varepsilon\left[\delta(\varepsilon-\varepsilon_k+\mu)\,G^A(\mathbf{k},\varepsilon+\hbar\omega)+\delta(\varepsilon+\hbar\omega-\varepsilon_k+\mu)\,G^R(\mathbf{k},\varepsilon)\right]\\
&=\frac{2ie^2\hbar^2}{3m^{*2}\omega\Omega}\sum_k\frac{k^2\tau}{\tau_0}\int_{-\infty}^{\infty}d\varepsilon\;\delta(\varepsilon-\varepsilon_k+\mu)\left[f(\varepsilon)\,G^A(\mathbf{k},\varepsilon+\hbar\omega)+f(\varepsilon-\hbar\omega)\,G^R(\mathbf{k},\varepsilon-\hbar\omega)\right]\\
&=\frac{2ie^2}{3m^*\omega}\int_{-\infty}^{\infty}\rho(\varepsilon)\,d\varepsilon\,\frac{\varepsilon\tau}{\tau_0}\left[\frac{f(\varepsilon-\mu)}{\hbar\omega-i\gamma}+\frac{f(\varepsilon-\hbar\omega-\mu)}{-\hbar\omega+i\gamma}\right]\\
&=\frac{2ie^2}{3m^*\omega}\int_{-\infty}^{\infty}\rho(\varepsilon)\,d\varepsilon\,\frac{\varepsilon\tau}{\tau_0}\,\frac{f(\varepsilon-\mu)-f(\varepsilon-\hbar\omega-\mu)}{\hbar\omega-i\gamma},\end{aligned} \tag{7.135}$$

[8] For example, at the upper branch we take $G(\mathbf{k},z+i\delta)$, z is real. Then $G(\mathbf{k},z+i\delta)=G^R(\mathbf{k},z)$. Generally, the Green's function $G(\mathbf{k},z)$ with real z has the analytical continuation to complex z-plane as retarded function for $\text{Im}\,z>0$ and advanced function for $\text{Im}\,z<0$.

and in the limit $\omega \to 0$ we finally obtain static conductivity:

$$\sigma = \frac{2e^2}{3m^*} \int_{-\infty}^{\infty} d\varepsilon \; \rho(\varepsilon) \, \varepsilon\tau \left(-\frac{\partial f}{\partial \varepsilon} \right), \tag{7.136}$$

where we used $\gamma = \hbar / 2\tau_0$. This result coincides with that obtained with the kinetic equation and is valid in both nondegenerate and degenerate semiconductors. In the degenerate case, Eq. (7.136) gives $\sigma = ne^2\tau / m^*$.

7.14 Quantum Hall Effect: Hall Conductivity as a Berry Phase: Thouless-Kohmoto-Nightingale-Nijs (TKNN) Theory

As follows from Eq. (7.19), transverse (Hall) voltage E_y is proportional to the longitudinal current, j_x, and perpendicular magnetic field $\boldsymbol{B}$, as $E_y = RBj_x$, where $R = 1/ne$ is the Hall coefficient and n is the electron density. Electron spectrum in two-dimensional electron gas subjected perpendicular magnetic field consists of a series of Landau levels, $E_{n_L} = (n_L + 1/2)\,\hbar\omega_c$, where $\omega_c = eB/m^*$ is the cyclotron frequency and $n_L = 1,2,\ldots$ (see Chapter 5). All these levels are degenerate since there are many electron states with the same n_L but different momenta determining the locations of the orbit center (for definiteness, we use the description based on the Landau gauge). Thus, there are $eB/2\pi\hbar$ states with the same n_L per unit area (see Eq. (5.28) in Chapter 5). If the first ν levels are occupied, the total number of electrons at these levels is $n = eB\nu / 2\pi\hbar$. Correspondingly, the value of Hall coefficient in a 2D semiconductor with fully occupied ν levels is

$$R = \frac{h}{e^2 B\nu} \tag{7.137}$$

and the Hall voltage as a function of current is quantized:

$$E_y = \frac{h}{\nu e^2} j_x, \tag{7.138}$$

which means quantization of off-diagonal resistivity, $\rho_{xy} = h/\nu e^2$.

Such quantization was observed by Klitzing et al. in the measurement of ρ_{xy} as a function of a magnetic field B. When B varies, Landau levels move relative to the Fermi level. Once B reaches the value with exactly ν levels filled, the magnitude of ρ_{xy} becomes $h/\nu e^2$ with high precision. When every subsequent level becomes filled, the off-diagonal resistivity jumps by h/e^2. At the same values of the magnetic field, the longitudinal ohmic resistivity ρ_{xy} becomes zero that is not surprising as electrons cannot scatter in the filled Landau levels. The quantum leaps in ρ_{xy} and zeroes in ρ_{xx} have been called *Quantum Hall Effect* (QHE).

What is surprising is the precision of ρ_{xy} quantization. Indeed, in real 2D-materials, the Landau levels have nonzero width $\sim \hbar/\tau_0$ due to impurities and defects. The interaction of electrons with phonons can also affect the off-diagonal conductivity. Another surprise is that, as described above, the quantized values in $\rho_{xy}(B)$ and zeroes in ρ_{xx} should have appeared at definite values of magnetic field $B = hn\nu/e\nu$, ν is an integer. The experiment, however, shows plateaus in $\rho_{xy}(B)$ and corresponding finite ranges of the magnetic field where $\rho_{xx} = 0$. The origin of plateaus comes from the disorder and electron localization in the Landau subbands. Thus, the explanation of the QHE requires a more sophisticated approach.

In the quantum transport theory, the Hall conductivity, σ_H, can be calculated based on the model of 2D electrons in a periodic lattice potential and perpendicular magnetic field:

$$H = -\frac{\hbar^2}{2m_0} \left[\left(\nabla_x - \frac{ieA_x}{\hbar} \right)^2 + \nabla_y^2 \right] + V_c(\boldsymbol{r}), \tag{7.139}$$

where m_0 is the free-electron mass, $V_c(\boldsymbol{r})$ is the lattice potential, $\boldsymbol{A}(\boldsymbol{r})$ is the vector potential, and we use the Landau gauge $\boldsymbol{A}(\boldsymbol{r})=(-yB,0,0)$, in which only the x-component of vector potential is nonzero.

The Hamiltonian is translation-invariant along axis x. Since we consider a 2D crystal of area $L\times L$, by using the periodic boundary condition for the wave function $\psi(x,y)=\psi(x+L,y)$, we do not break translational invariance. The situation with translational symmetry along axis y is not so simple. Nevertheless, it turns out that if the ratio of magnetic flux BL^2 to magnetic flux quantum $\hbar/|e|$ is a rational number p/q, one can use a generalized boundary condition to provide the translational symmetry along axis y. For more detail, see Ref. [8]. It results in a possibility to introduce as quantum numbers the components of momentum k_x, k_y labeling the eigenstates of Hamiltonian (7.139). Correspondingly, the eigenfunctions are the Bloch functions, $u_{kn}(\boldsymbol{r})\,e^{i\boldsymbol{k}\cdot\boldsymbol{r}}$, the Hamiltonian in this representation is a matrix $H_{\boldsymbol{k}}$, and the operator of the electron velocity has the form,

$$\hat{\mathbf{v}}=\frac{1}{\hbar}\frac{dH_{\boldsymbol{k}}}{d\boldsymbol{k}}. \tag{7.140}$$

That is the starting point for the TKNN theory of QHE [9].

Due to Landau quantization, the electronic structure consists of many Landau levels. We assume chemical potential μ is in the gap between levels so that all i-states have energy $\varepsilon_i<\mu$ whereas the j-states have energies $\varepsilon_j>\mu$. For brevity, we denote by i the set of quantum numbers $i=(\boldsymbol{k},n)$ defining the electron states of 2D electrons in the perpendicular magnetic field and periodic potential $V_c(\boldsymbol{r})$.

Following the Onsager symmetry principle (7.64), we define the Hall conductivity σ_H as an antisymmetric part of the conductivity tensor:

$$\begin{aligned}
\sigma_H&=\frac{1}{2}(\sigma_{xy}-\sigma_{yx})=\frac{e^2}{2\omega L^2}\int\frac{d\varepsilon}{2\pi}\sum_{i,j}\Big\{\big[(\mathrm{v}_x)_{ij}(\mathrm{v}_y)_{ji}-(\mathrm{v}_y)_{ij}(\mathrm{v}_x)_{ji}\big]G_j(\varepsilon+\hbar\omega)\,G_i(\varepsilon)\\
&\quad+\big[(\mathrm{v}_x)_{ji}(\mathrm{v}_y)_{ij}-(\mathrm{v}_y)_{ji}(\mathrm{v}_x)_{ij}\big]G_i(\varepsilon+\hbar\omega)\,G_j(\varepsilon)\Big\}\\
&=\frac{e^2}{2\omega L^2}\int\frac{d\varepsilon}{2\pi}\sum_{i,j}\Bigg\{\frac{(\mathrm{v}_x)_{ij}(\mathrm{v}_y)_{ji}-(\mathrm{v}_y)_{ij}(\mathrm{v}_x)_{ji}}{[\varepsilon+\hbar\omega-\varepsilon_j+\mu+i\gamma\ \mathrm{sign}\,(\varepsilon+\hbar\omega)]\,[\varepsilon-\varepsilon_i+\mu+i\gamma\ \mathrm{sign}\ \varepsilon]}\\
&\quad+\frac{(\mathrm{v}_x)_{ji}(\mathrm{v}_y)_{ij}-(\mathrm{v}_y)_{ji}(\mathrm{v}_x)_{ij}}{[\varepsilon+\hbar\omega-\varepsilon_i+\mu+i\gamma\ \mathrm{sign}\,(\varepsilon+\hbar\omega)]\,[\varepsilon-\varepsilon_j+\mu+i\gamma\ \mathrm{sign}\ \varepsilon]}\Bigg\}\\
&=\frac{ie^2}{2\omega L^2}\sum_{i,j}\left\{\frac{(\mathrm{v}_x)_{ij}(\mathrm{v}_y)_{ji}-(\mathrm{v}_y)_{ij}(\mathrm{v}_x)_{ji}}{\varepsilon_i+\hbar\omega-\varepsilon_j}+\frac{(\mathrm{v}_x)_{ji}\ (\mathrm{v}_y)_{ij}-(\mathrm{v}_y)_{ji}(\mathrm{v}_x)_{ij}}{\varepsilon_i-\hbar\omega-\varepsilon_j}\right\}\\
&=\frac{ie^2}{2\omega L^2}\sum_{i,j}\big[(\mathrm{v}_x)_{ij}(\mathrm{v}_y)_{ji}-(\mathrm{v}_y)_{ij}(\mathrm{v}_x)_{ji}\big]\left(\frac{1}{\varepsilon_i+\hbar\omega-\varepsilon_j}-\frac{1}{\varepsilon_i-\hbar\omega-\varepsilon_j}\right).
\end{aligned} \tag{7.141}$$

In the limit $\omega\to 0$, we obtain

$$\sigma_H=-\frac{ie^2\hbar}{L^2}\sum_{i,j}\frac{(\mathrm{v}_x)_{ij}(\mathrm{v}_y)_{ji}-(\mathrm{v}_y)_{ij}(\mathrm{v}_x)_{ji}}{(\varepsilon_i-\varepsilon_j)^2}. \tag{7.142}$$

Now we use Eq. (7.140) to calculate the matrix elements of velocity in Eq. (7.142):

$$
\begin{aligned}
(\mathrm{v}_x)_{ij} &= \frac{1}{\hbar}\int d^2r\ u_i^* e^{-i\boldsymbol{k}\cdot\boldsymbol{r}} \left(\frac{\partial H}{\partial k_x}\right) u_j e^{i\boldsymbol{k}\cdot\boldsymbol{r}} \\
&= \frac{1}{\hbar}\int d^2r\ u_i^* e^{-i\boldsymbol{k}\cdot\boldsymbol{r}} \left(\frac{\partial}{\partial k_x} H - H\frac{\partial}{\partial k_x}\right) u_j e^{i\boldsymbol{k}\cdot\boldsymbol{r}} \\
&= \frac{1}{\hbar}\left(\varepsilon_i - \varepsilon_j\right)\int d^2r\ u_i^* \frac{\partial}{\partial k_x} u_j .
\end{aligned}
\tag{7.143}
$$

From Eq. (7.142), one finds

$$
\sigma_H = -\frac{ie^2}{\hbar}\sum_n \int d^2k\ d^2r \left(\frac{\partial u_{kn}^*}{\partial k_x}\frac{\partial u_{kn}}{\partial k_y} - \frac{\partial u_{kn}^*}{\partial k_y}\frac{\partial u_{kn}}{\partial k_x}\right),
\tag{7.144}
$$

where we explicitly separated the sum over i into a sum over n and integral over $\boldsymbol{k}$ in the whole Brillouin zone. Hall conductivity σ_H includes the Berry curvature

$$
\mathcal{B}(\boldsymbol{k}) = i\int d^2r \left(\frac{\partial u_{kn}^*}{\partial k_x}\frac{\partial u_{kn}}{\partial k_y} - \frac{\partial u_{kn}^*}{\partial k_y}\frac{\partial u_{kn}}{\partial k_x}\right),
\tag{7.145}
$$

which is by Eq. (7.87)

$$
\mathcal{B}(\boldsymbol{k}) = \nabla_{\boldsymbol{k}} \times \mathcal{A}(\boldsymbol{k}) = \epsilon_{\alpha\beta}\frac{\partial}{\partial k_\alpha}\mathcal{A}_\beta(\boldsymbol{k}),
\tag{7.146}
$$

$$
\mathcal{A}(\boldsymbol{k}) = i\int d^2r\ u_{kn}^* \frac{\partial}{\partial \boldsymbol{k}} u_{kn},
\tag{7.147}
$$

where $\alpha, \beta = x, y$, and $\epsilon_{\alpha\beta}$ is the 2D unit antisymmetric tensor.

Using Stokes' theorem, we replace the integral over the interior of the 2D Brillouin zone (7.144) with the integral over its boundary:

$$
\begin{aligned}
\sigma_H &= -\frac{e^2}{\hbar}\sum_n \oint dk_\alpha\ \mathcal{A}_\alpha(\boldsymbol{k}) \\
&= -\frac{ie^2}{\hbar}\sum_n \oint dk_\alpha \int d^2r \left(u_{kn}^* \frac{\partial u_{kn}}{\partial k_\alpha} - \frac{\partial u_{kn}^*}{\partial k_\alpha} u_{kn}\right),
\end{aligned}
\tag{7.148}
$$

where the contour integral runs around the Brillouin zone. The integrand in Eq. (7.148) equals $2\left(d\varphi_{kn}/dk_\alpha\right)$, where φ_{kn} is the phase of Bloch amplitude u_{kn}. Since $u_{kn}(\boldsymbol{r})$ is a single-valued function of $\boldsymbol{k}$, the integral over closed contour equals $2\pi\nu_i$, where ν_i is an integer. Finally, this results in

$$
\sigma_H = \frac{e^2}{h}\sum_i \nu_i,
\tag{7.149}
$$

which means that σ_H is quantized. So, the quantization of Hall conductivity originates from the topology of electron energy states, in other words, the behavior of Bloch amplitudes in the Brillouin zone.

7.15 Laughlin and Halperin Explanation of QHE

The Nobel Prize in Physics 1998 was awarded to Robert Laughlin for his theoretical explanation of QHE [10]. To discuss Laughlin's arguments, let us consider the electron gas in a 2D ribbon of width a in a perpendicular magnetic field. To ensure that periodic boundary conditions hold, one makes a large loop by connecting ends of the ribbon. The magnetic field is assumed to be perpendicular to the surface at each point, as shown in Figure 7.5.

The electron energy spectrum consists of a series of Landau levels, while impurity scattering makes the DOS a series of peaks of finite width (Landau subbands). As discussed in Chapter 3, the electronic states in the DOS tail are localized being delocalized near the DOS peak. If the chemical potential is in the gap between Landau subbands, the ribbon is an insulator. The voltage E applied to the edges of the ribbon generates the Hall current along the loop.

The Hamiltonian of electrons in conductive states has the form,

$$H = -\frac{\hbar^2}{2m^*}\left[\left(\nabla_x - \frac{ieA_x}{\hbar}\right)^2 + \nabla_y^2\right] + eEy, \tag{7.150}$$

where $\boldsymbol{A} = \left(-yB, 0, 0\right)$ is the vector potential related to magnetic field $\boldsymbol{B}$ along axis z and E is the electric field in the y-direction. The eigenfunction of this Hamiltonian is

$$\psi_{k_x}(\boldsymbol{r}) = \frac{1}{\sqrt{L}}\, e^{ik_x x}\, \phi_n\left(y - y_0\right), \tag{7.151}$$

where $\phi_n(y)$ is the eigenfunction of a linear oscillator in parabolic potential,

$$U(y) = \frac{1}{2}\, m^*\omega_c^2 \left(y - y_0\right)^2, \tag{7.152}$$

where $\omega_c = eB/m^*$ and

$$y_0 = -\frac{\hbar k_x}{m^*\omega_c} - \frac{eE}{m^*\omega_c^2} \tag{7.153}$$

being the center of orbit. The eigenvalue of Hamiltonian (7.150),

$$\begin{aligned} \varepsilon_n(k_x) &= \hbar\omega_c\left(n + \frac{1}{2}\right) - \frac{\hbar k_x eE}{m^*\omega_c} - \frac{e^2E^2}{2m^*\omega_c^2} \\ &= \hbar\omega_c\left(n + \frac{1}{2}\right) + eEy_0 + \frac{e^2E^2}{2m^*\omega_c^2}, \end{aligned} \tag{7.154}$$

as well as the center of the orbit y_0 depends on electric field E.

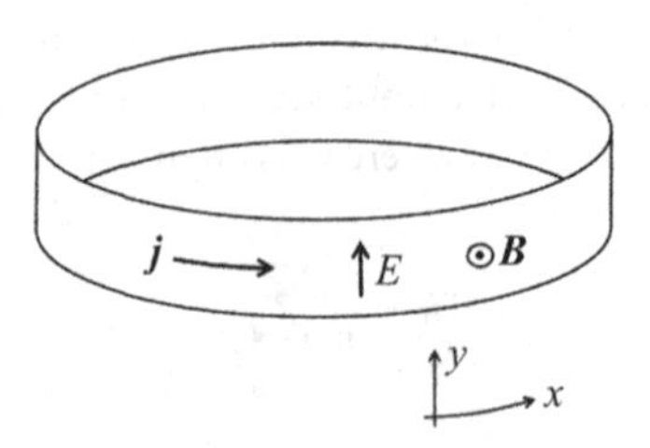

FIGURE 7.5 The geometry of Laughlin's experimental setting.

The energy change associated with the displacement of an electron on distance a in electric field E equals $\Delta\varepsilon = eEa = eU$, U is the potential difference between the points.

One can calculate the electron velocity v_x in the state (7.151) by using (7.154):

$$v_x = \frac{1}{\hbar}\frac{\partial \varepsilon_n}{\partial k_x} = -\frac{eE}{m^*\omega_c} = -\frac{E}{B}, \tag{7.155}$$

that is the velocity the electron is transmitting electric charge in the x-direction. The electric current flowing along the ribbon is the current transmitted by all electrons provided that each electron state contributes as ev_x / L:

$$j_x = -\frac{1}{L}\frac{\partial \varepsilon}{\partial A_x}, \tag{7.156}$$

where ε is the total electron energy.

To calculate the derivative, we can use the gauge invariance. One can add to A_x a small constant value δA and calculate the corresponding variation of the energy. Due to gauge invariance, the change $A_x \to A_x + \delta A$ is equivalent to $k_x \to k_x - e\,\delta A / \hbar$, which results in the shift in the phase of the wavefunction (7.151) by $\delta\varphi(x) = ex\,\delta A / \hbar$. On the other hand, this is compatible with the periodicity condition in the closed loop: $\delta\varphi(x) = \delta\varphi(x+L)$, so that $eL\,\delta A / \hbar = 2\pi n$, where n is an integer. Thus, one can use only quantized $\delta A_n = nh / eL$. For the calculation of current, we take the value

$$\delta A = \frac{h}{eL} = -\frac{\phi_0}{L} \tag{7.157}$$

where $\phi_0 = h/|e|$ is the magnetic flux quantum. From Eq. (7.156), the electric current becomes the variation of total electron energy per flux quantum,

$$j_x = \frac{\delta\varepsilon}{\phi_0}, \tag{7.158}$$

Due to the gauge invariance, the variation of vector potential $\delta A = -\phi_0 / L$ is equivalent to $k_x \to k_x - 2\pi/L$. So, the wavevector transforms into the equivalent one, and the system is again in the same state. So, electron transfer around the loop occurs adiabatically, meaning not changing the quantum state. Therefore, the change of the energy is associated only with the net transfer of the number of electrons n from one edge to the other one, $\delta\varepsilon = neU$, which gives us finally

$$j_x = \frac{ne^2U}{\hbar}, \tag{7.159}$$

that is the quantization of current in QHE. Since n is the integer number, it is called the *Integer Quantum Hall Effect*. Laughlin's explanation of QHE implies adiabatic electron transmission, cyclic boundary conditions, and gauge invariance. These general conditions are not sensitive to impurity scattering.

Explanation of precision in quantization relies on the chemical potential lying in the energy gap between Landau levels. In this case, there are no free electrons and holes, and the semiconductor is an insulator. It is not quite clear how the Hall current can be nonzero in an insulator. The answer was given by Halperin [11], who pointed out that the averaged potential created by the ions in crystal lattice sites is disappearing continuously at the edge leading to the bending of Landau levels and making them conductive at the edge of the sample. The edge states in a 2D-sample are one-dimensional and conductive even in the presence of impurities. In conclusion, the Hall current in the quantum Hall insulator implies the electron transport along the sample edges.

7.16 Fractional QHE

It was found experimentally two years after von Klitzing's work that in extremely clean samples, there appear additional plateaus of conductivity, $\sigma = \nu e^2 / h$, $\nu = p / q$, with fractional values $\nu = 1/3, 2/5, 3/7$, etc [12]. Here, the denominator is an odd number, but even denominators also were found in later experiments. In the explanation of this *Fractional Quantum Hall Effect* (FQHE), the essential role plays electron–electron interaction. The other important point is that the FQHE plateaus correspond to the chemical potential within the energy range of conductive states, i.e., near the peak of DOS.

Among several explanations to fractional ν, the most successful was Laughlin's theory [13]. The main point is that we deal with a strongly correlated electron system in a magnetic field, which should be described by the many-electron wave function. Under a magnetic field, the electrons are not so strongly repulsive but prefer to be at a certain distance, which is determined by the magnetic field. Basing on this consideration, Laughlin proposed an approximation for the wave function for the ground state of n electrons, corresponding to the state with fractional filling $1/q$

$$\psi_q(z_1, z_2, \ldots, z_n) = \text{Const}\, e^{-\left(|z_1|^2 + |z_2|^2 + \cdots + |z_n|^2\right)/S_0} (z_1 - z_3)^q (z_2 - z_3)^q \times (z_1 - z_2)^q \ldots (z_{n-1} - z_n)^q, \quad (7.160)$$

where $z_i = x_i + iy_i$ is the complex coordinate of ith electron and $S_0 = \phi_0 / B$. Since the wave function should be antisymmetric to transposition in each electron pair, the value of q should be an odd number.

Laughlin explained how the excitations appear when the magnetic flux through 2D-plane changes. When the increasing flux reaches the value ϕ_0, an excitation appears, which has the fractional charge $e^* = e/q$ or $-e/q$ (depending on the sign of magnetic flux).

The statistics of these excitations are neither Fermi nor Bose. In other words, under the permutation of two identical quasiparticles, the phase of wavefunction changes on neither 0 (Bose) nor π (Fermi), but on *statistical angle* $\theta = \pi / q$. Therefore, these excitations are usually called *anyons*.

There is another approach to the explanation of FQHE, which assumes the strong coupling of electrons to magnetic vortices leading to the creation of the *composite fermions* [14,15]. In this picture, one understands the FQHE as an integer QHE for the composite quasiparticles. In particular, the filling factors $n = 1, 2, 3$ in the integer QHE correspond to the fractional values $\nu = 1/3, 2/5, 3/7$. The reason is that the effective field acting on the composite fermions differs from the external magnetic field as part of magnetic fluxes are captured into composite fermions.

7.17 Anderson Localization

From the transport properties of semiconductors, we know that the disorder potential can create localized states in the gap and affect the band states by the electron momentum relaxation. The latter means that the electron scattering changes the direction of wavevector $\boldsymbol{k}$, which manifests itself as an imaginary part of the energy $i\gamma = i\hbar / 2\tau$, where τ is the momentum relaxation time.

All of the above is correct only for a weak disorder. It was assumed by Philip Anderson that in the strong disorder potential, all electron states become localized and thus the conductivity turns zero at $T = 0$. That is the case when the magnitude of randomly fluctuating short-range potential V is of the order of magnitude or larger than the width of the conduction band. If the disorder is not that strong, bands states are partially localized, even though the characteristic localization length L of these states might be large compared to the lattice constant. One indentifies the localization of electron state i by time-dependent probability density $|\psi_i(r,t)|^2$: for the localized state, it does not depend on time and $|\psi_i(r,t)|^2 \sim e^{-r/L}$ at large r, whereas for the nonlocalized state, this function is diffusing in space with time. Anderson's assumption has been verified numerically by using different models (like, for example, the model with a random onsite potential).

The condition of strong localization has been given as Ioffe-Regel-Mott criterium: an electron cannot propagate in space if the wavelength λ is larger than the mean free path $\ell\,(k\ell<1)$. Equivalently, $\varepsilon\tau/\hbar<1$, where k and ε are the electron wavevector and energy, respectively.

7.18 Theory of Weak Localization

Substantial progress in the understanding of Anderson localization is related to the theory of weak localization [16]. This theory calculates small *localization corrections to the conductivity* and their dependence on such parameters as temperature, magnetic field, spin-orbit interaction, etc. Even though these corrections are relatively small, they were observed in experiments.

In the weak localization theory, the analysis employs Feynman diagrams for conductivity presented in Section 7.13. Hamiltonian of an electron in a random field has the form,

$$H=-\frac{\hbar^2\nabla^2}{2m^*}+W(\boldsymbol{r}) \tag{7.161}$$

where $W(\boldsymbol{r})$ is the random potential, which we assume to be short-ranged and having the following properties:

$$\langle W(\boldsymbol{r})\rangle=0,\quad \langle W(\boldsymbol{r})\,W(\boldsymbol{r}')\rangle=\xi_0\,\delta(\boldsymbol{r}-\boldsymbol{r}'). \tag{7.162}$$

Here we denote $\xi_0=n_iV_0^2$, n_i is the impurity density, and the angular brackets mean averaging over realizations of the random potential. This type of randomness is called *white noise* because it corresponds to the uncorrelated in space random potential.

The diagrams for the averaged Green's functions and the Kubo loop diagram (see Figure 7.2) look like those for scattering on impurities in Born approximation, where each impurity line is ξ_0. Note that the vertex corrections are absent as the disorder potential is short-ranged. However, the ladder-type diagrams can appear in the loops, making them grow with the increasing parameter $\varepsilon\tau/\hbar$. In degenerate semiconductors at low temperatures, $k_BT\ll\mu$, the main parameter that controls localization corrections is $\mu\tau/\hbar\ll 1$.

There are two types of ladder diagrams corresponding to the renormalization of the impurity-mediated interaction between electrons. Diagrams, rendered in Figure 7.6, are fundamental elements of the theory of weak localization: the one with the antiparallel Green's function lines called *diffuson*, and the second one with parallel lines – the *cooperon*. As shown below, they carry information on elementary excitations in the disordered system as having simple poles for ω at definite wavevectors $\boldsymbol{q}$. In other words, they determine the spectrum of particle-like excitations $\omega=f(\boldsymbol{q})$.

The sum of all ladder diagrams with antiparallel lines gives the equation for diffuson,

$$D(\boldsymbol{q},\omega)=\xi_0+\xi_0\sum_{\boldsymbol{k}_1}G(\boldsymbol{k}_1+\boldsymbol{q},\varepsilon+\hbar\omega)\,G(\boldsymbol{k}_1,\varepsilon)\,D(\boldsymbol{q},\omega), \tag{7.163}$$

where

$$G(\boldsymbol{k},\varepsilon)=\frac{1}{\varepsilon-\varepsilon_{\boldsymbol{k}}+\mu+i\gamma\ sign\ \varepsilon},$$

$$\varepsilon_k=\frac{\hbar^2k^2}{2m^*},\quad \gamma=\pi\xi_0\rho(\mu). \tag{7.164}$$

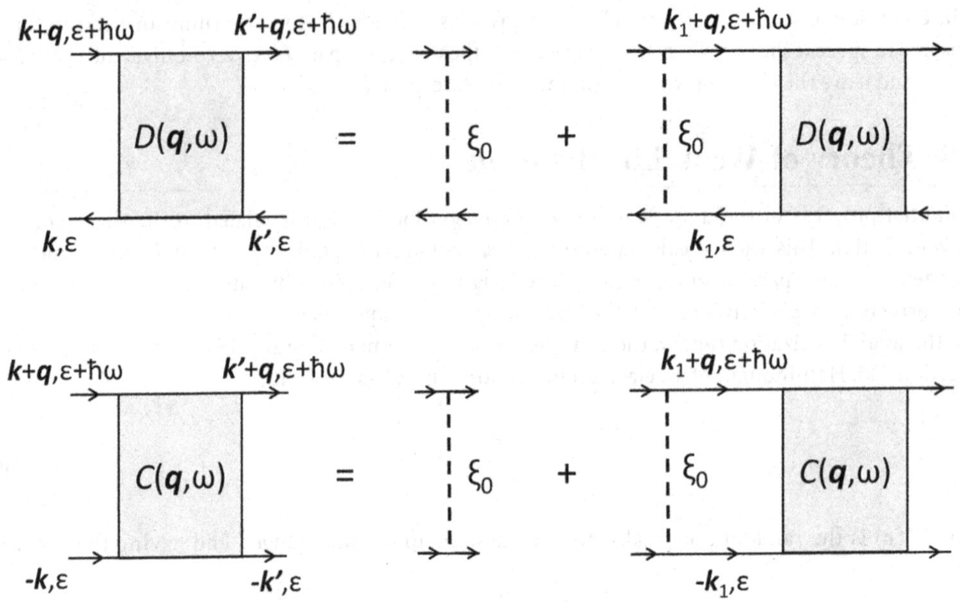

FIGURE 7.6 The equations for diffuson (upper row) and cooperon (lower row).

The solution to (7.163) is

$$D(\boldsymbol{q},\omega)=\frac{\xi_0}{1-\xi_0\,\Pi_1(\boldsymbol{q},\omega)}, \tag{7.165}$$

where

$$\begin{aligned}\Pi_1(\boldsymbol{q},\omega)&=\frac{1}{\Omega}\sum_{\boldsymbol{k}}\frac{1}{\left[\varepsilon+\hbar\omega-\varepsilon_{k+q}+\mu+i\gamma\ sign\ (\varepsilon+\hbar\omega)\right]\left[\varepsilon-\varepsilon_k+\mu+i\gamma\ sign\ \varepsilon\right]}\\&\simeq\int\frac{\rho(\varepsilon_k)\,d\varepsilon_k}{\left[\varepsilon_k-\varepsilon-\hbar\omega-\mu-i\gamma\ sign\ (\varepsilon+\hbar\omega)\right]\left[\varepsilon_k-\varepsilon-\mu-i\gamma\ sign\ \varepsilon\right]}\\&\quad-\frac{\hbar^2q^2}{2m^*}\int\frac{\rho(\varepsilon_k)\,d\varepsilon_k}{\left[\varepsilon_k-\varepsilon-\hbar\omega-\mu-i\gamma\ sign\ (\varepsilon+\hbar\omega)\right]^2\left[\varepsilon_k-\varepsilon-\mu-i\gamma\ sign\ \varepsilon\right]}\\&\quad+\frac{2\hbar^2q^2}{3m^*}\int\frac{\varepsilon_k\ \rho(\varepsilon_k)\,d\varepsilon_k}{\left[\varepsilon_k-\varepsilon-\hbar\omega-\mu-i\gamma\ sign\ (\varepsilon+\hbar\omega)\right]^3\left[\varepsilon_k-\varepsilon-\mu-i\gamma\ sign\ \varepsilon\right]}.\end{aligned} \tag{7.166}$$

In the limit of small ω and q, we obtain,

$$\Pi_1(\boldsymbol{q},\omega)\simeq\frac{\pi\ \rho(\mu)}{\gamma}+\frac{i\pi\hbar\omega\ \rho(\mu)}{2\gamma^2}-\frac{\pi\mu\ \rho(\mu)}{2\gamma^3}\frac{\hbar^2q^2}{3m^*}=\frac{1}{\xi_0}+\frac{i\hbar\omega}{2\xi_0\gamma}-\frac{\mu}{2\xi_0\gamma^2}\frac{\hbar^2q^2}{3m^*}. \tag{7.167}$$

Substituting (7.167) to (7.165), we obtain

$$D(\boldsymbol{q},\omega)=\frac{\xi_0/\tau}{-i\omega+Dq^2}, \tag{7.168}$$

where $D = v_F^2\tau/3$ is the diffusion coefficient in a 3D semiconductor. Thus, the diffuson determines the excitation spectrum $\omega = iDq^2$, the spectrum corresponding to the diffusion equation $\left(\frac{\partial}{\partial t} + iDq^2\right) D(q, t-t') = \delta(t-t')$.

Below follows the equation for cooperon,

$$C(q,\omega) = \xi_0 + \xi_0 \sum_{k_1} G(k_1 + q, \varepsilon + \hbar\omega)\, G(-k_1, \varepsilon)\, C(q,\omega), \tag{7.169}$$

and its solution in the form,

$$C(q,\omega) = \frac{\xi_0}{1 - \xi_0\, \Pi_2(q,\omega)}, \tag{7.170}$$

where

$$\Pi_2(q,\omega) = \frac{1}{\Omega} \sum_k \frac{1}{\left[\varepsilon + \hbar\omega - \varepsilon_{k+q} + \mu + i\gamma\ sign\,(\varepsilon + \hbar\omega)\right]\left[\varepsilon - \varepsilon_{-k} + \mu + i\gamma\ sign\ \varepsilon\right]} \tag{7.171}$$

Since $\varepsilon_k = \varepsilon_{-k}$, we obtain formally the same result, $\Pi_2(q,\omega) = \Pi_1(q,\omega)$, as for the diffuson. However, there is one essential detail. The method demonstrated above implies the coherence of electron wave function, meaning that the phase of moving electron depends only on the propagation path. The coherence breaks at finite temperatures. The temperature-induced decoherence is describing by introducing the *phase relaxation time* τ_φ. As a result, we have to write for the cooperon

$$C(q,\omega) = \frac{\xi_0/\tau}{-i\omega + Dq^2 + 1/\tau_\varphi}. \tag{7.172}$$

As the diffuson (7.168) determines the propagation of electron density, $\langle \psi^\dagger(r)\, \psi(r) \rangle$, it is not affected by the decoherence. The point is that the propagator (7.168), the diffuson, is the Green's function for diffusion equation, which links to particle conservation, thus being not sensitive to the electron phase. The detailed analysis shows that cooperon (7.172) is very sensitive to spin–orbit interactions and magnetic field, so it depends on parameters of the spin-flip scattering – spin relaxation time τ_s and spin–orbit relaxation time τ_{so}.

Now, one can use ladder diagrams in Figure 7.6 to calculate quantum corrections to conductivity, as illustrated in Figure 7.7. As we already know (see the end of Section 7.12.1), the vertex correction from the diffuson does not affect the conductivity if the impurity potential is short-ranged, $V(q) = V_0$. Thus, the main effect comes from the cooperon included in the Kubo diagram in Figure 7.7. The analytical

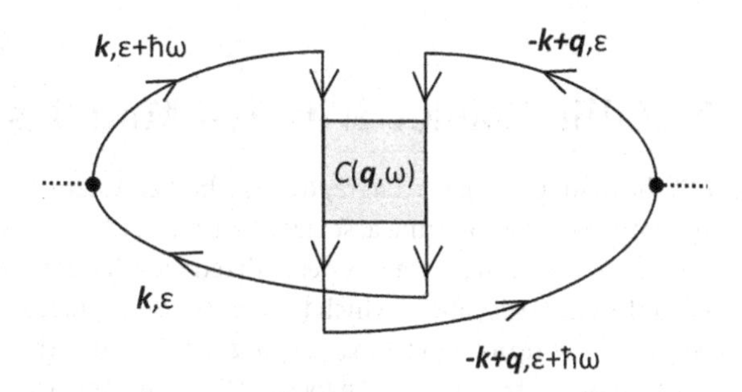

FIGURE 7.7 Quantum correction to the conductivity.

expression for this correction follows (we performed the same calculation steps as in Section 7.13, see Eq. (7.128)):

$$\Delta\sigma = \lim_{\omega\to 0}\frac{2e^2\hbar^2}{3m^{*2}\omega\Omega^2}\int\frac{d\varepsilon}{2\pi}\sum_{k,q}k\cdot(-k+q)\ G(k,\varepsilon+\hbar\omega)\,G(k,\varepsilon)$$

$$\times C(q,\omega)\ G(-k+q,\varepsilon+\hbar\omega)\,G(-k+q,\varepsilon)$$

$$\approx -\frac{e^2\hbar}{3\pi m^*\Omega}\int\frac{\varepsilon_k\,\rho(\varepsilon_k)\,d\varepsilon_k}{(\varepsilon_k-\mu-i\gamma)^2\,(\varepsilon_k-\mu+i\gamma)^2}\sum_q\frac{\xi_0}{Dq^2\tau+\tau/\tau_\varphi}$$

$$\approx -\frac{e^2\hbar\mu\,\rho(\mu)}{3m^*\gamma^3\Omega}\sum_q\frac{\xi_0}{Dq^2\tau+\tau/\tau_\varphi} = -\frac{2e^2}{\pi\hbar\Omega}\sum_q\frac{1}{q^2+1/\ell_\varphi^2}, \tag{7.173}$$

where $\ell_\varphi=\sqrt{D\tau_\varphi}$ is the *phase coherence length*. As we assume phase relaxation time τ_φ is much larger than momentum relaxation time τ, the value of ℓ_φ might become longer than the sample size (for example, in a thin semiconductor film, $\ell_\varphi \gg d$, where d is the film thickness). In such samples, the electron transport is effectively two-dimensional or one-dimensional, even though from the viewpoint of usual size quantization, this is a 3D semiconductor. The samples, which are effectively low-dimensional, are called *mesoscopic*. Bearing this in mind, one can calculate from Eq. (7.173) the quantum correction to conductivity for the samples of different effective dimensionality:

$$\Delta\sigma^{(1D)} = -\frac{e^2\,\ell_\varphi}{\pi\hbar}, \tag{7.174}$$

$$\Delta\sigma^{(2D)} = -\frac{e^2}{2\pi^2\hbar}\log\frac{\tau_\varphi}{\tau}, \tag{7.175}$$

$$\Delta\sigma^{(3D)} = -C+\frac{e^2}{2\pi^2\hbar\,\ell_\varphi}. \tag{7.176}$$

In the 2D case, we have cut off the integral over q at $q_{max}=1/\ell$, assuming that the quantum correction should disappear when $\tau_\varphi\to\tau$. In the 3D case, we cannot integrate at large q as used the cooperon calculated for $q\ll 1/\ell$. Therefore, we hide the contribution from large q in constant $-C$, where $C>0$.

In all dimensions, the quantum correction to conductivity is negative, being an indicator of weak localization. The correction magnitude does not depend on σ_0, so the relative value of quantum correction increases at smaller σ_0. The temperature dependence of localization correction is determining by $\tau_\varphi(T)$. In many cases, one can take $\tau_\varphi\sim T^{-p}$ with p depending on the mechanism of phase relaxation.

7.19 Minimum Metallic Conductivity and Mott Transition

As mentioned in Section 7.17, the main parameter that regulates whether a state with energy ε is localized is $\varepsilon\tau/\hbar$, where τ is the momentum relaxation time associated with electron scattering by impurities: if $\varepsilon\tau/\hbar\ll 1$, the electron state is localized, otherwise it is delocalized (conductive). Thus, one can roughly determine the critical value of the energy ε_c, above which the states are conductive. This critical value of energy is the *localization edge*, and the transition from $\varepsilon<\varepsilon_c$ to $\varepsilon>\varepsilon_c$ is called the *Mott transition*. It can be considered an insulator-metal phase transition. The localization length L, which is finite at $\varepsilon<\varepsilon_c$, is increasing and diverges at $\varepsilon\to\varepsilon_c$.

For the scattering from short-range impurities, we have $\tau = \hbar / 2\pi n_i V_0^2 \rho(\varepsilon)$. Then taking $\rho(\varepsilon) = \rho_0 \varepsilon^{1/2}$ $\left[\rho_0 = (2m^*)^{3/2}/2\pi^2\hbar^3\right]$ we find for the localization edge $\varepsilon_c = (2\pi n_i V_0^2 \rho_0)^2$. The localization edge ε_c increases with the impurity density n_i.

The conductivity at the metallic side has the finite value:

$$\sigma_{min} = \frac{n(\mu)\, e^2 \tau(\mu)}{m^*} = \frac{8e^2 m^{*2} n_i V_0^2}{3\pi^3 \hbar^5}, \tag{7.177}$$

where we substituted the chemical potential $\mu = \varepsilon_c + \delta$ $(\delta \to 0)$. That is, the minimum metallic conductivity at the metal-insulator transition.

Note that in close vicinity to the Mott transition the fluctuations of conductivity can be strong so that the simple picture of insulator-metal transition as a jump from $\sigma = 0$ to $\sigma = \sigma_{min}$ at $\varepsilon = \varepsilon_c$ is not well justified. The enhanced probability of fluctuations of conductivity near the Mott transition is typical for phase transitions near the critical point.

The main reason for localization, discussed in this chapter, is the scattering by impurities. It turns out that Coulomb interaction between electrons is also important for localization. The theory of the interaction-induced contribution to weak localization was developed by Boris Altshuler and Arkady Aronov [17].

References

1. E.M. Lifshitz and L.P. Pitaevskii, *Physical Kinetics* (Elsevier, Amsterdam, 1981).
2. S.M. Sze, *Physics of Semiconductor Devices* (John Wiley & Sons, New York, 1981).
3. M.C. Chang and Q. Niu, "Berry phase, hyperorbits, and the Hofstadter spectrum: Semiclassical dynamics in magnetic Bloch bands", *Phys. Rev. B* **53**, 7010 (1996).
4. G. Sundaram and Q. Niu, "Wave-packet dynamicsin slowly perturbed crystals: Gradient correction and Berry-phase effects", *Phys. Rev. B* **59**, 14915 (1999).
5. A.L. Fetter and J.D. Walecka, *Quantum Theory of Many-Particle Systems* (Dover, New York, 2003).
6. A.A. Abrikosov, L.P. Gorkov, and I.E. Dzyaloshinskii, *Methods of Quantum Field Theory in Statistical Physics* (Dover, New York, 1963).
7. G.D. Mahan, *Many-Particle Physics*, Chapter 2. (Kluwer, New York, 2000).
8. J. Zak, "Magnetic translation group. II. Irreducible representations". *Phys. Rev.* **134**, A1607 (1964).
9. D.J. Thouless, M. Kohmoto, M.P. Nightingale, and M. de Nijs, "Quantized Hall conductance in a two-dimensional periodic potential". *Phys. Rev. Lett.* **49**, 405 (1982).
10. R.B. Laughlin, "Quantized Hall conductivity in two dimensions". *Phys. Rev. B* **23**, 5632 (1981).
11. B.I. Halperin, "Quantized Hall conductance, current-carrying edge states, and the existence of extended states in a two-dimensional disordered potential". *Phys. Rev. B* **25**, 2185 (1982).
12. D.C. Tsui, H.L. Störmer, and A.C. Gossard, "Two-dimensional magnetotransport in the extreme quantum limit". *Phys. Rev. Lett.* **48**, 1559 (1982).
13. R.B. Laughlin, "Anomalous quantum Hall effect: An incompressible quantum fluid with fractionally charged excitations". *Phys. Rev. Lett.* **50**, 1395 (1983).
14. J.K. Jain, "Composite-fermion approach for the fractional quantum Hall effect", *Phys. Rev. Lett.* **63**, 199 (1989).
15. B.I. Halperin, P.A. Lee, and N. Read, "Theory of the half-filled Landau level". *Phys. Rev. B* **47**, 7312 (1993).
16. B.L. Altshuler, A.G. Aronov, D.E. Khmelnitskii, and A.I. Larkin, "Coherent effects in disordered conductors", In: *Quantum Theory of Solids*, Ed. by I.M. Lifshits (MIR Publisher, Moscow, 1982), pp. 130–237.
17. B.L. Altshuler and A.G. Aronov, "Electron-electron interaction in disordered conductors", In: *Electron-Electron Interactions in Disordered Systems*, Ed. by A.L. Efros and M. Pollak (Elsevier, Amsterdam, 1985), pp. 1–153.

8

Impurity Band Conductivity

8.1 Low-Temperature Conductivity and Electron Hopping

In Chapter 7, we discussed conductivity in the conduction and valence bands. In intrinsic semiconductors (negligible density of dopants) and high temperatures, the carrier density is generated mostly by the direct thermal excitation of electron–hole pairs: $n, p \sim N_{c,v} \exp\left(-E_g/2k_BT\right)$, E_g is the bandgap (see Chapter 4). The conductivity repeats strong exponential temperature dependence as mobilities μ_n, μ_p weakly depend on T: $\sigma = q\left(n\mu_n + p\mu_p\right) \sim \exp\left(-E_g/2k_BT\right)$.

At low temperatures, $k_BT << E_g$, the electrons or holes are mostly due to the activation of impurity levels. Let us consider an n-type semiconductor in which conduction electrons come from the thermal activation of donors. For low donor density, one can neglect the effect of disorder on the conduction band (see the discussion in Chapter 3). Then the concentration of electrons depends on temperature as $n \sim \exp\left(-\Delta E_d/2k_BT\right)$ for $k_BT < \Delta E_d$ and $n \simeq const$ for $k_BT >> \Delta E_d$ (here ΔE_d is the donor activation energy) and thus determines the temperature dependence of conductivity.

When the concentration of donors increases, the discrete impurity level becomes an impurity band of finite width. The impurity band manifests itself as a smeared peak of the density of states in the bandgap. This peak may merge into the conduction band upon a further increase in donor density. We discussed the formation of the impurity band in Section 3.10 of Chapter 3.

If spatially decaying impurity electron states have a localization radius less than the average inter-donor distance, electron states in the impurity band are localized and not conductive. Upon increasing donor density, the wavefunctions become overlapped, which does not result in conduction in a disordered impurity system. In a pair of neighboring impurities with equal energy of localized electrons, the overlap repels the levels, transforming them into states with different energies. These bonding and antibonding two-impurity states do not mix to the state at a third impurity as long as the distances in different pairs are not the same. As a result, we obtain three-level electron states, in which one state is separated from two others, belonging to the pair of nearest impurities. At a moderate impurity density, the impurity band consists of energy levels with electrons localized at different sites. In other words, at $T = 0$, an electron cannot hop from site to site, meaning a lack of conductivity in the impurity band. However, at $T \neq 0$, an electron can move a long distance by phonon-assisted jumps between different energy states. The phonon-assisted motion of electrons, most of the time residing at donors or acceptors, is called *hopping conductivity*. At low temperatures, hopping is the dominant mechanism of conductivity in doped semiconductors. Note that when k_BT is of the order of the impurity band width, the electron hopping occurs mainly between the nearest neighbors.

At lower temperatures, only the hopping between the energy states separated by k_BT is possible. We discuss this case in the following section.

DOI: 10.1201/9780429285929-8

8.2 Variable-Range Hopping

The theory of variable-range hopping conductivity was developed in 1971 [1]. In a pair of impurities at a distance $R_{ij} = |\boldsymbol{R}_i - \boldsymbol{R}_j|$ apart, the donor energy splitting falls with R_{ij}; thus, the higher R_{ij}, the lower energy splitting. The theory assumes that the dominant contribution to conductivity comes from jumps between sites at distances corresponding to the level splittings less than parameter $\varepsilon_0 \sim k_B T$.

The main assumption of the theory is that the dominating contribution to conductivity comes from jumps between sites located at distance R_{ij} corresponding to energy levels separation smaller than a certain value proportional to T. At low temperatures, impurities that contribute much to hopping conductivity are at large distances.

Let us consider the hopping probability between an arbitrary pair of localized impurity states E_i and E_j at a distance R_{ij} apart. Hopping rate between localized states, $\gamma_{ij} \sim \exp(-2\alpha R_{ij})$, where α is the inverse localization radius, the scale of spatial fall-off of impurity wavefunction. We assume that the localized states are separated by a long distance, $\alpha R_{ij} \gg 1$.

The detailed balance principle [2] relates the probabilities of electron transitions $i \to j$ and $j \to i$ between the states of different energies:

$$\gamma_{ij} = \gamma_{ji} \exp\left(\frac{E_i - E_j}{k_B T}\right), \left[\mathrm{s}^{-1}\right]. \tag{8.1}$$

That ensures the balance between transitions $i \to j$ and $j \to i$ in the equilibrium. With distance dependence included, the hopping probability takes the form,

$$\gamma_{ij} = \gamma_0 \begin{cases} \exp\left(-2\alpha R_{ij} - \dfrac{E_j - E_i}{k_B T}\right), & E_j > E_i, \\ \exp\left(-2\alpha R_{ij}\right), & E_i > E_j, \end{cases} \tag{8.2}$$

where γ_0 relates to electron-phonon coupling. As the transition can only occur from occupied state i to empty state j, the inclusion of occupation factors brings the transition rate to the form:

$$\Gamma_{ij} = n_i\left(1 - n_j\right)\gamma_{ij}, \tag{8.3}$$

where the occupation factor $n_i = \left[\exp(E_i - \mu)/k_B T + 1\right]^{-1}$ and μ is the chemical potential.

In what follows, we assume that the main contribution to hopping conductivity is due to transitions within an energy range of $2\varepsilon_0$ around the Fermi energy, as shown in Figure 8.1: $|E_i - E_j| < 2\varepsilon_0$, $\mu - \varepsilon_0 < E_i, E_j < \mu + \varepsilon_0$, where ε_0 is a parameter to be determined later. We also assume that $\varepsilon_0 \gg k_B T$, and, correspondingly, $|E_i - \mu|, |E_j - \mu| \gg k_B T$.

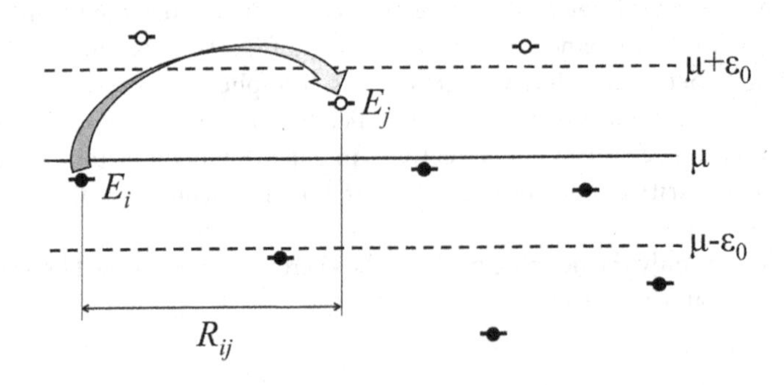

FIGURE 8.1 Variable-range hopping of electrons in the impurity band.

Using the first row in (8.2) and $n_i \simeq 1, n_j \simeq 0$, the transition rate (8.3) can be written as

$$\Gamma_{ij} = \gamma_0 \exp\left[-2\alpha R_{ij} - \frac{1}{2k_B T}\left(|E_i - \mu| + |E_j - \mu| + |E_i - E_j|\right)\right]. \tag{8.4}$$

Indeed, if $E_j > \mu$ and $E_i < \mu$, then

$$\frac{1}{2}\left(|E_i - \mu| + |E_j - \mu| + |E_i - E_j|\right) = E_j - E_i \,, \tag{8.5}$$

and one obtains

$$\Gamma_{ij} = \gamma_0 \exp\left(-2\alpha R_{ij} - \frac{E_j - E_i}{k_B T}\right), \quad \left(E_i < \mu, E_j > \mu\right), \tag{8.6}$$

Note that the transition rate (8.4) satisfies the detailed balance principle, $\Gamma_{ij} = \Gamma_{ji}$.

An electric field E applied along R_{ij} breaks the equilibrium. Chemical potentials at $\boldsymbol{R}_i$ and $\boldsymbol{R}_j$ points are not the same, $\mu_i = \mu_0 + \delta\mu_i$ and $\mu_j = \mu_0 + \delta\mu_j$, respectively ($\mu_0$ is the equilibrium value). The local variations in chemical potential are related by the condition $\delta\mu_j - \delta\mu_i = qU_{ij}$, where $U_{ij} = ER_{ij}$ is the applied voltage. Besides, the electric field changes the energy of levels E_i, E_j at different points. In linear to electric field approximation, the current has the form,

$$I_{ij} = q\left(\Gamma_{ij} - \Gamma_{ji}\right) = G_{ij} U_{ij} \,. \tag{8.7}$$

Here, we introduced the conductance G_{ij} of (ij) element of a network consisting of all possible impurity pairs providing the transfer of electrons by the hopping:

$$G_{ij} = \frac{q^2 \Gamma_{ij}^0}{k_B T}, \tag{8.8}$$

where Γ_{ij}^0 is the equilibrium transition rate. Such a network of (ij) elements is known as the *Miller–Abrahams resistance network* [3] and schematically shown in Figure 8.2.

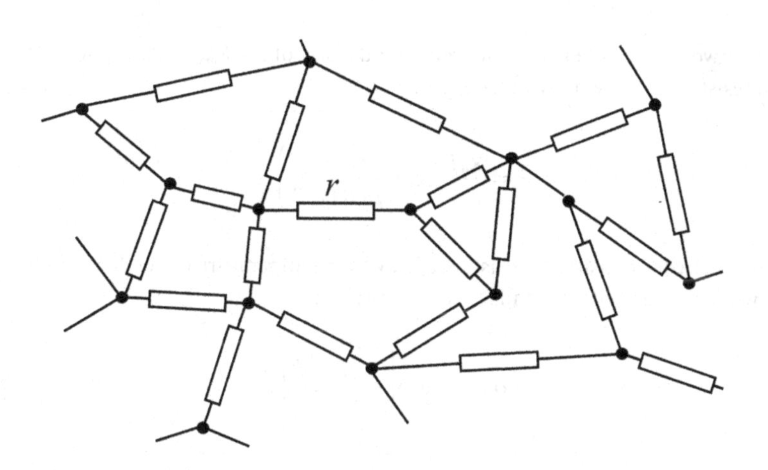

FIGURE 8.2 Miller–Abrahams resistance network.

Numerical simulation of the Miller–Abrahams resistance network shows that once conductance of every single element is less than critical, $G_{ij} < r_c^{-1}$, the whole network is not conductive. For more information on the Miller–Abrahams network, as well as the percolation theory approach to the problem, see Ref. [4].

Using Eq. (8.4), we find the following condition of nonzero hopping conductivity:

$$\gamma_0 \exp\left[-2\alpha R_{ij} - \frac{1}{2k_BT}\left(|E_i - \mu| + |E_j - \mu| + |E_i - E_j|\right)\right] > \frac{k_BT}{q^2 r_c}, \tag{8.9}$$

or in the other form,

$$\frac{R_{ij}}{R_{\max}} + \frac{|E_i - \mu| + |E_j - \mu| + |E_i - E_j|}{2E_{\max}} < 1, \tag{8.10}$$

where we denote

$$R_{\max} = \frac{1}{2\alpha}\frac{E_{\max}}{k_BT}, \quad E_{\max} = k_BT \log\left(\frac{\gamma_0 q^2 r_c}{k_BT}\right). \tag{8.11}$$

Equation (8.10) holds if $R_{ij} < R_{\max}$ and $|E_i - \mu|, |E_j - \mu| < E_{\max}$. Thus, we have found $\varepsilon_0 = E_{\max}$.

To satisfy the condition $R_{ij} < R_{\max}$, the density of impurities with energies $|E_i - \mu| < \varepsilon_0$ should not be too small. Assuming the density of states in the impurity band is nearly constant $\rho(\varepsilon) \simeq \rho(\mu)$ in the narrow energy range $\mu - \varepsilon_0 < \varepsilon < \mu + \varepsilon_0$, we find the volume concentration of impurities with energies in the interval $2\varepsilon_0$:

$$\Delta n = 2\rho(\mu)\varepsilon_0. \tag{8.12}$$

The average distance between these impurities is $\bar{R}_{ij} \simeq (\Delta n)^{-1/3}$. Correspondingly, the condition for hopping on long distances (percolation) $R_{ij} < R_{\max}$ can be written as $\Delta n\, R_{\max}^3 > 1$. A more sophisticated percolation theory gives

$$\Delta n\, R_{\max}^3 > \nu_c \tag{8.13}$$

with $\nu_c \simeq 4$.

The critical network corresponds to the impurity density of $\Delta n_c R_{\max}^3 = \nu_c$. Using (8.11) and (8.12), we obtain critical resistance of the network element,

$$r_c = \frac{k_BT}{\gamma_0 q^2} \exp\left(\frac{4\alpha^3 \nu_c}{\rho(\mu) k_BT}\right)^{1/4}. \tag{8.14}$$

Thus, the critical network consists of resistances r_c, which temperature dependence follows from (8.14). The whole network conductivity depends on temperature as

$$\sigma(T) \sim \exp\left[-\left(\frac{T_0}{T}\right)^{1/4}\right], \tag{8.15}$$

where

$$T_0 = \frac{4\alpha^3 \nu_c}{\rho(\mu)\, k_B}. \tag{8.16}$$

Equation (8.15) is *Mott's law of conductivity.*

In two-dimensional systems, Eq. (8.13) is to be replaced by $\Delta n_{2D}\, R_{\max}^2 > \nu_c^{2D}$, where Δn_{2D} is the sheet density of impurities with the energies in $\mu - \varepsilon_0 < \varepsilon < \mu + \varepsilon_0$ range. After calculating the critical resistance, we obtain the temperature dependence of 2D conductivity,

$$\sigma_{2D}(T) \sim \exp\left[-\left(\frac{T_0^{2D}}{T}\right)^{1/3}\right], \tag{8.17}$$

where $T_0^{2D} = 2\alpha^2 \nu_c^{2D} / \rho_{2D}(\mu)\, k_B$ and $\rho_{2D}(\mu)$ is the 2D density of states at the Fermi level.

References

1. V. Ambegaokar, B.I. Halperin, and J.S. Langer, "Hopping conductivity in disordered systems", *Phys. Rev. B.* **4**, 2612 (1971).
2. L.D. Landau and E.M. Lifshitz, *Quantum Mechanics*, Chapter 3 (Pergamon, New York, 1977).
3. A. Miller and E. Abrahams, "Impurity conduction at low concentrations", *Phys. Rev.* **120**, 745 (1960).
4. B.I. Shklovskii and A.L. Efros, *Electronic Properties of Doped Semiconductors* (Springer, Berlin, 1984).

9

Spin-Resolved Transport in Semiconductors

9.1 Spin Transport and Spin Current

An electron carries charge and spin. Therefore, electron transport means the transfer of electric charge (electric current) and spin (spin current). Let us first remind arguments of quantum mechanics [1] concerning the current density. We consider the time derivative of the local charge of an electron $e|\psi(\boldsymbol{r},t)|^2$, where $\psi(\boldsymbol{r},t)$ is the scalar wavefunction normalized to the crystal volume Ω: $\int_\Omega d^3r|\psi|^2=1$,

$$
\begin{aligned}
e\frac{\partial}{\partial t}|\psi|^2 &= \frac{ie}{\hbar}\left[\left(H_0\psi^*\right)\psi-\psi^*\left(H_0\psi\right)\right] \\
&= -\frac{i\hbar e}{2m^*}\left[\left(\Delta\psi^*\right)\psi-\psi^*\left(\Delta\psi\right)\right]=-\nabla_i j_i,
\end{aligned}
\tag{9.1}
$$

where we used the wave equation $\left(i\hbar\frac{\partial}{\partial t}-H_0\right)\psi(\boldsymbol{r},t)=0$ with Hamiltonian $H_0=-\hbar^2\Delta/2m^*$ and defined the current density $j_i(\boldsymbol{r},t)$ as

$$
j_i=\frac{ie\hbar}{2m^*}\left[\left(\nabla_i\psi^*\right)\psi-\psi^*\nabla_i\psi\right].
\tag{9.2}
$$

Equation (9.1) becomes a condition for charge conservation (the continuity equation):

$$
\frac{\partial}{\partial t}\,e|\psi|^2+\operatorname{div}\mathbf{j}=0.
\tag{9.3}
$$

If one includes spin into consideration, the Hamiltonian should be a 2×2 matrix in spin space, while wavefunction $\psi(\boldsymbol{r},t)$ is the two-component spinor (see Section 1.11, Chapter 1). Let us assume that the Hamiltonian does not explicitly contain spin so that H_0 is the diagonal matrix. We consider local spin density $\mathbf{S}(\boldsymbol{r},t)=\psi^\dagger(\boldsymbol{r},t)\sigma\psi(\boldsymbol{r},t)$ and calculate the time derivative:

$$
\begin{aligned}
\frac{\partial}{\partial t}\left(\psi^\dagger\sigma_\mu\psi\right) &= \frac{i}{\hbar}\left[\left(H_0\psi^\dagger\right)\sigma_\mu\psi-\psi^\dagger\sigma_\mu\left(H_0\psi\right)\right] \\
&= -\frac{i\hbar}{2m^*}\left[\left(\Delta\psi^\dagger\right)\sigma_\mu\psi-\psi^\dagger\sigma_\mu\left(\Delta\psi\right)\right]=-\nabla_i J_i^\mu,
\end{aligned}
\tag{9.4}
$$

DOI: 10.1201/9780429285929-9

where we define the spin current $J_i^{\mu}(\boldsymbol{r},t)$ as the spin density flux – the flux in the i-direction of the μ-component of the spin density vector,

$$J_i^{\mu} = \frac{i\hbar}{2m^*}\left[\left(\nabla_i\psi^{\dagger}\right)\sigma_{\mu}\psi - \psi^{\dagger}\sigma_{\mu}\nabla_i\psi\right]. \tag{9.5}$$

Correspondingly, one can consider Eq. (9.4) as a spin conservation law

$$\frac{\partial S_{\mu}}{\partial t} + \nabla_i J_i^{\mu} = 0. \tag{9.6}$$

Quantum-field-theory (QFT) method of introducing the spin current relies on symmetry to rotations in spin space. If the Hamiltonian does not contain spin operator $\boldsymbol{\sigma}$, then it is invariant under unitary transformation $T = \exp(i\alpha_{\mu}\sigma_{\mu})$, where α_{μ} is the angle of spin rotation around axis r_{μ}: $TH_0T^{\dagger} = H_0$. We use notation, which implies summation over repeated indices. If α_{μ} does not depend on coordinate, it is called *global symmetry.*

Once we require *local invariance* to spin rotations on angle $\alpha_{\mu}(\boldsymbol{r},t)$ that is to operation $T(\boldsymbol{r},t) = \exp(i\alpha_{\mu}(\boldsymbol{r},t)\sigma_{\mu})$, additional terms – gauge fields – appear in the Lagrangian (and Hamiltonian). For more detail, let us consider the Lagrangian:

$$L = \int d^3r\ \psi^{\dagger}\left(i\hbar\frac{\partial}{\partial t} - H_0\right)\psi, \tag{9.7}$$

where $\psi(\boldsymbol{r},t)$ is the spinor wave function. Time and space derivatives of the operator $T^{\dagger}(\boldsymbol{r},t)$ have the form,

$$\partial_t T^{\dagger} = \partial_t \exp(-i\alpha_{\mu}\sigma_{\mu}) = \exp(-i\alpha_{\mu}\sigma_{\mu})\left[\partial_t - i\left(\partial_t\alpha_{\mu}\right)\sigma_{\mu}\right],$$

$$\nabla_i T^{\dagger} = \nabla_i \exp(-i\alpha_{\mu}\sigma_{\mu}) = \exp(-i\alpha_{\mu}\sigma_{\mu})\left[\nabla_i - i\left(\nabla_i\alpha_{\mu}\right)\sigma_{\mu}\right]. \tag{9.8}$$

Then after transformation $\psi' = T\psi$ and $\left(i\hbar\partial_t - H_0\right) \to T\left(i\hbar\partial_t - H_0\right)T^{\dagger}$, we obtain

$$L' = \int d^3r\,\psi'^{\dagger}\left[i\hbar\left(\frac{\partial}{\partial t} - i\phi\right) + \frac{\hbar^2}{2m^*}(\nabla_i - iA_i)^2\right]\psi', \tag{9.9}$$

where $\phi = \phi^{\mu}\sigma_{\mu} = \left(\partial_t\alpha_{\mu}\right)\sigma_{\mu}$ and $A_i = A_i^{\mu}\sigma_{\mu} = \left(\nabla_i\alpha_{\mu}\right)\sigma_{\mu}$. Functions $\phi(\boldsymbol{r},t)$ and $\boldsymbol{A}(\boldsymbol{r},t)$ are matrix-valued gauge potentials. Lagrangians L and L' and, correspondingly, wave functions ψ and $\psi' = \exp(i\alpha_{\mu}(r,t)\sigma_{\mu})\psi$ are related by gauge transformation and physically equivalent as they describe the same observables. So, to provide local gauge invariance of the Lagrangian, we have to introduce potentials $\phi(\boldsymbol{r},t)$ and $\boldsymbol{A}(\boldsymbol{r},t)$ (spin potentials). If we do so, an arbitrary local rotation of spinor fields $\psi \to T\psi$ is accompaned by the transformation of potentials $\phi \to -iT\phi T^{\dagger} + T\partial_t T^{\dagger}$ and $\boldsymbol{A} \to -iT\boldsymbol{A}T^{\dagger} + T\nabla T^{\dagger}$. For more information about the spin fields, see Ref. [2].

Now we define the spin current as a variational derivative,

$$J_i^{\mu} = \frac{1}{\hbar}\frac{\delta L}{\delta A_i^{\mu}} = \frac{\hbar}{2m^*}\psi^{\dagger}\left[(\nabla_i - iA_i)\sigma_{\mu} + \sigma_{\mu}(\nabla_i - iA_i)\right]\psi = \frac{1}{2}\psi^{\dagger}\left\{\hat{v}_i,\sigma_{\mu}\right\}\psi, \tag{9.10}$$

where

$$\hat{v}_i = -\frac{i\hbar}{m^*}(\nabla_i - iA_i) \tag{9.11}$$

is the velocity operator, $\{\hat{a},\hat{b}\} = \hat{a}\hat{b} + \hat{b}\hat{a}$ is the anticommutator of operators $\hat{a}$ and $\hat{b}$. In the absence of external spin fields, we omit A_i in Eq. (9.11), as the one eliminated by a suitable gauge transformation.

The condition of minimum action with respect to the variation of $\psi^\dagger$ (zero variation of $\mathcal{A} = i\int dt\, L$) results in the usual Schrödinger equation for ψ. Variation for spin-rotation angle α_μ gives us the spin conservation equation (9.6). Thus, in (9.6), we come to the previous result (9.5) using the QFT method.

Spin current definition (9.5) related to spin conservation (9.6) cannot be generalized to the case when Hamiltonian includes spin. Conservation is a consequence of the Hamiltonian not depending on spin variables. Once the Hamiltonian includes spin–orbit interaction, it becomes nondiagonal in spin space, the spin density conservation breaks, and additional term – torque density – appears on the right-hand side of Eq. (9.6). Therefore, we will be using (9.5) and (9.6) when the spin–orbit interaction is absent or negligibly small. The definition of torque is discussed below in Section 9.6.

9.2 Anomalous Hall Effect

Several perspective spintronics applications one expects from the *anomalous Hall effect* (AHE). This effect allows the measuring of internal magnetization – main characteristics of magnetic semiconductors. Besides, quantized AHE could serve as a standard of resistance complementary to the existing quantum-Hall-based standard used in metrology.

In a typical setting, we direct electric current $\boldsymbol{j}$ along the x-axis perpendicular to magnetization $\boldsymbol{M}$ (in Figure 9.1, vector $\boldsymbol{M}$ is along axis z). Hall electric field E_y appears along axis y. The nonzero transverse electric field (or transverse current) in the ferromagnet is the AHE. At first glance, the AHE is similar to the ordinary Hall effect (Chapter 7 Section 7.5). The difference is that there is no external magnetic field in AHE, and instead, internal magnetization $\boldsymbol{M}$ causes the effect (compare Figures 9.1 and 7.1). In contrast to the ordinary Hall effect, the Lorentz force does not play a role because the magnetic induction associated with internal magnetization is too weak to give quantitative agreement with experiments.

Different physical mechanisms cause AHE, and all of them are due to the spin–orbit interaction. In other words, the AHE is the manifestation of SO interaction, which magnitude follows from measuring anomalous Hall voltage.

The general expression for spin–orbit interaction appears from relativistic quantum mechanics on taking the nonrelativistic limit (see Eq. (1.81), Chapter 1),

$$H_{so} = -\frac{i\lambda_0^2}{4}\left[\sigma \times \left(\nabla W(r)\right)\right] \cdot \nabla, \tag{9.12}$$

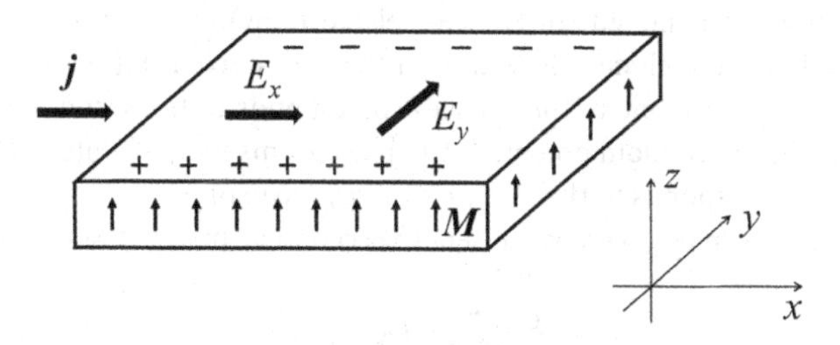

FIGURE 9.1 The geometry of the AHE.

where $\lambda_0 = \lambda_c/2\pi$, λ_c being the Compton wavelength, $W(\mathbf{r}) = e\varphi(\mathbf{r})$, and $\varphi(\mathbf{r})$ is the electrostatic potential. This form of SO interaction can also be used for band electrons in semiconductors, providing the coefficient in (9.12) replaced with material- and structure-dependent coupling constant g, which can be several orders of magnitude larger than the corresponding constant for a free electron.

One can observe AHE in both clean and doped metals and semiconductors. If the main contribution is related to impurities, $W(\mathbf{r})$ is the impurity potential. It is a case of an *extrinsic mechanism* of AHE. In clean materials, we call it the *intrinsic mechanism*, $W(\mathbf{r})$ is the lattice-periodic potential or the interface potential (in a semiconductor layer on a substrate). When analyzing experimental data, one should account for all possible mechanisms, and the dominating mechanism depends on material parameters.

9.3 Mechanism of AHE Related to the Spin-Orbit Scattering from Impurities

If impurity scattering plays the leading role, we deal with the extrinsic mechanism of AHE. The Hamiltonian describes an electron in a doped magnetic crystal,

$$H = -\frac{\hbar^2 \Delta}{2m^*} + M\sigma_z + W(\mathbf{r}) - ig\left[\sigma \times \left(\nabla W(\mathbf{r})\right)\right] \cdot \nabla, \tag{9.13}$$

where the first term is the electron kinetic energy, term $M\sigma_z$ is the spin-split electron energy due to magnetization, $W(\mathbf{r})$ is random impurity potential, and fourth term is the spin–orbit impurity scattering. The spin-split parameter M relates to magnetization m by $M = Jm$, where J is the exchange coupling constant.

We assume that the random potential is Gaussian with the following properties:

$$\langle W(\mathbf{r})\rangle = 0, \quad \langle W(\mathbf{r})W(\mathbf{r}')\rangle = \xi_0 \delta(\mathbf{r} - \mathbf{r}'), \tag{9.14}$$

and then all higher-order correlators are zero. Here the brackets $\langle \ldots \rangle$ mean averaging over all realizations of the random potential $W(\mathbf{r})$.

In what follows, we calculate the electrical current using Green's function method discussed in Section 7.13 (Chapter 7). To calculate current, we include the interaction with the electromagnetic field by taking into account gauge invariance (see details in Chapter 12),

$$\nabla \to \nabla - \frac{ie}{\hbar}\mathbf{A}(t), \quad \mathbf{A}(t) = \mathbf{A}_\omega \exp(-i\omega t). \tag{9.15}$$

Defining the operator of electric current through the velocity operator, $\hat{\mathbf{j}} = e\hat{\mathbf{v}}$, one obtains,

$$\hat{\mathbf{j}} = -\frac{\partial H}{\partial \mathbf{A}} = -\frac{ie\hbar}{m^*}\left(\nabla - \frac{ie\mathbf{A}}{\hbar}\right) + \frac{eg}{\hbar}\left[\sigma \times \left(\nabla W(\mathbf{r})\right)\right] \tag{9.16}$$

The second term in the right-hand side of (9.16) is related to SO impurity scattering. As AHE originates from SO interaction, we include this term into the Hall current represented by Kubo diagrams in Figure 9.2. In these diagrams, one of the vertices corresponds to the second term in (9.16), and the dashed line is the correlator of random potential $W(\mathbf{r})$. Diagrams with SO interaction in the vertex and the Gaussian correlator corresponds to the *side-jump mechanism* of AHE.

Using the Fourier transform of $W(\mathbf{r})$, we present the second term as the SO vertex function,

$$\mathbf{j}_{so} = \frac{eg}{\hbar\Omega}\sum_q e^{iq\mathbf{r}}(\sigma \times \mathbf{q})W(\mathbf{q}), \tag{9.17}$$

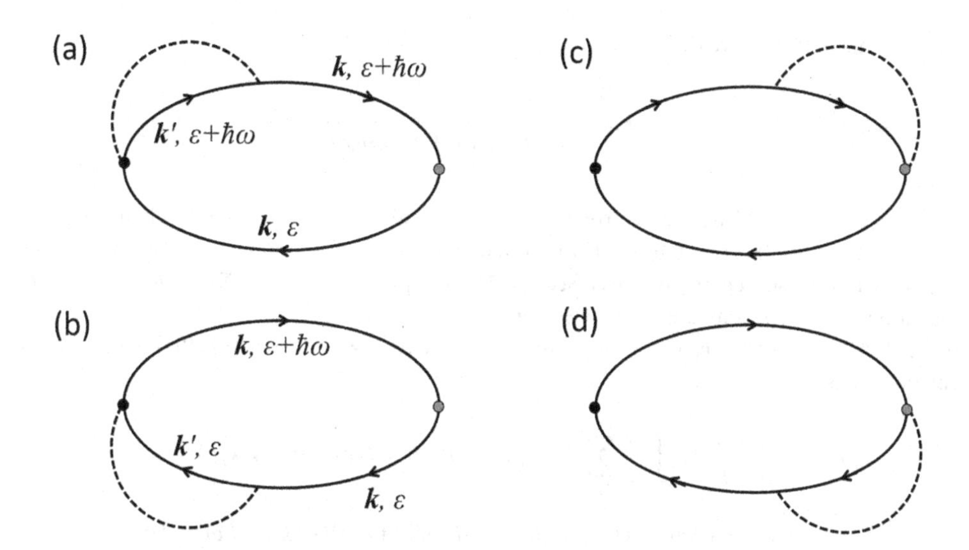

FIGURE 9.2 (a–d) Feynman diagrams for the calculation of transverse current.

where Ω is the sample volume.

Two diagrams with the left SO vertex (9.17) and the right vertex corresponding to electromagnetic field perturbation

$$H_{int} = \frac{ie\hbar}{m^*} \mathbf{A}(t) \cdot \nabla, \tag{9.18}$$

are presented in Figure 9.2a and b. The Fourier transform of Gaussian correlator is given as follows:

$$\langle W(\mathbf{q}) W(\mathbf{q}') \rangle = \xi_0 \delta(\mathbf{q} - \mathbf{q}'). \tag{9.19}$$

The diagrams in Figure 9.2c and d differ from (a) and (b) by the transposition of vertices and give the same contribution. Then the analytic expression for the sum of all four diagrams gives the current density,

$$\begin{aligned} j_i(\omega) = &-\frac{2e^2 g\xi_0}{m^*\Omega^2} \in_{ijl} A_{m\omega}\, Tr\,\sigma_j \\ &\times \int \frac{d\varepsilon}{2\pi} \sum_{kk'} \left[(k_l - k_l')\, k_m\, \hat{G}(\mathbf{k}', \varepsilon + \hbar\omega) \hat{G}(\mathbf{k}, \varepsilon + \hbar\omega) \hat{G}(\mathbf{k}, \varepsilon) + (k_l' - k_l) k_m\, \hat{G}(\mathbf{k}, \varepsilon + \hbar\omega) \hat{G}(\mathbf{k}, \varepsilon) \hat{G}(\mathbf{k}', \varepsilon) \right] \\ = &-\frac{2e^2 g\xi_0}{3m^*\Omega^2} \in_{ijl} A_{l\omega}\, Tr\,\sigma_j \\ &\times \int \frac{d\varepsilon}{2\pi} \sum_{kk'} k^2 \left[\hat{G}(\mathbf{k}', \varepsilon + \hbar\omega) \hat{G}(\mathbf{k}, \varepsilon + \hbar\omega) \hat{G}(\mathbf{k}, \varepsilon) - \hat{G}(\mathbf{k}, \varepsilon + \hbar\omega) \hat{G}(\mathbf{k}, \varepsilon) \hat{G}(\mathbf{k}', \varepsilon) \right], \end{aligned} \tag{9.20}$$

where $\in_{ijl}$ is the unit antisymmetric tensor and $\hat{G}(\mathbf{k}, \varepsilon)$ is the diagonal matrix

$$\hat{G} = \begin{pmatrix} G_\uparrow & 0 \\ 0 & G_\downarrow \end{pmatrix}. \tag{9.21}$$

Matrix elements in (9.21) have the form,

$$G_{\uparrow,\downarrow}(\boldsymbol{k},\varepsilon)=\frac{1}{\varepsilon-\varepsilon_{\uparrow,\downarrow k}+\mu+i\Gamma_{\uparrow,\downarrow}\,sign\,\varepsilon}, \tag{9.22}$$

where $\varepsilon_{\uparrow,\downarrow k}=\hbar^2k^2/2m^*\pm M$ are the electron magnetic subbands, we count ε from the chemical potential μ. Green's function (9.22) corresponds to Hamiltonian (9.13) without SO interaction. The random potential is included in the self-energy part (see Section 7.12, Chapter 7), $\Gamma_{\uparrow,\downarrow}=\left|\mathrm{Im}\,\Sigma_{\uparrow,\downarrow}\right|=\hbar/2\tau_{\uparrow,\downarrow}$, and $\tau_{\uparrow,\downarrow}$ is the relaxation time of spin-up and -down electrons.

For the electric field $\boldsymbol{E}(t)=\boldsymbol{E}_\omega\exp(-i\omega t)$ directed along axis x, we find from (9.20) the transverse Hall current along axis y:

$$\begin{aligned} j_y(\omega)=-\frac{2e^2g\xi_0}{3m^*\Omega^2}A_{x\omega}\int\frac{d\varepsilon}{2\pi}\sum_{kk'}k^2[&G_\uparrow(\boldsymbol{k}',\varepsilon+\hbar\omega)G_\uparrow(\boldsymbol{k},\varepsilon+\hbar\omega)G_\uparrow(\boldsymbol{k},\varepsilon)\\ &-G_\uparrow(\boldsymbol{k},\varepsilon+\hbar\omega)G_\uparrow(\boldsymbol{k},\varepsilon)G_\uparrow(\boldsymbol{k}',\varepsilon)-G_\downarrow(\boldsymbol{k}',\varepsilon+\hbar\omega)G_\downarrow(\boldsymbol{k},\varepsilon+\hbar\omega)G_\downarrow(\boldsymbol{k},\varepsilon)\\ &+G_\downarrow(\boldsymbol{k},\varepsilon+\hbar\omega)G_\downarrow(\boldsymbol{k},\varepsilon)G_\downarrow(\boldsymbol{k}',\varepsilon)]. \end{aligned} \tag{9.23}$$

Using relation $i\omega\boldsymbol{A}_\omega=\boldsymbol{E}_\omega$, we find from (9.23) the off-diagonal component of the conductivity tensor:

$$\begin{aligned} \sigma_{yx}(\omega)=\frac{2ie^2g\xi_0}{3m^*\Omega^2\omega}\int\frac{d\varepsilon}{2\pi}\sum_{kk'}k^2[&G_\uparrow(\boldsymbol{k}',\varepsilon+\hbar\omega)G_\uparrow(\boldsymbol{k},\varepsilon+\hbar\omega)G_\uparrow(\boldsymbol{k},\varepsilon)\\ &-G_\uparrow(\boldsymbol{k},\varepsilon+\hbar\omega)G_\uparrow(\boldsymbol{k},\varepsilon)G_\uparrow(\boldsymbol{k}',\varepsilon)-G_\downarrow(\boldsymbol{k}',\varepsilon+\hbar\omega)G_\downarrow(\boldsymbol{k},\varepsilon+\hbar\omega)G_\downarrow(\boldsymbol{k},\varepsilon)\\ &+G_\downarrow(\boldsymbol{k},\varepsilon+\hbar\omega)G_\downarrow(\boldsymbol{k},\varepsilon)G_\downarrow(\boldsymbol{k}',\varepsilon)]. \end{aligned} \tag{9.24}$$

The main contribution in (9.24) comes from the poles of Green's functions in complex ε-plane. The poles reside in different half-planes for $\varepsilon+\hbar\omega>0$ and $\varepsilon<0$. Then in the limit of $\omega\to0$, we obtain,

$$\sigma_{yx}=\frac{2ie^2g\xi_0\hbar}{3\pi m^*\Omega^2}\sum_{kk'}k^2\left\{G_\uparrow^R(\boldsymbol{k}')\left[G_\uparrow^R(\boldsymbol{k})-G_\uparrow^A(\boldsymbol{k})\right]G_\uparrow^A(\boldsymbol{k})-G_\downarrow^R(\boldsymbol{k}')\left[G_\downarrow^R(\boldsymbol{k})-G_\downarrow^A(\boldsymbol{k})\right]G_\downarrow^A(\boldsymbol{k})\right\}, \tag{9.25}$$

where G^R and G^A are the retarded and advanced Green's functions at the Fermi level, $\varepsilon=0$, respectively. Note that (9.25) has the contributions of opposite signs from spin-up and spin-down electrons, that means that the off-diagonal conductivity vanishes in the absence of magnetization. To simplify the Hall conductivity, one can use the relation

$$G_\sigma^R(\boldsymbol{k})-G_\sigma^A(\boldsymbol{k})=2i\,\mathrm{Im}\,G_\sigma^R(\boldsymbol{k})\simeq-2\pi i\,\delta\left[\varepsilon_\sigma(\boldsymbol{k})-\mu\right]=-\frac{2\pi im^*}{\hbar^2k_{F\sigma}}\delta(k-k_{F\sigma}), \tag{9.26}$$

where $k_{F\sigma}$ is the Fermi vector for electrons in σ-subband. Then after calculating integrals over $\boldsymbol{k}$ and $\boldsymbol{k}'$ in (9.25), we finally obtain

$$\sigma_{yx}=\frac{2e^2g\xi_0m^*}{3\pi\hbar^4}\left(k_{F\uparrow}^4\tau_\uparrow-k_{F\downarrow}^4\tau_\downarrow\right). \tag{9.27}$$

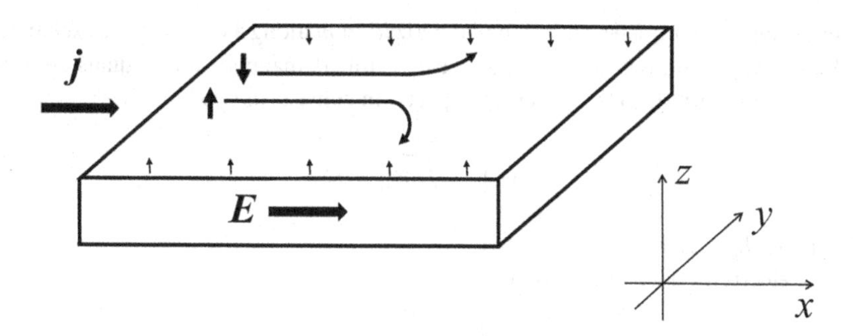

FIGURE 9.3 Separation of spin-up and spin-down currents due to impurity spin-orbit scattering.

As follows from (9.24) to (9.27), the off-diagonal conductivity σ_{yx} (AHE) has two contributions corresponding to spin-up and spin-down electrons, which are of opposite signs. In other words, SO scattering depends on the electron spin as schematically shown in Figure 9.3: moving along x spin-up electrons are preferably scattered to the right, whereas spin-down to the left. A net transverse current is not zero if the numbers of spin-up and down electrons are different due to the magnetization field, meaning the transverse spin accumulation occurs that will be described later as a transverse spin current.

For short-ranged impurities, $W(\boldsymbol{q}) \simeq W_0 = const$ (see Section 7.18, Chapter 7), one may use the explicit forms of correlator (9.19),

$$\xi_0 = n_i \; W_0^2, \tag{9.28}$$

and the relaxation time,

$$\tau_\sigma = \frac{\hbar}{2\pi\rho_\sigma n_i W_0^2}, \tag{9.29}$$

where n_i is the impurity density, $\rho_\sigma = m^* k_{F\sigma}/2\pi\hbar^2$ is the density of states of electrons with spin σ at the Fermi level. Substituting (9.28) and (9.29) into (9.27), we obtain

$$\sigma_{yx} = \frac{2e^2 g}{3\pi\hbar}\left(k_{F\uparrow}^3 - k_{F\downarrow}^3\right). \tag{9.30}$$

Surprisingly, σ_{yx} does not contain information about impurity potential even though we considered AHE due to the impurity SO scattering. Cancellation of W_0^2 in the product $\tau_\sigma \xi_0$ is the property of short-ranged impurity potential.

There is another mechanism of AHE related to impurity SO scattering called *skew scattering*. The diagrams for skew scattering does not include SO interaction in the vertex part while including it in the 3rd order correlator of non-Gaussian random potential. Which extrinsic mechanism of AHE prevails (side-jump or skew scattering) depends on material parameters.

9.4 Intrinsic Mechanism of AHE: Quantization of AHE

The intrinsic mechanism of AHE relies on the spin-orbit part of the lattice-periodic potential. We consider the 2D electrons under perpendicular magnetization, so the k-p model for Bloch amplitudes has the form of the Dirac Hamiltonian,[1]

$$H_k = M\sigma_z + \mathrm{v}_0 \boldsymbol{\sigma} \cdot \boldsymbol{k}, \tag{9.31}$$

[1] The Hamiltonian also describes surface electrons in topological insulators with perpendicular magnetization field.

where $M = Jm$ is the gap parameter related to magnetization $\boldsymbol{m}$ along axis z, J is the exchange coupling parameter, $\boldsymbol{k} = (k_x, k_y)$ is the two-dimensional wave vector. If magnetization equals zero, parameter $\mathrm{v}_0/\hbar$ is the electron velocity. The electron energy spectrum consists of two branches

$$\varepsilon_{1,2}(\boldsymbol{k}) = \pm\sqrt{M^2 + \mathrm{v}_0^2 k^2} \tag{9.32}$$

separated by energy $E_g = 2|M|$.

We define the electron velocity operator as[2]

$$\mathbf{v} = \frac{1}{\hbar}\frac{dH_k}{d\boldsymbol{k}} = \frac{\mathrm{v}_0}{\hbar}\sigma. \tag{9.33}$$

To calculate current along y-axis induced by x-directed electric field $\boldsymbol{E}(t) = \boldsymbol{E}_\omega \exp(-i\omega t)$, we use the Kubo expression (see Section 7.12, Chapter 7):

$$j_y(\omega) = -\frac{ie\mathrm{v}_0}{\hbar S} Tr\int \frac{d\varepsilon}{2\pi}\sum_{\boldsymbol{k}} \sigma_y\, \hat{G}(\boldsymbol{k}, \varepsilon + \hbar\omega) H_{int} \hat{G}(\boldsymbol{k}, \varepsilon), \tag{9.34}$$

where S is the sample area,

$$H_{int} = -\frac{e\mathrm{v}_0}{\hbar}\sigma_x A_{x\omega} \tag{9.35}$$

is the electromagnetic field-induced perturbation, and

$$\hat{G}(\boldsymbol{k}, \varepsilon) = \frac{\varepsilon + \mu + M\sigma_z + \mathrm{v}_0\sigma\cdot\boldsymbol{k}}{\left(\varepsilon - \varepsilon_1(\boldsymbol{k}) + \mu + i\delta\, sign\,\varepsilon\right)\left(\varepsilon - \varepsilon_2(\boldsymbol{k}) + \mu + i\delta\, sign\,\varepsilon\right)}$$
$$= \frac{\varepsilon + \mu + M\sigma_z + \mathrm{v}_0\sigma\cdot\boldsymbol{k}}{\varepsilon_1(\boldsymbol{k}) - \varepsilon_2(\boldsymbol{k})}\left(\frac{1}{\varepsilon - \varepsilon_1(\boldsymbol{k}) + \mu + i\delta\,\mathrm{sign}\,\varepsilon} - \frac{1}{\varepsilon - \varepsilon_2(\boldsymbol{k}) + \mu + i\delta\,\mathrm{sign}\,\varepsilon}\right) \tag{9.36}$$

is the casual Green's function corresponding to Hamiltonian (9.31), $\hat{G}(\boldsymbol{k}, \varepsilon) = \left(\varepsilon - H_k + \mu + i\delta\, sign\,\varepsilon\right)^{-1}$. In the following, we assume that the chemical potential μ is in the gap. For $T = 0$, there are no carriers, so we take δ in (9.34) an infinitesimally small to provide the correct location of poles.

Substituting (9.35) and (9.36) into (9.34), we obtain

$$j_y(\omega) = \frac{e^2\mathrm{v}_0^2 E_{x\omega}}{\hbar^2\omega S} Tr\int \frac{d\varepsilon}{2\pi}\sum_{\boldsymbol{k}} \sigma_y \frac{\varepsilon + \hbar\omega + \mu + M\sigma_z + \mathrm{v}_0\sigma\cdot\boldsymbol{k}}{\varepsilon_1(\boldsymbol{k}) - \varepsilon_2(\boldsymbol{k})}$$
$$\times\left(\frac{1}{\varepsilon + \hbar\omega - \varepsilon_1(\boldsymbol{k}) + \mu + i\delta} - \frac{1}{\varepsilon + \hbar\omega - \varepsilon_2(\boldsymbol{k}) + \mu - i\delta}\right)\sigma_x \frac{\varepsilon + \mu + M\sigma_z + \mathrm{v}_0\sigma\cdot\boldsymbol{k}}{\varepsilon_1(\boldsymbol{k}) - \varepsilon_2(\boldsymbol{k})}$$
$$\times\left(\frac{1}{\varepsilon - \varepsilon_1(\boldsymbol{k}) + \mu + i\delta} - \frac{1}{\varepsilon - \varepsilon_2(\boldsymbol{k}) + \mu - i\delta}\right). \tag{9.37}$$

[2] This definition is in agreement with the previously used current operator $\boldsymbol{v} = j/e = -\,\partial H/e\partial \boldsymbol{A}$, where $\boldsymbol{A}$ is the vector potential introduced with substitution $\boldsymbol{k} \to \boldsymbol{k} - e\boldsymbol{A}/\hbar$.

After calculating the integral over ε and trace over spin, in the limit of $\omega \to 0$, we find the off-diagonal conductance

$$\sigma_{yx} = \frac{4e^2 v_0^2 M}{\hbar} \int \frac{d^2k}{(2\pi)^2} \frac{1}{\left[\varepsilon_1(\boldsymbol{k}) - \varepsilon_2(\boldsymbol{k})\right]^3} = \frac{e^2 v_0^2 M}{4\pi\hbar} \int_0^\infty \frac{kdk}{\left(M^2 + v_0^2 k^2\right)^{3/2}} = \frac{e^2}{4\pi\hbar} \, sign\, M. \tag{9.38}$$

Hence, we obtained the nonzero Hall current in the insulating state. Moreover, the value of σ_{yx} equals $\frac{1}{2}$ in universal conductance units e^2/h, i.e., does not depend on material parameters, and jumps between $\pm e^2/2h$ once M changes sign. This means that the Hall conductance is quantized. Such AHE is usually called the *topological contribution* to the AHE.

The quantized AHE takes place only if the chemical potential lies in the gap, and thus Hall current flows along the sample boundary. If the chemical potential falls in the conduction or valence band, there appear bulk conduction resulting in nontopological (and thus nonquantized) contributions to the AHE. That is the metallic state, in which σ_{yx} depends on material parameters, impurities, and carrier density. Topological contribution to AHE in this state also exists, being related to boundary conduction and masked by diagonal conductivity on the Fermi surface. For more information on the theory of AHE in metals, see Ref. [3].

To better understand the topological origin of (9.38), we present calculation a bit differently [4], that is to express the Hamiltonian (9.31) as

$$H_k = \varepsilon_1(\boldsymbol{k})\boldsymbol{\sigma} \cdot \mathbf{n}(\boldsymbol{k}) \tag{9.39}$$

where $\boldsymbol{n}(\boldsymbol{k})$ is the unit vector with components

$$n_{x,y}(\boldsymbol{k}) = \frac{v_0 k_{x,y}}{\varepsilon_1(\boldsymbol{k})}, \quad n_z(\boldsymbol{k}) = \frac{M}{\varepsilon_1(\boldsymbol{k})}. \tag{9.40}$$

It means that we have the vector field $\boldsymbol{n}(\boldsymbol{k})$ defined in each point of 2D k-plane, realizing a mapping of k-plane (k-space) to the unit Berry sphere (image space), see Figure 9.4. For each point at the k-plane, there is a corresponding point on the Berry sphere.

Now using (9.40) for the calculation of current, we obtain the expression for conductance (for brevity, we omit the dependence of $\varepsilon_{1,2}$ and $\boldsymbol{n}$ on $\boldsymbol{k}$):

$$\begin{aligned}
\sigma_{yx}(\omega) = {} & \frac{e^2}{\hbar^2 \omega S} Tr \int \frac{d\varepsilon}{2\pi} \sum_{\boldsymbol{k}} \sigma_\mu \left(\frac{\partial \varepsilon_1}{\partial k_y} n_\mu + \varepsilon_1 \frac{\partial n_\mu}{\partial k_y} \right) \frac{\varepsilon + \hbar\omega + \mu + \varepsilon_1 \boldsymbol{\sigma} \cdot \boldsymbol{n}}{\varepsilon_1 - \varepsilon_2} \\
& \times \left(\frac{1}{\varepsilon + \hbar\omega - \varepsilon_1 + \mu + i\delta} - \frac{1}{\varepsilon + \hbar\omega - \varepsilon_2 + \mu - i\delta} \right) \sigma_\nu \left(\frac{\partial \varepsilon_1}{\partial k_x} n_\nu + \varepsilon_1 \frac{\partial n_\nu}{\partial k_x} \right) \\
& \times \frac{\varepsilon + \mu + \varepsilon_1 \boldsymbol{\sigma} \cdot \boldsymbol{n}}{\varepsilon_1 - \varepsilon_2} \left(\frac{1}{\varepsilon - \varepsilon_1 + \mu + i\delta} - \frac{1}{\varepsilon - \varepsilon_2 + \mu - i\delta} \right) \\
= {} & -\frac{e^2}{2\hbar^2 \omega} \epsilon_{\mu\nu\lambda} \int \frac{d^2k}{(2\pi)^2} \frac{\partial n_\mu}{\partial k_y} \frac{\partial n_\nu}{\partial k_x} n_\lambda \varepsilon_1 \left(\frac{\hbar\omega - \varepsilon_2}{\varepsilon_2 + \hbar\omega - \varepsilon_1} + \frac{\varepsilon_2}{\varepsilon_2 - \varepsilon_1} \right)
\end{aligned} \tag{9.41}$$

and for $\omega \to 0$ we obtain

$$\sigma_{yx} = \frac{e^2}{8\pi^2\hbar} \epsilon_{\mu\nu\lambda} \int d^2k \frac{\partial n_\mu}{\partial k_y} \frac{\partial n_\nu}{\partial k_x} n_\lambda. \tag{9.42}$$

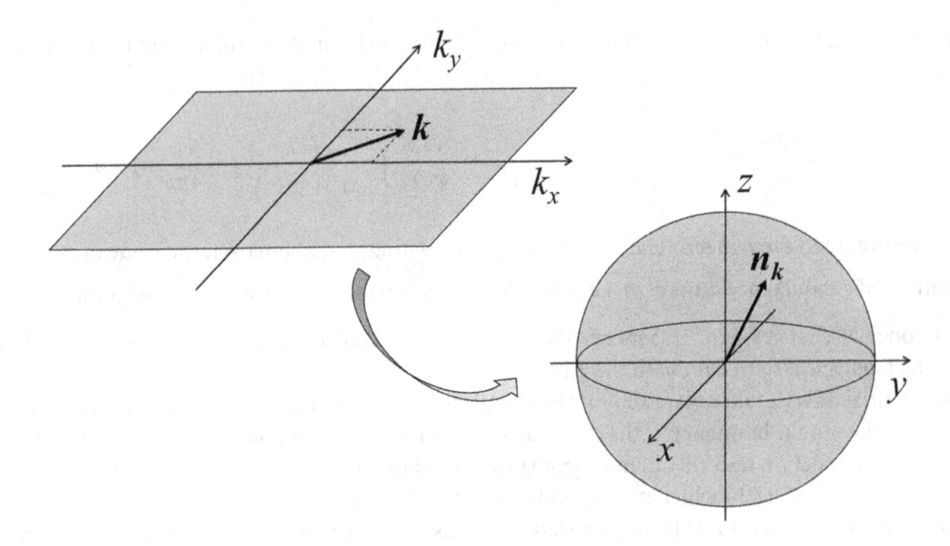

FIGURE 9.4 Mapping of the k-plane to the surface of the Berry sphere.

Integral (9.42) contains the *topological invariant*:

$$Q = \frac{1}{4\pi} \in_{\mu\nu\lambda} \int d^2k \frac{\partial n_\mu}{\partial k_y} \frac{\partial n_\nu}{\partial k_x} n_\lambda. \tag{9.43}$$

Here Q is the winding number – the number of times the Berry sphere is covered while $\boldsymbol{k}$ runs over the 2D Brillouin zone. To see it, we use spherical angles $\theta(\boldsymbol{k})$ and $\varphi(\boldsymbol{k})$ determining the unit vector $\boldsymbol{n}(\boldsymbol{k})$,

$$n_x = \sin\theta\cos\varphi, \quad n_y = \sin\theta\sin\varphi, \quad n_z = \cos\theta. \tag{9.44}$$

As follows from definition (9.40), the whole k-plane is mapping to the half-sphere of $\boldsymbol{n}$-points, and the corresponding spherical angle is 2π.

Using (9.44), we calculate the function under integral (9.42),

$$\in_{\mu\nu\lambda} \frac{\partial n_\mu}{\partial k_y} \frac{\partial n_\nu}{\partial k_x} n_\lambda = \sin\theta \left(\frac{\partial\theta}{\partial k_y} \frac{\partial\varphi}{\partial k_x} - \frac{\partial\theta}{\partial k_x} \frac{\partial\varphi}{\partial k_y} \right) = \sin\theta \frac{\partial(\theta,\varphi)}{\partial(k_y,k_x)}, \tag{9.45}$$

which is the spherical angle at Berry sphere corresponding to the unit area at k_x, k_y plane. Correspondingly, integration over the whole k-plane gives us the angle 2π. Substituting it to (9.42) we come to the result (9.38).

The intrinsic mechanism had been studying by using the model of 2D-metal with Rashba spin-orbit interaction. It was shown that nontopological contributions suppress the topological term in AHE [5]. For a review of AHE, its history, various theoretical approaches to understanding this effect, see Ref. [6].

9.5 Spin Hall Effect

As follows from Section 9.3, the AHE presents an algebraic sum of spin-up and spin-down currents flowing in opposite directions (see, e.g., Eq. (9.23)). Correspondingly, the effect vanishes if there is no magnetization field, and thus currents cancel each other. However, the spin Hall current is not zero even without magnetization as spin currents of spin-up and -down electrons flowing in the opposite direction have the same sign. To confirm this, we calculate the spin current using the operator introduced in Section 9.1:

$$\hat{j}_i^\mu = \frac{1}{2}\{\hat{v}_i, \sigma_\mu\}. \tag{9.46}$$

We calculate spin current in the model of Section 9.3, assuming that the impurity SO scattering is weak, and thus spin nonconservation negligible. From (9.46) it follows that simultaneous change signs of the velocity and spin do not affect $\hat{j}_i^\mu$.

In analogy with (9.17), we can present the expression for SO vertex contributing to the spin current,

$$J_{SO}^\mu = \frac{eg}{2\hbar\Omega}\sum_q e^{iq\cdot r}\{(\sigma \times \mathbf{q}), \sigma_\mu\} W(\mathbf{q}). \tag{9.47}$$

Then performing the same calculations as we did in Section 9.3, we come to (cf. with (9.20))

$$J_i^\mu(\omega) = -\frac{e^2 g\xi_0}{3m^*\Omega^2} \in_{ijl} A_{l\omega}\, Tr\{\sigma_j, \sigma_\mu\}$$

$$\int \frac{d\varepsilon}{2\pi}\sum_{kk'} k^2\left[\hat{G}(\mathbf{k}',\varepsilon+\hbar\omega)\hat{G}(\mathbf{k},\varepsilon+\hbar\omega)\hat{G}(\mathbf{k},\varepsilon) - \hat{G}(\mathbf{k},\varepsilon+\hbar\omega)\hat{G}(\mathbf{k},\varepsilon)\hat{G}(\mathbf{k}',\varepsilon)\right]. \tag{9.48}$$

Correspondingly, for J_y^z, we obtain (cf. with Eq. (9.23))

$$J_y^z(\omega) = -\frac{2e^2 g\xi_0}{3m^*\Omega^2} A_{x\omega}\int \frac{d\varepsilon}{2\pi}\sum_{kk'} k^2[G_\uparrow(\mathbf{k}',\varepsilon+\hbar\omega)G_\uparrow(\mathbf{k},\varepsilon+\hbar\omega)G_\uparrow(\mathbf{k},\varepsilon)$$

$$-G_\uparrow(\mathbf{k},\varepsilon+\hbar\omega)G_\uparrow(\mathbf{k},\varepsilon)G_\uparrow(\mathbf{k}',\varepsilon) + G_\downarrow(\mathbf{k}',\varepsilon+\hbar\omega)G_\downarrow(\mathbf{k},\varepsilon+\hbar\omega)G_\downarrow(\mathbf{k},\varepsilon)$$

$$-G_\downarrow(\mathbf{k},\varepsilon+\hbar\omega)G_\downarrow(\mathbf{k},\varepsilon)G_\downarrow(\mathbf{k}',\varepsilon)], \tag{9.49}$$

which in the static limit $\omega \to 0$ becomes

$$J_y^z = \frac{2e^2 g\xi_0 m^* E_x}{3\pi\hbar^4}\left(k_{F\uparrow}^4\tau_\uparrow + k_{F\downarrow}^4\tau_\downarrow\right). \tag{9.50}$$

Correspondingly, if the magnetization M is zero $(\tau_\uparrow = \tau_\downarrow \equiv \tau)$, we obtain *pure spin Hall current* (here, "pure" means that the transverse charge current j_y is zero):

$$J_y^z = \frac{4e^2 g\xi_0 m^* k_F^4 \tau}{3\pi\hbar^4}E_x. \tag{9.51}$$

This phenomenon is called the *spin Hall effect* (SHE). Like in the case of AHE, there are several different mechanisms of SHE. Here we discussed the side-jump mechanism.

Even though the spin current is theoretically ill-defined in systems with spin–orbit interaction (since spin is not conserved), SHE leads to spin accumulation at the sample boundaries and can be observed experimentally.

9.6 Current-Induced Spin Polarization and Spin Torque

The main aim of semiconductor spintronics is the possibility to manipulate electrically by magnetic moments. One way of converting electric current into a spin accumulation in the transverse to current direction is the spin-Hall effect. Another possibility is to use the intrinsic electron spin polarization induced by electric current flowing through a system with Rashba spin-orbit interaction – the Edelstein-Rashba effect [7].

We consider the 2D Rashba Hamiltonian (see Chapter 2.4)

$$H = H_0 + W(\mathbf{r}) = \frac{\hbar^2 k^2}{2m^*} + \alpha\left(\sigma_x k_y - \sigma_y k_x\right) + W(\mathbf{r}), \tag{9.52}$$

where α is the Rashba spin–orbit interaction, and $W(r)$ is the disorder potential related to impurities. We assume $\langle W(\mathbf{r})\rangle = 0$ and $\langle W(\mathbf{r})W(\mathbf{r}')\rangle = \xi_0\,\delta(\mathbf{r}-\mathbf{r}')$.

The Kubo diagram in Figure 9.5a (see also Section 7.12, Chapter 7) represents the current-induced spin density, where the left vertex $\tilde{\sigma}$ accounts for disorder-induced renormalization of the bare vertex σ (see Figure 9.5b), and the right one is due to the SO-dependent part of electromagnetic perturbation,

$$H_{int}^{so} = -\frac{e\alpha}{\hbar}\left(\sigma_x A_y - \sigma_y A_x\right). \tag{9.53}$$

In what follows, we assume that the electric field $\boldsymbol{E}(t) = \boldsymbol{E}_\omega \exp(-i\omega t)$ is along axis x and, correspondingly, $\boldsymbol{A}(t) = \boldsymbol{E}_\omega \exp(-i\omega t)/i\omega$.

The analytic expression for the spin density corresponding to diagram Figure 9.5a is

$$s_i(\omega) = -\frac{i}{S} Tr \int \frac{d\varepsilon}{2\pi} \sum_k \tilde{\sigma}_i(\omega)\, \hat{G}(\boldsymbol{k}, \varepsilon + \hbar\omega)\, H_{int}^{so}(\omega)\, \hat{G}(\boldsymbol{k}, \varepsilon), \tag{9.54}$$

where S is the sample area,

$$\hat{G}(\boldsymbol{k},\varepsilon) = \left(\varepsilon - H_0 + \mu + i\delta\, sign\,\varepsilon\right)^{-1} = \frac{\varepsilon + \mu - \varepsilon_k + \alpha\left(\sigma_x k_y - \sigma_y k_x\right)}{\left[\varepsilon - \varepsilon_1(\boldsymbol{k}) + \mu + i\Gamma_1\, sign\,\varepsilon\right]\left[\varepsilon - \varepsilon_2(\boldsymbol{k}) + \mu + i\Gamma_2\, sign\,\varepsilon\right]} \tag{9.55}$$

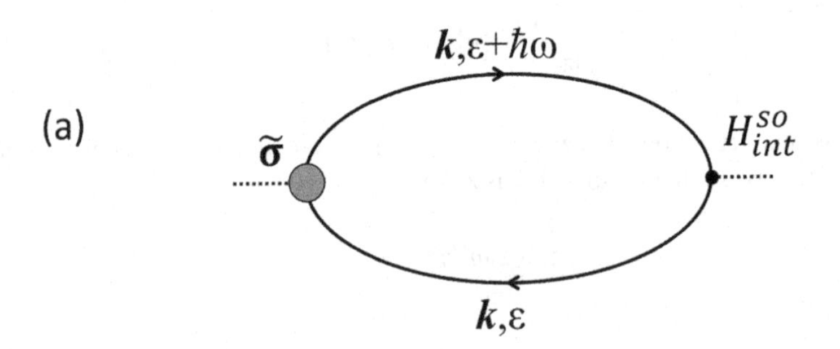

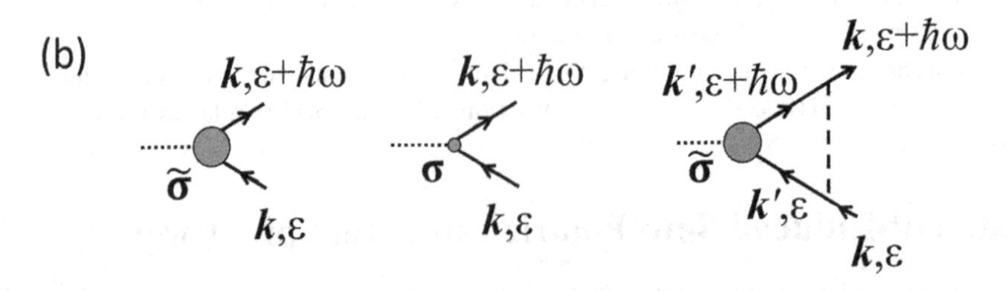

FIGURE 9.5 Feynman diagram for the current-induced spin polarization (a) and the ladder equation for the vertex part (b).

is the electron Green's function, $\varepsilon_{1,2}(\mathbf{k}) = \varepsilon_k \pm \alpha k$ is the electron energy in two subbands, $\varepsilon_k = \hbar^2 k^2 / 2m^*$, and $\Gamma_{1,2} = \hbar / 2\tau_{1,2}$, where $\tau_{1,2}$ are the electron relaxation time in subbands 1 and 2.

First, we calculate the vertex function $\tilde{\sigma}(\omega)$. The ladder equation for $\tilde{\sigma}$ is (see Figure 9.5b)

$$\tilde{\sigma}(\omega) = \sigma + \frac{\xi_0}{S} \sum_{k} \hat{G}(\mathbf{k}, \varepsilon + \hbar\omega) \tilde{\sigma}(\omega) \hat{G}(\mathbf{k}, \varepsilon). \tag{9.56}$$

We look for a solution to this equation in the form,

$$\tilde{\sigma}_i(\omega) = a_i^{\mu}(\omega) \sigma_{\mu}, \tag{9.57}$$

where $a_i^{\mu}(\omega)$ are some coefficients to be determined. Substituting (9.55) and (9.57) to (9.56), we get three equations for the vector components with $i = x, y, z$ (here we omit the explicit dependence of $\varepsilon_{1,2}$ on k)

$$a_i^{\mu}\sigma_{\mu} = \sigma_i + \frac{\xi_0}{S} \sum_{k} \frac{\varepsilon + \hbar\omega + \mu - \varepsilon_k + \alpha\left(\sigma_x k_y - \sigma_y k_x\right)}{\left[\varepsilon + \hbar\omega - \varepsilon_1 + \mu + i\Gamma_1\, sign(\varepsilon + \hbar\omega)\right]\left[\varepsilon + \hbar\omega - \varepsilon_2 + \mu + i\Gamma_1\, sign(\varepsilon + \hbar\omega)\right]}$$
$$\times \frac{a_i^{\mu}\sigma_{\mu}\left[\varepsilon + \mu - \varepsilon_k + \alpha\left(\sigma_x k_y - \sigma_y k_x\right)\right]}{\left(\varepsilon - \varepsilon_1 + \mu + i\Gamma_1\, sign\,\varepsilon\right)\left(\varepsilon - \varepsilon_2 + \mu + i\Gamma_2\, sign\,\varepsilon\right)}, \tag{9.58}$$

After averaging over $\mathbf{k}$-directions in (9.58), $a_i^{\mu}\sigma_{\mu}$ simplifies to

$$a_i^{\mu}\sigma_{\mu} = \sigma_i + \frac{\xi_0}{S} \sum_{k} \frac{\varepsilon + \hbar\omega + \mu - \varepsilon_k}{\left[\varepsilon + \hbar\omega - \varepsilon_1 + \mu + i\Gamma_1\, sign(\varepsilon + \hbar\omega)\right]\left[\varepsilon + \hbar\omega - \varepsilon_2 + \mu + i\Gamma_2\, sign(\varepsilon + \hbar\omega)\right]}$$
$$\times \frac{a_i^{\mu}\sigma_{\mu}(\varepsilon + \mu - \varepsilon_k) - a_i^{z}\sigma_z \alpha^2 k^2}{\left(\varepsilon - \varepsilon_1 + \mu + i\Gamma_1\, sign\,\varepsilon\right)\left(\varepsilon - \varepsilon_2 + \mu + i\Gamma_2\, sign\,\varepsilon\right)}. \tag{9.59}$$

It can be written as three equations for the coefficients before σ_x, σ_y, and σ_z. Then taking $i = x$ or $i = y$ we find $a_x^y = a_x^z = a_y^x = a_y^z = 0$ and

$$a_x^x = a_y^y = \left(1 - \frac{\xi_0}{S} \sum_{k} \frac{(\varepsilon + \hbar\omega + \mu - \varepsilon_k)(\varepsilon + \mu - \varepsilon_k)}{\left[\varepsilon + \hbar\omega - \varepsilon_1 + \mu + i\Gamma_1\, sign(\varepsilon + \hbar\omega)\right]\left[\varepsilon + \hbar\omega - \varepsilon_2 + \mu + i\Gamma_2\, sign(\varepsilon + \hbar\omega)\right]}\right.$$
$$\left. \times \frac{1}{\left(\varepsilon - \varepsilon_1 + \mu + i\Gamma_1\, sign\,\varepsilon\right)\left(\varepsilon - \varepsilon_2 + \mu + i\Gamma_2\, sign\,\varepsilon\right)}\right)^{-1}. \tag{9.60}$$

Choosing $i = z$, we get $a_z^x = a_z^y = 0$ and

$$a_z^z = \left(1 - \frac{\xi_0}{S} \sum_{k} \frac{(\varepsilon + \hbar\omega + \mu - \varepsilon_k)(\varepsilon + \mu - \varepsilon_k) - \alpha^2 k^2}{\left[\varepsilon + \hbar\omega - \varepsilon_1 + \mu + i\Gamma_1\, sign(\varepsilon + \hbar\omega)\right]\left[\varepsilon + \hbar\omega - \varepsilon_2 + \mu + i\Gamma_2\, sign(\varepsilon + \hbar\omega)\right]}\right.$$
$$\left. \times \frac{1}{\left(\varepsilon - \varepsilon_1 + \mu + i\Gamma_1\, sign\,\varepsilon\right)\left(\varepsilon - \varepsilon_2 + \mu + i\Gamma_2\, sign\,\varepsilon\right)}\right)^{-1}. \tag{9.61}$$

Correspondingly, the renormalized by impurities vertex function is

$$\tilde{\sigma}_x = a_x^x \sigma_x, \quad \tilde{\sigma}_y = a_y^y \sigma_y, \quad \tilde{\sigma}_z = a_z^z \sigma_z . \tag{9.62}$$

Next, we calculate the vertex $\tilde{\sigma}(\omega)$ for electrons at the Fermi energy ($\varepsilon \to 0$) in the limit of $\omega \to 0$ and $\alpha \to 0$, assuming $\varepsilon < 0$ and $\varepsilon + \hbar\omega > 0$. As we see later, this is what we need to calculate the static spin density $\boldsymbol{s}$ from Eq. (9.54). The main contributions to integrals[3] in (9.60) and (9.61) are related to poles of integrands. Then we obtain (note that $\Gamma_1 = \Gamma_2 = \Gamma$ at $\alpha \to 0$):

$$\begin{aligned}
&\int \frac{d^2k}{(2\pi)^2} \frac{(\mu - \varepsilon_k)^2}{(\mu - \varepsilon_1 + i\Gamma)(\mu - \varepsilon_2 + i\Gamma)(\mu - \varepsilon_1 - i\Gamma)(\mu - \varepsilon_2 - i\Gamma)} \\
&= -2i\pi \int \frac{d^2k}{(2\pi)^2} (\mu - \varepsilon_k)^2 \left(\frac{\delta(\mu - \varepsilon_1)}{-2i\Gamma(\varepsilon_1 - \varepsilon_2)^2} + \frac{\delta(\mu - \varepsilon_2)}{-2i\Gamma(\varepsilon_1 - \varepsilon_2)^2} \right) \\
&= \frac{\pi\tau}{\hbar} \int \frac{d^2k}{(2\pi)^2} \delta(\varepsilon_k - \mu) = \frac{m^*\tau}{2\hbar^3},
\end{aligned} \tag{9.63}$$

where we kept $\Gamma \ll \hbar\omega$, which is justified for low impurity density and weak scattering.

On the other hand, from Chapter 7, the relaxation time τ can be calculated from the self-energy, see Eq. (7.107). In the limit $\alpha \to 0$, τ is the relaxation time of 2D electrons due to the scattering from impurities,

$$\frac{\hbar}{2\tau} = \xi_0 \operatorname{Im} \int \frac{d^2k}{(2\pi)^2} \frac{1}{\mu - \varepsilon_k - i\Gamma} = \pi\xi_0 \int \frac{d^2k}{(2\pi)^2} \delta(\varepsilon_k - \mu) = \frac{m^*\xi_0}{2\hbar^2} \tag{9.64}$$

Finally,

$$\tau = \frac{\hbar^3}{m^*\xi_0}. \tag{9.65}$$

Substituting (9.63) and (9.65) in (9.60) and (9.61), we obtain $a_x^x = a_y^y = 2$ and correspondingly,

$$\tilde{\sigma}_{x,y}(0) = 2\sigma_{x,y}. \tag{9.66}$$

Performing similar calculations, we find in the same limit $a_z^z = 1$ and $\tilde{\sigma}_z(0) = \sigma_z$.

To calculate spin polarization, we substitute (9.53) and (9.55) into (9.54):

$$\begin{aligned}
s_i(\omega) = &-\frac{e\alpha E_{x\omega}}{\hbar S} Tr \int \frac{d\varepsilon}{2\pi} \sum_k \frac{\tilde{\sigma}_i \left[\varepsilon + \hbar\omega + \mu - \varepsilon_k + \alpha(\sigma_x k_y - \sigma_y k_x) \right]}{\left[\varepsilon + \hbar\omega - \varepsilon_1 + \mu + i\Gamma_1 \, sign(\varepsilon + \hbar\omega) \right]\left[\varepsilon + \hbar\omega - \varepsilon_2 + \mu + i\Gamma_1 \, sign(\varepsilon + \hbar\omega) \right]} \\
&\times \frac{\sigma_y \left[\varepsilon + \mu - \varepsilon_k + \alpha(\sigma_x k_y - \sigma_y k_x) \right]}{(\varepsilon - \varepsilon_1 + \mu + i\Gamma_1 \, sign\,\varepsilon)(\varepsilon - \varepsilon_2 + \mu + i\Gamma_2 \, sign\,\varepsilon)} \\
= &-\frac{4e\alpha a_y^y \delta_{iy} E_{x\omega}}{\hbar\omega S} \int \frac{d\varepsilon}{2\pi} \sum_k \frac{\varepsilon + \hbar\omega + \mu - \varepsilon_k}{\left[\varepsilon + \hbar\omega - \varepsilon_1 + \mu + i\Gamma_1 \, sign(\varepsilon + \hbar\omega) \right]\left[\varepsilon + \hbar\omega - \varepsilon_2 + \mu + i\Gamma_1 \, sign(\varepsilon + \hbar\omega) \right]} \\
&\times \frac{\varepsilon + \mu - \varepsilon_k}{(\varepsilon - \varepsilon_1 + \mu + i\Gamma_1 \, sign\,\varepsilon)(\varepsilon - \varepsilon_2 + \mu + i\Gamma_2 \, sign\,\varepsilon)}.
\end{aligned} \tag{9.67}$$

---

[3] Sum over $\boldsymbol{k}$ is a different notation for integral, $\sum_k \to S \int d^2k/(2\pi)^2$.

In the limit of $\omega \to 0$, the integral over ε is not zero when $\varepsilon + \hbar\omega > 0$ and $\varepsilon < 0$. Then we get

$$s_y \simeq -\frac{4e\alpha E_x}{\pi} \int \frac{d^2k}{(2\pi)^2} \frac{(\mu - \varepsilon_k)^2}{(\mu - \varepsilon_1 + i\Gamma)(\mu - \varepsilon_2 + i\Gamma)(\mu - \varepsilon_1 - i\Gamma)(\mu - \varepsilon_2 - i\Gamma)} = -\frac{2e\alpha m^* \tau E_x}{\pi\hbar^3}, \tag{9.68}$$

where we used (9.63) and the limit of small α.

Using the standard formula for conductivity (Chapter 7, Eq. (7.9)), one calculates current density j_x at $\alpha = 0$:

$$j_x = \frac{e^2 n\tau E_x}{m^*} = \frac{e^2 \mu\tau E_x}{\pi\hbar^2}, \tag{9.69}$$

where $n = m^*\mu/\pi\hbar^2$ is the 2D electron density. After substituting E_x from Eq. (9.69) into (9.68), we finally find the current-induced spin density

$$s_y = -\frac{2\alpha m^*}{e\hbar\mu} j_x. \tag{9.70}$$

Thus, we demonstrated the spin-charge conversion in a 2D spin-orbit electron gas.

If a soft magnetic layer with magnetic moment $\boldsymbol{M}$ is on the top of the 2D Rashba system, the current-induced spin polarization (9.70) affects the $\boldsymbol{M}$-orientation. Indeed, the energy of coupling reads $U = J\,\boldsymbol{M}\cdot\boldsymbol{s}$, where J is the constant of the exchange interaction between electron spin and a local spin. Variation of that energy caused by the rotation of vector $\boldsymbol{M}$ at small angle $\delta\varphi$ is $\delta U = J\,(\delta\varphi \times \boldsymbol{M})\cdot\boldsymbol{s}$. Correspondingly, the torque $\boldsymbol{K}$ acting on the moment $\boldsymbol{M}$ is

$$\boldsymbol{K} = -\frac{\partial U}{\partial \varphi} = -J\,\boldsymbol{M} \times \boldsymbol{s}. \tag{9.71}$$

In this case, it is called *spin torque*. So, in the system with spin–orbit interaction, there is a possibility to manipulate magnetic moments by electric current. The physical mechanism of this effect is related to the current-induced spin polarization of electrons.

References

1. L.D. Landau and E.M. Lifshitz, *Quantum Mechanics*, Chapter III, §19 (Elsevier, Amsterdam, 1977).
2. G. Tatara, "Effective gauge field theory of spintronics", *Physica E*. **106**, 208 (2019).
3. N.A. Sinitsyn, A.H. MacDonald, T. Jungwirth, V.K. Dugaev, and J. Sinova. "Anomalous Hall effect in a two-dimensional Dirac band: The link between the Kubo-Streda formula and the semiclassical Boltzmann equation approach". *Phys. Rev. B*. **75**, 045315 (2007).
4. G.E. Volovik, "An analog of the quantum Hall effect in a superfluid ^{3}He film". *Sov. Phys. JETP*. **67**, 1804 (1988).
5. T.S. Nunner, N.A. Sinitsyn, M.F. Borunda, V.K. Dugaev, A.A. Kovalev, Ar. Abanov, C. Timm, T. Jungwirth, J. Inoue, A.H. MacDonald, and J. Sinova, "Anomalous Hall effect in a two-dimensional electron gas", *Phys. Rev. B*. **76**, 235312 (2007).
6. N. Nagaosa, J. Sinova, S. Onoda, A.H. MacDonald, and N.P. Ong. "Anomalous Hall effect", *Rev. Mod. Phys.* **82**, 1539 (2010).
7. V.M. Edelstein, "Spin polarization of conduction electrons induced by electric current in two-dimensional asymmetric electron systems", *Sol. State Communs.* **73**, 233 (1990).

10

Electron Scattering

Electrical and thermal conductivities in solids are the kinetic coefficients defined as a response to the external perturbation, which creates the nonequilibrium state in a lattice. Nonequilibrium state in the electron system means that electron distribution differs from the Fermi function (see Chapter 4). After perturbation turns off, the relaxation to thermal equilibrium goes by scattering on lattice imperfections (vacancies, interstitials, dislocations), impurities, phonons, as well as electron–electron scattering. Relaxation processes determine the kinetic properties of solids. In particular, electron scattering in solids is at the origin of ohmic conductivity and the electron part of the thermal conductivity. Below we discuss the concept of scattering and introduce related physical parameters.

10.1 Elements of Scattering Theory

Electron scattering may occur on various agents such as atoms, electrons, phonons, impurities, etc. If the particles do not change their energy states in the course of the process, the scattering is elastic. Colliding particles have the same total kinetic energy before and after the collision. An example of such an event is the electron collision with charged or neutral impurity. Inelastic scattering implies that part of the total energy goes to the change in their internal energy states. Examples of inelastic scattering are electron–atom collisions accompanied by the intraatomic transition between ground and excited states, also, the electron scattering on acoustic phonons.

We suppose that scattering potential $V(\boldsymbol{r})$ depends on the relative coordinate between an electron and the scattering center, $\boldsymbol{r} = \boldsymbol{r}_1 - \boldsymbol{r}_2$. Indexes 1 and 2 relate to colliding particles. We use the center of mass reference frame, in which the problem comes to the motion of one particle of mass $\mu = m_1 m_2/(m_1 + m_2)$ in the field $V(\boldsymbol{r})$. We also suppose that the potential acts within finite distance a so that $V(r > a) = 0$.

Free electron elastic scattering is the quantum transition from initial state $\psi_{0i} = \exp(i\boldsymbol{k}_i \cdot \boldsymbol{r})$ to final state $\psi_{0f} = \exp(i\boldsymbol{k}_f \cdot \boldsymbol{r})$, both belonging to a continuous spectrum, $\boldsymbol{k}_i$, $\boldsymbol{k}_f$ are electron wavevectors. Wave functions are normalized so that the quantum-mechanical current is equal to electron velocity:

$$j_i = -\frac{i\hbar}{2\mu}\left(\psi_{0i}^* \nabla \psi_{0i} - \psi_{0i} \nabla \psi_{0i}^*\right) = \frac{\hbar \boldsymbol{k}_i}{\mu}. \tag{10.1}$$

If N is the electron volume density, Nj_i gives the incident flux density far from the scattering center (number of electrons per unit square per second).

If the Hamiltonian includes scattering potential, the solution to the Schrödinger equation is Born series (3.32) (Chapter 3). The wave function in the *first Born approximation* is calculated with the first two terms in series (3.32), in other words, by replacement of T-matrix (3.33) by interaction potential. At large distance $r >> a$ the solution is a sum of incident and diverging from the center scattered waves:

$$\psi_i(\boldsymbol{r}) = \psi_{0i}(\boldsymbol{r}) + \psi_s(\boldsymbol{r}),$$

DOI: 10.1201/9780429285929-10

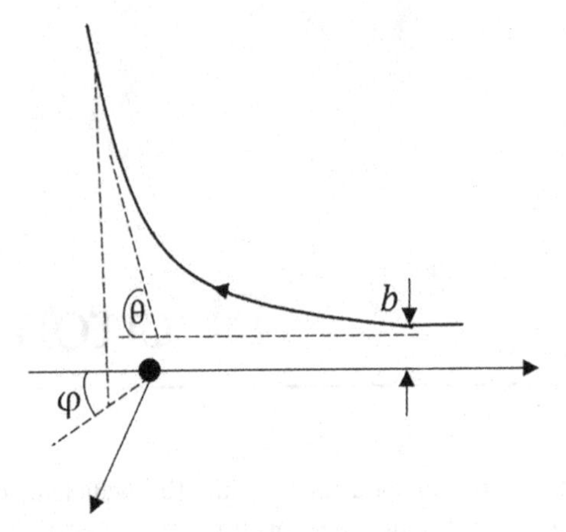

FIGURE 10.1 Particle incident at impact parameter b scatters at angles θ, φ by an immobile impurity center (black circle).

$$\psi_s(\mathbf{r}) = A_{fi} \frac{\exp(ikr)}{r},$$

$$A_{fi} = -\frac{\mu}{2\pi\hbar^2} \int d^3\mathbf{r}' \psi_{0f}^*(\mathbf{r}') V(\mathbf{r}') \psi_{0i}(\mathbf{r}') \equiv -\frac{\mu}{2\pi\hbar^2} \langle \psi_{0f} | V | \psi_{0i} \rangle. \quad (10.2)$$

The term $\psi_s(\mathbf{r})$ in (10.2) is the spherically propagating scattered wave in which factor A_{fi} is the *scattering amplitude*. The amplitude depends on the energy of relative motion, scattering potential, and scattering angles θ and φ.

In the classical picture of a collision, the scattering angles depend on the impact distance, as illustrated in Figure 10.1 For the sake of simplicity, illustration implies the scattering center immobile.

The radial current of scattered particles

$$j_s = -\frac{i\hbar}{2\mu}\left(\psi_s^* \nabla_r \psi_s - \psi_s \nabla_r \psi_s^*\right) = \frac{\hbar k_f}{\mu r^2} |A_{fi}|^2. \quad (10.3)$$

The number of electrons per second scattered in the sold angle $do = \sin\theta \, d\theta \, d\varphi$ and passed through area $r^2 do$ is $N j_s r^2 do$. Now we are ready to define the *differential scattering cross-section* $d\sigma$ as a ratio of scattered and incident fluxes:

$$d\sigma = \frac{k_f}{k_i} |A_{fi}|^2 \, do. \quad (10.4)$$

For elastic scattering, $k_i = k_f \equiv k$,

$$d\sigma = \frac{\mu^2}{4\pi^2\hbar^4} |\psi_{0f}|V|\psi_{0i}|^2 \, do, \; \left[\mathrm{m}^2\right]. \quad (10.5)$$

Result (10.5) follows from the first term of power series on scattering potential (see Chapter 3, Eq. (3.32)), that is, corresponds to the *first Born approximation*. For elastic scattering on spherically-symmetric potential ($k_i = k_f = k$), the validity condition for the approximation reads [1,2]:

$$\left| \int_0^\infty V(r)\left[e^{2ikr} - 1\right] dr \right| \ll \hbar v, \quad (10.6)$$

$v = \hbar k/\mu$ is the electron velocity. In the limits of slow and fast incident particles, the condition follows:

$$\begin{cases} \bar{V} \ll \dfrac{\hbar^2}{2\mu a^2}, & ka \ll 1, \\ \tilde{V} \ll \dfrac{\hbar^2 k}{\mu a}, & ka \gg 1, \end{cases} \tag{10.7}$$

where constants $\bar{V}$ and $\tilde{V}$ relate to the potential:

$$\bar{V} = \frac{1}{4\pi a^2}\left|\int \frac{V(r)}{r} d^3 r\right|, \quad \tilde{V} = \frac{1}{a}\left|\int V(r) dr\right|. \tag{10.8}$$

Inspecting the first line of (10.7), one finds that $\hbar^2/2\mu a^2$ is the minimum depth of the spherical potential well $(V(r) < 0)$ that provides for the first bound state, meaning that the first Born approximation is valid unless the potential captures a slow electron. For fast electrons (the second row in (10.7) the condition is equivalent to $\tilde{V}a << \hbar v$. In the screened Coulomb field,

$$V(r) = \frac{Z_1 Z_2 e^2}{4\pi\varepsilon_0 r} \exp\left(-\frac{r}{a}\right), \tag{10.9}$$

where ε_0 is the vacuum permittivity, $Z_{1,2}$ are the charges of colliding particles. Parameter $\tilde{V}a$ in (10.8) weakly (logarithmically) depends on a, and thus, the validity of Born approximation for Coulomb potential (both screened and unscreened) reads as $Z_1 Z_2 e^2/\varepsilon_0 \ll v\hbar$. The larger the charges, the higher the electron velocity that satisfies the first Born approximation.

The differential cross-section for spherically symmetric potential (10.9) follows from (10.5) and is given by

$$d\sigma = \left[\frac{\mu Z_1 Z_2 e^2}{2\pi\varepsilon_0\left(4p^2 \sin^2\dfrac{\theta}{2} + \hbar^2 a^{-2}\right)}\right]^2 do \equiv \sigma(\theta)\, do, \quad p = \hbar k. \tag{10.10}$$

where the first term in the denominator is the momentum transfer square. The momentum transfer in the course of the scattering event is illustrated in Figure 10.2

It follows from (10.10) that for slow electrons, the cross-section weakly depends on angle. If screening is absent $(a \to \infty)$, Eq.(10.10) tends to the Rutherford cross-section:

$$\sigma(\theta) = \left(\frac{\mu Z_1 Z_2 e^2}{8\pi\varepsilon_0 p^2 \sin^2\dfrac{\theta}{2}}\right)^2. \tag{10.11}$$

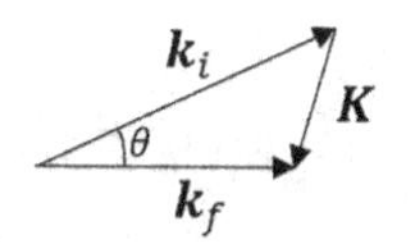

FIGURE 10.2 Momentum transfer in an elastic scattering event, $K = 2k\sin(\theta/2)$.

Divergence in the cross-section at small angles means that the particle scatters no matter how far from the center it moves. It is a consequence of slow fall-off with distance in the Coulomb potential.

In the range $\theta > \theta_0$, cross-section (10.10) strongly depends on θ:

$$2k\sin\frac{\theta_0}{2} \approx \frac{1}{a}, \tag{10.12}$$

Equation (10.12) equals the momentum transfer during a collision, and $\Delta k \approx 1/a$ is the width of the wave packet of size a. Representation of incident and scattered waves by plane waves implies both waves exist in the entire space, including the interaction region. Instead, one should deal with wave packets, all the more the experiments operate with a collimated beam of particles well spatially localized. For a large packet size, $R >> a$, the uncertainty in momentum inside the packet $\Delta k \approx 1/R$ is small compared to momentum transfer $\sim 1/a$ in the course of the scattering event. If this condition holds, the approximation with plane waves is acceptable.

It is instructive to note that on the transition from the screened to unscreened Coulomb scattering, that is, from (10.10) to (10.11), Planck's constant drops out from $\sigma(\theta)$. The original Rutherford formula, obtained in classical collision theory, stays the same if obtained from the quantum theory – the property specific to Coulomb interaction.

10.2 Electron Scattering in Solids

Transport characteristics of solids follow from the solution of the kinetic Boltzmann equation (see Chapter 7). The scattering cross-section is a solution to a mechanical problem of the two-particle collision. It is related to the main parameter in kinetic theory – transition rate in a quasi-discrete electron spectrum. To establish the relation, we start from the quantum-mechanical expression of transition rate (transition probability per second):

$$W_{\lambda\lambda'} = \frac{2\pi}{\hbar}\left|\langle\lambda|V(r)|\lambda'\rangle\right|^2 \delta(E_\lambda - E_{\lambda'}), \ \left[\mathrm{s}^{-1}\right],$$

$$\langle\lambda|V(\boldsymbol{r})|\lambda'\rangle = \int \psi_\lambda^* V(\boldsymbol{r})\psi_{\lambda'} d^3r, \ \int \psi_\lambda^* \psi_{\lambda'} d^3r = \delta_{\lambda\lambda'}, \tag{10.13}$$

where ψ_λ is the Bloch wave function in an ideal crystal lattice, $\lambda = \{\boldsymbol{k}, j, \sigma\}$ are wavevector, band index, and spin, respectively, $V(\boldsymbol{r})$ is the perturbation, that is the deviation of the potential energy from that in an ideal crystal. The delta-function guarantees energy conservation in the collision event. In what follows, we neglect spin and consider $\boldsymbol{k} \to \boldsymbol{k}'$ transitions in a single-band model.

Escape rate from initial state $\boldsymbol{k}_i$ (the inverse time during which an electron remains in state $\boldsymbol{k}_i$) reads as

$$\sum_{k'} W_{k_i k'} = \frac{2\pi}{\hbar}\frac{\Omega}{(2\pi)^3}\int\left|\langle\boldsymbol{k}_i|V|\boldsymbol{k}'\rangle\right|^2 \delta\left(E_{k_i} - E_{k'}\right)k'^2\, dk'\, do, \tag{10.14}$$

where Ω is the crystal volume, sum over discrete wavevector is replaced by integral, and $do = \sin\theta\, d\theta\, d\varphi$. For spherically symmetric potential, the scattering rate in solid angle do depends on θ only and takes the form:

$$W(\theta) = \frac{2\pi}{\hbar}\frac{\Omega}{(2\pi)^3}\int\left|\langle\boldsymbol{k}_i|V|\boldsymbol{k}'\rangle\right|^2 \delta\left(E_{k_i} - E_{k'}\right)k'^2\, dk' \tag{10.15}$$

Assuming parabolic electron dispersion, $E=\hbar^2k^2/2m$, and using relation $\delta(E_k-E_{k'})=\delta(k-k')/\hbar v'$, one obtains transition rate from initial $\boldsymbol{k}_i$ to final state $\boldsymbol{k}_f$ in solid angle *do*:

$$W(\theta)=\frac{\Omega m^2 v_f}{(2\pi)^2\hbar^4}\left|\langle \boldsymbol{k}_i|V|\boldsymbol{k}_f\rangle\right|^2,\quad k_i=k_f. \tag{10.16}$$

Note that expression (10.13) is more general than (10.16) as Eq. (10.13) does not necessarily imply elastic electron scattering $(k_i=k_f)$. Once the energy of quantum state comprises electron and phonon (or other quasiparticles) contributions, the energy conservation allows inelastic processes in which electron energy changes.

If N electrons move with velocity v, the flux – the number of electrons per second moving through a unit area normal to the velocity – is $N\mathrm{v}/\Omega$, Ω is the volume. The number of electrons per second striking the center of cross-section $\sigma(\theta)$ is $N\mathrm{v}\,\sigma(\theta)/\Omega$, [s^{-1}]. On the other hand, this number is the probability per second of scattering to angle θ, $W(\theta)$, multiplied by the number of electrons N:

$$W(\theta)=\frac{\mathrm{v}\sigma(\theta)}{\Omega}. \tag{10.17}$$

With (10.16) and (10.17), the cross-section takes the form:

$$\sigma(\theta)=\frac{\Omega^2 m^2}{4\pi^2\hbar^4}\left|\langle \boldsymbol{k}_i|V|\boldsymbol{k}_f\rangle\right|^2, \tag{10.18}$$

which coincides with that followed from the first Born approximation (10.5). Additional factor Ω^2 in (10.18) comes from the normalization of wave functions: $|\boldsymbol{k}\rangle\sim 1/\sqrt{\Omega}$, so the cross-section does not depend on volume.

10.3 Impurity Scattering: Momentum Relaxation Time

The transition rate $W(\theta)$ relates electron scattering to electrical conductivity and other kinetic coefficients (see Chapter 7). In particular, the collision integral in the Boltzmann equation can be approximated and then expressed through the *momentum relaxation time* τ. For elastic scattering the relaxation time reads as

$$\frac{1}{\tau}=\int W(\theta)(1-\cos\theta)do=\int_0^{2\pi}d\varphi\int_0^{\pi}W(\theta)(1-\cos\theta)\sin\theta\,d\theta=\frac{2\pi\mathrm{v}}{V}\int_0^{\pi}\sigma(\theta)(1-\cos\theta)\sin\theta\,d\theta. \tag{10.19}$$

Cross-section $\sigma(\theta)$ describes collision on a single impurity. In a sample with impurity density n_I and the electron wavelength less than l, all scattering events are independent. Then, the total collision rate is

$$\frac{1}{\tau}=2\pi\mathrm{v}n_I\int_0^{\pi}\sigma(\theta)(1-\cos\theta)\sin\theta\,d\theta. \tag{10.20}$$

Initial nonequilibrium electrons experience collisions on the way to the thermal equilibrium. Formally, thermalization means that in time τ the electron distribution function comes to its equilibrium value. Parameter τ determines the *mean free path* $l=\mathrm{v}\tau$, conductivity, and other transport characteristics, and thus also called the *transport relaxation time*.

10.3.1 Electron Scattering by a Screened Coulomb Potential

For a screened Coulomb potential, one substitutes cross-section (10.10) into (10.20) and obtain

$$\frac{1}{\tau}=2\pi v n_I\left(\frac{mZ_1Z_2a^2e^2}{2\pi\hbar^2\varepsilon_0}\right)^2\int_0^{\pi}\frac{(1-\cos\theta)\sin\theta}{\left(4k^2a^2\sin^2\frac{\theta}{2}+1\right)^2}d\theta=\frac{m\,n_I}{8\pi(\hbar k)^3}\left(\frac{Z_1Z_2e^2}{\varepsilon_0}\right)^2\left[\log\left(1+4k^2a^2\right)-\frac{4k^2a^2}{1+4k^2a^2}\right] \tag{10.21}$$

the *Brooks-Herring* expression for relaxation rate related to ionized impurity scattering (we used m and $v=\hbar k/m$ as the electron mass and velocity, respectively). As a first Born approximation, the Brooks-Herring result is applicable, providing conditions (10.7) hold. Besides, the use of cross-section on a single impurity (10.10) implies that the screening radius is less than the average interimpurity distance, $an_I^{1/3}\ll 1$. Otherwise, the scattering event involves many impurities, and thus may not be considered a two-body interaction.

According to the electron energy distributions (see Chapter 4), the energy of most electrons is $E\approx k_BT$ in nondegenerate semiconductors (Maxwell distribution) and $E\approx E_F$ in metals (E_F is the Fermi energy). At ambient temperatures in semiconductors, $k^2a^2\approx 2mk_BTa^2/\hbar^2\gg 1$, and the relaxation time (10.21) is insensitive to screening length as it contains $\log(ka)$.

10.3.2 Unscreened Coulomb Potential

If screening is ineffective ($a\to\infty$), Equation (10.21) diverges logarithmically and fails to describe the collision rate. Once $an_I^{1/3}\gg 1$, collision to a single impurity occurs on an unscreened Coulomb potential, and thus, one must use cross-section (10.11), divergent at small angles. To obtain a finite result, one has to cut off the integral (10.20) by introducing a minimum scattering angle θ_{min}. The point is to justify the physical meaning of θ_{min}. For this, one can use the classical theory of scattering, which, as mentioned earlier, is valid concerning Coulomb scattering. The scattering angle decreases with an increase in impact distance b (see Figure 10.1). If an electron happens to move in between two impurities, the scattering effect disappears. Thus, the scattering angle reaches the minimum when the impact distance is half a distance between impurities. More exactly, $2b_{max}=n_I^{-1/3}$, where $n_I^{-1/3}$ is the mean inter-impurity distance.

The relation between b_{max} and θ_{min} follows from electron dynamics in the course of elastic scattering [3]. Below we consider the dynamics based on the geometry of the hyperbolic trajectory. The trajectory of the charged particle colliding to unscreened Coulomb potential is depicted in Figure 10.3.

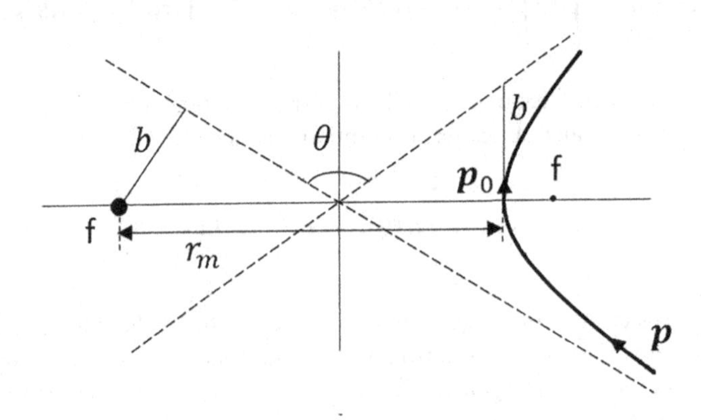

FIGURE 10.3 The hyperbolic trajectory of the particle scattered on impurity center, r_m is the minimum distance to the center, b is the impact distance.

When an incident electron passes the vertex of the hyperbola, the energy and angular momentum conservation conditions read as

$$\frac{p^2}{2m} = \frac{p_0^2}{2m} + \frac{Ze^2}{4\pi\varepsilon_0 r_m},$$

$$pb = p_0 r_m, \tag{10.22}$$

where $p = mv$ is the momentum of an electron as a classical particle. As follows from the geometry of the trajectory,

$$r_m = b\frac{1+\sin\dfrac{\theta}{2}}{\cos\dfrac{\theta}{2}}. \tag{10.23}$$

From (10.22) and (10.23), one obtains

$$b = R\cot\frac{\theta}{2}, \quad R \equiv \frac{mZe^2}{4\pi\varepsilon_0 p^2}. \tag{10.24}$$

The minimum angle corresponds to maximum impact distance $b_{\max}$:

$$\cot\frac{\theta_{\min}}{2} = \frac{n_I^{-1/3}}{2R} = \frac{2\pi\varepsilon_0 p^2}{mZe^2\, n_I^{1/3}}. \tag{10.25}$$

Now, using the Rutherford cross-section (10.11) in (10.20), one calculates the *Conwell-Weisskopf relaxation rate*:

$$\frac{1}{\tau} = \frac{n_I m\left(Z_1 Z_2 e^2\right)^2}{32\varepsilon_0^2 \pi p^3} I,$$

$$I = \int_{\theta_{\min}}^{\pi} \frac{(1-\cos\theta)\sin\theta}{\sin^4\dfrac{\theta}{2}} d\theta = -\log\left(\sin\frac{\theta_{\min}}{2}\right) = \frac{1}{2}\log\left[1+\left(\frac{2\pi\varepsilon_0 p^2}{mZe^2\, n_I^{1/3}}\right)^2\right]. \tag{10.26}$$

In (10.26) we used identity $\sin\left(arc\cot(x)\right) = \left(1+x^2\right)^{-1/2}$.

Relaxation times (10.21) and (10.26) are solutions to the mechanical problem of electron motion in the Coulomb field of ionized impurity (screened and unscreened), and they depend on the electron energy. Neglecting logarithmic factors, one concludes that both relaxation times have energy dependence, $E^{3/2}$. Since electrons of various energies take part in transport, the observable kinetic coefficients include energy integration. That is why E-dependence of τ is very important.

10.4 Neutral Impurity Scattering

Neutral impurity potential is short-ranged. If one presented it as screened Coulomb interaction in the limit $ka \ll 1$, cross-section (10.10) would not depend on energy and resulted in relaxation time $\tau \sim 1/\sqrt{E}$. However, expression (10.10) implies that an incident electron does not affect the field of the scattering center. Short-ranged interaction allows the electron to get close to impurity that makes (10.10) inapplicable for several reasons. First, the incident electron may excite intraimpurity electron transitions that result in inelastic scattering. Second, the electron may polarize the neutral atom modifying the potential from screened Coulomb to the dipole field. Third, when two indistinguishable particles (incident electron and electron in an atom) get close, the exchange effect may come into play. In semiconductors, the generally accepted model includes polarization and exchange effects in the course of elastic collisions with a neutral hydrogen atom. As a result, the relaxation time is determined by the Erginsoy formula:

$$\tau = \frac{m^2 e^2}{80\pi\varepsilon_0 \varepsilon \hbar^3 n_I}. \tag{10.27}$$

Scattering on neutral impurities determines electrical resistivity in semiconductors at low temperatures when only a low impurity number is ionized.

10.5 Phonon Scattering

Correction to electron energy due to interaction with longitudinal acoustical vibrations (6.108) (Chapter 6) can be written as interaction Hamiltonian in second quantization representation (see the procedure applied to Fröhlich Hamiltonian in Section 6.6.2):

$$H = \sqrt{\frac{\hbar}{2MN}} \sum_{kq} \frac{\sigma i q}{\sqrt{\omega(q)}} \left(b_{qj} c^{+}_{k+q} c_k - b^{+}_{qj} c^{+}_{k-q} c_k \right), \tag{10.28}$$

where σ is the acoustic deformation potential, $\omega(\boldsymbol{q})$ is the frequency of longitudinal phonon branch, M is the total mass of the unit cell, and N is the number of unit cells in a sample.

Interaction (10.28) induces transitions between quantum states which without interaction is just a product of Bloch wavefunction and phonon number state n_q (see Appendix 1), https://www.routledge.com/Modern-Semiconductor-Physics-and-Device-Applications/Dugaev-Litvinov/p/book/9780367250829#:

$$\left|\boldsymbol{k}, n_q\right\rangle = \left|\boldsymbol{k}\right\rangle \left|n_q\right\rangle = c^{+}_{k} \frac{1}{\sqrt{n_q!}} \left(b^{+}_{q}\right)^{n_q} \left|0\right\rangle. \tag{10.29}$$

The energy of the state (10.29) is the sum of electron energy and vibrational energy of the lattice: $E(\boldsymbol{k}, n_q) = E_k + \hbar\omega(\boldsymbol{q})(n_q + 1/2)$.

As Hamiltonian (10.28) is linear in operators b and b^{+}, nonzero matrix elements exist only between states in which phonon numbers differ by unity. Two terms in (10.28) have matrix elements,

$$\begin{aligned} &\left\langle \boldsymbol{k}', n'_q \middle| b_{qj} c^{+}_{k+q} c_k \middle| \boldsymbol{k}, n_q \right\rangle = \sqrt{n_q} \left\langle n'_q \middle| n_q - 1 \right\rangle \left\langle \boldsymbol{k}' \middle| \boldsymbol{k} + \boldsymbol{q} \right\rangle = \sqrt{n_q}\, \delta_{k', k+q} \delta_{n'_q, n_q - 1}, \\ &\left\langle \boldsymbol{k}', n'_q \middle| b^{+}_{qj} c^{+}_{k-q} c_k \middle| \boldsymbol{k}, n_q \right\rangle = \sqrt{n_q + 1}\, \delta_{k', k-q} \delta_{n'_q, n_q + 1}, \end{aligned} \tag{10.30}$$

where Kronecker symbols follow from orthonormality of electron and phonon states. So, matrix elements that enter transition rate (10.13) have the form:

$$W^{+}_{kq} = \frac{2\pi}{\hbar} \left| \left\langle \boldsymbol{k} - \boldsymbol{q}, n_q + 1 \middle| H \middle| \boldsymbol{k}, n_q \right\rangle \right|^2 \delta\left(E(\boldsymbol{k} - \boldsymbol{q}) - E(\boldsymbol{k}) + \hbar\omega(\boldsymbol{q})\right) \equiv \frac{\pi\sigma^2 q^2}{MN\omega(\boldsymbol{q})} (n_q + 1)\, \delta\left(L^{+}_{kq}\right),$$

$$W^{-}_{kq} = \frac{2\pi}{\hbar} \left| \left\langle \boldsymbol{k} + \boldsymbol{q}, n_q - 1 \middle| H \middle| \boldsymbol{k}, n_q \right\rangle \right|^2 \delta\left(E(\boldsymbol{k} + \boldsymbol{q}) - E(\boldsymbol{k}) - \hbar\omega(\boldsymbol{q})\right) \equiv \frac{\pi\sigma^2 q^2}{MN\omega(\boldsymbol{q})} n_q\, \delta\left(L^{-}_{kq}\right),$$

$$L_{kq}^{\pm} \equiv \frac{\hbar^2}{2m}\left[q^2 \pm (Q - 2kqt)\right],\quad Q \equiv \frac{2m\omega(\boldsymbol{q})}{\hbar} = \frac{2m\mathrm{v}_s q}{\hbar}, \tag{10.31}$$

where t is the cosine of the angle between $\boldsymbol{k}$ and $\boldsymbol{q}$, v_s is the sound velocity: $\omega(\boldsymbol{q}) = \mathrm{v}_s q$.

Energy conservation in (10.31) implies parabolic electron dispersion in $L_{kq}^{\pm}$.

The transition rate with phonon absorption W_{kq}^{-} is proportional to the Bose factor n_q (phonon mode occupation number), in other words, to the number of real phonons in the crystal. The rate with phonon emission W_{kq}^{+} is proportional to $n_q + 1$ and thus nonzero even if there are no real phonons in the system, meaning that *spontaneous phonon emission* occurs.

Full transition rate with phonon emission is the sum over all phonon modes:

$$W_k^{+} = \sum_q W_{kq}^{+} = \frac{\pi\sigma^2}{MN\mathrm{v}_s}\sum_q q(n_q + 1)\delta(L_{kq}^{+}), \tag{10.32}$$

Assuming the Debye model for phonon spectrum and replacing sum with integral (see Eq. (6.79), Chapter 6), one obtains phonon emission rate,

$$W_k^{+} = \frac{\sigma^2\Omega_0}{4\pi M\mathrm{v}_s}\int_{-1}^{1} dt \int_0^{q_D} q^3\, dq (n_q + 1)\delta(L_{kq}^{+}), \tag{10.33}$$

where q_D is the Debye wavevector, Ω_0 is the unit cell volume.

At low temperatures, $n_q \approx 0$, one finds spontaneous emission rate calculating (10.33) with the help of identity,

$$\int F(x)\delta(L(x)) = \sum_i \frac{F(x_i)}{(\partial L/\partial x)_{x_i}},\quad L(x_i) = 0. \tag{10.34}$$

Root of equation $L_{kq}^{+}(q) = 0$ should not be negative: $kt \geq m\mathrm{v}_s/\hbar$. The result of integration has the form:

$$W_k^{+} = \begin{cases} \dfrac{2\sigma^2 m\Omega_0}{3\pi\hbar^2 Mk\mathrm{v}_s}\left(k - \dfrac{m\mathrm{v}_s}{\hbar}\right)^3, & k > m\mathrm{v}_s/\hbar, \\ 0, & k < m\mathrm{v}_s/\hbar. \end{cases} \tag{10.35}$$

As follows from (10.35) at the minimum threshold electron velocity, which equals to sound velocity, an electron spontaneously emits a phonon in the same direction. The meaning of the result is similar to the Cherenkov effect, in which a charged particle moving in a medium emits photons once its velocity exceeds the speed of light. Phonon absorption during a collision with an electron occurs without a threshold. It follows from the fact that condition $L_{kq}^{-}(q) = 0$ holds at any $\boldsymbol{k}$.

The time introduced as $\tau^{-1} = W_k^{+}$ presents the electron lifetime in state $\boldsymbol{k}$. The lifetime is finite even if real phonons are absent as it manifests electron interaction with zero-mode phonon vibrations (spontaneous phonon emission). The lifetime is not the same as transport relaxation time to be discussed in the next section.

10.5.1 Transport Relaxation Time

Momentum relaxation time that determines transport properties is given in (10.19):

$$\frac{1}{\tau_k} = \sum_q W_{kq}(1 - \cos\theta),$$

$$W_{kq} = W_{kq}^{+} + W_{kq}^{-} = \frac{\pi\sigma^2 q}{MN\text{v}_s}\left[\left(n_q+1\right)\delta\left(L_{kq}^{+}\right) + n_q\delta\left(L_{kq}^{-}\right)\right], \tag{10.36}$$

where θ is the angle between $\boldsymbol{k}$ and $\boldsymbol{k}' = \boldsymbol{k} + \boldsymbol{q}$.

Collisions with phonons in semiconductors are elastic as long as phonon energy is much less than the average electron energy. In metals and degenerate semiconductors, the electron energy is close to the Fermi energy that exceeds the phonon energy. In nondegenerate semiconductors, the elastic regime is a good approximation at $T > T_D$ no matter acoustical or optical phonons take part in the scattering process. Assuming scattering is elastic, one may neglect $\hbar\omega(q)$ in $L_{kq}^{\pm}$:

$$L_{kq}^{+} \approx L_{kq}^{-} = \frac{\hbar^2}{2m}\left[q^2 + 2kqt\right],$$

$$W_{kq} = \frac{\pi\sigma^2 q\left(2n_q+1\right)}{MN\text{v}_s}\delta\left(\frac{\hbar^2}{2m}\left[q^2 + 2kqt\right]\right). \tag{10.37}$$

Note that W_{kq} depends on t, the cosine of the angle between $\boldsymbol{k}$ and $\boldsymbol{q}$, while τ_k (10.36) contains $\cos\theta$, the angle between $\boldsymbol{k}$ and $\boldsymbol{k}' = \boldsymbol{k} + \boldsymbol{q}$, so one has to express factor $(1 - \cos\theta)$ through t:

$$1 - \cos\theta = 1 - \frac{\boldsymbol{k}\cdot\boldsymbol{k}'}{kk'} = 1 - \frac{k^2 + kqt}{k^2} = -\frac{qt}{k}. \tag{10.38}$$

In Eq. (10.38), we used the fact the scattering is elastic: $k = k'$. Now, the relaxation time becomes

$$\frac{1}{\tau_k} = -\frac{\pi\sigma^2}{MN\text{v}_s k}\sum_q q^2 t\left(2n_q+1\right)\delta\left(\frac{\hbar^2}{2m}\left[q^2 + 2kqt\right]\right). \tag{10.39}$$

When the temperature is higher than the Debye temperature, the Bose factor becomes $n_q \approx k_B T/\hbar\omega(q) \gg 1$. In this limit, (10.39) takes the form

$$\frac{1}{\tau_k} = -\frac{2\pi\sigma^2 k_B T}{MN\hbar \text{v}_s^2 k}\sum_q qt\delta\left(\frac{\hbar^2}{2m}\left[q^2 + 2kqt\right]\right), \tag{10.40}$$

which is valid as well in non-degenerate semiconductors where electron energy $\simeq k_B T$.

Summation in (10.40) goes the way described in (10.33) and (10.34):

$$\frac{1}{\tau_k} = \frac{mk\sigma^2 k_B T\Omega_0}{\pi\hbar^3 M\text{v}_s^2}. \tag{10.41}$$

As $k \sim \sqrt{E}$, the momentum relaxation time depends on energy as $\sim 1/\sqrt{E}$. With electron velocity $\text{v} = \hbar k/m$, the mean free path,

$$l = \text{v}\tau = \frac{\pi\hbar^4 M\text{v}_s^2}{m^2\sigma^2 k_B T\Omega_0}, \tag{10.42}$$

does not depend on the electron energy and drops as temperature increases.

In Chapter 6 we have considered interaction with deformation acoustic (DA, (6.108)), deformation optical (nonpolar optical) (PO, (6.112)), polar optical (PO, (6.122)), and polar acoustic (piezoelectric interaction, PA, (6.125)) phonons. Detailed calculation given above for DA phonons, results in Eq. (10.41).

For all other mechanisms, one should repeat calculations (10.37)–(10.41) but with corresponding q-dependent matrix elements and phonon energy spectrum. Results are listed below:

$$\begin{array}{ccc} \text{Interaction matrix element} \sim & \text{Relaxation time, } \tau_k \sim & \text{Phonons} \\ \sqrt{q} & \dfrac{1}{k_B T\sqrt{E}} & DA \\ q^0 & \dfrac{1}{k_B T\sqrt{E}} & DO \\ \dfrac{1}{q} & \dfrac{\sqrt{E}}{k_B T} & PO \\ \dfrac{1}{\sqrt{q}} & \dfrac{\sqrt{E}}{k_B T} & PA \end{array} . \tag{10.43}$$

At low temperatures $k_B T \ll \hbar\omega_0$, spontaneous emission is forbidden due to low electron energy. An electron absorbs and then emits a phonon (induced emission), in which process the electron energy remains almost unchanged due to weak phonon dispersion. That is one more region where elastic scattering is a good approximation.

The concept of momentum relaxation time has sense if scattering is elastic. In this case, the Boltzmann equation allows an analytical solution, and the relaxation time fully determines the non-equilibrium distribution function. Generally, the electron–phonon scattering is not elastic as it causes electron energy relaxation and thus brings initially disturbed electron system into thermal equilibrium with lattice. Under specific conditions, scattering is approximately elastic since the momentum thermalization happens much faster than energy relaxation. For example, in metals, in the course of electron–phonon collisions, the initial and final electron momenta stay on the Fermi surface, $E = E' = E_F$. In many instances, the elastic collisions well describe the experimental findings in the transport properties of solids.

10.6 Simultaneous Action of Several Scattering Mechanisms

Often several scattering mechanisms act simultaneously. The scattering rate is additive, meaning that the total relaxation time equals the sum of terms following from all relaxation mechanisms:

$$\frac{1}{\tau} = \sum_i \frac{1}{\tau_i}. \tag{10.44}$$

As an example, we consider simultaneous scattering on ionized impurities and acoustic phonons. Corresponding momentum relaxation times are from (10.21) and (10.41)

$$\tau_I = \tau_{0I} E^{3/2}, \quad \tau_{ph} = \frac{\tau_{0ph}}{k_B T\sqrt{E}}. \tag{10.45}$$

Effective relaxation time becomes

$$\tau = \frac{\tau_{0I}\tau_{0ph} E^{3/2}}{\tau_{0ph} + \tau_{0I} E^2 k_B T}. \tag{10.46}$$

If one of the mechanisms becomes ineffective, the corresponding $\tau_0 \to \infty$, dropping out from (10.46).

10.7 Spin-Dependent Scattering and Spin Relaxation Time

The exclusion of the spin variable from calculations in previous sections means we studied collisions that conserve spin. However, account for spin adds new scattering channels, which cause spin-flip and, consequently, spin relaxation. Moreover, in crystals with no spatial inversion symmetry, momentum and spin relaxation channels depend on each other in the sense that momentum scattering induces spin-flip processes. Spin relaxation is the parameter that determines the lifetime of spin-polarized electrons in the course of optical spin orientation in semiconductors, and also a spin injection in all-electrical spintronic devices. There are several mechanisms behind the origin of spin-flip collisions. The most obvious one is the electron scattering on magnetic impurities, the spin–spin interaction which conserves the total spin (an electron and a magnetic atom). This type of scattering is to discuss in the next section. Another reason for spin-flip collisions is the spin–orbit interaction, which makes scattering by phonons and (or) nonmagnetic impurities spin-dependent.

Spin–orbit interaction has the form:

$$H_{so} = \frac{\hbar}{4m_0^2c^2}\boldsymbol{\sigma} \times \nabla V(\boldsymbol{r}) \cdot \boldsymbol{p}, \tag{10.47}$$

where m_0 is the free-electron mass, c is the speed of light in free space, $\boldsymbol{\sigma} = \{\sigma_x, \sigma_y, \sigma_z\}$ is the spin operator represented by Pauli matrices. So, H_{so} is the 2×2 matrix in the space of 2-spinors, $|\uparrow\rangle = \begin{pmatrix} 1 \\ 0 \end{pmatrix}$ and $|\downarrow\rangle = \begin{pmatrix} 0 \\ 1 \end{pmatrix}$, where spin-up and -down directions imply spin quantizing axis. To arrange the axis, one may apply a magnetic field and then reduce its strength to zero. So, neither Zeeman splitting nor Landau quantization affects electron motion.

When considering elastic spin-flip scattering, we imply the electron energies in spin-up and -down states are degenerate. That is the case in centrosymmetric crystals such as Ge, II–VI semiconductors, and cubic modification of GaN, where spin–orbit interaction does not split spin subbands. Even in noncentrosymmetric semiconductors such as zinc-blende III-V materials, the spectrum calculated in the *kp*-perturbation scheme remains spin-degenerate up to k^2 -terms (for low-energy electrons). The Dresselhaus spin-splitting in III–V materials starts with k^3-terms.

The total electron energy is a sum of the lattice-periodic field $V(\boldsymbol{r})$ and potentials violating periodicity such as impurity potential and (or) lattice vibrations $U(\boldsymbol{r})$. With the spin–orbit part taken into account, the total potential energy reads as

$$W(\boldsymbol{r}) = V(\boldsymbol{r}) + H_{so}^V + U(\boldsymbol{r}) + H_{so}^U, \tag{10.48}$$

where $H_{so}^{V,U}$ is the spin–orbit contributions (10.47) from periodic potential $V(\boldsymbol{r})$ and defect potential $U(\boldsymbol{r})$, respectively. So, the first two terms in (10.48) correspond to an ideal lattice. Two others are the potential and spin–orbit fields that scatter electrons. If H_{so}^V is weak, as compared to $V(\boldsymbol{r})$, the spin–orbit interaction does not affect Bloch wave functions and electron energy bands, and thus, spin-flip collisions originate from H_{so}^U-term only.

In crystals with a strong spin–orbit coupling, the Bloch amplitudes become a mixture of the initial spin states $|\uparrow\rangle$ and $|\downarrow\rangle$; therefore, spin-flip processes may also originate from scattering on the potential part of defect potential $U(\boldsymbol{r})$. The potential itself can represent electron-impurity, electron–phonon, or Coulomb electron–hole interaction.

The spin-flip transitions determine the lifetime of an electron in state $|\boldsymbol{k}\uparrow\rangle$ relatively to the loss of spin orientation, *spin scattering time*:

$$\frac{1}{\tau_s(\mathbf{k})} = N_I \sum_{k'} W_{k\uparrow k'\downarrow}$$

$$W_{k\uparrow k'\downarrow} = \frac{2\pi}{\hbar} \left| \langle \mathbf{k} \uparrow | U(\mathbf{r}) | \mathbf{k}' \downarrow \rangle \right|^2 \delta(E_k - E_{k'}), \tag{10.49}$$

where N_I is the number of scattering centers.

We define the spin relaxation rate as $T_1 = \tau_s/2$, which is the time needed for the average spin component parallel to the magnetic field to increase from zero to a maximum value. There is another characteristic time that controls transverse components of the average spin. If the spin vector deviates in the direction perpendicular to the magnetic field, transverse components appear, and then, during the time T_2, relax to zero. Parameters T_1 and T_2 are called longitudinal and transverse spin relaxation times, respectively.

Relaxation time (10.49) is due to scattering on a spin-independent agent in centrosymmetric crystals with strong spin–orbit interaction and corresponds to the *Elliott-Yafet* (EY) mechanism of spin relaxation. The time is proportional to momentum relaxation time and tends to infinity when the spin–orbit parameter (spin–orbit valence band splitting) goes to zero.

In noncentrosymmetric materials, the conduction band energy experiences zero-magnetic field momentum-dependent spin splitting. In zinc-blende materials, the splitting is proportional to $\mathbf{k}^3$ (Dresselhaus spectrum) while in wurtzite crystals to $\mathbf{k}$ (Rashba spectrum).

In both cases, it is equivalent to spin splitting in an effective magnetic field, the direction of which depends on $\mathbf{k}$. When elastic collisions scatter electrons, initial spin orientation changes, and this new channel of spin relaxation is called the *Dyakonov-Perel* mechanism (DP). If ΔE_k is the magnitude of spin-splitting in point $\mathbf{k}$, and $\Delta E_k \tau \ll \hbar$ (τ is the momentum relaxation time), the spin relaxation time is $T_1 \approx \hbar^2/(\Delta E_k)^2 \tau$.

So, the EY mechanism dominates in inversion-symmetric materials, while DP mechanism requires the lack of spatial inversion. Spin relaxation times T_1 in EY and DP mechanisms depend on τ differently: direct (EY) and inverse (DP) proportionality. They both tend to infinity (spin relaxation becomes ineffective) when the valence band spin–orbit splitting goes to zero.

One more channel of the spin relaxation, called the Bir-Aronov-Pikus (BAP) mechanism, originates from the exchange part of Coulomb interaction between an electron and a hole.

In many semiconductors, the spin–orbit scattering rate originated from the impurity potential is much less than the one induced by the spin–orbit interaction in an ideal crystal. For more detail on the relative importance among various mechanisms, see Refs. [4,5].

10.8 Kondo Effect

Electrons interact with ionized nonmagnetic impurities via Coulomb interaction. If impurity has a magnetic moment, it scatters an electron via direct and exchange parts of Coulomb interaction. The exchange part can be presented as spin–spin interaction (see also Chapter 3):

$$H_{ex}(\mathbf{r}) = -\frac{J}{n} \sum_i \mathbf{S}_i \cdot \boldsymbol{\sigma} \delta(\mathbf{r} - \mathbf{R}_i), \tag{10.50}$$

where J is the exchange constant of energy dimension and n is the density of host atoms. The interaction is nonzero in the spatial region where wave functions of localized and band electrons overlap, so we use contact interaction as an approximation. Minus sign makes the parallel orientation of electron and impurity spins energetically favorable (ferromagnetic coupling).

In the first Born approximation, the scattering amplitude is proportional to $A \sim -(J/n)\mathbf{S}_i \cdot \boldsymbol{\sigma}$. In metals, the second-order term in amplitude gives

$$A \sim \left(\frac{J}{n}\right)^2 \rho(E_F)\log\left(\frac{E_F}{\max\{k_BT, E-E_F\}}\right)\mathbf{S}_i \cdot \boldsymbol{\sigma}, \tag{10.51}$$

where $\rho(E_F)$ is the density of electron states (number per volume per energy interval) at the Fermi level. The logarithmic divergence in scattering amplitude causes an anomaly in electrical resistivity:

$$r = r_0 + r_B\left(1 - \frac{2J\rho(E_F)}{n}\log\frac{E_F}{k_BT}\right), \tag{10.52}$$

where r_0 is the resistivity due to scattering on phonons and non-magnetic impurities, r_B stems from the first Born approximation to magnetic scattering term (10.50) and depends on the density of magnetic atoms. Expression (10.52) is the perturbation series in J and thus valid as long as the second term in the bracket is less than unity. If $J<0$, the divergent part of resistivity logarithmically rises when temperature decreases. As r_0 monotonically decreases at $T \to 0$, total resistivity (10.52) approaches a minimum at low temperatures. The minimum, illustrated in Figure 10.4, is called the *Kondo effect*.

The sign of the exchange constant J is essential. The Kondo effect originates from spin-flip scattering which conserves the total spin of interacting particles. If z is the quantizing axis, the electron spin has two components $s_z = \pm 1/2$ while for impurity spin S, $S_z = S, S-1, \ldots$. If $J>0$ spins of electron and impurity are parallel, the total z-projection equals a maximum value $S+1/2$. This value stays before and after a collision, so the spin-flip transition is prohibited. If $J<0$, spins of electron and impurity are antiparallel, the total z-projection is $S-1/2$, and it may become $S-1+1/2$ after a collision. In this process, the total spin is conserved while impurity spin projection changes and electron spin flips.

As (10.52) followed from the perturbation approach, the expression becomes invalid when the anomalous term increases to become close in magnitude to the first term. It happens when temperature decreases below the *Kondo temperature*:

$$k_BT_k \approx E_F \exp\left(-\frac{n}{|J|\rho(E_F)}\right), \quad J<0. \tag{10.53}$$

Since the Kondo temperature follows from relative magnitudes of terms in bracket (10.52), it does not coincide with the minimum of resistivity as the minimum varies with the density of magnetic atoms hidden in coefficient r_B.

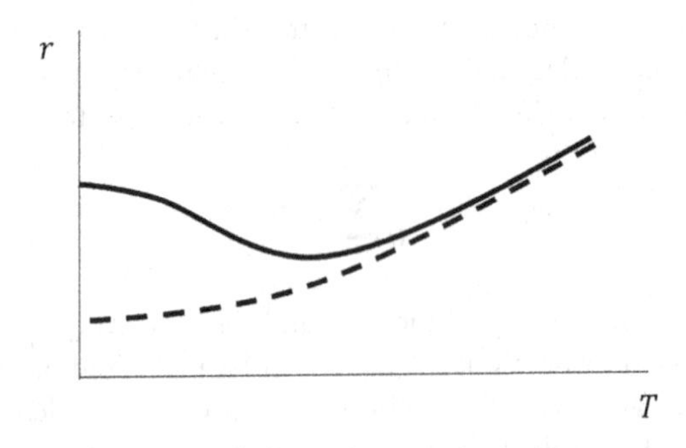

FIGURE 10.4 Temperature dependence of resistivity in metals. Scattering on phonons and nonmagnetic impurities (dashed line). Magnetic impurities and Kondo effect (solid line).

In experiments, there is no singularity at $T < T_k$, but instead, the resistivity shows plateaus. This discrepancy is called the Kondo problem. The problem was solved by nonperturbative methods, which for $T < T_k$ show reducing to zero the magnetic moment of impurity due to screening by Fermi electrons. In other words, the system which comprises the un-perturbed Fermi sea of electrons and the local magnetic moment becomes unstable toward the formation of a collective state, so-called *Abrikosov-Suhl resonance* [6–8]. The resonance level appears at the Fermi energy manifesting the singlet state of local magnetic moment and electrons. The state of size $\hbar v_F / k_B T_k$ effectively scatters electrons contributing to the resistance. Formation of the singlet state screens the magnetic moment similar to a screening of an electric charge inside the electron gas. For electrically charged magnetic impurity, both types of screening occur at the same time. For an exact solution to the Kondo problem, see Ref. [9].

References

1. D. Bohm, *Quantum Theory* (Prentice Hall Inc., New York, 1952).
2. A.S. Davydov, *Quantum Mechanics* (International Series of Monographs in Natural Philosophy, ed. D. ter Haar, Pergamon Press Ltd., Headington Hill Hall, Oxford, UK, 1965).
3. M. Born, *Atomic Physics* (2nd ed., Blackie & Son, Glasgow, UK, 1937).
4. V.F. Gantmaher and Y.B. Levinson, *Carrier Scattering in Metals and Semiconductors* (Modern Problems in Condensed Matter Sciences, v.19, Elsevier Science Publishers, Amsterdam, 1987).
5. G.E. Pikus and A.N. Titkov, Chapter 3 "Spin Relaxation under Optical Orientation in Semiconductors", In: *Optical Orientation*, Ed. by F. Meyer and B.P. Zakharchenya (Elsevier Science Publishers, 1984).
6. H. Suhl, "Dispersion theory of the Kondo effect", *Phys. Rev.* **138**, A515 (1965); "Paramagnetic impurities in metals at finite temperatures", *Physics* **2**, 39 (1965).
7. Y. Nagaoka, "Self-consistent treatment of Kondo's effect in dilute alloys", *Phys. Rev.* **138**, A1112 (1965).
8. A.A. Abrikosov, "Electron scattering on magnetic impurities in metals and anomalous resistivity effects", *Physics*. **2**, 5 (1965).
9. A.C. Hewson, *The Kondo Problem to Heavy Fermions* (Cambridge University Press, Cambridge, UK, 1993).

11

Magnetic Semiconductors

Materials that combine magnetic and semiconductor properties are being explored and implemented in various spintronic applications. Binary chalcogenides such as EuTe and MnTe(Se) are magnetic semiconductors. However, effective electrical manipulation of magnetic behavior requires high-electron-mobility structures, fabricated as *p–n* junctions, the main elements of any device. That is why much attention attracts *diluted magnetic semiconductors*, which are well-developed III-V, IV-VI, and II-VI semiconductor templates doped with magnetic impurities – rare earth or transition metal atoms [1].

Use of the same semiconductor host, nonmagnetic and magnetically doped, enables effective injection of spin-polarized electrons from magnetic to nonmagnetic material. In comparison, spin injection from the ferromagnetic metal contact into a semiconductor is suppressed by a large conductivity mismatch on both sides of the junction.

At a small content of magnetic component, semiconductors are paramagnetic. The solubility of magnetic impurity atoms can be up to several percent, so, as content increases, magnetic dopants might form ferro-, antiferro-, or spin-glass- order, depending on the detail of the coupling between magnetic atoms and their spatial distribution in the host lattice. Magnetic coupling enters the Hamiltonian as the Heisenberg interaction term $H_{int} = J_{ij}\ \mathbf{S}_i \cdot \mathbf{S}_j$, where $\mathbf{S}_i$ is the impurity spin operator localized in site *i*, J_{ij} is the exchange coupling constant. The negative sign of J_{ij} favors the parallel spin orientation of two impurities or ferromagnetic ground state across the whole crystal of a magnetic semiconductor. Positive J_{ij} corresponds to an antiferromagnetic coupling, which makes the total magnetic moment zero. Spatially oscillating exchange interaction appears positive or negative in the pairs of different distances apart, so it calls for more complicated magnetic structures discussed below.

11.1 Direct and Indirect Interactions between Magnetic Impurities

Direct exchange coupling between two electrons stems from Coulomb interaction and the overlap of their wave functions. In a single atom, the direct exchange is always negative, $J_{ij}<0$. It is related to the antisymmetry of the two-electron wave function to permutation of electrons, which makes the triplet state ($S_{tot} = 1$) energetically favorable since the antisymmetric combination of orbitals prevents two electrons from being close, thus suppressing Coulomb energy (Hund's rule [2]). For two electrons residing at neighboring impurities at distance $R \simeq a_0$ (a_0 is the lattice parameter), the direct exchange can be either sign depending on the structure of overlapping wavefunctions. For example, in a hydrogen molecule, two electrons in the ground state have zero total spin or antiferromagnetic sign of exchange interaction (see Figure 6.2, Chapter 6). The main difference between single and different atoms the pair of electrons belong to is that atomic orbitals in a single atom are orthogonal and result in the Hund's rule, or ferromagnetic sign of exchange. The orbitals belonging to different sites are not orthogonal, making the sign of direct exchange coupling dependent on wavefunctions. In simple metals, the direct exchange is ferromagnetic, $J_{ij} < 0$.

DOI: 10.1201/9780429285929-11

If the magnetic atoms are not neighbors and separated by a nonmagnetic atom in the host lattice, magnetic polarization of the host atom creates a *superexchange* interaction mechanism resulting in $J_{ij} > 0$. In oxides, the superexchange mechanism favors the antiferromagnetic interaction of magnetic moments.

The density of magnetic atoms in magnetic semiconductors is low. For example, in GaAs, the maximum concentration of magnetic ions of Mn is about 6%. Therefore, the probability of finding two moments at the distance of lattice parameter a_0 is small.

One can estimate this probability by using the Poisson distribution, which is justified if the probability of a single random event $p \ll 1$. In our problem, p is the probability for one impurity to be within an arbitrary chosen volume $4\pi a_0^3/3$ in the sample. Correspondingly, we have $p = 4\pi a_0^3/3\Omega \ll 1$, where Ω is the sample volume. The probability of getting m same results in a series of N trials is expressed by Poisson distribution,

$$P(m) = \frac{\lambda^m}{m!} e^{-\lambda}, \tag{11.1}$$

where $\lambda = Np$ is called the expectation value. In our case, N is the total number of magnetic impurities. Hence, one can present the probability to find m impurities in volume $4\pi a_0^3/3$ as

$$P(m) = \frac{\left(4\pi n_i a_0^3/3\right)^m}{m!} \exp\left(-\frac{4\pi n_i a_0^3}{3}\right), \tag{11.2}$$

where we denoted by n_i the density of magnetic impurities, $n_i = N/\Omega$.

Let us choose a single impurity in the crystal and find the probability that no more impurity comes to the volume of radius a_0 around the chosen one. The probability of finding impurity isolated within distance $r \sim a_0$ is given by Eq. (11.2) with $m = 0$. Taking $n_i = x a_0^{-3}$, we find that for $x = 0.01$, the corresponding probability $P(0) \simeq 0.96$, and for $x = 0.06$, this quantity falls to $P(0) \simeq 0.8$. In other words, for $x < 0.01$, one can neglect the existence of pairs at a distance $r \simeq a_0$. Correspondingly, for $P(0) \simeq 1$, the direct exchange interaction is negligible.

The primary mechanism of magnetic interaction in semiconductors is the indirect exchange. It is related to the magnetic polarization of electron density around a magnetic atom. In contrast to the direct exchange interaction, this type of coupling is long-ranged and may cause the magnetic order of randomly distributed magnetic moments.

11.2 RKKY Interaction

We consider the electron gas coupled to impurity spin $\mathbf{S}_i$, located at $\mathbf{r} = 0$, and assume the spin classical, $S \gg 1$. The Hamiltonian of this model reads

$$H = \int d^3r\, \psi^\dagger(\mathbf{r}) \left[-\frac{\hbar^2 \nabla^2}{2m^*} + J(\boldsymbol{\sigma} \cdot \mathbf{S}_i)\delta(\mathbf{r}) \right] \psi(\mathbf{r}), \tag{11.3}$$

where $\psi(\mathbf{r})$ is the electron spinor field, $\boldsymbol{\sigma}$ is the Pauli matrix acting on spinors, the last term describes the point interaction of electron and impurity spin $\mathbf{S}_i$, and J is the exchange constant.

The electron spin density in a point $\mathbf{r}$ has the form,

$$\mathbf{s}(\mathbf{r}) = \langle \psi^\dagger(\mathbf{r}) \boldsymbol{\sigma} \psi(\mathbf{r}) \rangle = -i\, Tr\, \boldsymbol{\sigma} G(\mathbf{r},t;\mathbf{r},t+\delta)|_{\delta\to+0} = -i\, Tr \int \frac{d\varepsilon}{2\pi} \boldsymbol{\sigma} G(\mathbf{r},\mathbf{r};\varepsilon) e^{i\varepsilon\delta} \Big|_{\delta\to+0}, \tag{11.4}$$

where $G(\boldsymbol{r},\boldsymbol{r}';\varepsilon)$ is the causal Green's function of electrons corresponding to Hamiltonian (11.3). The trace operation is the sum of matrix diagonal elements. The exponent $\exp(i\varepsilon\delta)$ allows calculating the integral in $(-\infty,\infty)$ limits by closing the integration path in the upper half-plane of complex ε.

The second term in Eq. (11.3), taken in the first order of perturbation, gives the Green's function as

$$G(\boldsymbol{r},\boldsymbol{r};\varepsilon)\simeq G_0(\boldsymbol{r},\boldsymbol{r};\varepsilon)+J\,G_0(\boldsymbol{r},0;\varepsilon)(\boldsymbol{\sigma}\cdot\boldsymbol{S}_i)G_0(0,\boldsymbol{r};\varepsilon), \tag{11.5}$$

where $G_0(\boldsymbol{r},\boldsymbol{r}';\varepsilon)$ is the Green's function of a free electron.

Using expansion (11.5) and Eq. (11.4), one can write the expression for impurity-induced electron spin density,

$$\begin{aligned} s_\alpha(\boldsymbol{r})&=-iJS_{i\beta}\,Tr\int\frac{d\varepsilon}{2\pi}\,\sigma_\alpha G_0(\boldsymbol{r},0;\varepsilon)\sigma_\beta G_0(0,\boldsymbol{r};\varepsilon)\\ &=-2iJS_{i\alpha}\int\frac{d\varepsilon}{2\pi}G_0(\boldsymbol{r},0;\varepsilon)\,G_0(0,\boldsymbol{r};\varepsilon), \end{aligned} \tag{11.6}$$

The Green's function of a free electron in $(\boldsymbol{k},\varepsilon)$ representation is (see Chapter 7, Eq. (7.103))

$$G_0(\boldsymbol{k},\varepsilon)=\frac{1}{\varepsilon-\varepsilon_{\boldsymbol{k}}+\mu+i\delta\,sign\,\varepsilon}, \tag{11.7}$$

where $\varepsilon_{\boldsymbol{k}}=\hbar^2k^2/2m^*$. Then we get

$$\begin{aligned} G_0(\boldsymbol{r},\boldsymbol{r}',\varepsilon)&=\int\frac{d^3k}{(2\pi)^3}\,e^{i\boldsymbol{k}\cdot(\boldsymbol{r}-\boldsymbol{r}')}G_0(\boldsymbol{k},\varepsilon)=\int\frac{d^3k}{(2\pi)^3}\frac{e^{i\boldsymbol{k}\cdot(\boldsymbol{r}-\boldsymbol{r}')}}{\varepsilon-\varepsilon_{\boldsymbol{k}}+\mu+i\delta\ sign\ \varepsilon}\\ &=\frac{1}{4\pi^2}\int_0^\infty k^2dk\int_{-1}^{1}\frac{e^{ik|\boldsymbol{r}-\boldsymbol{r}'|t}\,dt}{\varepsilon-\varepsilon_{\boldsymbol{k}}+\mu+i\delta\ sign\ \varepsilon}=-\frac{i}{4\pi^2|\boldsymbol{r}-\boldsymbol{r}'|}\int_0^\infty kdk\frac{e^{ik|\boldsymbol{r}-\boldsymbol{r}'|}-e^{-ik|\boldsymbol{r}-\boldsymbol{r}'|}}{\varepsilon-\varepsilon_{\boldsymbol{k}}+\mu+i\delta\ sign\ \varepsilon}\\ &=-\frac{i}{4\pi^2|\boldsymbol{r}-\boldsymbol{r}'|}\int_{-\infty}^{\infty}\frac{e^{ik|\boldsymbol{r}-\boldsymbol{r}'|}\,kdk}{\varepsilon-\varepsilon_{\boldsymbol{k}}+\mu+i\delta\,sign\,\varepsilon}=\frac{im^*}{2\pi^2\hbar^2|\boldsymbol{r}-\boldsymbol{r}'|}\int_{-\infty}^{\infty}\frac{e^{ik|\boldsymbol{r}-\boldsymbol{r}'|}\,kdk}{k^2-2m^*(\varepsilon+\mu+i\delta\,sign\,\varepsilon)/\hbar^2} \end{aligned} \tag{11.8}$$

If $\varepsilon>-\mu$, Eq. (11.8) can be presented as

$$G_0(\boldsymbol{r},\boldsymbol{r}',\varepsilon)=\frac{im^*}{2\pi^2\hbar^2|\boldsymbol{r}-\boldsymbol{r}'|}\int_{-\infty}^{\infty}\frac{e^{ik|\boldsymbol{r}-\boldsymbol{r}'|}\,kdk}{\left[k-k_0(\varepsilon+i\delta\,sign\,\varepsilon)\right]\left[k+k_0(\varepsilon+i\delta\,sign\,\varepsilon)\right]}, \tag{11.9}$$

where $k_0(z)=\sqrt{2m^*(z+\mu)}/\hbar$. Since for complex z, the multivalued function $k_0(z)$ has a branch point at $z=-\mu$, we make a cut along the real z-axis from $z=-\mu$ to $z=\infty$. Then we get $k_0(\varepsilon+i\delta\,sign\,\varepsilon)=k_0(\varepsilon)sign\,\varepsilon+i\delta$. Integral (11.9) is calculated along the contour in the upper halfplane of complex k. As a result, we obtain

$$G_0(\boldsymbol{r},\boldsymbol{r}';\varepsilon)=-\frac{m^*}{2\pi\hbar^2|\boldsymbol{r}-\boldsymbol{r}'|}\exp\left(i\left[\,k_0(\varepsilon)\ sign\ \varepsilon+i\delta\right]|\boldsymbol{r}-\boldsymbol{r}'|\right),\quad \varepsilon+\mu>0. \tag{11.10}$$

When $\varepsilon<-\mu$ we get

$$G_0(\boldsymbol{r},\boldsymbol{r}',\varepsilon)=\frac{im^*}{2\pi^2\hbar^2|\boldsymbol{r}-\boldsymbol{r}'|}\int_{-\infty}^{\infty}\frac{e^{ik|\boldsymbol{r}-\boldsymbol{r}'|}\,kdk}{(k-i\kappa)(k+i\kappa)}=-\frac{m^*e^{-\kappa|\boldsymbol{r}-\boldsymbol{r}'|}}{2\pi\hbar^2|\boldsymbol{r}-\boldsymbol{r}'|},\quad \varepsilon+\mu<0. \tag{11.11}$$

where $\kappa(\varepsilon)=\sqrt{2m^*|\varepsilon+\mu|}/\hbar$.

If $\mu>0$ (degenerate semiconductor), we present integration over ε as integration in complex z-plane along the paths shifted to complex z

$$s_\alpha(r)=-\frac{iJS_{i\alpha}m^{*2}}{4\pi^3\hbar^4r^2}\left(\int_{-\infty-i\delta}^{-i\delta}+\int_{i\delta}^{\infty+i\delta}\right)dz\quad\exp\left(\frac{2ir\sqrt{2m^*(z+\mu)}}{\hbar}\right)$$

$$=-\frac{iJS_{i\alpha}m^{*2}}{4\pi^3\hbar^4r^2}\left(\int_{-\infty-i\delta}^{-i\delta}+\int_{-\infty+i\delta}^{\infty+i\delta}-\int_{-\infty+i\delta}^{i\delta}\right)dz\ \exp\left(\frac{2ir\sqrt{2m^*(z+\mu)}}{\hbar}\right) \tag{11.12}$$

The second integral from $(-\infty+i\delta)$ to $(\infty+i\delta)$ gives zero after calculating it by contour integration in the upper halfplane. The other two integrals compensate each other at the segment $(-\infty,-\mu)$ because the function $k_0(z)$ does not depend on the shift sign if $z<-\mu$ (remember that the function $k_0(z)$ has opposite signs only for $z>-\mu$, i.e., at the different edges of the cut). As a result, we get

$$s_\alpha(r)=-\frac{iJS_{i\alpha}m^{*2}}{4\pi^3\hbar^4r^2}\left(\int_{-\mu-i\delta}^{-i\delta}-\int_{-\mu+i\delta}^{i\delta}\right)dz\quad\exp\left(\frac{2ir\sqrt{2m^*(z+\mu)}}{\hbar}\right)$$

$$=\frac{iJS_{i\alpha}m^{*2}}{2\pi^3\hbar^4r^2}\int_{-\mu}^{0}d\varepsilon\quad\exp\left(\frac{2ir\sqrt{2m^*(\varepsilon+\mu)}}{\hbar}\right)=\frac{JS_{i\alpha}m^*}{8\pi^3\hbar^2r^4}\left[2k_Fr\cos(2k_Fr)-\sin(2k_Fr)\right],$$

$$k_F\equiv\frac{\sqrt{2m^*\mu}}{\hbar}. \tag{11.13}$$

Function $s_\alpha(r)$ is oscillating at large distances as $s_\alpha(r)\sim\cos(2k_Fr)/r^3$ for $2k_Fr\gg1$, and this is called *Friedel oscillations* of the spin density. At small distances, $2k_Fr\ll1$, from Eq. (11.13), we have $s_\alpha(r)=-JS_{i\alpha}m^*k_F^3/3\pi^2\hbar^2r$. This function is formally divergent at $r\to0$ because we used the model with point perturbation. Note that at small distances, the electron spins align ferromagnetically if $J<0$.

The interaction between two spins $\boldsymbol{S}_i$ and $\boldsymbol{S}_j$ separated by vector $\boldsymbol{r}$ has the form,

$$W_{ji}(\boldsymbol{r})=J\ \boldsymbol{S}_j\cdot\boldsymbol{s}(\boldsymbol{r}), \tag{11.14}$$

where $\boldsymbol{s}(\boldsymbol{r})$ is the electron spin density induced by $\boldsymbol{S}_i$. Then using Eq. (11.13), we find the interaction energy

$$W_{ji}(r)=\frac{J^2m^*\left(\boldsymbol{S}_i\cdot\boldsymbol{S}_j\right)}{8\pi^3\hbar^2r^4}\left[2k_Fr\cos(2k_Fr)-\sin(2k_Fr)\right] \tag{11.15}$$

This formula is usually called the *RKKY interaction* (Ruderman-Kittel-Kasuya-Yosida). RKKY mechanism describes well indirect magnetic interaction in metals and degenerate semiconductors.

11.3 Indirect Exchange in Dielectrics: Bloembergen–Rowland Mechanism

Within the one-band model discussed above, we find that for $\mu < 0$ (empty conduction band in a nondegenerate semiconductor), $s_\alpha(\boldsymbol{r}) = 0$, meaning that in the absence of free electrons, no impurity-induced spin polarization exists. However, more adequate semiconductor models are based on the k-p perturbation approach (see Kane or Dirac models, Chapter 1, Sections 1.12–1.14). Below we consider a two-band semiconductor within the Dirac model. The model describes two bands and allows calculating indirect exchange interaction in an insulating state when the chemical potential lies in the bandgap. It turns out that magnetic polarization is nonzero due to *vacuum polarization* (excitation of virtual electron–hole pairs). It is the *Bloembergen–Rowland* mechanism of indirect exchange interaction we discuss below. The mechanism was proposed and discussed in Refs. [3,4] within a simplistic semiconductor model with parabolic electron spectrum.

The two-band semiconductor model–the isotropic Dirac Hamiltonian has the matrix form by band index:

$$H = \int d^3r\, \psi^\dagger(\boldsymbol{r}) \left[\begin{pmatrix} \Delta & -iv\,\boldsymbol{\sigma}\cdot\nabla \\ -iv\,\boldsymbol{\sigma}\cdot\nabla & -\Delta \end{pmatrix} + JS_{i\alpha}\,\delta(\boldsymbol{r}) \begin{pmatrix} \sigma_\alpha & 0 \\ 0 & \sigma_\alpha \end{pmatrix} \right] \psi(\boldsymbol{r})$$

$$\equiv \int d^3r\, \psi^\dagger(\boldsymbol{r}) \left[\Delta\tau_z - iv\,\tau_x\, \boldsymbol{\sigma}\cdot\nabla + JS_{i\alpha}\,\delta(\boldsymbol{r})\,\sigma_\alpha \right] \psi(\boldsymbol{r}), \tag{11.16}$$

where $\psi(\boldsymbol{r})$ is the bispinor field, $\Delta = E_g/2$ is the bandgap parameter, τ_x and τ_z are the Pauli matrices acting in band (electron and hole) space, and σ_α are the spin Pauli matrices, v is the parameter, which is related to electron velocity v/$\hbar$.

One can calculate the interaction between two spins starting from expression,

$$W_{ji}(r) = -iS_{j\alpha}S_{i\beta}\; Tr \int \frac{d\varepsilon}{2\pi}\, \hat{J}\sigma_\alpha\, G_0(\boldsymbol{r},0;\varepsilon)\, \hat{J}\sigma_\beta G_0(0,\boldsymbol{r};\varepsilon). \tag{11.17}$$

Such a direct method is useful for a multiband model of a semiconductor, which can include different exchange coupling constants for conduction and valence bands (in this case, the exchange coupling $\hat{J}$ would be a matrix in the band space).

The Green's function of free Dirac electron in $(\boldsymbol{k},\varepsilon)$ representation is

$$G_0(\boldsymbol{k},\varepsilon) = \frac{\varepsilon + \mu + \Delta\tau_z + v\tau_x\, \boldsymbol{\sigma}\cdot\boldsymbol{k}}{(\varepsilon - E_k + \mu + i\delta\; sign\; \varepsilon)(\varepsilon + E_k + \mu + i\delta\; sign\; \varepsilon)} \tag{11.18}$$

where $E_k = \sqrt{\Delta^2 + v^2k^2}$ is the electron spectrum in the conduction band. In the following, we take $\mu = 0$ (i.e., chemical potential in the energy gap). Then the Green's function of free electrons in $(\boldsymbol{r},\varepsilon)$ representation is

$$G_0(\boldsymbol{r},\boldsymbol{r}',\varepsilon) = \int \frac{d^3k}{(2\pi)^3}\, e^{i\boldsymbol{k}\cdot(\boldsymbol{r}-\boldsymbol{r}')}\, \frac{\varepsilon + \tau_z\Delta + v\tau_x\, \boldsymbol{\sigma}\cdot\boldsymbol{k}}{(\varepsilon + i\delta\, sign\, \varepsilon)^2 - E_k^2} \tag{11.19}$$

Calculation of the integrals over angles gives us

$$\int \frac{d^3k}{(2\pi)^3} e^{ik(r-r')} \ldots = \frac{1}{4\pi^2} \int_0^\infty k^2 dk \int_{-1}^{1} dt \; e^{ik|r-r'|t} \ldots$$

$$= -\frac{i}{4\pi^2 |r-r'|} \int_0^\infty kdk \left(e^{ik|r-r'|} - e^{-ik|r-r'|} \right) \ldots = -\frac{i}{4\pi^2 |r-r'|} \int_{-\infty}^{\infty} kdk e^{ik|r-r'|} \ldots \tag{11.20}$$

$$\int \frac{d^3k}{(2\pi)^3} e^{ik(r-r')} \ldots = \frac{1}{4\pi^2} \frac{\boldsymbol{r}-\boldsymbol{r}'}{|\boldsymbol{r}-\boldsymbol{r}'|} \int_0^\infty k^3 dk \int_{-1}^{1} tdte^{ik|r-r'|t} \ldots$$

$$= -\frac{1}{4\pi^2} \frac{\boldsymbol{r}-\boldsymbol{r}'}{|\boldsymbol{r}-\boldsymbol{r}'|^3} \int_0^\infty kdk \left[ik|\boldsymbol{r}-\boldsymbol{r}'| \left(e^{ik|r-r'|} + e^{-ik|r-r'|} \right) - \left(e^{ik|r-r'|} - e^{-ik|r-r'|} \right) \right] \cdots$$

$$= -\frac{1}{4\pi^2} \frac{\boldsymbol{r}-\boldsymbol{r}'}{|\boldsymbol{r}-\boldsymbol{r}'|^3} \int_{-\infty}^{\infty} kdk \left(ik|\boldsymbol{r}-\boldsymbol{r}'| - 1 \right) e^{ik|r-r'|} \ldots \tag{11.21}$$

Then using Eqs. (11.20) and (11.21), we present Eq. (11.19) as

$$G_0(\boldsymbol{r},\boldsymbol{r}',\varepsilon) = -\frac{i}{4\pi^2 |\boldsymbol{r}-\boldsymbol{r}'|} \int_{-\infty}^{\infty} kdk \; e^{ik|r-r'|} \frac{\varepsilon + \tau_z \Delta}{(\varepsilon + i\delta \; sign \; \varepsilon)^2 - E_k^2}$$

$$- \frac{\mathrm{v}\tau_x \, \boldsymbol{\sigma} \cdot (\boldsymbol{r}-\boldsymbol{r}')}{4\pi^2 |\boldsymbol{r}-\boldsymbol{r}'|^3} \int_{-\infty}^{\infty} kdk \; e^{ik|r-r'|} \frac{ik|\boldsymbol{r}-\boldsymbol{r}'| - 1}{(\varepsilon + i\delta \, sign \varepsilon)^2 - E_k^2}$$

$$= \frac{i}{4\pi^2 \mathrm{v}^2 |\boldsymbol{r}-\boldsymbol{r}'|} \int_{-\infty}^{\infty} kdk \; e^{ik|r-r'|} \frac{\varepsilon + \tau_z \Delta}{k^2 - k_0^2 (\varepsilon + i\delta \; sign \varepsilon)}$$

$$+ \frac{\tau_x \, \boldsymbol{\sigma} \cdot (\boldsymbol{r}-\boldsymbol{r}')}{4\pi^2 \mathrm{v} |\boldsymbol{r}-\boldsymbol{r}'|^3} \int_{-\infty}^{\infty} kdk \; e^{ik|r-r'|} \frac{ik|\boldsymbol{r}-\boldsymbol{r}'| - 1}{k^2 - k_0^2 (\varepsilon + i\delta \, sign \varepsilon)}, \tag{11.22}$$

where $k_0(z) = \sqrt{z^2 - \Delta^2}/\mathrm{v}$. The function $k_0(z)$ of complex variable z has two branch points $z = \pm\Delta$. Therefore, we make two cuts along the Rez-axis from $-\infty$ to $-\Delta$ and from Δ to ∞. Then for $|\varepsilon| > \Delta$ we obtain $k_0(\varepsilon + i\delta \; sign \; \varepsilon) = \sqrt{\varepsilon^2 - \Delta^2}/\mathrm{v} + i\delta$, and for $|\varepsilon| < \Delta$ we get $k_0(\varepsilon + i\delta \; sign \; \varepsilon) = i\sqrt{\Delta^2 - \varepsilon^2}/\mathrm{v}$.

After calculating the integrals in Eq. (11.22) along the contour in the upper halfplane of complex k, we obtain

$$G_0(\boldsymbol{r},\boldsymbol{r}',\varepsilon) = \left\{ -\frac{\varepsilon + \tau_z \Delta}{4\pi \mathrm{v}^2 |\boldsymbol{r}-\boldsymbol{r}'|} + \frac{i\tau_x \, \boldsymbol{\sigma} \cdot (\boldsymbol{r}-\boldsymbol{r}')}{4\pi \mathrm{v} |\boldsymbol{r}-\boldsymbol{r}'|^3} \left[ik_0(\varepsilon) |\boldsymbol{r}-\boldsymbol{r}'| - 1 \right] \right\} e^{i[k_0(\varepsilon) + i\delta] |r-r'|}. \tag{11.23}$$

Then the expression for interaction to the spin $\boldsymbol{S}_j$ located at the point $\boldsymbol{r}$ has the following form:

$$W_{ji}(\boldsymbol{r}) = -iJ^2 S_{j\alpha} S_{i\beta}\, Tr \int_{-\infty}^{\infty} \frac{d\varepsilon}{2\pi}\, \sigma_\alpha\, G_0(\boldsymbol{r},0;\varepsilon)\, \sigma_\beta\, G_0(0,\boldsymbol{r};\varepsilon)$$

$$= -\frac{iJ^2 S_{j\alpha} S_{i\beta}}{32\pi^5 \mathrm{v}^4 r^2}\, Tr \int_{-\infty}^{\infty} d\varepsilon\; e^{2ir[k_0(\varepsilon)+i\delta]}\, \sigma_\alpha \left[\varepsilon + \tau_z \Delta - \frac{i\mathrm{v}\tau_x\, \boldsymbol{\sigma}\cdot\boldsymbol{r}}{r^2}\left(irk_0(\varepsilon)-1\right)\right]\sigma_\beta$$

$$\times \left[\varepsilon + \tau_z \Delta + \frac{i\mathrm{v}\tau_x\, \boldsymbol{\sigma}\cdot\boldsymbol{r}}{r^2}\left(irk_0(\varepsilon)-1\right)\right]$$

$$= -\frac{iJ^2 S_{j\alpha} S_{i\beta}}{8\pi^5 \mathrm{v}^4 r^2}\left(\int_{-\infty-i\delta}^{-i\delta} + \int_{i\delta}^{\infty+i\delta}\right) dz e^{2irk_0(z)}$$

$$\times \left[\left(z^2+\Delta^2\right)\delta_{\alpha\beta} + \frac{\mathrm{v}^2\left(2r_\alpha r_\beta - r^2\delta_{\alpha\beta}\right)}{r^4}(irk_0(z)-1)^2\right]. \tag{11.24}$$

By substitution $z = i\xi$, we shift the integration path in Eq. (11.24) to the imaginary axis in the complex z-plane:

$$W_{ji}(\boldsymbol{r}) = -\frac{iJ^2 S_{j\alpha} S_{i\beta}}{4\pi^5 \mathrm{v}^4 r^2}\int_0^{i\infty} dz e^{2irk_0(z)}\left[\left(z^2+\Delta^2\right)\delta_{\alpha\beta} + \frac{\mathrm{v}^2\left(2r_\alpha r_\beta - r^2\delta_{\alpha\beta}\right)}{r^4}\left(ik_0(z)+1\right)^2\right]$$

$$= \frac{J^2 S_{j\alpha} S_{i\beta}}{4\pi^5 \mathrm{v}^4 r^2}\int_0^{\infty} d\xi e^{-2r\sqrt{\xi^2+\Delta^2}/\mathrm{v}}\left[\left(\Delta^2-\xi^2\right)\delta_{\alpha\beta} + \frac{\mathrm{v}^2\left(2r_\alpha r_\beta - r^2\delta_{\alpha\beta}\right)}{r^4}\left(\frac{r\sqrt{\xi^2+\Delta^2}}{\mathrm{v}}+1\right)^2\right], \tag{11.25}$$

At large distances, $r \gg \mathrm{v}/\Delta$, we find

$$W_{ji}(\boldsymbol{r}) = \frac{J^2 S_{j\alpha} S_{i\beta}}{4\pi^5 \mathrm{v}^4 r^2}\int_0^{\infty} d\xi\; e^{-\frac{2r\sqrt{\xi^2+\Delta^2}}{\mathrm{v}}}\left[-2\xi^2\delta_{\alpha\beta} + \frac{2r_\alpha r_\beta}{r^2}\left(\xi^2+\Delta^2\right)\right]. \tag{11.26}$$

Here the integrals over ξ can be calculated using the substitution $\xi = \Delta \sinh t$.

$$\int_0^{\infty} d\xi \exp\left(-\frac{2r\sqrt{\xi^2+\Delta^2}}{\mathrm{v}}\right) = \Delta\int_0^{\infty}\cosh t \exp\left(-\frac{2r\Delta\cosh t}{\mathrm{v}}\right) dt = \Delta\; K_1(2r\Delta/\mathrm{v}), \tag{11.27}$$

$$\int_0^{\infty}\xi^2 d\xi \exp\left(-\frac{2r\sqrt{\xi^2+\Delta^2}}{\mathrm{v}}\right) = \Delta^3\int_0^{\infty}\sinh^2 t \cosh t \exp\left(-\frac{2r\Delta\cosh t}{\mathrm{v}}\right) dt$$

$$= \frac{\mathrm{v}\Delta^2}{2r}\int_0^{\infty}\cosh(2t)\exp\left(-\frac{2r\Delta\cosh t}{\mathrm{v}}\right) = \frac{\mathrm{v}\Delta^2}{2r} K_2(2r\Delta/\mathrm{v}), \tag{11.28}$$

where we used an integral representation of the modified Bessel function

$$K_n(z)=\int_0^\infty \cosh(nz)e^{-z\cosh z}\,dz. \tag{11.29}$$

Then using Eqs. (11.27) and (11.28) and the asymptotics of Bessel functions for large z, $K_n(z)\simeq\left(\pi/\sqrt{2z}\right)e^{-z}$ we find the dominating interaction at large r:

$$W_{ji}(\boldsymbol{r})\simeq\frac{J^2(\boldsymbol{S}_j\cdot\boldsymbol{r})(\boldsymbol{S}_i\cdot\boldsymbol{r})\Delta^3}{2\pi^5 v^4 r^4}K_1(2r\Delta/v)=\frac{J^2(\boldsymbol{S}_j\cdot\boldsymbol{r})(\boldsymbol{S}_i\cdot\boldsymbol{r})\Delta^{5/2}}{4\pi^4 v^{7/2} r^{9/2}}\exp\left(-\frac{2r\Delta}{v}\right). \tag{11.30}$$

Thus, the exchange interaction is nonzero due to the vacuum polarization (no carriers), as follows from Hamiltonian (11.17). In this model, the interaction is highly anisotropic, and orienting the pair of impurity spins antiferromagnetically along the line connecting two impurities.

11.4 Ferromagnetic Ordering of Magnetic Impurities

The oscillating character of pair interaction (11.14) between impurity spins in the degenerate semiconductor leads to ferromagnetic ordering for sufficiently large spin density. If most of the randomly distributed spins are at distances $2k_F r \ll 1$, RKKY exchange is ferromagnetic. It is correct if $n_i^{-1/3} \ll r \ll k_F^{-1}$ (here n_i is the impurity density).

Let us assume now that all impurity spins align ferromagnetically. Then the spin density created in point $r=0$ by all impurities except the one at $r=0$ is

$$s_\alpha(0)\simeq\frac{JS_{i\alpha}m^*n_i}{2\pi^2\hbar^2}\int_{a_0}^\infty\frac{dr}{r^2}\left[2k_F r\cos(2k_F r)-\sin(2k_F r)\right]$$

$$=\frac{JS_{i\alpha}m^*n_ik_F}{\pi^2\hbar^2}\int_{2k_Fa_0}^\infty\frac{dx}{x^2}(x\cos x-\sin x)=-\frac{JS_{i\alpha}m^*n_ik_F}{\pi^2\hbar^2}\frac{\sin(2k_Fa_0)}{2k_Fa_0}, \tag{11.31}$$

where a_0 is the lattice parameter. From Eq. (11.31) follows that for the case of $2k_Fa_0 \ll 1$, $s_\alpha(0)\simeq -JS_{i\alpha}m^*n_ik_F/\pi^2\hbar^2$. Correspondingly the energy of spin $\boldsymbol{S}_0$ located at $r=0$ becomes

$$\varepsilon_{int}=-\frac{J^2\,(\boldsymbol{S}_0\cdot\boldsymbol{S}_i)\,m^*n_ik_F}{\pi^2\hbar^2}, \tag{11.32}$$

so the mean-field acting on spin $\boldsymbol{S}_0$ at $r=0$ makes the ferromagnetic state energetically favorable. The critical temperature of ferromagnetic ordering can be estimated as $k_BT_c \approx |\varepsilon_{int}|$:

$$T_c\simeq\frac{J^2\,S(S+1)\,m^*n_ik_F}{\pi^2\hbar^2k_B}, \tag{11.33}$$

where S is the spin value and $\boldsymbol{S}^2=S(S+1)$ is the eigenvalue of a spin square. The method of calculating T_c corresponds to the *mean-field approximation* in the theory of phase transition.

In a more formalized version of mean-field theory, we start with the partition function of the Heisenberg spin system,

$$Z=Tr\exp\left(-\frac{1}{2k_BT}\sum_{ij}\hat{S}_{i\alpha}\,g(\boldsymbol{R}_i-\boldsymbol{R}_j)\,\hat{S}_{j\alpha}\right), \tag{11.34}$$

where $g(\boldsymbol{r}-\boldsymbol{r}')$ is the range function, which enters RKKY interaction (11.15) as $W_{ij}(\boldsymbol{r}-\boldsymbol{r}')=(\boldsymbol{S}_i\cdot\boldsymbol{S}_j)g(\boldsymbol{r}-\boldsymbol{r}')$, and the trace implies summation over all possible states of the spin system.

The standard trick which makes possible the calculation of trace is using the Hubbard–Stratonovich decoupling method resulting in

$$Z=Tr\int Ds(\boldsymbol{r})\exp\left[\sum_i\sum_r \hat{S}_{i\alpha}\,\delta(\boldsymbol{R}_i,\boldsymbol{r})\,s_\alpha(\boldsymbol{r})+\frac{k_BT}{2}\sum_{rr'}s_\alpha(\boldsymbol{r})g^{-1}(\boldsymbol{r}-\boldsymbol{r}')\,s_\alpha(\boldsymbol{r}')\right], \tag{11.35}$$

where $\delta(\boldsymbol{R}_i,\boldsymbol{r})$ is the Kronecker symbol, functional integration runs in each point of space over a dimensionless auxiliary field $\boldsymbol{s}(\boldsymbol{r})$, which is proportional to the electron spin density field: $\int Ds(\boldsymbol{r})\equiv\int\cdots\int\prod_r ds_r$. The Hubbard–Stratonovich method, which helps to get Eq. (11.35) from Eq. (11.34), is based on an arbitrary shift of variable in the multidimensional Gauss integral:

$$\begin{aligned}I&=\int d^nx\,\exp\left(-\frac{1}{2}\boldsymbol{x}A^{-1}\boldsymbol{x}\right)=\int d^nx\,\exp\left[-\frac{1}{2}(\boldsymbol{x}-\boldsymbol{y}A)\,A^{-1}\,(\boldsymbol{x}-A\boldsymbol{y})\right]\\&=\int d^nx\exp\left(-\frac{1}{2}\,\boldsymbol{x}A^{-1}\boldsymbol{x}+\boldsymbol{x}\cdot\boldsymbol{y}-\frac{1}{2}\,\boldsymbol{y}A\boldsymbol{y}\right),\end{aligned} \tag{11.36}$$

where $\boldsymbol{x}$ and $\boldsymbol{y}$ are the n-component vectors and A is the $n\times n$ positive definite matrix. Note that integral I depends only on the choice of matrix A.[1] From Eq. (11.36) follows

$$\exp\left(\frac{1}{2}\boldsymbol{y}A\boldsymbol{y}\right)I=\int d^nx\,\exp\left(-\frac{1}{2}\,\boldsymbol{x}A^{-1}\boldsymbol{x}+\boldsymbol{x}\cdot\boldsymbol{y}\right). \tag{11.37}$$

In our case, $s_\alpha(\boldsymbol{r})$ corresponds to the vector $\boldsymbol{x}$ with components $\boldsymbol{x}_r$ in the points of discrete $\boldsymbol{r}$-space, $s_\alpha(\boldsymbol{r})=\boldsymbol{x}_r$. Correspondingly, function $-\frac{g(\boldsymbol{r}-\boldsymbol{r}')}{k_BT}\equiv-\frac{g_{rr'}}{k_BT}=A_{rr'}$, vector $\boldsymbol{y}_r=\sum_i\hat{S}_{i\alpha}\,\delta(\boldsymbol{r},\boldsymbol{R}_i)$, and function $g^{-1}(\boldsymbol{r}-\boldsymbol{r}')\equiv g^{-1}_{rr'}$. In these notations, the expression in Eq. (11.35) is similar to the right-hand part of (11.37). After using I in Eq. (11.34) and comparing it with (11.37), the formula for Z in Eq. (11.34) becomes (11.35) (inserting a constant does not affect the partition function Z).

Now we calculate the trace over spin states of impurity $\boldsymbol{S}_i$ (we omit index i and denote $\boldsymbol{s}=\boldsymbol{s}(\boldsymbol{R}_i)$)

$$\begin{aligned}Tr\exp(\boldsymbol{S}\cdot\boldsymbol{s})&=\sum_{S_z=-S}^{S_z=S}e^{S_zs}=e^{sS}\left(1+e^{-s}+e^{-2s}+\cdots+e^{-2s}\right)\\&=\left(e^{sS}-e^{-s(S+1)}\right)\left(1+e^{-s}+e^{-2s}+\cdots\right)=\frac{e^{sS}-e^{-s(S+1)}}{1-e^{-s}}=\frac{\sinh\left[s(S+1/2)\right]}{\sinh(s/2)},\end{aligned} \tag{11.38}$$

where we assume the orientation of vector $\boldsymbol{s}$ along axis z. After calculating the trace over states of all impurities, we obtain

[1] After diagonalization of A, one finds $I=(2\pi)^{n/2}(\det A)^{-1/2}$.

$$Z = \int Ds(\boldsymbol{r}) \exp\left(\sum_i \sum_r \delta(\boldsymbol{r}, \boldsymbol{R}_i) \log \frac{\sinh[s(\boldsymbol{r})(S+1/2)]}{\sinh[s(\boldsymbol{r})/2]} + \frac{k_B T}{2} \sum_{rr'} s(\boldsymbol{r})\, g^{-1}(\boldsymbol{r}-\boldsymbol{r}')\, s(\boldsymbol{r}') \right). \tag{11.39}$$

The mean-field value of $s(\boldsymbol{r})$ corresponds to the minimum in the exponent,

$$\sum_i \delta(\boldsymbol{r}, \boldsymbol{R}_i) \left\{ (S+1/2) \coth[s(\boldsymbol{r})(S+1/2)] - \frac{1}{2} \coth[s(\boldsymbol{r})/2] \right\} = -k_B T \sum_{r'} g^{-1}(\boldsymbol{r}-\boldsymbol{r}')\, s(\boldsymbol{r}'). \tag{11.40}$$

Using the matrix form of this equation we can rewrite it as

$$s(\boldsymbol{r}) = -\frac{1}{k_B T} \sum_{r'} \sum_i g(\boldsymbol{r}-\boldsymbol{r}')\, \delta(\boldsymbol{r}', \boldsymbol{R}_i) \left\{ \left(S+\frac{1}{2}\right) \coth\left[s(\boldsymbol{r}')\left(S+\frac{1}{2}\right) \right] - \frac{1}{2} \coth\left[\frac{s(\boldsymbol{r}')}{2} \right] \right\}. \tag{11.41}$$

Then, assuming a weak variation of functions $s(\boldsymbol{r})$ and $g(\boldsymbol{r}-\boldsymbol{r}')$ at the mean distance between impurities $n_i^{-1/3}$, we can average over impurity locations, and in the limit of $s(\boldsymbol{r}) \to 0$ we finally get the equation for mean field polarization

$$s(\boldsymbol{r}) = -\frac{n_i S(S+1)}{3 k_B T} \int d^3 r' g(\boldsymbol{r}-\boldsymbol{r}')\, s(\boldsymbol{r}'). \tag{11.42}$$

Assuming $s(\boldsymbol{r})$ constant, we find the temperature of ferromagnetic ordering as a solution to (11.42):

$$T_c = -\frac{n_i S(S+1)}{3 k_B} \int d^3 r' g(\boldsymbol{r}-\boldsymbol{r}') = -\frac{n_i S(S+1)}{3 k_B} \int d^3 r' g(\boldsymbol{r}'). \tag{11.43}$$

Here we use $g(\boldsymbol{r}-\boldsymbol{r}') = g(|\boldsymbol{r}-\boldsymbol{r}'|)$ so that the integral over the whole space does not depend on coordinates. The critical temperature corresponds to rough estimation (11.33).

11.5 Magnetic Order and Percolation

The necessary condition for using approximation (11.33) is $n_i \gg k_F^3$: the mean distance between impurity spins $n_i^{-1/3}$ should be much smaller than the Fermi wavelength of electrons $\lambda_F = 2\pi / k_F$).[2] It means that there is a large number of spins in the volume $\sim k_F^{-3}$.

If $n_i < k_F^3$, the impurity density fluctuates so that only relatively small clusters of impurity spins, which are at the distance $r < k_F^{-1}$, are ferromagnetically ordered, but there is no ordering between moments belonging to different clusters because the interaction sign is oscillating with r for $r > k_F^{-1}$. Hence, a degenerate semiconductor with a small density of impurity spins is not a ferromagnet.

Another condition of using exchange interaction in form of Eq. (11.15) is a relatively weak effect of disorder, which affects the wave functions and eigenstates of an electron. Indeed, the wave vector is a good quantum number if an electron propagates on a distance smaller than the mean free path l related to scattering at impurities and defects. Account for disorder introduces additional factor $\exp(-r/l)$ in Eq. (11.15) [5]. Besides, it is assumed inequality $l > k_F^{-1}$ holds, otherwise the electron states are localized.

---

[2] Note that this assumption makes sense only for semiconductors with low carrier density. In metals, $k_F \simeq a_0^{-1}$ so that inequality $r < k_F^{-1}$ implies distances shorter than the lattice constant a_0.

Thus, the exchange interaction in a "dirty" degenerate semiconductor has a finite range $R_0 \simeq l$. So, it makes sense to consider the model with ferromagnetic coupling $W(r)$ of finite range R_0. At a low density of randomly distributed impurity spins, $n_i \ll R_0^{-3}$, only in a small number of clusters the inter-spin distances $r < R_0$ make clusters ferromagnetic. Ferromagnetic clusters are growing in size with the increasing density of spins so that at critical n_i all ferromagnetically coupled clusters merge and penetrate the whole sample (*percolation point*). Critical spin density n_i depends on temperature as impurities can align spins only if the ferromagnetic coupling is large enough, $W(r) \gg k_B T$, otherwise, temperature fluctuations break ferromagnetic order. Distance r_c determines *percolation* point: $W(r_c) = k_B T_c$, T_c is the critical temperature of percolation transition. Critical distance r_c relates to the mean distance between impurity spins as $r_c = \beta n_i^{-1/3}$, where $\beta \simeq 0.87$ was found by computer simulations [6]. As a result, we get

$$k_B T_c = W\left(\beta n_i^{-1/3}\right). \tag{11.44}$$

Although $\beta \approx 1$, the critical temperature of the percolation transition can differ significantly from the mean-field result. The difference is due to strong dependence (e.g., exponential decay) of the coupling energy on a distance at $r \sim R_0$.

11.6 Spin Glass

If the electron Fermi vector is large, $k_F \gg n_i^{1/3}$, strongly oscillating RKKY interaction (11.15) of randomly distributed impurity spins looks like aleatory in sign. Even a small change of a distance between two impurities changes the sign of interaction. The mean-field consideration of Section 11.3 is not applicable in this case. The effective field, acting on an arbitrarily chosen spin, depends on the specific distribution of surrounding spins and appears random in sign and direction. At the same time, the magnitude of interaction can be sufficiently large. In this situation, there arises randomly distributed and oriented strongly coupled impurity spins – the *spin glass*.

In the model, we consider impurity spins homogeneously distributed in space with small random deviations Δr from fictitious regular positions $\bar{R}_i$. The distance between neighboring spins is almost the same, equal to average distance $\bar{r} = n_i^{-1/3}$, up to the random deviation Δr and that $k_F^{-1} \ll \Delta r \ll \bar{r}$. Then the averaging of pair interaction $g_{ij}(r) = g_{0ij}(r)\cos(2k_F r)$ of the spins at distance r over random deviation Δr is, in fact averaging over variable q_{ij}, which determines fluctuations of $\cos(2k_F r)$ due to random r-variations. To calculte the average, we introduce distribution function $w(q_{ij})$, characterizing disorder in positions of impurity spins.

To describe the transition to spin-glass order, we start from the partition function (11.34):

$$Z = Tr\ \exp\left(-\frac{1}{2k_B T}\sum_{ij} q_{ij}\hat{S}_{i\mu} g_{0ij}\hat{S}_{j\mu}\right) \tag{11.45}$$

where random variables q_{ij} account for fluctuations of the interaction between spins $\mathbf{S}_i$ and $\mathbf{S}_j$ at a distance $|\mathbf{R}_i - \mathbf{R}_j|$.

Our goal is to get average free energy $F = -k_B T\langle \log Z\rangle$ of the spin system over random realizations of pair interaction. For averaging $\log Z$, we use the trick based on the expression for logarithm and called *replica method* [7,8]:

$$\log x = \lim_{n \to 0} \frac{1}{n}\left(x^n - 1\right). \tag{11.46}$$

With this expression, the averaged free energy becomes

$$F = -k_BT \lim_{n\to 0} \frac{1}{n}\int dq_{ij}\, w(q_{ij})\left\{Tr \exp\left(-\frac{1}{2k_BT}\sum_{ij} q_{ij}\, \hat{S}^{\alpha}_{i\mu}\, g_{0ij}\, \hat{S}^{\alpha}_{j\mu}\right)-1\right\}, \tag{11.47}$$

where $\alpha = 1,2,\ldots,n$ is the replica index. This means that we consider now n equivalent replicas of the system, and the trace in Eq. (11.47) is the sum that runs over spin states in each replica.

In what follows, we use the Gauss distribution function for random variable q_{ij},

$$w(q_{ij}) = \frac{a}{\sqrt{\pi}} e^{-a^2 q_{ij}^2}, \tag{11.48}$$

where $a = 1/\sigma\sqrt{2}$, and σ is the variance (fluctuations magnitude). Then after integrating Eq. (11.47) over q_{ij} we get

$$F = -k_BT \lim_{n\to 0}\frac{1}{n}\left\{Tr \exp\left[\frac{1}{16a^2k_B^2T^2}\sum_{ij} g_{0ij}^2\left(\hat{S}^{\alpha}_{i\mu}\hat{S}^{\beta}_{i\mu}\right)\left(\hat{S}^{\beta}_{j\nu}\hat{S}^{\alpha}_{j\nu}\right)\right]-1\right\}. \tag{11.49}$$

The replica method allows the introduction of the order parameters responsible for the spin-glass state, off-diagonal correlators $Q_i^{\alpha\beta} = \left\langle S^{\alpha}_{i\mu} S^{\beta}_{i\mu}\right\rangle, \alpha \neq \beta$. Here <...> means both thermodynamic averaging and the averaging over random q_{ij}. The correlators appear as a result of the decoupling of spin product in the exponent:

$$F = -k_BT \lim_{n\to 0}\frac{1}{n}\left\{\int DQ_i^{\alpha\beta}\; Tr \exp\left[\frac{nS^2(S+1)^2}{16a^2k_B^2T^2}\sum_{ij} g_{0ij}^2 + \sum_i Q_i^{\alpha\beta}\hat{S}^{\beta}_{i\mu}\hat{S}^{\alpha}_{i\mu}\right.\right.$$

$$\left.\left. -4a^2k_B^2T^2\sum_{ij}\frac{1}{g_{0ij}^2}\, Q_i^{\alpha\beta}Q_j^{\beta\alpha}\right]-1\right\}, \tag{11.50}$$

where we use the functional integral notation $\int DQ_i^{\alpha\beta} \equiv \int\cdots\int\prod_i dQ_i^{\alpha\beta}$.

Assuming replica space isotropic, $Q_i^{\alpha\beta} = Q_i$ (all off-diagonal components are equal), we get

$$F = -k_BT \lim_{n\to 0}\frac{1}{n}\left\{\int DQ_i\; Tr \exp\left[\frac{nS^2(S+1)^2}{16a^2k_B^2T^2}\sum_{ij} g_{0ij}^2 + \sum_{i\alpha\beta} Q_i\hat{S}^{\beta}_{i\mu}\hat{S}^{\alpha}_{i\mu}\right.\right.$$

$$\left.\left. -nS(S+1)\sum_i Q_i - 4n(n-1)a^2k_B^2T^2\sum_{ij}\frac{Q_iQ_j}{g_{0ij}^2}\right]-1\right\} \tag{11.51}$$

and after using Hubbard–Stratonovich decoupling in the second term of Eq. (11.51), we obtain

$$F=-k_BT\lim_{n\to 0}\frac{1}{n}\left\{\int DQ_i\,Dm_i^{\mu}\,Tr\exp\left[\frac{nS^2(S+1)^2}{16a^2k_B^2T^2}\sum_{ij}g_{0ij}^2+\sum_i\left(\sum_{\alpha}\sqrt{2Q_i}m_{i\mu}\hat{S}_{i\mu}^{\alpha}-\frac{m_i^2}{2}\right)\right.\right.$$

$$\left.\left.-nS(S+1)\sum_i Q_i-4n(n-1)a^2k_B^2T^2\sum_{ij}\frac{Q_iQ_j}{g_{0ij}^2}\right]-1\right\}$$

$$=-k_BT\lim_{n\to 0}\frac{1}{n}\left\{\int DQ_i\,Dm_i\,\exp\left[\frac{nS^2(S+1)^2}{16a^2k_B^2T^2}\sum_{ij}g_{0ij}^2+n\sum_i\log\frac{\sinh\left[m_i\sqrt{2Q_i}(S+1/2)\right]}{\sinh\left[m_i\sqrt{2Q_i}/2\right]}\right.\right.$$

$$\left.\left.-\sum_i\frac{m_i^2}{2}-nS(S+1)\sum_i Q_i+4na^2k_B^2T^2\sum_{ij}\frac{Q_iQ_j}{g_{0ij}^2}\right]-1\right\}$$

$$=-k_BT\int DQ_i\,d^3m_i\,e^{-\sum_i(m_i^2/2)}\left[\frac{S^2(S+1)^2}{16a^2k_B^2T^2}\sum_{ij}g_{0ij}^2+\sum_i\log\frac{\sinh\left[m_i\sqrt{2Q_i}(S+1/2)\right]}{\sinh\left[m_i\sqrt{2Q_i}/2\right]}\right.$$

$$\left.-S(S+1)\sum_i Q_i+4a^2k_B^2T^2\sum_{ij}\frac{Q_iQ_j}{g_{0ij}^2}\right]. \tag{11.52}$$

The saddle-point equation $\partial F/\partial Q_i=0$ reads

$$\int_0^{\infty}m_i^2\,dm_i\,e^{-m_i^2/2}\left(\frac{m_i(S+1/2)}{\sqrt{2Q_i}}\coth\left(m_i\sqrt{2Q_i}(S+1/2)\right)-\frac{m_i}{\sqrt{2Q_i}}\coth\left(m_i\sqrt{2Q_i}/2\right)\right.$$

$$\left.-S(S+1)+8a^2k_B^2T^2\sum_j\frac{Q_j}{g_{0ij}^2}\right)=0. \tag{11.53}$$

In the limit of $Q_i\to 0$, we can use expansion,

$$\coth x\simeq\frac{1}{x}+\frac{x}{3}-\frac{x^3}{45}+\cdots \tag{11.54}$$

Then using Eq. (11.54) and integrating over m_i in Eq. (11.53), we come to

$$\frac{2}{3}\left[S^2(S+1)^2+\frac{1}{2}S(S+1)\right]Q_i=8a^2k_B^2T^2\sum_j\frac{Q_j}{g_{0ij}^2}, \tag{11.55}$$

the linear algebraic equation for vector with components Q_i. For convenience, one can express the equation as

$$\left[S^2(S+1)^2+\frac{1}{2}S(S+1)\right]\sum_j g_{0ij}^2\,Q_j=12a^2k_B^2T^2\,Q_i. \tag{11.56}$$

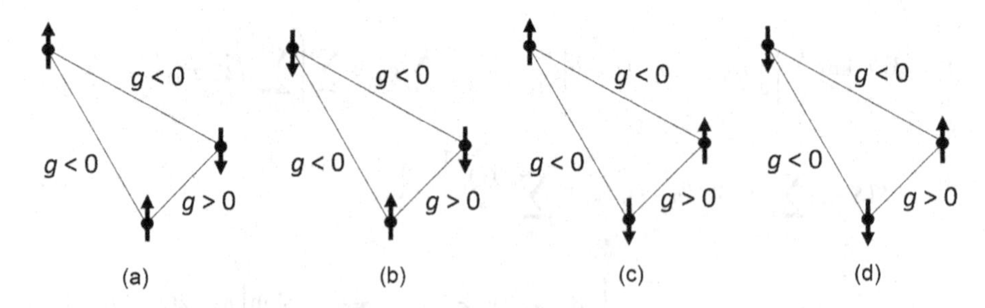

FIGURE 11.1 An example of frustration in the three-spin system. The energies of all states are the same, which makes possible jumps between these states.

Assuming spin-glass state homogeneous, $Q_i = Q$, we obtain the temperature of the spin-glass transition:

$$T_{SG} = \frac{\tilde{g}}{2\sqrt{3}ak_B}\left[S^2(S+1)^2 + \frac{1}{2}S(S+1)\right]^{1/2}, \tag{11.57}$$

where

$$\tilde{g} \simeq \left(n_i \int d^3r\, g_0^2(r)\right)^{1/2} = \frac{J^2 m^* k_F}{4\pi^3\hbar^2}\left(4\pi n_i \int_{a_0}^{\infty} \frac{dr}{r^4}\right)^{1/2} = \frac{J^2 m^* k_F}{4\pi^3\hbar^2}\left(\frac{4\pi n_i}{3a_0^3}\right)^{1/2} \tag{11.58}$$

and as g_0 we used the nonoscillating part of pair interaction from Eq. (11.15) for $2k_F r \gg 1$.

The transition from sum over j in Eq. (11.56) to the integral over whole space in Eq. (11.58) is justified if the spin distribution is close to homogeneous, so that the number of spins in the element of volume $d\Omega = 4\pi r^2 dr$ is equal to $n_i d\Omega$. Constant a determines the magnitude of fluctuations in pair interaction and enters Eq. (11.57): the smaller a, the larger fluctuations, and larger T_{SG}: disorder favors the transition to the spin-glass state. In this state, there exists a spin order with zero average magnetization. The spin-glass ground state is degenerate as after reversing the directions of several spins (without changing the total spin of the system), the total energy does not change. Schematically, it is illustrated in Figure 11.1 for three spins and corresponding signs of pair interaction g.

Figure 11.1 illustrates spin frustration: there is no way to satisfy the spin configuration required by given interaction signs on the bonds. That is the source of spin disorder in spin-glass.

For more details about the spin-glass state in specific materials, see Ref. [9].

11.7 Diluted Magnetic Semiconductors

If temperature T is above the critical temperature of magnetic ordering, the magnetically doped semiconductor is paramagnetic. When we turn the external magnetic field on, the localized magnetic moments align parallel to the external field and enhance the Zeeman field acting on electron spin. The physical mechanism of this effect is related to the exchange coupling of electrons and localized magnetic moments. To demonstrate it, we consider a model with electrons and homogeneously distributed localized spins in magnetic field $\boldsymbol{B}$. The Hamiltonian of this system reads

$$H = \sum_r \psi^\dagger(r)\left(-\frac{\hbar^2\Delta}{2m^*} + g\mu_B\, \boldsymbol{\sigma}\cdot\boldsymbol{B} + J\sum_i \boldsymbol{\sigma}\cdot\boldsymbol{S}_i\, \delta(\boldsymbol{R}_i, r)\right)\psi(r) + g\mu_B \sum_i \boldsymbol{S}_i\cdot\boldsymbol{B} \tag{11.59}$$

where m^* is the electron effective mass, $\mu_B = e\hbar / 2m^*$ is the Bohr magneton, g is the Landé factor, here $\psi^\dagger(\boldsymbol{r})$ and $\psi(\boldsymbol{r})$ are dimensionless electron fields – the spinor electron creation and annihilation operators. The free energy of this system is

$$F = -k_BT\log Z = -k_BT\ \log Tr\int D\psi\ D\psi^\dagger\ \exp\left(-\frac{H}{k_BT}\right) = -k_BT\log Tr\int D\psi\ D\psi^\dagger$$

$$\times\exp\left\{-\frac{1}{k_BT}\left[\sum_r \psi^\dagger\left(-\frac{\hbar^2\Delta}{2m^*} + g\mu_B\ \sigma\cdot\boldsymbol{B} + J\sum_i \sigma\cdot\boldsymbol{S}_i\ \delta(r,R_i)\right)\psi + g\mu_B\sum_i \boldsymbol{S}_i\cdot\boldsymbol{B}\right]\right\} \quad (11.60)$$

where the path integral over Fermi fields $\psi(\boldsymbol{r})$ and $\psi^\dagger(\boldsymbol{r})$ implies integration in each point of space, $\int D\psi\, D\psi^\dagger \equiv \int\cdots\int \prod_r d\psi(\boldsymbol{r})\ d\psi^\dagger(\boldsymbol{r})$. The trace runs over the spin states of impurities.

Integration over Fermi fields in quantum field theory uses anticommuting variables, which are called the *Grassmann fields*. The anticommutation property $\{\theta_i, \theta_j\} \equiv \theta_i\theta_j + \theta_j\theta_i = 0$ of any two Grassmann variables θ_i and θ_j and the simple integration rules with Grassmann variables

$$\int d\theta_i = 0 \quad \int \theta_i\ d\theta_i = 1, \quad (11.61)$$

lead to the formula

$$\int d\theta\ d\theta^*\ e^{a\theta^*\theta} = \int d\theta\ d\theta^*(1 + a\theta^*\theta) = a. \quad (11.62)$$

Correspondingly, we get the expression for Gauss integration as

$$\int D\theta_i\ D\theta_i^\dagger\ e^{\theta_i^\dagger A_{ij}\theta_j} = \det A \quad (11.63)$$

More details about using Grassmann fields in quantum field theory can be found in Ref. [10].

The external field $\boldsymbol{B}$ aligns electron and impurity spins. One can calculate magnetization in the impurity spin system neglecting electrons. Taking axis z in the direction of $\boldsymbol{B}$, we obtain (here H_S is the last term in Eq. (11.59) and $Z_S = Tr\ \exp(-H_S/k_BT)$),

$$S_{iz} \equiv \hat{S}_{iz} = \frac{1}{Z_S}\ Tr\ \hat{S}_{iz}\exp\left(-\frac{H_S}{k_BT}\right) = -\frac{k_BT}{g\mu_B}\frac{\partial}{\partial B}\log Tr\ \exp\left(-\frac{g\mu_B}{k_BT}\ \hat{S}_{iz}B\right)$$

$$= -\frac{k_BT}{g\mu_B}\frac{\partial}{\partial B}\log\frac{\sinh\left[\frac{g\mu_BB(S+1/2)}{k_BT}\right]}{\sinh\left(\frac{g\mu_BB}{2k_BT}\right)} = -S\ B_S\left(\frac{Sg\mu_BB}{k_BT}\right), \quad (11.64)$$

where $B_S(x)$ is the Brillouin function

$$B_S(x) = \frac{2S+1}{2S}\coth\left(\frac{(2S+1)\,x}{2S}\right) - \frac{1}{2S}\coth\left(\frac{x}{2S}\right). \quad (11.65)$$

Then we use magnetization in Eq. (11.59) and obtain electron Hamiltonian where magnetic field acts both directly and through the impurity mean-field:

$$H_e = \sum_r \psi^{\dagger}(r)\left(-\frac{\hbar^2\Delta}{2m^*} + g_{eff}\mu_B\, \sigma\cdot B\right)\psi(r), \tag{11.66}$$

where we introduced the effective g factor

$$g_{\text{eff}} = g - \frac{Jn_iS}{\mu_B B} B_S\left(\frac{Sg\mu_B B}{k_B T}\right), \tag{11.67}$$

and n_i is the density of localized spins. Using the parameters of a diluted magnetic semiconductor like CdTe doped with Mn, one can find that the second term in Eq. (11.64) can be much larger than $g = 2$, which means Zeeman field enhancement by impurity contribution.

11.8 GaMnAs Magnetic Semiconductor

Manganese is the magnetic impurity, which has a solubility in GaAs up to 6 at. %. Bulk Mn-doped GaAs (GaMnAs) crystals were identified, theoretically and experimentally, as ferromagnetic materials below ~150 K. High-temperature ferromagnetism in III-V semiconductors makes them materials of choice in spintronic applications. For example, GaMnAs can be used to inject spin-polarized electrons into a nonmagnetic semiconductor. It is anticipated that the material would enable the implementation of a spin transistor – the main component of spin-based electronics.

At a large Mn content, the exchange interaction stems from holes as the chemical potential is in the valence band. At high hole density, the magnetic order stems from the RKKY mechanism discussed in Section 11.3. Hole-mediated RKKY interaction is more effective than the electron-mediated one due to a large effective mass: $m_h^* \gg m_e^*$.[3]

Strictly speaking, classical RKKY-interaction does not apply to the hole-mediated interaction as the valence band has a complex structure described by the Kane model (see Chapter 1). At a high concentration of Mn impurities, it is convenient to calculate Curie temperature evaluating the total free energy of the magnetically ordered state. Introducing magnetization field $\boldsymbol{M}$ into the Kane model, one can calculate energy levels and free energy of electron system F_e. The free energy of localized spins F_S can be found by using the susceptibility of the local spin magnetization $\boldsymbol{M}$ to the magnetic field $\boldsymbol{B}$: $\boldsymbol{B}(\boldsymbol{M}) = -\delta F_S / \delta \boldsymbol{M}$. Then, the equation for $\boldsymbol{M}$ follows from the minimum total free energy with respect to $\boldsymbol{M}$, and the non-zero solution determines the ferromagnetic critical temperature T_c. This approach ends with numerical calculations of critical temperature in GaMnAs and other magnetically doped semiconductors [11].

Likewise to magnetic interaction in dielectrics (see Section 11.3), there is also a possibility of exchange coupling between the moments via electrons occupying the localized states, which form an impurity band in the vicinity of the valence band edge. Indeed, the Mn impurity is an acceptor in GaAs creating the acceptor level near the edge of the valence band (see Chapter 3.6). If the concentration of Mn is very high the impurity band merges the valence band so that the electron states at the Fermi level are delocalized. But in the case of small or moderate doping, the impurity band consists of mostly localized states with some hopping of electrons between these states. In this case, the Fermi level is located in the impurity band. Such a transformation of strongly localized states to impurity band and then to the band states with increasing concentration of impurities has been already discussed in Chapter 3. Experimental evidence of impurity band in GaAs(Mn) and in-the-gap position of Fermi level has been discussed in Refs. [12,13]. For impurity-band related mechanism of indirect exchange interaction, see Refs. [14,15].

[3] The electron mass in GaAs is $m_e^* \simeq 0.066 m_0$, whereas the mass of heavy holes $m_h \simeq 0.54 m_0$.

11.9 Stoner Ferromagnetism

The possibility of magnetic ordering in semiconductors is related to the ordering of impurity spins as we described it before. Another mechanism (usually called the *itinerant magnetism*) describes ferromagnetism in the system of itinerant electrons and appears to be the leading mechanism of ferromagnetism in magnetic metals and alloys with transition-metal atoms. The magnetization relies on the spin-dependent part of electron–electron interaction.

To understand Stoner magnetism, we consider the Hamiltonian of electrons with exchange electron-electron interaction in a continuous model:

$$H=\int d^3r\,\psi^\dagger(\boldsymbol{r})\left[\varepsilon(-i\nabla)-\mu\right]\psi(\boldsymbol{r})$$
$$+\frac{1}{2}\int d^3r\,d^3r'\left[\psi^\dagger(\boldsymbol{r})\,\sigma_\mu\,\psi(\boldsymbol{r})\right]\mathrm{v}(\boldsymbol{r}-\boldsymbol{r}')\left[\psi^\dagger(\boldsymbol{r}')\,\sigma_\mu\,\psi(\boldsymbol{r}')\right] \tag{11.68}$$

where $\varepsilon(\boldsymbol{k})$ is the electron energy spectrum, and $\mathrm{v}(\boldsymbol{r}-\boldsymbol{r}')$ is the interaction function we assume short-ranged, $\psi^\dagger(\boldsymbol{r}), \psi(\boldsymbol{r})$ are the spinor operators of electron creation and annihilation in point $\boldsymbol{r}$ at certain time t (for shortness we do not explicitly indicate the dependence on t), $\psi(\boldsymbol{r})=\Omega^{-1/2}\sum_{\boldsymbol{k}}\chi(\boldsymbol{k})e^{i\boldsymbol{k}\cdot\boldsymbol{r}}$, $\chi(\boldsymbol{k})$ is the Grassmann field. The second term in Eq. (11.68) is the spin–spin-density interaction.

The partition function is the path integral,

$$Z=\int D\chi\,D\chi^\dagger\,\exp\left\{\frac{i}{\hbar}\int dt\left(\int d^3r\,\psi^\dagger(\boldsymbol{r})\,i\hbar\partial_t\,\psi(\boldsymbol{r})-H\right)\right\}. \tag{11.69}$$

The functional integration in Eq. (11.69) is defined as $\int D\chi\,D\chi^\dagger\equiv\int\ldots\prod_{\boldsymbol{k}}d\chi(\boldsymbol{k})\,d\chi^\dagger(\boldsymbol{k})$.

One can transform the four-electron interaction by introducing the Gaussian integration over auxiliary field $\boldsymbol{M}(\boldsymbol{r})$, which has the meaning of an exchange field at point $(\boldsymbol{r},t)$

$$Z=\int D\chi\,D\chi^\dagger\,DM\,\exp\left\{\frac{i}{\hbar}\int dt\,d^3r\left[\psi^\dagger(\boldsymbol{r})\left(i\hbar\,\partial_t-\varepsilon(-i\nabla)-\boldsymbol{\sigma}\cdot\boldsymbol{M}(\boldsymbol{r})+\mu\right)\psi(\boldsymbol{r})\right.\right.$$
$$\left.\left.+\frac{1}{2}\int d^3r\,M_\mu(\boldsymbol{r})\,\mathrm{v}^{-1}(\boldsymbol{r}-\boldsymbol{r}')\,M_\mu(\boldsymbol{r}')\right]\right\}$$
$$=\int D\chi\,D\chi^\dagger\,DM\,\exp\left\{i\int\frac{d\varepsilon}{2\pi}\,d^3r\left[\psi^\dagger(\boldsymbol{r})\left(\varepsilon-\varepsilon(-i\nabla)-\boldsymbol{\sigma}\cdot\boldsymbol{M}(\boldsymbol{r})+\mu\right)\psi(\boldsymbol{r})\right.\right.$$
$$\left.\left.+\frac{1}{2}\int d^3r'\,M_\mu(\boldsymbol{r})\,\mathrm{v}^{-1}(\boldsymbol{r}-\boldsymbol{r}')\,M_\mu(\boldsymbol{r}')\right]\right\}, \tag{11.70}$$

where

$$\mathrm{v}^{-1}(\boldsymbol{r}-\boldsymbol{r}')=\int\frac{d^3q}{(2\pi)^3}\,\frac{e^{i\boldsymbol{k}\cdot(\boldsymbol{r}-\boldsymbol{r}')}}{\mathrm{v}(q)} \tag{11.71}$$

and we used the Fourier transformation $\psi(\boldsymbol{r},t)=(1/2\pi)\int d\varepsilon\,\exp(-i\varepsilon t/\hbar)\psi(\boldsymbol{r},\varepsilon)$.

Now we can calculate the Gauss integral over the Fermi field in Eq. (11.70) by using Eq. (11.63)

$$\int D\chi\, D\chi^{\dagger} \exp\left\{ i \int \frac{d\varepsilon}{2\pi} d^3r\; \psi^{\dagger}(r)\left[\varepsilon - \varepsilon(-i\nabla) - \boldsymbol{\sigma}\cdot\boldsymbol{M}(\boldsymbol{r}) + \mu\right]\psi(\boldsymbol{r}) \right\}$$
$$= \det\left[\varepsilon - \varepsilon(-i\nabla) - \boldsymbol{\sigma}\cdot\boldsymbol{M}(\boldsymbol{r}) + \mu\right] \equiv \prod_{r,\varepsilon}\left[\varepsilon - \varepsilon(-i\nabla) - \boldsymbol{\sigma}\cdot\boldsymbol{M}(\boldsymbol{r}) + \mu\right]$$
$$= \exp\left\{ Tr \int \frac{d\varepsilon}{2\pi} d^3r\; \log\left(\varepsilon - \varepsilon(-i\nabla) - \boldsymbol{\sigma}\cdot\boldsymbol{M}(\boldsymbol{r}) + \mu\right) \right\}. \tag{11.72}$$

Substituting Eq. (11.72) to (11.68) we obtain

$$Z = \int DM \exp\left\{ i \int \frac{d\varepsilon}{2\pi} d^3r \left[Tr \log\left(\varepsilon - \varepsilon(-i\nabla) - \boldsymbol{\sigma}\cdot\boldsymbol{M}(\boldsymbol{r}) + \mu\right) \right.\right.$$
$$\left.\left. + \frac{1}{2}\int d^3r'\; M_{\mu}(\boldsymbol{r})\, \mathrm{v}^{-1}(\boldsymbol{r}-\boldsymbol{r}')\, M_{\mu}(\boldsymbol{r}') \right]\right\}. \tag{11.73}$$

Calculating the derivative over $\boldsymbol{M}(\boldsymbol{r})$ in the exponent of (11.71) and equating it to zero, we find the saddle-point equation for $\boldsymbol{M}(\boldsymbol{r})$

$$-i\, Tr \int \frac{d\varepsilon}{2\pi}\, \boldsymbol{\sigma}\, G(\boldsymbol{r},\boldsymbol{r};\varepsilon) + \int d^3r'\; \mathrm{v}^{-1}(\boldsymbol{r}-\boldsymbol{r}')\, \boldsymbol{M}(\boldsymbol{r}') = 0, \tag{11.74}$$

where $G(\boldsymbol{r},\boldsymbol{r}';\varepsilon)$ is the Green's function of electron in magnetization field $\boldsymbol{M}(\boldsymbol{r})$. Green's function obeys equation

$$\left[\varepsilon - \varepsilon(-i\nabla) - \boldsymbol{\sigma}\cdot\boldsymbol{M}(\boldsymbol{r}) + \mu\right] G(\boldsymbol{r},\boldsymbol{r}';\varepsilon) = \delta(\boldsymbol{r}-\boldsymbol{r}'). \tag{11.75}$$

In the mean-field approximation, the exchange field is a constant, $\boldsymbol{M}(\boldsymbol{r}) \to \boldsymbol{M}$. Then, using Eq. (11.74), we find the equation for $\boldsymbol{M}$:

$$\boldsymbol{M} = i\mathrm{v}_0\; Tr \int \frac{d\varepsilon}{2\pi} \frac{d^3k}{(2\pi)^3}\, \boldsymbol{\sigma}\, G(\varepsilon,\boldsymbol{k}), \tag{11.76}$$

where (we use the notation $\varepsilon_k \equiv \varepsilon(\boldsymbol{k})$)

$$G(\varepsilon,\boldsymbol{k}) = (\varepsilon - \varepsilon_k - \boldsymbol{\sigma}\cdot\boldsymbol{M} + \mu + i\delta\; sign\; \varepsilon)^{-1} \tag{11.77}$$

and

$$\mathrm{v}_0 = \int d^3r'\; \mathrm{v}(\boldsymbol{r}-\boldsymbol{r}') = \mathrm{v}(\boldsymbol{q}=0). \tag{11.78}$$

Taking z the quantization axis, we present Eq. (11.76) as

$$1 = 2i\mathrm{v}_0 \int \frac{d\varepsilon}{2\pi} \frac{d^3k}{(2\pi)^3} \frac{1}{\varepsilon_1 - \varepsilon_2}\left(\frac{1}{\varepsilon - \varepsilon_1 + \mu + i\delta\; sign\; \varepsilon} - \frac{1}{\varepsilon - \varepsilon_2 + \mu + i\delta\; sign\; \varepsilon} \right)$$
$$= -2\mathrm{v}_0 \int \frac{d^3k}{(2\pi)^3} \frac{f(\varepsilon_1) - f(\varepsilon_2)}{\varepsilon_1 - \varepsilon_2}, \tag{11.79}$$

where $f(\varepsilon)$ is the Fermi-Dirac function at $T=0$[4] and we denoted $\varepsilon_{1,2}=\varepsilon(\mathbf{k})\pm M$.

In the limit of $M\rightarrow 0$, we get

$$1=2\mathrm{v}_0\int\frac{d^3k}{(2\pi)^3}\left(-\frac{\partial f(\varepsilon_k)}{\partial\varepsilon_k}\right)=\mathrm{v}_0\ \rho(\mu), \tag{11.80}$$

where $\rho(\mu)$ is the electron density of states at the Fermi level. Thus, we come to the condition for a spontaneous magnetic moment to exist: $\mathrm{v}_0\ \rho(\mu)=1$. As finite temperature destroys any order, the ferromagnetism at $T\neq 0$ requires

$$\mathrm{v}_0\ \rho(\mu)>1. \tag{11.81}$$

This is the *Stoner criterion* of ferromagnetism. The onset of itinerant ferromagnetism requires large enough interaction strength and density of states at the Fermi level.

In conclusion, itinerant ferrromagnetism relies on the spin-dependent Coulomb interaction. As such interaction one can use direct repusion between two electrons of opposite spins residing on the same lattice site (Hubbard model): $H=\sum_{ij}t_{ij}+U\sum_i n_{i\uparrow}n_{i\downarrow}$, t_{ij} is the hopping integral that corresponds to electron movement in the conduction band, U is the contact Coulomb interaction. Hubbard model is capable of explaining ferromagnetism in a narrow band, $U\gg t$ [16].

References

1. J.K. Furdyna, "Diluted magnetic semiconductors", *J. Appl. Phys.* **64**, R29 (1988).
2. L.D. Landau and E.M. Lifshitz, *Quantum Mechanics*, Section 62. (Pergamon, New York, 1977).
3. N. Bloembergen and T.J. Rowland, "Nuclear spin exchange in solids: Tl_{203} and Tl_{205} magnetic resonance in thallium and thallic oxide", *Phys. Rev.* **97**, 1680 (1955).
4. A.A. Abrikosov, "Spin-glass with a semiconductor as a host", *Low Temp. Phys.*, **39**, 217 (1980).
5. P.G. de Gennes, "Polarisation de charge (ou de spin) au voisinage d'une impureté dans un alliage", *J. Phys. Radium* **23**, 630 (1962).
6. I.Y. Korenblit and E.F. Shender, "Ferromagnetism of disorderd systems", *Sov. Phys. Uspekhi* **21**, 832 (1978).
7. M. Mezard and A. Montanari, *Information, Physics, and Computation* (Oxford University Press, Oxford, UK, 2009).
8. M. Mezard, G. Parisi, and M.A. Virasoro, *Spin-Glass Theory and Beyond* (World Scientific, Singapore, 1987).
9. K. Binder and A.P. Young, "Spin glasses: Experimental facts, theoretical concepts, and open questions", *Rev. Mod. Phys.* **58**, 801 (1986).
10. P. Coleman, *Introduction to Many-Body Physics*, Chapter 12. (Cambridge University Press, Cambridge, 2015).
11. T. Dietl, H. Ohno, F. Matsukura, J. Cibert, and D. Ferrands, "Zener model of ferromagnetism in zink-blende magnetic semiconductors", *Science* **287**, 1019 (2000).
12. S. Ohya, K. Takata, and M. Tanaka, "Universal valence-band picture of the ferromagnetic semiconductor GaMnAs", arXiv:1009.2235 [cond-mat.mtrl-sci] (2010).
13. M. Tanaka, S. Ohya, and P.N. Hai, "Recent progress in III-V based ferromagnetic semiconductors: Band structure, Fermi level, and tunneling transport", *Appl. Phys. Rev.* **1**, 011102 (2014).

[4] The rigorous finite-temperature calculation shows that the second line in Eq. (11.79) with Fermi-Dirac functions inserted is correct also for $T\neq 0$.

14. V.I. Litvinov and V.K. Dugaev, "Ferromagnetism in magnetically doped III-V semiconductors", *Phys. Rev. Lett.* **86**, 5593 (2001).
15. V.I. Litvinov, *Wide Bandgap Semiconductor Spintronics* (Pan Stanford, Singapore, 2016).
16. R.M. White, *Quantum Theory of Magnetism* (3rd ed., Springer, Berlin, Heidelberg, 2007).

12

Optical Properties

12.1 Coupling to Electromagnetic Field and Gauges

Explanation of many relevant optical effects requires quantum-mechanical language for semiconductors and a quasi-classical approach to electromagnetic field and interaction between the two. An electromagnetic field obeys Maxwell equations for magnetic field $\boldsymbol{H}$ and the electric field $\boldsymbol{E}$, the experimentally observable variables. As the system under study contains field sources – electric charge density ρ and current density $\boldsymbol{j}$ – the physical fields obey the Maxwell equations:

$$\begin{cases} \nabla \cdot \boldsymbol{H} = 0, \\ \nabla \times \boldsymbol{E} = -\mu_0 \dfrac{\partial \boldsymbol{H}}{\partial t}, \end{cases}$$
$$\begin{cases} \nabla \cdot \boldsymbol{E} = \dfrac{\rho}{\varepsilon_0}, \\ \nabla \times \boldsymbol{H} = \varepsilon_0 \dfrac{\partial \boldsymbol{E}}{\partial t} + j, \end{cases} \tag{12.1}$$

where $\varepsilon_0 = 8.85 \times 10^{-12}$ F/m (farad per meter) and $\mu_0 = 4\pi \times 10^{-7}$ H/m (henry per meter) are the vacuum electrical permittivity and magnetic permeability, respectively.

By applying the divergence operation to the last row in Eq. (12.1), one can use notation $\boldsymbol{B} = \mu_0 \boldsymbol{H}$ and relation $\nabla \cdot (\nabla \times \boldsymbol{B}) = 0$ to obtain the continuity equation in the form:

$$\frac{\partial \rho}{\partial t} + \nabla \cdot \boldsymbol{j} = 0. \tag{12.2}$$

As Eq. (12.1) contains six variables, $\boldsymbol{B}$ and $\boldsymbol{E}$, it is convenient to reduce the number of variables to four by introducing auxiliary fields – the vector and scalar potentials, $\boldsymbol{A}$ and V, respectively. The potentials determine the physical fields as follows:

$$\boldsymbol{B} = \nabla \times \boldsymbol{A}, \ \boldsymbol{E} = -\frac{\partial \boldsymbol{A}}{\partial t} - \nabla V. \tag{12.3}$$

Let us dwell on the relation between fields and potentials. In classical electrodynamics, the transition from $\boldsymbol{B}$ and $\boldsymbol{E}$ to potentials $\boldsymbol{A}$ and V may be a matter of convenience as both formulations have equal footing. On the contrary, in quantum physics, one must use potentials as the energy operator – Hamiltonian – contains potentials, not physical fields. Quantum states described by eigenfunctions

DOI: 10.1201/9780429285929-12

of the Hamiltonian depend on potentials, which influence electron motion even where physical fields vanish. That is the background for Aharonov–Bohm-type effects. The effect is quantum in nature since it is related to the phase of the wave functions and experimentally observed as the sensitive to potentials interference patterns of electrons moving through the region where physical fields are absent [1].

The choice of representation (12.3) automatically satisfies the first two equations (12.1), as verified by direct substitution. What remains relevant is the second pair of the Maxwell equations, which, using (12.3), is reduced to

$$\nabla^2 \boldsymbol{A} - \frac{1}{c^2}\frac{\partial^2 \boldsymbol{A}}{\partial t^2} - \nabla\left(\frac{1}{c^2}\frac{\partial V}{\partial t} + \nabla \cdot \boldsymbol{A}\right) = -\mu_0 \boldsymbol{j},$$

$$\frac{\partial}{\partial t}(\nabla \cdot \boldsymbol{A}) + \nabla^2 V = -\frac{\rho}{\varepsilon_0}, \tag{12.4}$$

where $c = 1/\sqrt{\varepsilon_0 \mu_0}$ is the speed of light in free space. Vector identities, $\nabla \times \nabla \times \boldsymbol{A} = \nabla(\nabla \cdot \boldsymbol{A}) - \nabla^2 \boldsymbol{A}$ and $\nabla \cdot (\nabla V) = \nabla^2 V$, have been used in Eq. (12.4).

The system (12.4) with the appropriate boundary conditions determines four potentials $\boldsymbol{A}$ and V. The price we pay for simplifying the problem is the ambiguity of the potentials: equations (12.4) and the physical fields resulting from them will not change if we shift the potentials by a *gauge transformation*,

$$\boldsymbol{A}' = \boldsymbol{A} + \nabla\, \chi(\boldsymbol{r},t), \quad V' = V - \frac{\partial \chi(\boldsymbol{r},t)}{\partial t}, \tag{12.5}$$

where χ is the arbitrary differentiable function of coordinates and time.

The electromagnetic field represented by potentials is an example of the gauge field. We have to deal with a gauge field because it is the potentials, not physical fields, that enter the Hamiltonian and determine the total energy. So, in quantum mechanics, we must use the Hamilton operator H, which determines the electron levels and allows dealing with an electron and the electromagnetic field within a common approach. Observables should be independent of the gauge choice – this condition determines the form of Hamiltonian for an electron in the electromagnetic field. In a quantum-mechanical context, the gauge invariance means that the potentials $\boldsymbol{A}$ and V enter the Hamiltonian in such a way that gauge transformation (12.5) does not change the observables. Since observables are proportional to $|\psi(\boldsymbol{r},t)|^2$, the only effect of the gauge transformation could be an additional phase factor in the wave function,

$$\psi'(\boldsymbol{r},t) = \psi(\boldsymbol{r},t)\exp\left(i\Lambda(\boldsymbol{r},t)\right) \tag{12.6}$$

The only way to get the phase shift in wave function while doing the gauge transformation (12.5) is to choose the Hamiltonian as

$$H = \frac{1}{2m}(\boldsymbol{p} - q\boldsymbol{A})^2 + qV,$$

$$i\hbar \frac{\partial \psi(\boldsymbol{r},t)}{\partial t} = H\psi(\boldsymbol{r},t), \tag{12.7}$$

where $\boldsymbol{p} = -i\hbar\nabla$ is the operator of canonical momentum, m is the electron mass, and q is the elementary charge. Applying the gauge transformation (12.5) to the Hamiltonian (12.7), we obtain the Schrödinger equation for $\psi'(\boldsymbol{r},t)$ providing $\Lambda = iq\chi/\hbar$. So, the Hamiltonian (12.7) is gauge-invariant in the sense that the gauge transformation (12.5) leads to the local phase shift in the wave function (12.6).

It is instructive to trace the origin of the guess for the Hamiltonian (12.7). It comes from the classical Lagrangian for a charged particle in an electromagnetic field:

$$L=\frac{1}{2}m\dot{r}^2+j\cdot A-qV, \tag{12.8}$$

where $j=q\dot{r}$, $\dot{r}$ is the electron velocity. The first term is the kinetic energy of an electron, while two other terms present the electron potential energy in a scalar field qV and dynamic coupling to a vector field, $-j\cdot A$ [2]. The Hamiltonian follows from Eq. (12.8) and by using canonical momentum $p=\partial L/\partial\dot{r}=m\dot{r}+qA$ takes the form:

$$H=p\cdot\dot{r}-L=p\cdot\dot{r}-\frac{1}{2}m\dot{r}^2-q\dot{r}\cdot A+qV=\frac{1}{2m}(p-qA)^2+qV. \tag{12.9}$$

Momentum $p_k=p-qA$ determines the electron kinetic energy in the electromagnetic field and is called *kinetic momentum*. As the potentials possess the gauge freedom (12.5), the interaction term looks differently depending on the gauge chosen for a particular problem. To reduce or remove the gauge freedom, one has to fix the gauge, that is, to impose additional restrictions on A and V. In what follows, we give several examples of gauge fixing, each adjusted to the task.

Lorentz gauge. One may simplify the first row in Eq. (12.4) by imposing the gauge as follows:

$$\frac{1}{c^2}\frac{\partial V}{\partial t}+\nabla\cdot A=0, \tag{12.10}$$

It transforms (12.4) into decoupled equations for potentials A and V:

$$\nabla^2 A-\frac{1}{c^2}\frac{\partial^2 A}{\partial t^2}=-\mu_0 j,$$
$$\nabla^2 V-\frac{1}{c^2}\frac{\partial^2 V}{\partial t^2}=-\frac{\rho}{\varepsilon_0}. \tag{12.11}$$

Solutions to Eq. (12.11) represent retarded potentials propagating with velocity c [2–4].

The Lorentz gauge allows freedom in the sense of (12.5) with the gauge function which obeys the wave equation,

$$\nabla^2\chi-\frac{1}{c^2}\frac{\partial^2\chi}{\partial t^2}=0. \tag{12.12}$$

In other words, if χ obeys (12.12), the gauge transformation (12.5) results in new potentials that satisfy the Lorentz condition (12.10). The gauge fixing condition (12.10) is Lorentz invariant that means it holds in any of the inertial reference frames (r,t), which are related by Lorentz transformation.

Coulomb gauge. If Lorentz invariance is not essential, for example, we deal with a specific reference frame pinned to a media- the system of particles – it is convenient to use the gauge $\nabla\cdot A=0$. Then, the system of equations (12.4) looks as follows:

$$\nabla^2 A-\frac{1}{c^2}\frac{\partial^2 A}{\partial t^2}=-\mu_0\left(j-\varepsilon_0\frac{\partial}{\partial t}\nabla V\right),$$
$$\nabla^2 V=-\frac{\rho}{\varepsilon_0}, \tag{12.13}$$

where the first row is the inhomogeneous wave equation and the second row is the Poisson equation. Let us take a look at how the Coulomb gauge decouples potentials and physical fields into transverse and longitudinal parts. Expansion of potentials in a series of plane wave modes gives

$$A(r,t)=\sum_k A_k(t)\exp(ik\cdot r),\quad V(r,t)=\sum_k V_k(t)\exp(ik\cdot r), \tag{12.14}$$

then the Coulomb gauge means

$$\nabla\cdot A=\sum_k ik\cdot A_k(t)\exp(ik\cdot r)=0. \tag{12.15}$$

From Eq. (12.15) follows $k\cdot A_k=0$ – all modes of vector-potential and thus electric field $E_t=-\partial A/\partial t$ and magnetic field $B=\nabla\times A$ are perpendicular to the wave vector and called transverse. Calculating another part of the electric field, one gets it as longitudinal:

$$E_l=-\nabla V=-i\sum_k k\, V_k(t)\exp(ik\cdot r) \tag{12.16}$$

This gauge is convenient in treating non-relativistic problems where the electric field can be partitioned to transverse and longitudinal: $E=E_l+E_t$. It allows considering the transverse field as a perturbation to the ground state, which accounts for longitudinal fields including Coulomb interaction. Because of field decomposition, the first line of Eq. (12.13) implies two separate equations: for longitudinal and transverse field components,

$$\begin{gathered}\nabla^2 A-\frac{1}{c^2}\frac{\partial^2 A}{\partial t^2}=-\mu_0 j_t,\\ \varepsilon_0\frac{\partial}{\partial t}\nabla V=j_l,\\ j_l\equiv\sum_k\frac{k}{k^2}(k\cdot j_k)\exp(ik\cdot r),\quad j=j_t+j_l,\end{gathered} \tag{12.17}$$

where j_k is the Fourier mode of the total current density. The first line in (12.17) is the wave equation, while the second line is nothing more than a continuity equation (12.2). So, the field decomposition in the Coulomb gauge transforms the Eq. (12.13) from the one containing potentials A and V into a wave equation for potential A only.

Landau gauge. One may fix the gauge by imposing conditions,

$$\begin{gathered}A_y=xB_z,\\ A_x=A_z=0,\quad V=0.\end{gathered} \tag{12.18}$$

Landau gauge (12.18) is convenient for solving the Schrödinger equation for an electron in a constant magnetic field, $B=\nabla\times A=B_z$.

Velocity gauge. Electron interaction with an electromagnetic field determines the optical properties of a solid. The interaction term follows from the Hamiltonian (12.7):

$$\begin{gathered}H=H_0+H_1,\\ H_0=\frac{p^2}{2m}+qV_0,\\ H_1=\frac{i\hbar q}{2m}(\nabla\cdot A+A\cdot\nabla)+qV+\frac{q^2A^2}{2m},\end{gathered} \tag{12.19}$$

where V_0 and $(\boldsymbol{A}, V)$ are static electric and electromagnetic potentials, respectively. The perturbation H_1 is a quantum operator so that the term in the bracket acts on an arbitrary scalar function $f(\boldsymbol{r})$ as $\nabla \cdot \boldsymbol{A} f + \boldsymbol{A} \cdot \nabla f = f\nabla \cdot \boldsymbol{A} + 2\boldsymbol{A} \cdot \nabla f$. In a transverse field, $\nabla \cdot \boldsymbol{A} = 0$, the operator simplifies to $\nabla \cdot \boldsymbol{A} + \boldsymbol{A} \cdot \nabla = 2\boldsymbol{A} \cdot \nabla$.

Perturbation H_1 associated with the electric field $\boldsymbol{E}(\boldsymbol{r},t)$ without sources, implying $\rho = 0$ in (12.1), can be represented by potentials in a *velocity gauge*,

$$H_1 = \frac{i\hbar q}{m} \boldsymbol{A}^{\mathrm{v}} \cdot \nabla,$$
$$V^{\mathrm{v}} = 0, \boldsymbol{A}^{\mathrm{v}}(\boldsymbol{r},t) = -\int_{-\infty}^{t} \boldsymbol{E}(\boldsymbol{r},t')\, dt', \tag{12.20}$$

This gauge belongs to the class of the Lorentz gauge subjected additional transverse condition $\nabla \cdot \boldsymbol{A}^{\mathrm{v}} = 0$. In integral (12.20), one implies the adiabatic approximation: the field increases from zero at $t' = -\infty$ to $\boldsymbol{E}$ at $t' = t$, not perturbing the electron state. We neglected the term $\sim \boldsymbol{A}^2$, thereby limiting our consideration by the weak field, meaning that the electron energy acquired during subcycle is much less than the semiconductor bandgap E_g: $q^2 A^2 \ll 2mE_g$. In other words, the field is weak as compared to the crystal field in a solid.

By introducing the current density operator, interaction (12.20) takes the form,

$$H_1 = -\Omega \boldsymbol{A}^{\mathrm{v}} \cdot \boldsymbol{j},\ \boldsymbol{j} = -\frac{iq\hbar}{\Omega m} \nabla, \tag{12.21}$$

where Ω is the crystal volume.

Example field is the plane linearly polarized electromagnetic wave propagating in k-direction:

$$\boldsymbol{E}(\boldsymbol{r},t) = \boldsymbol{E}_0 \cos(\boldsymbol{k} \cdot \boldsymbol{r} - \omega t),\ \boldsymbol{E}_0 \cdot \boldsymbol{k} = 0,$$
$$\boldsymbol{A}^{\mathrm{v}}(\boldsymbol{r},t) = -\frac{\boldsymbol{E}_0}{\omega} \sin(\boldsymbol{k} \cdot \boldsymbol{r} - \omega t), \tag{12.22}$$

where $\boldsymbol{k}$ and ω are the wavevector and angular frequency, respectively.

Length gauge. If the field $\boldsymbol{E}(\boldsymbol{r},t)$ varies on a spatial scale much larger than the characteristic distance in the electron system, for instance, the lattice constant in a solid, the field is almost homogeneous and time-dependent only, $\boldsymbol{A}(t)$. Being applied to the electromagnetic wave (12.22), it means the long-wavelength limit $qr \ll 1$, or dipole approximation, and reads as

$$\boldsymbol{A}(t) = \frac{\boldsymbol{E}_0}{\omega} \sin \omega t. \tag{12.23}$$

Using transformation (12.5) with the gauge function

$$\chi(\boldsymbol{r},t) = -\boldsymbol{r} \cdot \boldsymbol{A}(t), \tag{12.24}$$

one goes to the Göppert-Mayer or *length gauge*:

$$\boldsymbol{A}^{l} = \boldsymbol{A}^{\mathrm{v}} + \nabla\chi(\boldsymbol{r},t) = \boldsymbol{A}^{\mathrm{v}}(\boldsymbol{r},t) - \boldsymbol{A}(t)$$
$$V^{l} = V^{\mathrm{v}} - \frac{\partial \chi(\boldsymbol{r},t)}{\partial t} = \boldsymbol{r} \cdot \frac{\partial \boldsymbol{A}(t)}{\partial t} = -\boldsymbol{r} \cdot \boldsymbol{E}(t). \tag{12.25}$$

For a plane wave, we have

$$A^{l} = A^{v} + \nabla\chi(r,t) = \frac{E_0}{\omega}\left(\sin(k\cdot r - \omega t) + \sin\omega t\right),$$
$$V^{l} = V^{v} - \frac{\partial\chi(r,t)}{\partial t} = -r\cdot E_0\cos\omega t, \tag{12.26}$$

In the dipole approximation, $A^{v}(r,t)\to A(t)$, so the gauge (12.25) and the interaction (12.19) tend to

$$A^{l} = 0,\ V^{l} = -r\cdot E(t),$$
$$H_1 = qV^{l} = -d\cdot E(t) \tag{12.27}$$

where $d = qr$ is the electron dipole moment.

In semiconductor physics, both representations (12.20) and (12.27) being used to study optical effects, and a choice between the two is a matter of convenience in solving a particular problem.

12.2 Phenomenological Approach to Wave Propagation

Maxwell equations (12.1) describe electrodynamics in free space. As a solid comprises charged particles involved in fast motion that gets electric and magnetic fields to fluctuate in space and time, the electromagnetic theory has to deal with averaged fields and correlations between fields in different spatial and temporal points. The Maxwell equations are linear in fields and sources, so formally, they preserve the same form after the averaging procedure. However, vacuum fields now are to be replaced by the average fields, which depend on material parameters. There are two main approaches to the averaging procedure. The macroscopic approach implies averaging of the fields over the physically small volume that is, by definition, small enough to consider mean fields homogeneous inside the volume while large enough to neglect their fluctuations. This approach requires the wavelength of the electromagnetic field to be large compared to an interatomic distance in a solid. Macroscopic phenomenology misses the effect of local fields acting on a charged particle in a solid or, in other words, neglects the effects of spatial dependence (spatial dispersion) in material parameters. On the other hand, the microscopic electrodynamics is free from the restrictions above due to ensemble averaging, which is, in virtue of the ergodic hypothesis, the same as averaging over time in every local point of space. The microscopic approach works in high-frequency fields when the wavelength is comparable to an interatomic distance. The two approaches differ in the way they establish the relationship between currents and fields. The macroscopic theory divides currents into parts associated with free and bound charges, thus introducing, respectively, the macroscopic conductivity and polarization of the medium (dielectric and magnetic). The microscopic approach deals with fields of atomic spatial scale where the free and bound currents are indistinguishable. This approach allows treating nonlocal field effects, meaning that the material properties of solid in any spatial point are determined by the fields nearby, resulting in spatial dispersion of material parameters. Effects of spatial dispersion become relevant in short-wavelength fields (ultraviolet spectral range) and in studying coupling between the external electromagnetic field and collective excitations in solids such as phonons and (or) excitons. Unless we are interested in the effects of spatial dispersion, we may use the macroscopic phenomenology that works well for semiconductors interacting with electromagnetic fields from static to optical frequencies. For details on the effects of spatial dispersion, see Ref. [5].

12.2.1 Quasi-Static Fields

As a result of averaging, the Maxwell equations include average fields in a solid which are affected by the electrical and magnetic polarizations as well as currents induced by the fields. By neglecting spatial dispersion in a medium, we assume that the relationship between fields, polarization, and currents is local – in the same spatial point. Once the frequency of the electromagnetic field is much less than the frequency of particle motion in the medium, the time delay between the field and the medium response is negligible, namely, the polarization and currents are following the field instantaneously. In a quasi-static regime, time-dependent electrical and magnetic fields are proportional, respectively, to the electrical and magnetic response induced in media. The time delay between the field and the response is equivalent to frequency dispersion of material parameters and will be discussed later in the text.

The electrical and magnetic inductions $\boldsymbol{D}$ and $\boldsymbol{B}$, respectively, characterize the total fields in the media, and in isotropic materials can be expressed as

$$\begin{aligned} \boldsymbol{D} &= \varepsilon_0 \boldsymbol{E} + \boldsymbol{P} \equiv \varepsilon\varepsilon_0 \boldsymbol{E}, \\ \boldsymbol{B} &= \mu_0 \boldsymbol{H} + \boldsymbol{M} \equiv \mu\mu_0 \boldsymbol{H}, \end{aligned} \tag{12.28}$$

where $\boldsymbol{P}$ is the electric polarization or dipole moment per unit volume, $\boldsymbol{M}$ is the magnetization or magnetic moment per unit volume, ε and μ are the relative dielectric constant and magnetic permeability, respectively. In nonmagnetic semiconductors, $\mu \approx 1$. Additional relation is needed to close the system (12.29) – the Ohm's law, which introduces one more material parameter, conductivity: $\boldsymbol{j} = \sigma \boldsymbol{E}$.

In the quasi-static limit, one may consider material parameters μ, ε, σ as frequency independent and equal to their values in static fields. In dielectrics, it is a good approximation in the transparency region – the frequency of the external field is much less than in atomic motion, which happens to oscillate in the infrared. The quasi-static approximation means that we neglect the displacement current in Maxwell equations. It works if the displacement current is much less than the ohmic current, $\sigma \boldsymbol{E} \gg \varepsilon\varepsilon_0 \, \partial \boldsymbol{E} / \partial t \approx \varepsilon\varepsilon_0 \omega \boldsymbol{E}$. This condition restricts the frequency range, where the conductivity may be considered frequency independent and equal to static conductivity, $\omega \ll \sigma / \varepsilon\varepsilon_0$. Another restriction comes from the microscopic mechanism of conductivity. Conductivity is frequency-independent if $\omega\tau \ll 1$, where τ is the electron momentum relaxation time. In other words, the period of the electromagnetic oscillations has to be much larger than the average time between electron scattering events. Otherwise, a significant change in the field amplitude between two consecutive collisions would affect conductivity, thus introducing the frequency-dependent phase shift between an electric field and a current.

The Maxwell equations in non-dispersive media take the form:

$$\begin{cases} \nabla \cdot \boldsymbol{H} = 0, \\ \nabla \times \boldsymbol{E} = -\mu\mu_0 \dfrac{\partial \boldsymbol{H}}{\partial t}, \end{cases}$$
$$\begin{cases} \varepsilon\varepsilon_0 \nabla \cdot \boldsymbol{E} = \rho, \\ \nabla \times \boldsymbol{H} = \varepsilon\varepsilon_0 \dfrac{\partial \boldsymbol{E}}{\partial t} + \mathrm{j}, \end{cases} \tag{12.29}$$

Equations (12.29) are similar to vacuum equations (12.1) with a correction for local and instantaneous material parameters μ, ε. Once we consider an electromagnetic field in neutral and homogeneous dielectrics, $\rho = j = 0$ should be put in Eq. (12.29). The equation for vector-potential $\boldsymbol{A}$ follows from Eq. (12.29) the same way as Eq. (12.4) follows from vacuum Maxwell equations (12.1):

$$\begin{aligned} &\nabla^2 \boldsymbol{A} - \frac{1}{\mathrm{v}^2} \frac{\partial^2 \boldsymbol{A}}{\partial t^2} = 0, \quad \mathrm{v} = \frac{c}{\sqrt{\varepsilon\mu}}, \\ &\nabla \cdot \boldsymbol{A} = 0. \end{aligned} \tag{12.30}$$

Formally, Eq. (12.30) is a wave equation, which describes the monochromatic plane wave propagating with the phase velocity $v = c/n_r$, where $n_r = \sqrt{\varepsilon\mu}$ is the refractive index of a material. Presenting the solution in the form of a plane wave, from Eq. (12.30), we find the relation between the wave vector and frequency,

$$A(r,t) = A_0 \exp\left[i(k \cdot r - \omega t)\right],$$
$$k^2 = \frac{\omega^2 n_r^2}{c^2}. \tag{12.31}$$

Calculated with (12.3) and (12.31) electric and magnetic fields are expressed as

$$E = i\omega A,\ H = \frac{i}{\mu\mu_0} A \times k = \frac{1}{Z}(E \times n), \tag{12.32}$$

where $Z = \sqrt{\mu\mu_0 / \varepsilon\varepsilon_0}$ is the wave impedance of the material [ohms] and $n = k/k$ is the unit vector in the propagation direction. As follows from Eq. (12.32), in a transverse electromagnetic wave, E, H, and k are mutually orthogonal. Observable fields are real parts of E and H from Eq. (12.32). Until we operate with linear Maxwell equations and their time- and spatial-periodic solutions, it is safe to present fields as complex phasors such as (12.31) and then take real parts of solutions to get the observables. As discussed in the next section, the energy flux is a product of fields, and it requires real-valued electric and magnetic fields for calculation.

12.2.2 Energy Flux

Electromagnetic wave transfers energy. The energy flux follows from the Maxwell equations. Multiplying second and fourth equations (12.29) by H and E, respectively, and then subtracting one from the other, one gets

$$\frac{\partial}{\partial t}\left(\frac{\varepsilon\varepsilon_0 E^2 + \mu\mu_0 H^2}{2}\right) = -j \cdot E - \nabla \cdot (E \times H), \tag{12.33}$$

where the vector identity, $E \cdot (\nabla \times H) - H \cdot (\nabla \times E) = -\nabla \cdot (E \times H)$, was used. Terms in Eq. (12.33) are local characteristics of the field: $(\varepsilon\varepsilon_0 E^2 + \mu\mu_0 H^2)/2$ is the volume energy density of the electromagnetic field, $j \cdot E$ is the Joule power dissipating in the media, and $\nabla \cdot (E \times H)$ is the power density which goes with radiation. The equation integrated over a finite volume presents the energy conservation law in that volume. The integral of the last term in (12.33) is equals the flux of vector $P = E \times H$ through the closed surface that bounds this volume (Gauss theorem) [2]. Poynting vector P [W/m^2] is the energy flowing through a plane surface per unit time per unit square.

In the plane wave propagating in a homogeneous and isotropic media, fields (12.32) determine the Poynting vector as

$$P = E \times H = \frac{1}{Z} E \times (E \times n) = \frac{1}{Z} E^2 n. \tag{12.34}$$

The direction of the energy flux coincides with the direction of the wave propagation k. Note that in Eq. (12.34), E and H are real physical fields. Working with periodic fields expressed as complex-valued phasors, $E = E_m \exp(-i\omega t)$ and $H = H_m \exp(-i\omega t)$, one should use their real parts when calculating the Poynting vector. Substituting real parts into (12.34) and integrating over the period, we find the averaged Poynting vector in the form:

$$\bar{P}=\frac{1}{T}\int_0^T \mathrm{Re}[E]\times\mathrm{Re}[H]\,dt=\frac{1}{2}\mathrm{Re}\left[E_m\times H_m^*\right], \tag{12.35}$$

where $T=2\pi/\omega$ is the period of electromagnetic wave. In Eq. (12.35), the terms oscillating with doubled frequency turn zero after integrating over the period.

Within the quantum description, the field energy density averaged over the period can be expressed through the energy of particles – photons – of density N, each carries energy $\hbar\omega$:

$$\bar{W}=\frac{1}{2}\left(\varepsilon\varepsilon_0\overline{E^2}+\mu\mu_0\overline{H^2}\right)=\frac{1}{4}\varepsilon\varepsilon_0|E_m|^2+\frac{1}{4Z^2}\mu\mu_0|E_m|^2$$
$$=\frac{1}{2}\varepsilon\varepsilon_0|E_m|^2=N\hbar\omega. \tag{12.36}$$

Equation (12.36) relates photon density N to the amplitude of the electromagnetic field.

12.2.3 Electromagnetic Waves in a Conductive Media

At high enough frequency to violate the quasi-static approximation, we have to account for the time delay between the fields and induced by them polarizations in a material. The delay causes the frequency dispersion in material parameters. There is another reason why we include dispersion into consideration. In semiconductors, even far from absorption spectral bands, we have to account for the finite conductivity and thus for the ohmic loss of electromagnetic energy (term $j\cdot E$ in Eq. (12.33)). As we will see below, lossy materials are always frequency dispersive.

The time delay means that instantaneous polarization forms by fields in preceding moments,

$$D(t)=\varepsilon_0\left[E(t)+\int_0^\infty f(\tau)\,E(t-\tau)\,d\tau\right], \tag{12.37}$$

where $f(\tau)$ describes the polarization properties of a material. As we consider time-periodic fields, it is convenient to use the frequency domain by applying the Fourier transform to Eq. (12.37):

$$D(\omega)=\varepsilon_0\,\varepsilon(\omega)\,E(\omega),$$
$$\varepsilon(\omega)=1+\int_0^\infty f(\tau)\exp(i\omega\tau)d\tau, \tag{12.38}$$

So, dispersion in $\varepsilon(\omega)$ originates from the time delay in polarization current associated with bound charges.

As it follows from Eq. (12.38), at finite frequency $\varepsilon(\omega)$ has real and imaginary parts meaning that time delay causes the phase shift between the field and the induction as a response: $\varepsilon(\omega)=\varepsilon'(\omega)+i\varepsilon''(\omega)$. Following definition (12.38), $\varepsilon'(-\omega)=\varepsilon'(\omega)$, $\varepsilon''(-\omega)=-\varepsilon''(\omega)$, and thus in static limit $\varepsilon'(\omega\to 0)=\varepsilon$, $\varepsilon''(\omega\to 0)=0$. High-frequency limit gives $\varepsilon'(\omega\to\infty)=1$, $\varepsilon''(\omega\to\infty)=0$ as polarization cannot respond to rapidly alternating electric field. In the upper half-plane of complex-valued ω, function $\varepsilon(\omega)-1$ is analytical and decreases exponentially on a semicircle of radius $R\to\infty$. Analytical properties of $\varepsilon(\omega)-1$ are the result of causality principle which dictates integration limits in Eq. (12.37). Real and imaginary parts of the dielectric function obey the Kramers–Kronig dispersion relations briefly discussed in Section 12.3 (for details and derivation, see Refs. [6,7]).

One more material equation, Ohm's law, should also reflect the delay between an electric field and an electric current as a response. Electric current is proportional to electron velocity averaged with a non-equilibrium distribution function. Upon a fast change in the electric field, the electron distribution function responds with a time delay approximately equal to the momentum relaxation time. This delay creates the phase shift between an electric field and a current, making the conductivity a complex-valued function of frequency. The Drude model of conductivity (see Chapter 7) reads as:

$$\sigma(\omega)=\frac{\sigma}{1-i\omega\tau}\equiv\sigma'(\omega)+i\sigma''(\omega), \tag{12.39}$$

where $\sigma=q^2n_e\tau/m$, m is the electron mass, n_e is the volume electron density, and τ is the momentum relaxation time.

To study the wave propagation in a frequency dispersive media, we use the Maxwell equations (12.29) with frequency-dependent dielectric constant $\varepsilon(\omega)$ that originates from the time delay in polarization current associated with bound charges. In the electromagnetic field, the ohmic current induced by free carriers is also dispersive and described with $\sigma(\omega)$.

The fourth Maxwell equation (12.29) for monochromatic field $\boldsymbol{E}=\boldsymbol{E}_0(\boldsymbol{r})\exp(-i\omega t)$ takes the form,

$$\nabla\times\boldsymbol{H}=\left(-i\omega\,\varepsilon'\varepsilon_0+\omega\,\varepsilon_0\varepsilon''+\sigma'+i\sigma''\right)\boldsymbol{E}. \tag{12.40}$$

Applying curl operation to the second row in Eq. (12.29), using identity $\nabla\times\nabla\times\boldsymbol{E}=-\nabla^2\boldsymbol{E}$ and Eq. (12.40), one obtains the wave equation in a magnetically nondispersive $(\mu(\omega)\approx\mu)$ and conductive media:

$$\begin{aligned}
&\nabla^2\boldsymbol{E}-\xi\frac{\partial\boldsymbol{E}}{\partial t}-\frac{1}{\mathrm{v}^2}\frac{\partial^2\boldsymbol{E}}{\partial t^2}=0,\\
&\mathrm{v}^2=c^2\left[\mu\varepsilon'(\omega)-\frac{\mu\sigma''(\omega)}{\varepsilon_0\omega}\right]^{-1},\\
&\xi=\left[\frac{\mu\omega\varepsilon''(\omega)}{c^2}+\mu\mu_0\sigma'(\omega)\right].
\end{aligned} \tag{12.41}$$

General expressions (12.41) account for frequency dispersion in both the dielectric function and conductivity. As we consider the polarization and ohmic currents as separate terms in Maxwell equations, the dielectric dispersion $\varepsilon(\omega)$ originates from the polarization of bound charges only. In other words, $\varepsilon(\omega)$ is the lattice part of the dielectric function which does not contain a contribution from free electrons. Dispersion in $\varepsilon(\omega)$ starts at a frequency higher than optical, where the difference between bound and free charges disappears, and the spatial dispersion comes into effect. On the contrary, $\sigma(\omega)$ is dispersive in the frequency range of interest: $\omega\tau>1$. So, in the optical frequency range, one can preserve dispersion in $\sigma(\omega)$ while replacing $\varepsilon(\omega)$ with optical dielectric constant ε and assuming $\varepsilon''(\omega)=0$.

As will be discussed below, the first time derivative in Eq. (12.41) describes attenuation the wave experiences propagating in a dispersive material. Two sources of attenuation correspond to two terms in ξ (12.41). The first term is the dielectric loss, i.e., the imaginary part of the lattice dielectric function which originates from the time delay in polarization current and becomes relevant if the frequency of the field is close to ultraviolet. Absorption of the electromagnetic field and frequency dispersion in dielectrics always come along. This contribution fades out when $\omega\to0$. The second term in ξ is the ohmic loss caused by finite conductivity $\sigma'(\omega)$, in other words, by free carriers that accelerate by the electromagnetic field and then transfer their energy to the lattice by collisions to atomic vibrations and crystal imperfections.

Presenting the solution in the form of plane wave $\boldsymbol{E}=\boldsymbol{E}_0\exp\left(i\left(\boldsymbol{k}\cdot\boldsymbol{r}-\omega t\right)\right)$, $\boldsymbol{E}_0$ is the complex-valued amplitude of circularly polarized wave (12.41), we find

$$k^2=\frac{\omega^2}{v^2}+i\omega\xi=\frac{\mu\omega^2}{c^2}\left[\varepsilon+\frac{i}{\varepsilon_0\omega}\left(\sigma'(\omega)+i\sigma''(\omega)\right)\right]. \quad (12.42)$$

Square bracket (12.42) presents the complex-valued dielectric function and defines optical constants n_r and κ as

$$\varepsilon(\omega)\equiv\varepsilon'+i\varepsilon''=\varepsilon+\frac{i\sigma}{\omega\varepsilon_0},\quad \left(n_r+i\kappa\right)^2=\varepsilon(\omega)\,\mu,$$
$$n_r=\sqrt{\frac{\mu}{2}\left[\varepsilon'+\sqrt{\varepsilon'^2+\varepsilon''^2}\right]},\quad 2n_r\kappa=\mu\varepsilon''. \quad (12.43)$$

Within the Drude model (12.39),

$$\varepsilon'=\varepsilon-\frac{\sigma''(\omega)}{\omega\varepsilon_0}=\varepsilon-\frac{\sigma\omega\tau}{\omega\varepsilon_0\left(1+\omega^2\tau^2\right)}=\varepsilon\left[1-\frac{\omega_p^2\tau^2}{1+\omega^2\tau^2}\right],$$
$$\varepsilon''=\frac{\sigma'(\omega)}{\omega\varepsilon_0}=\frac{\sigma}{\omega\varepsilon_0\left(1+\omega^2\tau^2\right)}=\frac{\varepsilon\omega_p^2\tau}{\omega\left(1+\omega^2\tau^2\right)}, \quad (12.44)$$
$$\omega_p^2=\frac{q^2n_e}{m\varepsilon_0\varepsilon}.$$

The wave velocity in the second row (12.41) contains ε', the real part of the dielectric function renormalized with the imaginary part of conductivity. Renormalization part is negative and at $\omega\,\varepsilon_0\varepsilon(0)<\sigma''(\omega)$ the phase velocity becomes imaginary that makes wave propagation impossible and means full reflection of the wave incoming to the boundary between vacuum and the medium. If $\omega^2\tau^2\gg1$,

$$\varepsilon'\approx\varepsilon\left[1-\frac{\omega_p^2}{\omega^2}\right], \quad (12.45)$$

In the frequency point, $\varepsilon'(\omega)=0$, the Maxwell equations allow the wave solution which is different from the transverse electromagnetic wave. Let us assume $\boldsymbol{j}=0,\ \rho=0$. From the first and fourth equations (12.29) one gets $\boldsymbol{k}\times\boldsymbol{H}=0$ and $\boldsymbol{k}\cdot\boldsymbol{H}=0$ that means $\boldsymbol{H}=0$.

Providing $\boldsymbol{H}=0$, the second equation gives $\boldsymbol{k}\times\boldsymbol{E}=0$ meaning the only possible wave solution should have the longitudinal electric field parallel to the wave vector. That is called the plasma oscillations with frequency ω_p, in other words, oscillations of carrier density and thus the longitudinal electric field.

Excitation of plasma oscillations results in full reflection of the wave incoming from vacuum to the boundary of the medium. The plasma frequency ω_p determines the frequency region of full reflection: $\omega<\omega_p$.

In metals, ω_p is of the order of tens eV, that is why they reflect all optical frequencies and acquire shiny color. In semiconductors, the carrier density is low and $\omega_p\approx0.01\,\text{eV}$. Increase in carrier density in semiconductors by heavy doping is accompanied with increased scattering rate that violates condition $\omega^2\tau^2\gg1$ and narrows frequency interval of fully reflected waves as follows from ε' (12.44). Besides, if condition $\omega^2\tau^2\gg1$ does not hold, one has to account ε'', so plasma excitations would require two conditions $\varepsilon'(\omega)=0$, $\varepsilon''(\omega)=0$ which can be satisfied with complex-valued ω, what corresponds to damped plasma oscillations.

On the other hand, $\varepsilon'=1$ at $\omega^2=\omega_1^2$, where $\omega_1^2=\omega_p^2\varepsilon/(\varepsilon-1)$. At frequency $\omega\geq\omega_1$, $n_r=\sqrt{\varepsilon'\mu}$ is real meaning the media is almost transparent and supports wave propagation. The rise of reflection happens when frequency decreases in the region $\omega_p<\omega\leq\omega_1$. In semiconductors, $\varepsilon>1$ and thus ω_p and ω_1 are close to each other that leads to sharp plasmonic reflection edge observed in experiments.

12.2.4 Negative Refraction

Above conclusions regarding reflection and transparency have physical sense if magnetic permeability μ is real and positive. In magnetic materials including semiconductors, μ may experience frequency dispersion sometimes of resonant character. Similar to dielectric susceptibility in metals, μ could be negative in a finite frequency region. Frequency dispersive ε and μ are complex-valued functions of frequency: $\varepsilon=\varepsilon'+i\varepsilon''$, $\mu=\mu'+i\mu''$. If ε' and μ' have opposite signs, the refractive index is imaginary meaning the wave could not propagate in the material. However, once there is a frequency region where both ε' and μ' are negative, the electromagnetic wave can propagate. Square of refractive index is $n_r^2=\varepsilon\mu=\varepsilon'\mu'\left[1-\varepsilon''\mu''/\varepsilon'\mu'+i\left(\varepsilon''/\varepsilon'+\mu''/\mu'\right)\right]$. We assume the media is dissipative, $\varepsilon''>0$, $\mu''>0$, and dissipation is weak, $\varepsilon''\mu''<\varepsilon'\mu'$. The refractive index n_r is

$$n_r=\pm\sqrt{\varepsilon\mu}\approx\pm\sqrt{\varepsilon'\mu'}\left(1-\frac{\varepsilon''\mu''}{2\varepsilon'\mu'}+i\frac{\varepsilon''\mu'+\varepsilon'\mu''}{2\varepsilon'\mu'}\right). \tag{12.46}$$

As follows from Eq. (12.31), condition $\mathrm{Im}(n_r)>0$ guarantees the medium is dissipative, so if $\varepsilon'>0$, $\mu'>0$ one has to choose sign plus in Eq. (12.46), while if $\varepsilon'<0$, $\mu'<0$, the correct sign is minus. Often ignoring small dissipation, one could write refraction index as $n_r=-\sqrt{\varepsilon'\mu'}$.

In natural and artificial materials, $\varepsilon'>0$, $\mu'>0$ may be frequency independent in a broad spectral range, and thus one finds positive refraction index in transparent lossless media. On the contrary, lossless medium with negative refraction does not exist as conditions ε', $\mu'<0$ holds in a finite frequency range implying frequency dispersion that always leads to dissipation in passive media.

The possibility of wave propagation in materials with negative refraction index [8] creates a background for study and engineering of metamaterials with unusual electromagnetic properties (see [9,10] for a review).

12.2.5 Intraband Free Carrier Absorption

The wave vector in Eq. (12.42) is a complex-valued number $k\equiv k_r+ik_i$, where choice of the sign provides for the positive k_i that guarantees the amplitude in $\boldsymbol{E}$ exponentially decreasing along the propagation path, let say, z-direction: $\boldsymbol{E}=\boldsymbol{E}_0\exp(-k_iz)\exp(i(k_rz-\omega t))$.

Eq. (12.42) can be rewritten as

$$k_r^2-k_i^2=\frac{\omega^2}{\mathrm{v}^2},\quad 2k_rk_i=\omega\xi. \tag{12.47}$$

Solution to (12.47) is

$$k_i^2=\frac{\omega^2}{2\mathrm{v}^2}\left(\sqrt{1+\frac{\mathrm{v}^4\xi^2}{\omega^2}}-1\right),\ k_r^2=\frac{\omega^2}{2\mathrm{v}^2}\left(\sqrt{1+\frac{\mathrm{v}^4\xi^2}{\omega^2}}+1\right). \tag{12.48}$$

In the regime $\omega\tau\ll1$, $\sigma''\approx0$, $\sigma'\approx\sigma$, and the dispersion relation (12.42) takes the form

$$k^2 = \frac{\mu\omega^2}{c^2}\left[\varepsilon + i\frac{\sigma}{\omega\varepsilon_0}\right], \tag{12.49}$$

where σ is the static conductivity. So, even without dielectric loss, the attenuation of the electromagnetic wave comes from the finite static conductivity. More detailed information follows from Eq. (12.48):

$$k_i^2 \approx \begin{cases} \dfrac{\mu\mu_0\omega\sigma}{2}, & \dfrac{\sigma}{\varepsilon_0\varepsilon\omega} \gg 1, \\ \dfrac{\mu\mu_0\sigma^2}{4\varepsilon_0\varepsilon}, & \dfrac{\sigma}{\varepsilon_0\varepsilon\omega} \ll 1. \end{cases} \tag{12.50}$$

If conductivity is high (metals), the first row in Eq. (12.50) determines the skin depth $k_i \equiv L^{-1} \approx \sqrt{\omega\mu\mu_0\sigma/2}$, the distance the electric and magnetic fields could penetrate the metal. An example is $L = 0.65\,\mu\text{m}\ (2\,\mu\text{m})$ in cooper at 10 (1.0) GHz.

In the opposite limit of low conductivity (second row in Eq. (12.50)), the free carrier absorption causes an exponential decrease of the field amplitude on the distance $l = k_i^{-1} \approx 2\sqrt{\varepsilon\varepsilon_0 / \mu\mu_0}\,/\,\sigma$, so the absorption coefficient is $\alpha = 2k_i = Z\sigma$, where $Z = \sqrt{\mu\mu_0 / \varepsilon\varepsilon_0}$ is the wave impedance, and coefficient 2 comes from the spatial decrement in power which is proportional to E^2 and thus twice as large as the decrement in electric field. So, the free carrier absorption coefficient is proportional to the ohmic conductivity.

Power flux. Using Eq. (12.32) we express electric and magnetic fields as

$$\begin{aligned} &\boldsymbol{E} = \boldsymbol{E}_m \exp(-i\omega t), \quad \boldsymbol{H} = \boldsymbol{H}_m \exp(-i\omega t), \\ &\boldsymbol{E}_m = \boldsymbol{E}_0 \exp(-z/l)\exp(ik_r z), \\ &\boldsymbol{H}_m = \frac{1}{Z}(\boldsymbol{E}_0 \times \boldsymbol{n})\exp\left(-\frac{z}{l}\right)\exp(ik_r z), \end{aligned} \tag{12.51}$$

and then the average power flux (12.35):

$$\bar{\boldsymbol{P}} = \frac{1}{2}\mathrm{Re}\left[\boldsymbol{E}_m \times \boldsymbol{H}_m^*\right] = \frac{\boldsymbol{n}}{2Z}|E_0|^2 \exp(-\alpha z) = Nv\hbar\omega\exp(-\alpha z)\boldsymbol{n}, \tag{12.52}$$

where $v = c/n_r$, N is the photon density (12.36).

12.3 Interband Absorption in Semiconductors

In semiconductors, the dominant mechanism of absorption relies on electron transitions from valence to conduction band if the photon energy is close to the bandgap $\hbar\omega \geq E_g$. The phenomenological approach applied to free carrier absorption cannot be used as we deal with field-induced transitions between quantum electron states in a crystal, that requires an approach based on interaction Hamiltonian (12.20) or (12.27). With time-periodic perturbation $\boldsymbol{A} = \mathrm{Re}\left\{\boldsymbol{A}_0 \exp(-i\omega t + i\boldsymbol{q}\cdot\boldsymbol{r})\right\}$, the interaction term (12.20) can be presented in the form

$$\begin{aligned} &H_1 = F\exp(-i\omega t) + F^+\exp(i\omega t), \\ &F = \frac{i\hbar e}{2m}\exp(i\boldsymbol{q}\cdot\boldsymbol{r})\,\boldsymbol{A}_0\cdot\nabla, \end{aligned} \tag{12.53}$$

here e is the elementary electrical charge, $\boldsymbol{q}$ is the wave vector of the electromagnetic wave, and F^+ is the operator Hermitian conjugate to F.

In the framework of the time-dependent perturbation theory [11], periodic terms (12.53) induce electronic transitions in the two-level system $E_v < E_c$. One of these terms is resonant for upward electronic transition $E_v \to E_c$, while the other is rapidly oscillating and thus turns zero after time integration. For upward transition, the terms swap their roles. The probability of a resonant transition of electrons $E_v \to E_c$ per unit time is equal to

$$w_{vc} = \frac{2\pi}{\hbar}\left|\langle\varphi_c|F|\varphi_v\rangle\right|^2 \delta(E_c - E_v - \hbar\omega), \left[s^{-1}\right] \tag{12.54}$$

where $\langle\varphi_c|F|\varphi_v\rangle = \Omega^{-1}\int \varphi_c^*(\boldsymbol{r})\ F\ \varphi_v(\boldsymbol{r})\ d^3\boldsymbol{r}$ is the matrix element of perturbation calculated with unperturbed wave functions which correspond to levels $E_{c,v}$ and Ω is the crystal volume. Transition rate (12.54) is nonzero at resonance: $\hbar\omega = E_c - E_v$.

In semiconductors, levels are actually bands with dispersion $E_{c,v}(\boldsymbol{k})$ and $\boldsymbol{k}$ is the electron wavevector. The transition rate per unit volume is a sum over all quantum states including spin:

$$R_{vc} = \frac{4\pi}{\Omega\hbar}\sum_{\boldsymbol{k},\boldsymbol{k}'}\left|\varphi_{ck}\,|\,F\,|\,\varphi_{vk'}\right|^2 \delta\left(E_c(\boldsymbol{k}) - E_v(\boldsymbol{k}') - \hbar\omega\right), \tag{12.55}$$

where $\boldsymbol{k} - \boldsymbol{k}' = \boldsymbol{q}$. Expression (12.55) is valid at zero temperature when the valence band is filled with electrons while the conduction band is empty. The Pauli principle precludes transitions to a final state filled with an electron, so the transition probability should by multiplied by Fermi filling factors to ensure that electron excited from the initially filled valence state goes to the empty final state in the conduction band:

$$R_{vc} = \frac{4\pi}{\Omega\hbar}\sum_{\boldsymbol{k},\boldsymbol{k}'}\left|\langle\varphi_{ck}|F|\varphi_{vk'}\rangle\right|^2 \delta\left(E_c(\boldsymbol{k}) - E_v(\boldsymbol{k}') - \hbar\omega\right) f_v(\boldsymbol{k}')\left(1 - f_c(\boldsymbol{k})\right), \tag{12.56}$$

where $f_{c,v}(\boldsymbol{k}) = \left\{1 + \exp\left[\left(E_{c,v}(\boldsymbol{k}) - \mu\right)/k_BT\right]\right\}^{-1}$ is the Fermi distribution function or the filling factor in thermal equilibrium.

The electron transition from lower to the upper level, $E_v \to E_c$, occurs due to photon absorption. There is also stimulated emission, i.e., inverse transitions $E_c \to E_v$ accompanied by an emission of a photon and described by the second term in H_1 (12.53). The rate of resonant electron transition from a filled state in the upper level to empty one in the lower level is

$$R_{cv} = \frac{4\pi}{\Omega\hbar}\sum_{\boldsymbol{k},\boldsymbol{k}'}\left|\langle\varphi_{ck}|F|\varphi_{vk'}\rangle\right|^2 \delta\left(E_v(\boldsymbol{k}) - E_c(\boldsymbol{k}') + \hbar\omega\right) f_c(\boldsymbol{k}')\left(1 - f_v(\boldsymbol{k})\right). \tag{12.57}$$

Matrix elements in Eqs. (12.56) and (12.57) are equal: $\left|\langle\varphi_{ck}|F|\varphi_{vk'}\rangle\right| = \left|\langle\varphi_{vk}|F|\varphi_{ck'}\rangle\right|$.

Net transition rate per photon flux Nv (12.52) is called the absorption coefficient $\left[m^{-1}\right]$:

$$\alpha \equiv \frac{R_{vc} - R_{cv}}{Nv} = \frac{2\pi e^2\hbar^2\mu}{\Omega m^2\omega\, n_r\varepsilon_0 c}\sum_{\boldsymbol{k},\boldsymbol{k}'}\left|\langle\varphi_{ck}|e^{i\boldsymbol{q}\cdot\boldsymbol{r}}\boldsymbol{e}\cdot\nabla|\varphi_{vk'}\rangle\right|^2 \delta\left(E_v(\boldsymbol{k}) - E_c(\boldsymbol{k}') + \hbar\omega\right)\left[f_v(\boldsymbol{k}') - f_c(\boldsymbol{k})\right], \tag{12.58}$$

where $\boldsymbol{e}$ is the unit vector in the direction of electric field-polarization vector. The polarization vector is introduced as $\boldsymbol{E} = \mathrm{Re}\left[\boldsymbol{e}\ E_0\exp\left(i\left(\boldsymbol{q}\cdot\boldsymbol{r} - \omega t\right)\right)\right]$. In general, it is a complex-valued vector. In transverse electromagnetic wave propagating in the z-direction $\boldsymbol{e}$ is perpendicular to z, $\boldsymbol{e} = \boldsymbol{e}_x + i\boldsymbol{e}_y$, and gives the electric field in the form:

$$E = e_x\ E_0 \cos(q \cdot r - \omega t) - e_y\ E_0 \sin(q \cdot r - \omega t),$$

$$\frac{E_x^2}{(e_x E_0)^2} + \frac{E_y^2}{(e_y E_0)^2} = 1 \qquad (12.59)$$

Relation (12.59) means that the end of vector E as a function of time rotates following an ellipse in the x-y plane corresponding to right-handed elliptically polarized wave, which turns to circularly polarized if $e_x = e_y$, and to linear polarized along y or along x if either e_x or e_y equals zero. In the incoming right-handed polarized wave, the electric field rotates counterclockwise. Choice $e = e_x - ie_y$ corresponds to a rotation in opposite direction-left-handed polarization.

The absorption coefficient (12.58) does not account for finite electron lifetime in the state with particular energy. Scattering relaxes strong energy conservation condition, so in order to fit the experimental data one may replace δ-function with the Lorentzian: $\delta(E_v(k) - E_c(k') + \hbar\omega) \to \Gamma/2\left[(E_v(k) - E_c(k') + \hbar\omega)^2 + (\Gamma/2)^2\right]^{-1}$ where Γ describes inelastic scattering rate.

In semiconductors, the resonant inter-band absorption is dominant, so one can neglect the free carrier absorption $(\sigma \approx 0)$ and consider inter-band transitions as the only contribution to the wave attenuation. In the limit $\varepsilon''(\omega)/\varepsilon \ll 1$ the parameters of wave equation (12.41) and the attenuation factor (12.48) become

$$v^2 \approx \frac{c^2}{\mu\varepsilon},\ \xi \approx \frac{\mu\omega\varepsilon''(\omega)}{c^2},$$

$$q^2 \approx \frac{\mu\omega^2\varepsilon''(\omega)^2}{4c^2\varepsilon}. \qquad (12.60)$$

Using the definition given after Eq. (12.50), one can relate the absorption coefficient to the imaginary part of the dielectric function as follows:

$$\alpha = 2\gamma = \frac{\sqrt{\mu}\ \omega\ \varepsilon''(\omega)}{c\sqrt{\varepsilon}}. \qquad (12.61)$$

From Eqs. (12.58) and (12.61), one obtains

$$\varepsilon''(\omega) = \frac{2\pi e^2\hbar^2}{\Omega m^2\omega^2\ \varepsilon_0} \sum_{k,k'} \left|\langle \varphi_{ck} | e^{iq\cdot r} e \cdot \nabla | \varphi_{vk'} \rangle\right|^2 \delta(E_v(k) - E_c(k') + \hbar\omega)\left[f_v(k') - f_c(k)\right]. \qquad (12.62)$$

Once $\varepsilon''(\omega)$ is known, the real part $\varepsilon'(\omega)$ follows from the Kramers–Kronig relation:

$$\varepsilon'(\omega) = 1 + \frac{2}{\pi} P \int_0^\infty \frac{x\varepsilon''(x)}{x^2 - \omega^2} dx, \qquad (12.63)$$

where the principal value in Eq. (12.63) means the integration path, which excludes the pole at $x = \omega$. Integration with (12.62) gives

$$\varepsilon'(\omega) - 1 = \frac{4e^2\hbar^4}{\Omega m^2 \varepsilon_0} \sum_{k,k'} \left|\langle \varphi_{ck} | e^{iq\cdot r} e \cdot \nabla | \varphi_{vk'} \rangle\right|^2 \frac{f_v(k') - f_c(k)}{\left[E_c(k) - E_v(k')\right]\left\{\left[E_c(k) - E_v(k')\right]^2 - (\hbar\omega)^2\right\}}. \qquad (12.64)$$

Dielectric function expressed in Eqs. (12.62) and (12.64) acquires frequency dispersion from band-to-band electron transitions.

12.3.1 Momentum Conservation in Direct Interband Transitions

A photon-assisted transition between electron energy levels is just an electron–photon collision in which both energy and momentum conservation laws hold. Formally, energy conservation in Eq. (12.58) is explicitly expressed by the δ-function while the momentum conservation is hidden in the matrix element and requires more detailed consideration. The electron wave function in a crystal is $\varphi_{nk}(\boldsymbol{r}) = u_{nk}(\boldsymbol{r})\exp(i\boldsymbol{k}\cdot\boldsymbol{r})$, where $u_{nk}(\boldsymbol{r})$ is the Bloch amplitude – a function of coordinates with unit cell periodicity – n is the band index. In Eq. (12.54), we have chosen Bloch functions normalized on the crystal volume that makes Bloch amplitudes normalized on the unit cell volume: $\int\limits_{\Omega_0} u^*_{nk}u_{n'k'}\, d^3r = \Omega_0\, \delta_{kk'}\delta_{nn'}$.

Matrix element (12.58) calculated with Bloch functions has the form:

$$\langle \varphi_{ck} | e^{i\boldsymbol{q}\cdot\boldsymbol{r}} \boldsymbol{e}\cdot\nabla | \varphi_{vk'} \rangle = I_1 + I_2,$$

$$I_1 = \frac{1}{\Omega}\int\limits_{\Omega} e^{i(\boldsymbol{q}-\boldsymbol{k}+\boldsymbol{k}')\boldsymbol{r}} u^*_{ck}\, \boldsymbol{e}\cdot\nabla\, u_{vk'}\, d^3r, \tag{12.65}$$

$$I_2 = \frac{1}{\Omega}\int\limits_{\Omega} e^{i(\boldsymbol{q}-\boldsymbol{k}+\boldsymbol{k}')\boldsymbol{r}} u^*_{ck}\, \boldsymbol{e}\cdot\boldsymbol{k}\, u_{vk'}\, d^3r.$$

Integrands in $I_{1,2}$ are products of unit-cell periodic Bloch amplitudes and long-periodic exponential factors. It is convenient to express the integral over the crystal volume as a sum of integrals over the unit cells:

$$I_1 = \frac{1}{\Omega}\sum_{\boldsymbol{R}}\int\limits_{\Omega_0} e^{i(\boldsymbol{q}-\boldsymbol{k}+\boldsymbol{k}')(\boldsymbol{R}+\boldsymbol{r})} u^*_{ck}(\boldsymbol{r})\, \boldsymbol{e}\cdot\nabla\, u_{vk'}(\boldsymbol{r})\, d^3r, \tag{12.66}$$

where $\boldsymbol{R}$*'s* are the lattice vectors. As the exponential function slowly varies within the unit cell, one can take the approximate exponent ($\boldsymbol{R}+\boldsymbol{r} \approx \boldsymbol{R}$) out of the integral:

$$I_1 = \frac{1}{\Omega}\sum_{\boldsymbol{R}} e^{i(\boldsymbol{q}-\boldsymbol{k}+\boldsymbol{k}')\boldsymbol{R}} \int\limits_{\Omega_0} u^*_{ck}(\boldsymbol{R}+\boldsymbol{r})\, \boldsymbol{e}\cdot\nabla\, u_{vk'}(\boldsymbol{R}+\boldsymbol{r}) d^3r. \tag{12.67}$$

Unit-cell periodic Bloch amplitudes make the integrals the same in each of N unit cells, so the integral can be moved out of $\boldsymbol{R}$ -sum:

$$I_1 = \frac{1}{\Omega}\int\limits_{\Omega} u^*_{ck}(\boldsymbol{r})\, \boldsymbol{e}\cdot\nabla\, u_{vk'}(\boldsymbol{r})\, d^3r \sum_{\boldsymbol{R}} e^{i(\boldsymbol{q}-\boldsymbol{k}+\boldsymbol{k}')\boldsymbol{R}} = \delta_{q-k+k'}\, \frac{1}{\Omega_0}\int\limits_{\Omega_0} u^*_{ck}(\boldsymbol{r})\, \boldsymbol{e}\cdot\nabla\, u_{vk'}(\boldsymbol{r})\, d^3r. \tag{12.68}$$

where we used the lattice sum that gives the Kronecker symbol

$$\sum_{\boldsymbol{R}} \exp\left[i(\boldsymbol{q}-\boldsymbol{k}+\boldsymbol{k}')\cdot\boldsymbol{R}\right] = N\,\delta_{mG,\, q-k+k'}, \tag{12.69}$$

and formally expresses momentum conservation in the transition between states $(c,\mathbf{k})$ and $(\mathrm{v},\mathbf{k}')$, $\mathbf{G}$ is the reciprocal lattice vector, $m=0,1,2,\ldots$ Integral I_2 is calculated similarly and equal to zero as c- and v -Bloch amplitudes are orthogonal.

For electron wave vectors in the first Brillouin zone, $m=0$, and the conservation law reads as $\mathbf{q}-\mathbf{k}+\mathbf{k}'=0$. Usually, the photon wave vector $\mathbf{q}$ is small compared to the electron one that reduces the momentum conservation condition to $\mathbf{k}\approx\mathbf{k}'$. Photon-assisted interband transitions conserve the electron momentum and are called direct or vertical. Transitions between valence band maximum and conduction band minimum may occur at small values of $\mathbf{k},\mathbf{k}'$, which are of the same order of magnitude as $\mathbf{q}$. However, the phase volume of these transitions (part of the $\mathbf{k}$ -space where transitions occur) is small, the density of states, and therefore, optical absorption is negligible. Finally, absorption coefficient (12.58) takes the form,

$$\alpha=\frac{2\pi e^2\mu}{\Omega\, m^2\omega\epsilon_0 c\, n_r}\sum_{\mathbf{k}}\left|\mathbf{e}\cdot\mathbf{p}_{cv}\right|^2\,\delta\big(E_{\mathrm{v}}(\mathbf{k})-E_c(\mathbf{k})+\hbar\omega\big)\big[f_{\mathrm{v}}(\mathbf{k})-f_c(\mathbf{k})\big],$$

$$\mathbf{p}_{cv}(\mathbf{k})=-\frac{i\hbar}{\Omega_0}\int_{\Omega_0}u_{ck}^*(\mathbf{r})\,\nabla u_{\mathrm{v}k}(\mathbf{r})d^3r. \tag{12.70}$$

In Section 12.8, we discuss the symmetry of Bloch amplitudes and selection rules for matrix elements (12.70).

12.3.2 Absorption Due to Allowed and Forbidden Transitions

In a generic point of the Brillouin zone, $\mathbf{p}_{cv}$ depends on momentum. If transitions occur close to band extrema at $\mathbf{k}\approx\mathbf{k}_0$, one can expand the matrix element as $\mathbf{e}\cdot\mathbf{p}_{cv}(\mathbf{k})=\mathbf{e}\cdot\mathbf{p}_{cv}(\mathbf{k}_0)+\left[\frac{\partial}{\partial\mathbf{k}}\mathbf{e}\cdot\mathbf{p}_{cv}(\mathbf{k})\right]_{\mathbf{k}_0}\cdot(\mathbf{k}-\mathbf{k}_0)$. The optical selection rule reads as $\mathbf{e}\cdot\mathbf{p}_{cv}(\mathbf{k}_0)\neq0$ meaning the transitions are allowed, and it is sufficient to use k-independent $\mathbf{p}_{cv}(\mathbf{k}_0)$ in Eq. (12.70). If the selection rule is not satisfied, transitions, called forbidden, do not prevent absorption as, in this case, k -dependent terms in the matrix element come into play. Optical absorption due to forbidden transitions will be discussed further in the text.

Analytical calculation of α caused by allowed transitions is possible in a semiconductor with parabolic energy dispersion,

$$E_{c,\mathrm{v}}(\mathbf{k})=\pm\left(E_g/2+\hbar^2\mathbf{k}^2/2m_{c,\mathrm{v}}\right),$$

$$E_c(\mathbf{k})-E_{\mathrm{v}}(\mathbf{k})\equiv E_g+E(\mathbf{k}),\ E(\mathbf{k})\equiv\frac{\hbar^2\mathbf{k}^2}{2m_r}, \tag{12.71}$$

$$m_r=\frac{m_c m_{\mathrm{v}}}{m_c+m_{\mathrm{v}}}.$$

In a nondegenerate semiconductor at $T=0$, $f_{\mathrm{v}}(\mathbf{k})=1$, $f_c(\mathbf{k})=0$, and

$$\alpha=\frac{2\pi e^2\mu}{\Omega\, m^2\omega\epsilon_0 c\, n_r}\left|\mathbf{e}\cdot\mathbf{p}_{cv}(0)\right|^2\sum_{\mathbf{k}}\delta\big(E_g+E(\mathbf{k})-\hbar\omega\big), \tag{12.72}$$

Replacing the sum with integral, $\sum_{\mathbf{k}}(\ldots)\rightarrow\Omega\int(\ldots)\,d^3k/(2\pi)^3$ and going to integration over energy variable $E=\hbar^2k^2/2m_r$, we define the joint density of states as

$$4\pi\int_0^\infty (\ldots)\, k^2 dk = \int_0^\infty (\ldots)\, \rho(E)\, dE,$$
$$\rho(E) = \frac{2\pi(2m_r)^{3/2}}{\hbar^3}\sqrt{E}. \tag{12.73}$$

Calculating integral in Eq. (12.72), we have

$$\alpha = \frac{e^2\mu}{4\pi^2 m^2 \varepsilon_0 c\, n_r \omega}\left|\boldsymbol{e}\cdot\boldsymbol{p}_{cv}(0)\right|^2 \begin{cases} \rho(\hbar\omega - E_g), & \hbar\omega > E_g \\ 0, & \hbar\omega \le E_g \end{cases}. \tag{12.74}$$

Figure 12.1 illustrates a typical semiconductor band diagram, the absorption edge, and the density of states.

In anisotropic semiconductors, effective masses depend on direction: m_i, $i = x, y, z$. Reduced masses corresponding to each direction are $m_{ri}^{-1} = m_{ci}^{-1} + m_{vi}^{-1}$. The joint density of states (12.73) would contain $\sqrt{m_{rx}m_{ry}m_{rz}}$ instead of $m_r^{3/2}$.

If the selection rule forbids transitions in $\boldsymbol{k} = \boldsymbol{k}_0$ we have to account for k-dependent matrix element: $\boldsymbol{e}\cdot\boldsymbol{p}_{cv}(\boldsymbol{k}) = \left[\frac{\partial}{\partial\boldsymbol{k}}\boldsymbol{e}\cdot\boldsymbol{p}_{cv}(\boldsymbol{k})\right]_{\boldsymbol{k}_0}\cdot(\boldsymbol{k}-\boldsymbol{k}_0)$. Let us assume $\boldsymbol{k}_0 = 0$, denote $\left[\frac{\partial}{\partial\boldsymbol{k}}\boldsymbol{e}\cdot\boldsymbol{p}_{cv}(\boldsymbol{k})\right]_{\boldsymbol{k}=0} \equiv \hbar\boldsymbol{b}$, and use $\boldsymbol{e}\cdot\boldsymbol{p}_{cv}(\boldsymbol{k}) = \hbar\boldsymbol{b}\cdot\boldsymbol{k}$ for calculation in Eq. (12.70). The sum in Eq. (12.72) now includes additional factor $\left|\boldsymbol{e}\cdot\boldsymbol{p}_{cv}(\boldsymbol{k})\right|^2 = \hbar^2 b^2 k^2\cos^2\theta$, where θ is the angle between $\boldsymbol{k}$ and $\boldsymbol{b}$. In spherical coordinates, we choose $\boldsymbol{b}$ as a polar axis and present momentum summation in the form:

$$\sum_k \left|\boldsymbol{e}\cdot\boldsymbol{p}_{cv}(\boldsymbol{k})\right|^2 \delta(E_g + E(k) - \hbar\omega) = \frac{\Omega\hbar^2 b^2}{(2\pi)^3}\int_0^{2\pi} d\varphi \int_0^{\pi}\sin\theta\cos^2\theta \int_0^\infty k^4 dk\, \delta(E_g + E(k) - \hbar\omega).$$

After the calculation of integrals, one gets the absorption coefficient,

$$\alpha = \frac{2^{3/2} e^2 \mu\, b^2 m_r^{5/2} Z_0}{3\pi\hbar^3 m^2 n_r}\begin{cases} \dfrac{(\hbar\omega - E_g)^{3/2}}{\omega}, & \hbar\omega > E_g \\ 0, & \hbar\omega \le E_g. \end{cases} \tag{12.75}$$

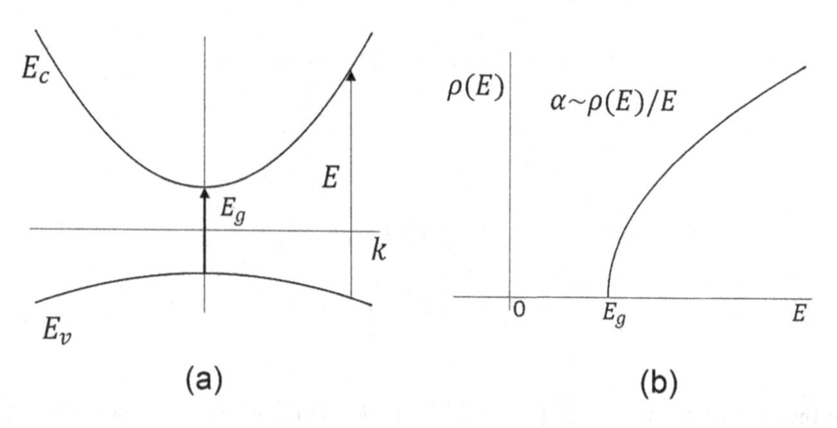

FIGURE 12.1 Absorption edge in a semiconductor with parabolic energy bands. (a) Band diagram. (b) Joint density of states and absorption caused by allowed transitions.

The signature of forbidden transitions is the absorption coefficient, which depends on frequency as $\alpha \sim \left(\hbar\omega - E_g\right)^{3/2} / \omega$.

12.3.3 Indirect Optical Transitions

Vertical optical transitions discussed in the previous section take place in semiconductors in which conduction and valence band extrema are at the same point in the Brillouin zone. In some materials, such as Si, Ge, and GaP, positions of the valence band maximum and the conduction band minimum in the $\boldsymbol{k}$-space do not coincide. In materials with an indirect bandgap, besides photon, another particle must present to ensure the conservation of the quasi-momentum in the course of optical transition. The third particle could be a phonon, an impurity, a lattice defect, as well as an electron. In the course of the impurity-assisted indirect optical transition, the electron changes the momentum and, at the same time, conserves the energy, as the impurity has a large mass and cannot assimilate the energy difference, thus making collision elastic. In moderately doped semiconductors, the probability of impurity scattering is small, as it is proportional to the impurity concentration.

In intrinsic semiconductors, the dominant mechanism of optical transitions relies on the emission or absorption of photons. The photon delivers the energy difference between the initial and final state, while the phonon provides momentum transfer.

In Figure 12.2, we show the diagram of the optical transition in the indirect-gap semiconductor Ge.

The indirect transition relies on two terms in the Hamiltonian: electron–photon and electron–phonon interactions. In the perturbation theory, the indirect transition $i \to f$ can be represented as a photon-assisted virtual vertical transition to an intermediate state m and following phonon-assisted virtual transition $m \to f$ with phonon emission or absorption (see Figure 12.2a). The alternative path in Figure 12.2b shows the same transition, $i \to f$, as a combination of vertical one $m' \to f$ and subsequent $i \to m'$ accompanied by phonon emission or absorption. Each transition respects the quasi-momentum conservation law as it occurs in a system with translational symmetry. For the real transition $i \to f$, the momentum conservation reads as $\boldsymbol{k}_f - \boldsymbol{k}_i = \pm\boldsymbol{q}$, where $\hbar\omega_q$ is the phonon energy.

The transition rate for the process rendered in Figure 12.2a includes matrix elements of electron–phonon, H_{ep}, and electron–photon, F, interactions. It is written below as a sum of two

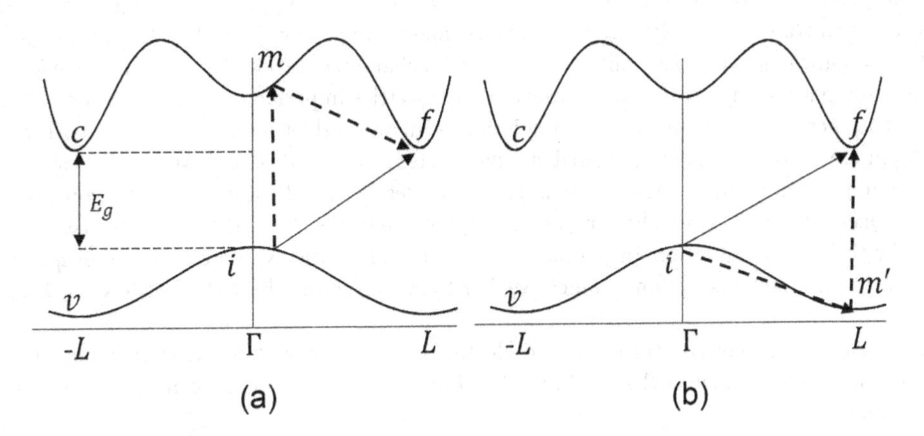

FIGURE 12.2 Two channels of an indirect optical transition in a multivalley semiconductor. i, f, and $m(m')$ are the initial, final, and intermediate states, respectively. Solid arrow – indirect transition; dashed arrows – virtual transitions through an intermediate state.

terms; each corresponds to emission or absorption of a phonon (we consider one-phonon processes only):

$$w_{fi}^{ind} = \frac{2\pi}{\hbar}\left\{\frac{\left|\langle s_f|H_{ep}^{+}|s_m\rangle\right|^2 \left|\langle \varphi_m|F|\varphi_i\rangle\right|^2}{(E_m - E_i - \hbar\omega)^2}\delta\left(E_f - E_i - \hbar\omega + \hbar\omega_q\right)\right.$$

$$\left. + \frac{\left|\langle s_f|H_{ep}^{-}|s_m\rangle\right|^2 \left|\langle \varphi_m|F|\varphi_i\rangle\right|^2}{(E_m - E_i - \hbar\omega)^2}\delta\left(E_f - E_i - \hbar\omega - \hbar\omega_q\right)\right\} \tag{12.76}$$

where s_n is the product of Bloch function φ_n and wave function of the lattice vibrational state, the state with a definite number of phonons. The electron–phonon interaction acts differently on the state s_n depending on whether it creates or annihilates a phonon:

$$\left|\langle s_f|H_{ep}^{\pm}|s_m\rangle\right|^2 \sim n_q + \frac{1}{2} \pm \frac{1}{2}. \tag{12.77}$$

Signs + and − correspond to phonon emission and absorption, respectively, and $n_q = \left[\exp\left(\hbar\omega_q / k_B T\right) - 1\right]^{-1}$ is the number of phonons in thermal equilibrium. At low temperature $n_q \to 0$, and electron–phonon processes can occur only with the emission of phonons.

Energy conservation expressed as an argument of δ-function in Eq. (12.76) involves the initial and final states only, as virtual transitions through intermediate states do not necessarily conserve energy.

The absorption coefficient is the transition rate summed over all intermediate states, all initial and final electron states as well as all phonon branches and looks similar to Eq. (12.72):

$$\alpha = A\sum_{sq}\sum_{k,k'}\frac{\left|\langle s_f|H_{ep}^{+}|s_m\rangle\right|^2 \left|\langle \varphi_m|F|\varphi_i\rangle\right|^2}{(E_m - E_{vk} - \hbar\omega)^2}\left\{(n_q+1)\,\delta\left(E_{ck'} - E_{vk} - \hbar\omega + \hbar\omega_{sq}\right)\right.$$

$$\left. + n_q\,\delta\left(E_{ck'} - E_{vk} - \hbar\omega - \hbar\omega_{sq}\right)\right\}. \tag{12.78}$$

As compared to Eq. (12.72), α includes additional factor $q = \left|\langle s_f|H_{ep}|s_m\rangle\right|^2 / (E_m - E_{vk} - \hbar\omega)^2$. As the vertical virtual transition implies $E_m = E_{ck}$, and we follow frequency lower than the direct gap $i-m$, the denominator in q is not resonant and estimates as characteristic electron excitation energy $\approx E_g^2$. The electron–phonon matrix element is nonzero due to lattice imperfections-ion vibrations. Phonons break lattice periodicity and introduce a deviation from an ideal lattice atomic positions. The disorder amplitude is proportional to the adiabatic parameter $\beta = \sqrt[4]{m/M}$, m and M are the masses of an electron and a lattice ion, respectively. In a static and ideally periodic lattice, the electron would not scatter. An estimated electron–phonon matrix element equals the characteristic electron energy multiplied by the relative vibration amplitude: $\left|\langle s_f|H_{ep}|s_m\rangle\right| \sim \beta E_g$. The estimate gives factor $q \sim \sqrt{m/M}$ small, so the indirect absorption is much weaker than in allowed direct transitions not involving phonons.

Once we consider allowed transitions and electrons interacting with acoustic or nonpolar optical phonons, the combined matrix element in Eq. (12.78) is not k-dependent and can be kept constant out of the sum:

$$\alpha = A\sum_{s}\sum_{k,k'}\left\{(n_{sq}+1)\,\delta\left(E_{ck'}-E_{vk}-\hbar\omega+\hbar\omega_{sq}\right)+n_{sq}\,\delta\left(E_{ck'}-E_{vk}-\hbar\omega-\hbar\omega_{sq}\right)\right\}. \tag{12.79}$$

One can present the double sum in (12.79) as

$$\sum_{k,k'}(\ldots)\rightarrow\frac{\Omega^2}{(2\pi)^6}\int\int(\ldots)\,d^3k\,d^3k',$$

$$\int\int\delta\left(E_{ck'}-E_{vk}-\hbar\omega\pm\hbar\omega_{sq}\right)d^3k\,d^3k' = \int_{-\infty}^{\infty}dE\int d^3k'\delta\left(E_{ck'}-E\right)\int d^3k\delta\left(E-E_{vk}-\hbar\omega\pm\hbar\omega_{sq}\right)$$

$$= \int_{-\infty}^{\infty}dE\rho_c(E)\,\rho_v\left(E-\hbar\omega\pm\hbar\omega_{sq}\right), \tag{12.80}$$

where $\rho_{c,v}$ are density of states in bands c, v:

$$\rho_c(E)\sim\begin{cases}\sqrt{E}\ , E>0\\ 0,\ E\le 0,\end{cases}\qquad \rho_v(E)\sim\begin{cases}\sqrt{-E-E_g}, & E<-E_g\\ 0, & E\ge -E_g.\end{cases} \tag{12.81}$$

Integral in Eq. (12.80) becomes

$$\int_{-\infty}^{\infty}dE\rho_c(E)\,\rho_v\left(E-\hbar\omega\pm\hbar\omega_{sq}\right)=\int_{0}^{\hbar\omega\mp\hbar\omega_{sq}-E_g}\sqrt{E\left(\hbar\omega\mp\hbar\omega_{sq}-E-E_g\right)}\,dE$$

$$=\frac{\pi}{8}\left(\hbar\omega\mp\hbar\omega_{sq}-E_g\right)^2. \tag{12.82}$$

The absorption edge is determined by condition $\hbar\omega - E_g \mp \hbar\omega_{sq} > 0$. Using Eqs. (12.72), (12.79), and (12.82), one obtains frequency dependence of the absorption coefficient as

$$\alpha\sim\frac{1}{\omega}\sum_{s}\left(\hbar\omega-\hbar\omega_{sq}-E_g\right)^2(n_{sq}+1)+n_{sq}\left(\hbar\omega+\hbar\omega_{sq}-E_g\right)^2. \tag{12.83}$$

Processes with phonon absorption allow optical transitions at $\hbar\omega < E_g$. So, the absorption coefficient as a function of frequency starts with weak indirect transitions at $\hbar\omega \le E_g$, then direct transitions turn on leading to a sharp increase in absorption at $\hbar\omega > E_g$.

Optical measurements are used to determine the bandgap. However, there are several reasons why the absorption edge does not necessarily coincide with the bandgap. One of them is that indirect transitions start with lower than the bandgap energy. Also, in disordered semiconductors the absorption does not reveal the sharp edge as the tail of density of states in the bandgap favors transitions with $\hbar\omega < E_g$. Another reason is that in degenerate semiconductors, the edge shifts to higher energy as final states in the conduction band are occupied up to the Fermi level ε_F. That is the Burstein–Moss shift illustrated in Figure 12.3. Absorption edge shifts as $\hbar\omega = E_g + \varepsilon_F(1 + m_c / m_v)$ for direct transitions and $\hbar\omega = E_g + \varepsilon_F$ for indirect transitions.

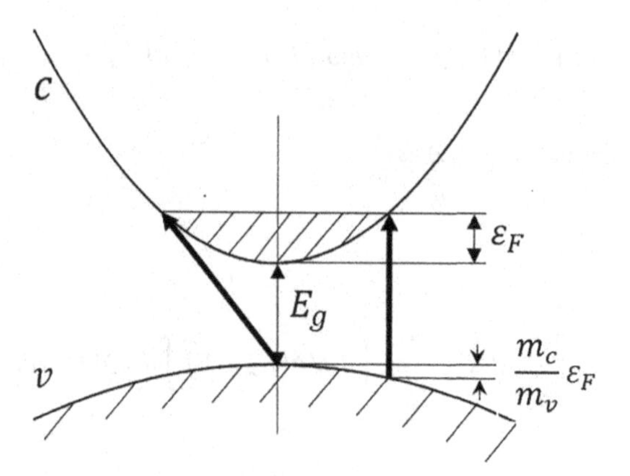

FIGURE 12.3 Moss–Burstein shift for direct and indirect transitions.

12.4 Exciton Transitions

In previous sections, band-to-band transitions have been discussing within the one-electron model of solids which does not account for the interaction between electrons. In the final state of interband transition, the electron in the conduction band and the hole in the valence band experience Coulomb attraction that bounds electron to hole, like an electron to the nucleus in the hydrogen atom, thus modifying the energy spectrum of a semiconductor. The bound state called Wannier–Mott exciton lowers the total energy of the electron–hole pair

$$E_n = E_g + \frac{p^2}{2(m_c + m_v)} - \frac{m_r e^4}{2n^2(\varepsilon\varepsilon_0\hbar)^2}, \quad m_r = \frac{m_c m_v}{m_c + m_v}. \tag{12.84}$$

where $\boldsymbol{p} = \hbar(\boldsymbol{k}_e + \boldsymbol{k}_h)$ is the center-of-mass momentum of the pair and $m_{c,v}$ are effective masses defined in Eq. (12.71). The first term is the energy of the unbound electron–hole pair, the second term is the kinetic energy of the pair, and the third term is the Coulomb binding energy. Index n numbers the hydrogen-like energy levels of the electron–hole pair which in the ground state $n = 1$ has effective Bohr radius $r_B = \varepsilon\varepsilon_0\hbar^2 / m_r e^2$. The energy spectrum E_n falls into the energy gap of a semiconductor as shown in Figure 12.4a.

Excitons affect the edge of optical absorption as shown in Figure 12.4b. If the exciton lifetime is long enough, the weak near-edge absorption coefficient becomes modified by strong exciton contribution.

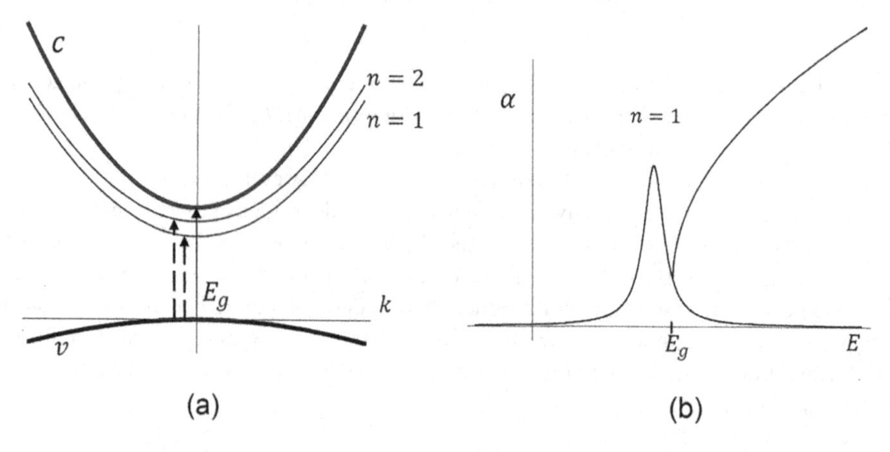

FIGURE 12.4 (a) Direct optical transition to exciton bands 1 and 2 (dashed arrows). (b) Optical absorption edge modified by ground state exciton.

In typical semiconductors, the binding energy of excitons is less than the bandgap, so one needs an optical electric field to create an exciton. Inspecting Eq. (12.84), one may conclude that narrow-gap semiconductors might be unstable with respect to the spontaneous generation of excitons providing the bandgap is less than the binding energy: $2E_g(\varepsilon\varepsilon_0\hbar)^2 < m_r e^4$. This exotic state in narrow-gap semiconductors and semimetals is called excitonic insulator [12,13].

12.5 Impurity-Band Optical Transitions

The feature of the absorption edge shown in Figure 12.4 takes place in pure (intrinsic) semiconductors. However, a similar structure of the absorption edge has been observed in samples, doped with donors and (or) acceptors. Optical absorption in extrinsic semiconductors involves transitions from the valence band on shallow donor levels (see Figure 12.5a) or from acceptor levels to the conduction band (see Figure 12.5b).

The absorption coefficient can be calculated similarly to the one obtained in Section 12.4. For the process rendered in Figure 12.5a, α is proportional to the density of ionized donors (see Eq. (4.32)) and the valence band density of states,

$$\alpha \sim n_d^+ \frac{\sqrt{\hbar\omega - E_g + E_D}}{\omega}, \quad n_d^+ = \frac{n_d}{1 + g\,\exp\left(\dfrac{\mu - E_D}{k_B T}\right)} \tag{12.85}$$

where n_d is the total donor concentration. For the process shown in Figure 12.5b, α relies on the density of occupied acceptors and the conduction band density of states. Experimental examples of impurity-band optical absorption can be found in Ref. [14].

12.6 Electroabsorption and Franz–Keldysh Effect

Constant electric field $\boldsymbol{E} \| z$ applied to a semiconductor affects its energy spectrum and wave functions of electrons in conduction and valence bands. In the xy-plane, a conduction electron is the Bloch particle with the energy $\varepsilon_\perp = E_c + p_\perp^2 / 2m_c$. The field breaks translation symmetry in a crystal that makes the electron wave function aperiodic in the z-direction, so the envelop function is no longer a plane wave but rather a solution to the Schrödinger equation,

$$\left(-\frac{\hbar^2}{2m_c}\nabla_z^2 - eEz\right)\varphi_c(z) = (\varepsilon - \varepsilon_\perp)\varphi_c(z) \tag{12.86}$$

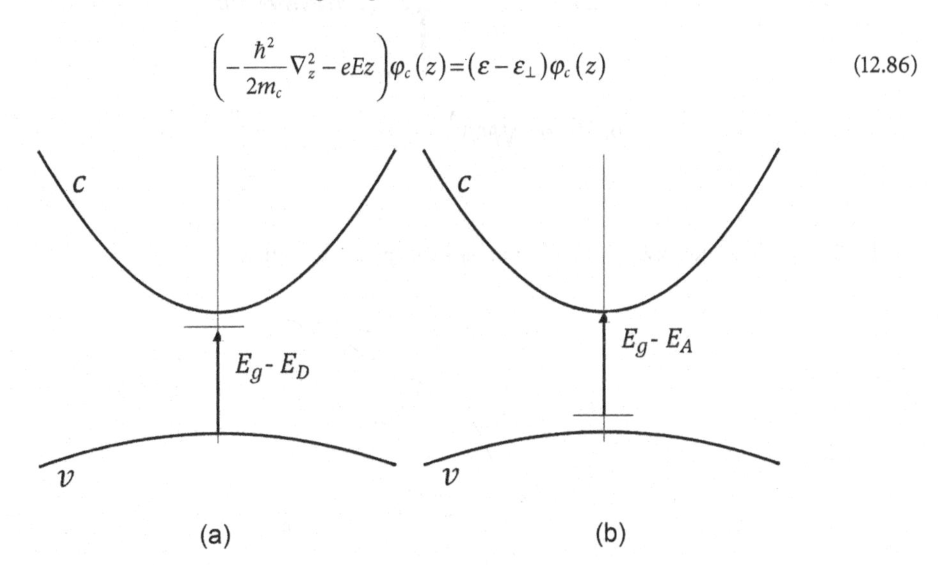

FIGURE 12.5 (a) Transition valence band-empty donor. (b) Occupied acceptor-conduction band.

Solution to (12.86) is expressed as

$$\varphi_c(\xi)=C\ Ai(\xi),\ \xi=\frac{\varepsilon_\perp-\varepsilon-eEz}{\Theta},\ \Theta=\left(\frac{e^2\hbar^2E^2}{2m_c}\right)^{1/3}, \tag{12.87}$$

where C is the normalization constant and $Ai(x)$ is the Airy function [15]. Figure 12.6 illustrates electron and hole wave functions in the electric field.

Absorption coefficient calculated with z-dependent envelops $\varphi_c(z)$ and $\varphi_v(z)$ gives [6,16]:

$$\alpha_E=R\sqrt{\hbar\omega_E}\ \pi\int_\beta^\infty\left[Ai(x)\right]^2dx=R\sqrt{\hbar\omega_E}\pi\ f(\beta),$$

$$f(\beta)=\left\{-\beta Ai^2(\beta)+\left[Ai'(\beta)\right]^2\right\}, \tag{12.88}$$

$$\hbar\omega_E=\left(\frac{e^2\hbar^2E^2}{2m_r}\right)^{1/3},\ \beta=\frac{E_g-\hbar\omega}{\hbar\omega_E},$$

where R is the coefficient before $\sqrt{\hbar\omega-E_g}$ in Eq. (12.74):

$$R=\frac{e^2\mu\ Z_0(2m_r)^{3/2}}{2\hbar^3\pi m^2n_r\omega}\left|e\cdot p_{cv}(0)\right|^2 \tag{12.89}$$

Function $f(\beta)$ shown in Figure 12.7 is nonzero at $\hbar\omega<E_g$ due to the electric field induced overlap of the wave functions in c- and v-bands (Franz–Keldysh effect).

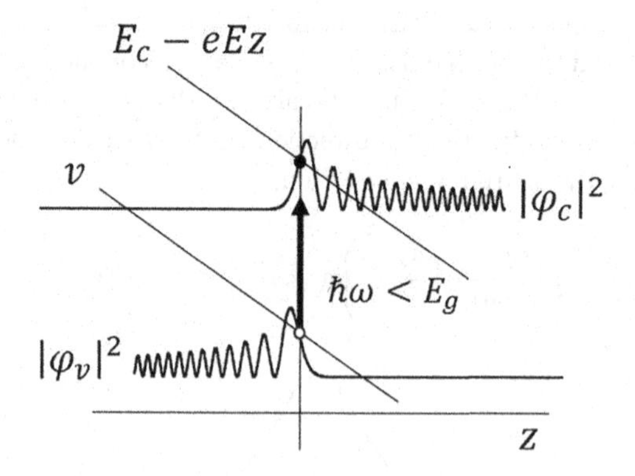

FIGURE 12.6 Franz–Keldysh effect: tunneling assisted optical absorption.

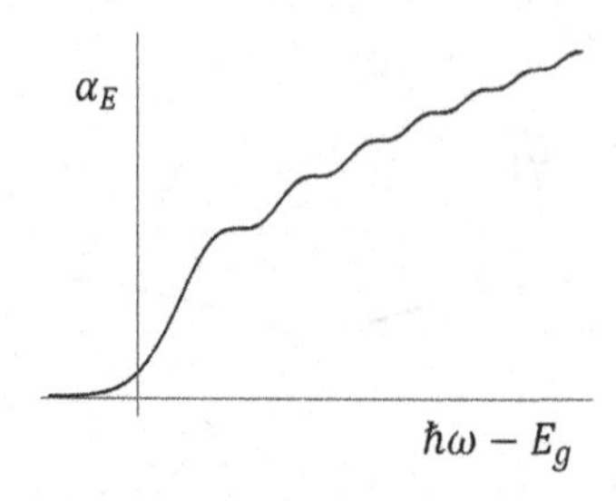

FIGURE 12.7 Electro-absorption coefficient for direct and allowed transitions.

Another effect of the electric field is a weak periodic modulation of the coefficient if photon energy exceeds the bandgap. Explicitly, the result shown in Figure 12.7 can be obtained by using the asymptotic expansion of Airy function $Ai(x) \approx \exp\left(-\frac{2}{3}x^{3/2}\right) \Big/ (2\sqrt{\pi}x^{1/4})$, $\beta \gg 1$, in (12.88) that gives the absorption coefficient

$$\alpha_E \approx R \frac{(\hbar\omega_E)^{\frac{3}{2}}}{8(E_g - \hbar\omega)} \exp\left[-\frac{4}{3}\left(\frac{E_g - \hbar\omega}{\hbar\omega_E}\right)^{\frac{3}{2}}\right], \quad \frac{E_g - \hbar\omega}{\hbar\omega_E} \gg 1, \tag{12.90}$$

The tail of the absorption coefficient in the bandgap is the exponential function of frequency and electric field. In the opposite limit $\beta = -t, \ t \gg 1$, we use asymptotic valid for negative arguments:

$$Ai(-x) \approx \frac{1}{\sqrt{\pi}x^{1/4}} \sin\left(\frac{2}{3}x^{3/2} + \frac{\pi}{4}\right). \tag{12.91}$$

Using (12.91) in $\alpha_E = R\sqrt{\hbar\omega_E}\ \pi \left\{qAi^2(-t) + \left[Ai'(-t)\right]^2\right\}$, one obtains

$$\alpha_E \approx \pi R\sqrt{\hbar\omega_E}\left[\frac{\left(4t^{3/2}\cos\left(\frac{\pi}{4} + \frac{2t^{3/2}}{3}\right) - \sin\left(\frac{\pi}{4} + \frac{2t^{3/2}}{3}\right)\right)^2}{16\pi\ t^{5/2}} + \frac{\sqrt{q}\sin\left(\frac{\pi}{4} + \frac{2t^{3/2}}{3}\right)^2}{\pi}\right]$$

$$= R\sqrt{\hbar\omega - E_g}\left[1 - \frac{(\hbar\omega_E)^{3/2}}{4\left(\hbar\omega - E_g\right)^{3/2}} \cos\left(\frac{4}{3}\left(\frac{\hbar\omega - E_g}{\hbar\omega_E}\right)^{3/2}\right)\right]. \tag{12.92}$$

The first term in Eq. (12.92) is the zero-field absorption coefficient, which coincides with (12.74). The second term is the weak oscillating modulation shown in Figure 12.7.

12.7 Lattice Absorption

If the photon energy becomes less than the semiconductor bandgap, the interband absorption tends to zero. However, photons may lose their energy by exciting the electrically active *TO* phonons. Phonon frequencies fall in the far-infrared range, so we deal with lattice absorption of photons of energy $\hbar\omega < E_g$. In semiconductors with a narrow bandgap, lattice absorption would coexist with interband transitions.

Transverse lattice displacements follow from Eq. (6.72) (Chapter 6):

$$\frac{\partial^2 \boldsymbol{x}_t}{\partial t^2} = -\omega_0^2\ \boldsymbol{x}_t + \omega_0\sqrt{\varepsilon_0\left(\varepsilon(0) - \varepsilon(\infty)\right)}\ \boldsymbol{E}_t, \tag{12.93}$$

where ω_0 is the frequency of TO mode $q = 0$. The second term is responsible for coupling between electromagnetic waves and lattice displacements. Polarization (Chapter 6, Eq. (6.67)) takes the form:

$$\mathbf{P} = \omega_0\sqrt{\varepsilon_0\left(\varepsilon(0) - \varepsilon(\infty)\right)}\ \boldsymbol{x}_t + \varepsilon_0\left(\varepsilon(\infty) - 1\right)\ \boldsymbol{E}_t \tag{12.94}$$

Since the external field $\boldsymbol{E}_t = \boldsymbol{E}_\omega \exp(-i\omega t)$ is time-periodic, we look for solution to (12.93) in the form $\boldsymbol{x}_t = \boldsymbol{x}_\omega \exp(-i\omega t)$:

$$x_\omega = \frac{\omega_0\sqrt{\varepsilon_0\left(\varepsilon(0)-\varepsilon(\infty)\right)}}{\left(\omega_0^2-\omega^2\right)}E_\omega, \tag{12.95}$$

Polarization $\boldsymbol{P} = \boldsymbol{P}_\omega \exp(-i\omega t)$ follows from Eqs. (12.94) and (12.95):

$$\boldsymbol{P}_\omega = \left[\frac{\omega_0^2\varepsilon_0\left(\varepsilon(0)-\varepsilon(\infty)\right)}{\omega_0^2-\omega^2}+\varepsilon_0\left(\varepsilon(\infty)-1\right)\right]\boldsymbol{E}_\omega \tag{12.96}$$

Using $\boldsymbol{P}_\omega = \varepsilon_0\left(\varepsilon(\omega)-1\right)\boldsymbol{E}_\omega$ and Lyddane-Sachs-Teller relation (6.73) (Chapter 6), one obtains

$$\varepsilon(\omega) = \frac{\varepsilon(\infty)\left(\omega_{LO}^2-\omega^2\right)}{\omega_{TO}^2-\omega^2}. \tag{12.97}$$

Dielectric function acquires dispersion due to coupling between an electromagnetic wave and transverse lattice vibrations. Equation $\varepsilon(\omega)=0$ determines the frequency of long-wavelength longitudinal phonons while $\varepsilon^{-1}(\omega)=0$ holds at transverse frequency.

In the frequency range $\omega_{TO}<\omega<\omega_{LO}$, dielectric function $\varepsilon(\omega)$ becomes negative so that electromagnetic wave cannot penetrate the sample. The calculation of reflection and absorption coefficients following from dielectric function (12.97) gives the relative magnitudes of reflected and absorbed power inside the nontransparency band. For details, see, for example, Ref. [14].

Polaritons. In the transparency region $\varepsilon(\omega)>0$, photons mix with transverse optical phonons to form coupled waves, polaritons. The wavevector of propagating waves relates to the refractive index as it follows from Eq. (12.31):

$$k^2 = \frac{\omega^2 n_r^2}{c^2} \tag{12.98}$$

Substituting (12.97) into (12.98) and using $n_r^2=\varepsilon(\omega)$, one gets the dispersion equation for polaritons:

$$\frac{c^2k^2}{\omega^2} = \frac{\varepsilon(\infty)\left(\omega_{LO}^2-\omega^2\right)}{\omega_{TO}^2-\omega^2}. \tag{12.99}$$

As a solution to the quadratic equation, the polariton spectrum consists of two branches, $\omega_1(k)$ and $\omega_2(k)$, illustrated in Figure 12.8.

In the long-wavelength limit, $k\to 0$, $\omega_1^2(k)\to c^2k^2/\varepsilon(0)$, $\omega_2^2(k)\to\omega_{LO}^2+c^2k^2/\varepsilon(\infty)$.

Polaritons are just electromagnetic waves in crystals with electrically active transverse phonons, or "dressed" photons, meaning they propagate together with lattice polarization.

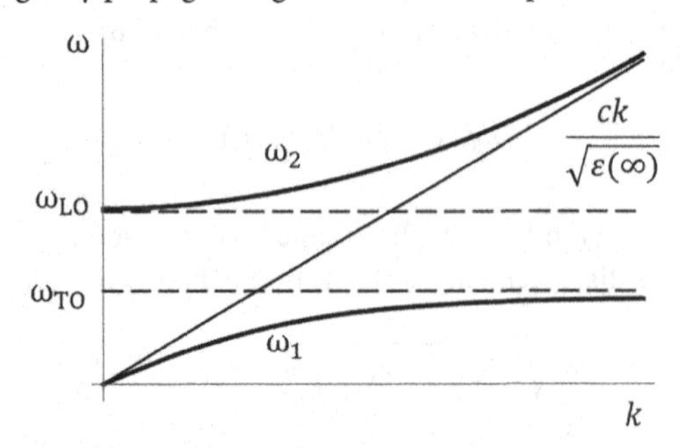

FIGURE 12.8 Polariton dispersion.

12.8 Optical Spin Orientation and Spin-Galvanic Effect

In previous sections, we discussed optical transitions using a simplistic model of a semiconductor which does not account for the spin–orbit interaction. In many practically important semiconductors, the spin–orbit interaction is large enough to form the energy spectrum of a material. Examples are zinc-blende GaAS, wurtzite GaN, and a variety of III-V ternary alloys. Illustrated in Figure 12.9 is the energy spectrum of GaAs in the vicinity of Γ-point that follows from the Kohn–Luttinger approximation.

Bloch amplitudes of the valence band in Γ-point are linear combinations of atomic p-orbitals, each corresponds to a certain total angular momentum and its z-component $|J, j\rangle$. Heavy and light valence bands correspond to total momentum $J = L + S = 3/2$, where L, S are the orbital and spin momentum, respectively. In spin–orbit split band $J = 1/2$. Bloch amplitudes of the conduction band in the center of the Brillouin zone are the s-type spherically symmetric functions $|iS\rangle|\uparrow\downarrow\rangle$ which correspond to $L = 0, S = 1/2, J = 1/2$. All Kohn–Luttinger Bloch amplitudes are listed below:

$$
\begin{aligned}
&\left|\frac{3}{2},\frac{3}{2}\right\rangle; \; u_{hh\uparrow} = -\frac{1}{\sqrt{2}}|(X+iY)\rangle|\uparrow\rangle, \\
&\left|\frac{3}{2},-\frac{3}{2}\right\rangle; \; u_{hh\downarrow} = \frac{1}{\sqrt{2}}|(X-iY)\rangle|\downarrow\rangle, \\
&\left|\frac{3}{2},\frac{1}{2}\right\rangle; \; u_{lh\uparrow} = -\frac{1}{\sqrt{6}}|(X+iY)\rangle|\downarrow\rangle + \sqrt{\frac{2}{3}}|Z\rangle|\uparrow\rangle, \\
&\left|\frac{3}{2},-\frac{1}{2}\right\rangle; \; u_{lh\downarrow} = \frac{1}{\sqrt{6}}|(X-iY)\rangle|\uparrow\rangle + \sqrt{\frac{2}{3}}|Z\rangle|\downarrow\rangle, \\
&\left|\frac{1}{2},\frac{1}{2}\right\rangle; u_{so\uparrow} = \frac{1}{\sqrt{3}}|(X+iY)\rangle|\downarrow\rangle + \frac{1}{\sqrt{3}}|Z\rangle|\uparrow\rangle, \\
&\left|\frac{1}{2},-\frac{1}{2}\right\rangle; u_{so\downarrow} = \frac{1}{\sqrt{3}}|(X-iY)\rangle|\uparrow\rangle - \frac{1}{\sqrt{3}}|Z\rangle\downarrow, \\
&\left|\frac{1}{2},\pm\frac{1}{2}\right\rangle; \; u_{c\uparrow\downarrow} = i|S\rangle|\uparrow\downarrow\rangle,
\end{aligned}
\tag{12.100}
$$

where wave functions comprise spherical harmonics multiplied by spin functions $|\uparrow\downarrow\rangle$.

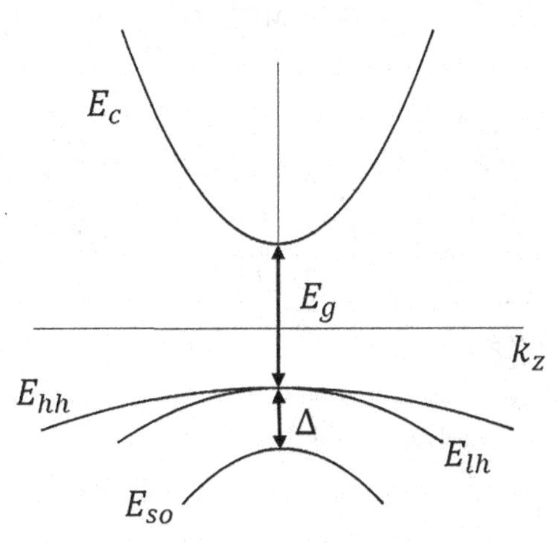

FIGURE 12.9 Four double-degenerate bands in GaAs and Δ is the spin–orbit splitting parameter.

Let us look at how the spin–orbit interaction $(\Delta \neq 0)$ affects optical transitions between bands shown in Figure 12.9. Transition rate is determined by matrix element $|\boldsymbol{e}\cdot\boldsymbol{p}_{cv}|^2$ (see Eq. (12.70)) which can be calculated using (12.100). Assuming circularly polarized wave propagates in the z -direction, we present a polarization unit vector as $\boldsymbol{e}=\boldsymbol{e}_x \pm i\boldsymbol{e}_y$, $\boldsymbol{e}_{x,y}$ are vectors in x and y directions, respectively. Dot product becomes $\boldsymbol{e}\cdot\boldsymbol{p}_{cv}=p_{cv}^x \pm i p_{cv}^y$, where signs $\pm$ label right- and left-handed polarization. Matrix element for a transition between states $u_{lh\downarrow}$ and $|S\uparrow\rangle$ follows:

$$M1^+ = |u_{lh\downarrow}\rangle p_x - ip_y|S\rangle|\uparrow\rangle = \frac{i}{\sqrt{6}}\left[\langle\uparrow|\uparrow\rangle\int d^3r\left(xp_x + yp_y\right)S(r)\right]$$

$$+i\sqrt{\frac{2}{3}}\langle\downarrow|\uparrow\rangle\int d^3r\left(z\,p_x + z\,p_y\right)S(r). \tag{12.101}$$

The second term in Eq. (12.101) is zero due to the spatial symmetry of wave functions: $p_y S \sim \partial S/\partial y \sim y$, and $p_x S \sim x$, so the integral turns zero. The last term is zero due to the orthogonality of spin functions: $\langle\uparrow|\uparrow\rangle=\langle\downarrow|\downarrow\rangle=1, \langle\uparrow|\downarrow\rangle=0$. What remains is

$$M1^+ = \frac{2i}{\sqrt{6}}\langle X|p_x|S\rangle. \tag{12.102}$$

Matrix element M1$^+$ describes $j=-1/2 \to j=1/2$ transition induced by a wave of circular polarization $\boldsymbol{\sigma}_+ = \boldsymbol{e}_x - i\boldsymbol{e}_y$. As that polarized photon carries angular momentum +1, $M1^+$ increases angular momentum by one, thus expressing angular momentum conservation in optical transition. All transitions, which increase angular momentum under $\boldsymbol{\sigma}_+$ illumination, are shown in Figure 12.10a.

Similarly, with polarization choice $\boldsymbol{\sigma}_- = \boldsymbol{e}_x + i\boldsymbol{e}_y$, all allowed transitions get the angular momentum decreased by -1 as illustrated in Figure 12.9b. The example matrix element is

$$M1^- = -\frac{2i}{\sqrt{6}}\langle X|p_x|S\rangle. \tag{12.103}$$

Once linear polarization is written as a sum $\boldsymbol{\sigma}_+ + \boldsymbol{\sigma}_-$, the matrix element becomes

$$M1^+ + M1^- = 0.$$

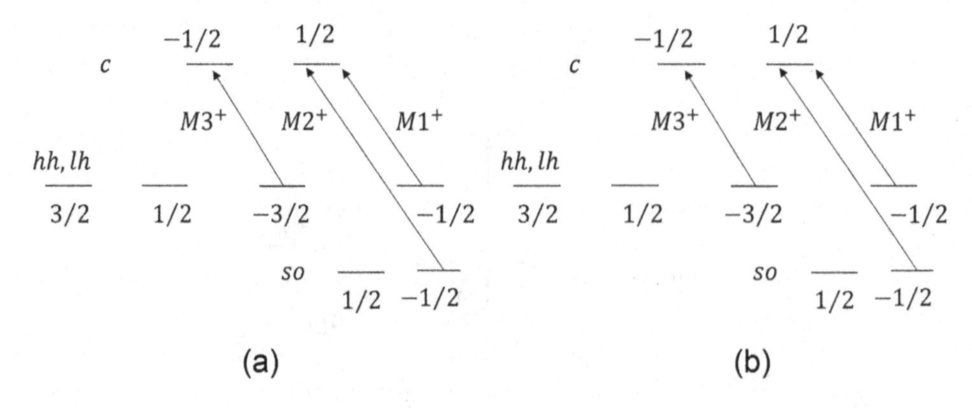

FIGURE 12.10 (a) Allowed transitions under σ_+-polarized illumination. (b) σ_--polarization.

Calculating matrix elements corresponding to transitions in Figure 12.10a one obtains

$$M2^+ = \frac{2i}{\sqrt{3}} \langle X | p_x | S \rangle, \quad M3^+ = \frac{2i}{\sqrt{2}} \langle X | p_x | S \rangle. \tag{12.104}$$

Transition rates are proportional to square matrix elements (12.102) and (12.104):

$$\begin{aligned} &I_1^+ \sim \left| M1^+ \right|^2 = \frac{2}{3} \left| \langle X | p_x | S \rangle \right|^2, \quad I_2^+ \sim \frac{4}{3} \left| \langle X | p_x | S \rangle \right|^2, \quad I_3^+ \sim 2 \left| \langle X | p_x | S \rangle \right|^2, \\ &I_1^+ : I_2^+ : I_3^+ = 1:2:3. \end{aligned} \tag{12.105}$$

We define the spin polarization of optically excited electrons as

$$P = \frac{n_\uparrow - n_\downarrow}{n_\uparrow + n_\downarrow} = \frac{I^+ - I^-}{I^+ + I^-}, \tag{12.106}$$

where $\uparrow\downarrow$ are spin states along the direction of the excitation beam and $n_{\uparrow\downarrow}$ are the numbers of excited conduction electrons which at the moment of creation are proportional to transition rates $I^\pm$. The degree of spin polarization depends on a wavelength. If all three processes indicated in Figure 12.10a $\left(\hbar\omega > E_g + \Delta\right)$ contribute to absorption, the spin polarization is zero as the conduction states with up-and down spins are equally populated: $n_\uparrow \sim 1+2$ (transitions $M1^+$ and $M2^+$) and $n_\downarrow \sim 3$ (transition $M3^+$). It also underlines the role of spin–orbit interaction: if spin splitting was absent ($\Delta = 0$), no spin polarization would exist in the process of interband absorption.

If $E_g + \Delta > \hbar\omega > E_g$, transitions $M1^+$ and $M3^+$ are involved in optical absorption and the degree of spin polarization along the excitation direction becomes

$$P_0 = \frac{I_1^+ - I_3^+}{I_1^+ + I_3^+} = \frac{1-3}{1+3} = -\frac{1}{2}. \tag{12.107}$$

Under continuous illumination, P_0 is the initial spin imbalance before spin relaxation processes bring spin polarization to its equilibrium value. The minus sign means that the spin orientation is opposite to the direction of incoming light. The inverse process is the interband recombination of electrons and holes. The degree of circular polarization in luminescence is the signature of spin polarization of nonequilibrium carriers. Counting transition rates for interband processes in Figure 12.10 with $E_g + \Delta > \hbar\omega > E_g$ and using (12.107), one gets the degree of circular polarization:

$$P_L^0 = \frac{I^+ - I^-}{I^+ + I^-} = \frac{3n_\downarrow + n_\uparrow - 3n_\uparrow - n_\downarrow}{3n_\uparrow + n_\downarrow + 3n_\downarrow + n_\uparrow} = -\frac{1}{2}\frac{n_\uparrow - n_\downarrow}{n_\uparrow + n_\downarrow} = \frac{1}{4}. \tag{12.108}$$

In a wider spectral range, $\hbar\omega > E_g + \Delta$, luminescence is not circularly polarized.

Expression (12.108) is valid if no spin relaxation occurs during the lifetime of excited electrons. The direction around which precession of oriented spins occurs may change randomly due to spin-orbit interaction or exchange interaction with magnetic impurities (see [17] for details). After spin relaxation time τ_S the initial axis of spin precession is completely forgotten, so a more realistic expression reads as

$$P_L = \frac{P_L^0}{1 + \tau / \tau_S}, \tag{12.109}$$

where τ is the lifetime of the nonequilibrium carrier with respect to recombination.

12.8.1 Spin Relaxation

The spin–orbit interaction couples spin of an electron to its momentum. If bands are split due to spin–orbit interaction, as it is in III-V semiconductors (see Figure 12.9), a potential that violates lattice periodicity (impurities or lattice vibrations) scatters electrons and also mixes states in c- and v-bands with opposite spins at the same momentum $\boldsymbol{k} \neq 0$. So, scattering is accompanied by spin relaxation (Elliot–Yafet (EY) mechanism). So, necessary ingredients for EY spin relaxation are the spin–orbit band splitting, an impurity or phonon scattering, and a finite electron momentum since the spin–orbit interaction is felt by a moving electron:

$$\frac{1}{\tau_S} \sim \frac{1}{\tau_p}\left(\frac{\Delta}{E_g + \Delta}\right)^2\left(\frac{E_k}{E_g}\right)^2, \tag{12.110}$$

where τ_p is the momentum relaxation time and E_k is the electron energy respective to the band extremum.

Another relaxation mechanism exists in wurtzite semiconductors as well as in a variety of structurally asymmetric quantum wells and heterostructures where inversion asymmetry and spin–orbit interaction lift spin degeneracy in each $\boldsymbol{k} \neq 0$ state, so the electron spectrum is Kramers degenerate: $E(\boldsymbol{k},\uparrow) = E(-\boldsymbol{k},\downarrow)$. The $\boldsymbol{k}$-dependent spin-splitting is equivalent to the effective momentum-dependent magnetic field. During the time between collisions, an electron moves in the effective magnetic field which randomly changes direction upon each act of scattering thus randomizing the axis of spin precession. If scattering frequency $1/\tau_p$ is larger than the spin precession frequency between collisions Ω, the regime presents the Dyakonov–Perel (DP) mechanism of spin relaxation:

$$\frac{1}{\tau_S} \sim \Omega^2 \tau_p, \tag{12.111}$$

Spin relaxations by EY and DP mechanisms can be discerned by their dependence on momentum relaxation time meaning different behavior with impurity concentration.

One more channel of spin dephasing is the Bir–Aronov–Pikus (BAP) mechanism, which relies on magnetic exchange interaction between electrons and holes. The interaction induces spin-flip scattering of electrons by holes providing the valence band is split by spin–orbit interaction. BAP may become a dominant mechanism in heavily p-doped III-V semiconductors. The relative importance of various regimes of spin relaxation is discussed in Refs. [17,18].

12.8.2 Spin-Galvanic Effect

As far as we deal with circularly polarized light, it is instructive to note that symmetry considerations do not exclude the relation between the current and the square electric field in an electromagnetic wave,

$$j_i = i\beta_{il}\left[\boldsymbol{E} \times \boldsymbol{E}^*\right]_l, \quad \boldsymbol{E} = \boldsymbol{e}\ \exp(-i\omega t), \tag{12.112}$$

where $\boldsymbol{e}$ is the complex-valued polarization vector. The right-hand side in (12.112) is proportional to the degree of circular polarization in a transverse electromagnetic wave: $i\left[\boldsymbol{E} \times \boldsymbol{E}^*\right] = \boldsymbol{n}E_0^2 P$, where n is the unit vector in the direction of propagation. Under inversion symmetry operation, the polar vector $\boldsymbol{j}$ changes sign while $\boldsymbol{E} \times \boldsymbol{E}^*$ does not, so tensor $\beta_{i\ln}$ is zero in centrosymmetric crystals and circular photogalvanic effect (12.112) may take place in systems lacking inversion symmetry. Under time inversion operation, both $\boldsymbol{j}$ and the cross-product in (12.112) change signs, so β_{il} remains untouched, meaning the effect is dissipationless. The photogalvanic effect has been theoretically described and experimentally demonstrated in various bulk materials and heterostructures (for a review, see [19,20]).

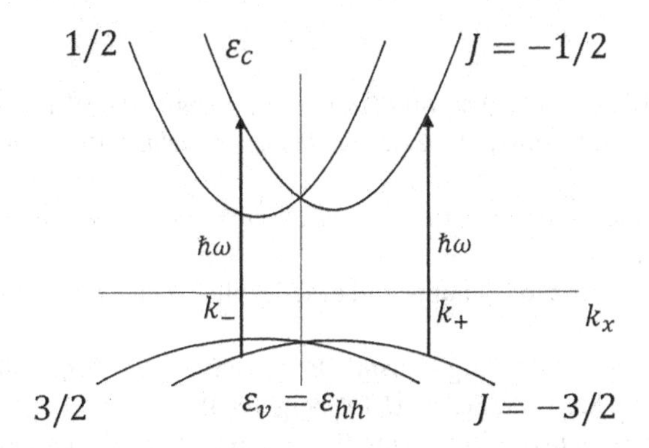

FIGURE 12.11 Optical transitions in the spin-split spectrum. Imbalance in the momentum of electrons $k_\pm$ creates the net electron flux in the x-direction. Transition corresponds to $M3$ shown in Figure 12.4a.

Expression (12.112) is the linear relation between polar vector $\boldsymbol{j}$ and axial-vector $\boldsymbol{E}\times\boldsymbol{E}^*$. This is possible in noncentrosymmetric crystals with gyrotropic point symmetry [21]. Spin polarization, discussed in the previous section, is also induced by $\boldsymbol{E}\times\boldsymbol{E}^*$. Thus, symmetry allows a similar relation between $\boldsymbol{j}$ and axial-vector of spin polarization $\boldsymbol{P}$:

$$j_i = Q_{ik}P_k, \tag{12.113}$$

Expression (12.113) is the phenomenological description of the spin-galvanic effect. Symmetry equivalence in Eqs. (12.112) and (12.113) means that the spin-galvanic current induced by circularly polarized light always occurs along with the photogalvanic effect. To illustrate the microscopic mechanism of the spin-galvanic effect, let us consider a noncentrosymmetric semiconductor system with k- linear spin splitting. Various implementations of the system include wurtzite bulk crystals, wurtzite quantum wells, as well as structurally asymmetric zinc-blende quantum wells.

As an example, we consider the $\boldsymbol{k}$-linear spin-split spectrum in conduction and heavy hole bands:

$$\varepsilon_c = E_g + \frac{\hbar^2 k_x^2}{2m_c} \pm \alpha_c k_x;\ \varepsilon_v = -\frac{\hbar^2 k_x^2}{2m_v} \mp \alpha_v k_x\ , \tag{12.114}$$

where $\alpha_{c,v}$ are the Rashba parameters for conduction and valence bands. The spectrum (12.114) and interband transitions are illustrated in Figure 12.11.

Electron momenta after excitation follow from condition $\varepsilon_c(-1/2) - \varepsilon_v(-3/2) = \hbar\omega$:

$$k_\pm = \frac{m_r}{\hbar^2}\left[\alpha_c + \alpha_v \pm \sqrt{(\alpha_c + \alpha_v)^2 + 2\hbar^2(\hbar\omega - E_g)/m_r}\right], \tag{12.115}$$

Group velocities of excited electrons $\hbar v_c = \partial\varepsilon_c / \partial k_x$ are expressed as

$$\hbar v_{c+} = \frac{\hbar^2 k_+}{m_c} - \alpha_c,\ \hbar v_{c-} = \frac{\hbar^2 k_-}{m_c} - \alpha_c,$$
$$v_{c+} + v_{c-} = 2\frac{m_v\alpha_v - \alpha_c m_c}{\hbar(m_c + m_v)}, \tag{12.116}$$

Net electric current is proportional to the group velocity (12.116), and it is nonzero unless condition $m_v\alpha_v = \alpha_c m_c$ holds.

For a more detailed microscopic description of photo-galvanic and spin-galvanic effects, see Ref. [22].

References

1. Y. Aharonov and D. Rohrlich, *Quantum Paradoxes* (Wiley-VCH, Weinheim, 2005).
2. L.D. Landau and E.M. Lifshitz, *The Classical Theory of Fields* (4th ed., Butterworth-Heinemann, Oxford, 1975).
3. W.K.H. Panofsky and M. Phillips, *Classical Electricity and Magnetism* (2nd ed., Addison-Wesley, Boston, MA, 1962).
4. J.D. Jackson, From Lorenz to Coulomb and other explicit gauge transformations, *Am. J. Phys.* **70**, 917 (2002).
5. V.M. Agranovich and V. Ginzburg, *Crystal Optics with Spatial Dispersion*, Springer Series Solid State Science (vol. 42, Springer, Berlin, Heidelberg, 1984).
6. A. Anselm, *Introduction to Semiconductor Theory* (Prentice-Hall, London, 1981).
7. L.D. Landau, E.M. Lifshitz, and L.P. Pitaevskii, *Electrodynamics of Continuous Media* (2nd ed., Pergamon Press, New York, 1984).
8. V. Veselago, "The electrodynamics of substances with simultaneously negative values of ε and μ", *Sov. Phys. Uspekhi*, **10**, 509 (1968).
9. J.B. Pendry, "Negative refraction", *Contemp. Phys.*, **45**, 191 (2004).
10. N. Engheta and R.W. Ziolkowski (Eds.) *Metamaterials: Physics and Engineering Explorations* (Wiley-IEEE Press, New York, 2006, 414 p).
11. L.D. Landau and E.M. Lifshitz, *Quantum Mechanics: Non-Relativistic Theory* (vol. 3, 3rd ed., Pergamon Press, New York, 1977).
12. D. Jérome, T.M. Rice, and W. Kohn, "Excitonic insulator", *Phys. Rev.*, **158**, 462 (1967).
13. J. Stajic, "Probing an excitonic insulator", *Science*, **358**(6370), 1552 (2017).
14. P.Y. Yu and M. Cardona, *Fundamentals of Semiconductors: Physics and Material Properties* (3rd ed., Springer, Berlin, 2001).
15. M. Abramovitz and I.E. Stegun (Eds), *Handbook of Mathematical Functions* (National Bureau of Standards, Gaithersburg, MD, 1964).
16. E. Rosencher and B. Vinter, *Optoelectronics* (Cambridge University Press, Cambridge, 2002).
17. M.I. Dyakonov (Ed.), *Spin Physics in Semiconductors*, Springer Series in Solid State Sciences (vol. 157, 2nd ed., Springer International Publishing, Berlin, Germany, 2017).
18. G.E. Pikus and A.N. Titkov, "Spin relaxation under optical orientation in semiconductors," In: *Modern Problems in Condensed Matter Science*, v. 8, Ed. by F. Meier and B.P. Zakharchenya (North-Holland, Amsterdam, 1984).
19. V.I. Belinicher and B.I. Sturman, "The photogalvanic effect in media lacking a center of Symmetry", *Sov. Phys. Uspekhi* **23**, 199 (1980).
20. E.L. Ivchenko and S.D. Ganichev, "Spin-dependent photogalvanic effects (a review)", arXiv:1710.09223v1 [cond-mat.mes-hall] (2017).
21. E.L. Ivchenko, *Optical Spectroscopy of Semiconductor Nanostructures* (Alpha Science, Harrow, 2005).
22. S.D. Ganichev and W. Prettl, "Spin photocurrents in quantum wells", *J. Phys. Condens. Matter* **15**, R935 (2003).

13

Nonequilibrium Electrons and Holes

In intrinsic semiconductors, at $T = 0$ and without external perturbations, there are no carriers in the conducting and valence bands. If at finite temperature an electronic system of a semiconductor is in thermal equilibrium with the lattice and the background electromagnetic field, the electron and hole densities n_0 and p_0, respectively, appear due to thermal generation. Recombination, the process in which electrons and holes collide and annihilate, balances the thermal generation providing for equilibrium densities of electrons and holes in a steady state.

In extrinsic semiconductors, conduction and valence bands, also impurity energy levels in the band-gap, are involved in the balance between thermal generation and recombination. The process includes band-to-band as well as band-to-impurity transitions.

In addition to thermal generation, an optical pumping or (and) electron injection from adjacent metal creates excess carriers δn and δp in the conduction and valence bands, respectively. Excess carriers recombine through various channels of two types: radiative and nonradiative.

Interband recombination occurs as a radiative process, direct or indirect (with phonons involved). Nonradiative band-to-band recombination is hardly probable, as it releases the energy larger than E_g and would require a large number of phonons to preserve energy conservation in the process. This type of nonradiative process becomes more probable in narrow-gap semiconductors.

Interband nonradiative recombination in intrinsic semiconductors occurs if the released energy of the order of E_g does not radiate as a photon but instead accelerates the third particle- electron or hole. The process is called *Auger recombination*. It involves three particles – two electrons and a hole or two holes and an electron-thereby providing energy and momentum conservation.

Other nonradiative processes are the band-to-impurity and the surface recombination. Some of the recombination mechanisms are discussed below in more detail.

13.1 Lifetime of Nonequilibrium Carriers: Phenomenological Approach

The concept of *lifetime* describes the time the particle dwells in the same quantum state. In solids, the quantum states of an electron discern by energy, momentum, and band index. Elastic scattering on crystal imperfections conserves electron energy while changes momentum: one says it determines the lifetime in a state with a particular momentum – the mean time interval between two elastic collision events. Inelastic scattering on phonons changes both energy and momentum and then determines the lifetime related to loss of coherence (phase) in the wavefunction. The time a carrier dwells in the band is finite because of various recombination processes. Recombination channels arise from electron interaction with light, impurities, phonons, and electron–electron Coulomb repulsion. On the time scale, the recombination time is of the order of 10^{-9} s, which is much longer than intraband scattering times $\left(\approx 10^{-12}\ \text{s}\right)$, so nonequilibrium carriers undergo fast intraband thermalization before they recombine.

In this chapter, we study recombination lifetime and population kinetics of carriers injected in a semiconductor.

DOI: 10.1201/9780429285929-13

13.1.1 Direct Band-to-Band Recombination

The lifetime parameter is introduced phenomenologically, in other words, regardless of the microscopic mechanisms of generation and recombination, and also whether the recombination is radiative or not. In band-to-band processes, generation and recombination mean the creation and annihilation of e–h pairs. Then, the rate of electron-hole recombination, the number of pairs annihilated per unit time per unit volume, is proportional to electron and hole concentrations n and p, respectively,

$$R = \gamma_R(T)np, \tag{13.1}$$

where $n = n_0 + \delta n$, $p = n_0 + \delta n$, $\gamma_R(T)$ is the temperature-dependent recombination coefficient $\left[\mathrm{m}^3\ \mathrm{s}^{-1}\right]$, which follows from microscopic theory discussed later. Generation rate comprises thermal G_t and excess carrier generation G. In equilibrium, $G = 0, n = n_0, p = p_0$ and balance between generation and recombination (13.1) casts to

$$G_t = \gamma_R(T)\, n_0 p_0. \tag{13.2}$$

In intrinsic semiconductors, $n_0 p_0 = n_i^2$ (see Eq. (4.23) Chapter 4), so the bandgap, temperature, and density of states at the edges of c- and v-bands determine the thermal generation rate.

Once excess carriers are generated, their density $\delta n = \delta p$ follow temporal evolution expressed as

$$\frac{\partial(\delta n)}{\partial t} = G_t + G - R. \tag{13.3}$$

Substituting (13.2) into (13.3) and turning off generation G at $t = 0$, one gets the equation that controls the relaxation process for weak deviation from equilibrium, $(\delta n)^2 \ll n_0 p_0$:

$$\frac{\partial(\delta n)}{\partial t} = \gamma_R(T)\left[n_0 p_0 - np\right] \approx -r(T)\, \delta n(n_0 + p_0), \tag{13.4}$$

with initial condition $t = 0, \delta n = \delta n(0)$. The solution has the form

$$\ln \frac{\delta n(t)}{\delta n(0)} = -\gamma_R(T)(n_0 + p_0)t,$$

or

$$\delta n(t) = \delta n(0) \exp\left(-\frac{t}{\tau_R}\right), \quad \tau_R = \frac{1}{\gamma_R(T)\left(n_0 + p_0\right)}. \tag{13.5}$$

Under the steady-state generation, excess carrier density does not depend on time, $\delta n(t) = \delta n(0)$, and from (13.3), one gets

$$G = \gamma_R(T)\delta n(0)(n_0 + p_0) = \frac{\delta n(0)}{\tau_R}. \tag{13.6}$$

When generation is turned on up to a steady-state level suddenly at $t=0$ the difference between $\delta n(0)=G\tau_R$ and (13.5) shows how the lifetime determines the temporal dynamics when excess carrier density tends to steady-state value:

$$\delta n(t)=G\tau_R\left[1-\exp\left(-\frac{t}{\tau_R}\right)\right]. \tag{13.7}$$

If the excitation is periodic, $G=G_0+G_1\sin(\omega t)$, Eq. (13.3) takes the form:

$$\frac{\partial(\delta n)}{\partial t}=\gamma_R(T)(n_0p_0-np)+G_0+G_1\sin(\omega t)\approx-\frac{\delta n}{\tau_R}+G_0+G_1\sin(\omega t). \tag{13.8}$$

Solution to (13.8) with initial condition $\delta n(t)=0$ is expressed as

$$\delta n(t)=G_0\tau_R\left[1-\exp\left(-\frac{t}{\tau_R}\right)\right]+G_1\tau_R\left[\frac{1}{2}\sin(2\varphi)\exp\left(-\frac{t}{\tau_R}\right)+\cos(\varphi)\sin(\omega t-\varphi)\right],$$

$$tg\ \varphi\equiv\omega\tau_R. \tag{13.9}$$

Since the moment of generation on ($t=0$), the lifetime regulates relaxation to the pure oscillating regime, which begins at $t\gg\tau_R$:

$$\delta n(t)=G_0\tau_R+G_1\tau_R\cos(\varphi)\sin(\omega t-\varphi). \tag{13.10}$$

Excess carrier density follows the periodic in time excitation with the phase shift φ, which corresponds to the delay due to finite lifetime. If $\omega\tau_R\ll 1$, then $\varphi\approx 0$ and density oscillates following the excitation with no time delay. In the opposite limit, $\omega\tau_R\gg 1, \varphi\approx\pi/2$, carriers cannot respond to high-frequency excitation, so the excess carrier density does not depend on time and relies on steady-state excitation component G_0.

For a high level of excitation, the relaxation process does not follow an exponential law, so that recombination lifetime is ill-defined. The concept of a lifetime and its role in population kinetics described above are not specific to band-to-band recombination and equally applicable to band-to-impurity processes of generation and capture of nonequilibrium carriers in extrinsic semiconductors.

13.1.2 Schokley–Read–Hall Recombination

In doped semiconductors, impurity centers with energy level E_t in the bandgap may capture carriers from c,v-bands and at finite temperature may release them to the bands. Capture and release (emission) correspond to the band-to-impurity and impurity to-band transitions, respectively. Rendered in Figure 13.1 are transitions that constitute Shockley–Read–Hall (SRH) processes.

Electrons and holes are captured independently by empty and filled traps, respectively. As empty and filled traps have different charge states, capture rates are different for electrons and holes. If an empty trap is neutral relative to the lattice (acceptor-like), it becomes negatively charged when filled with an electron. Thus, the capture of a hole is more probable than that of an electron because hole capture is assisted with Coulomb attraction while electron capture is not. Vice versa, if the filled trap is neutral (donor-like), the electron capture rate is larger than the hole capture rate. Different capture rates for electrons and holes imply separate kinetic equations similar to Eq. (13.3):

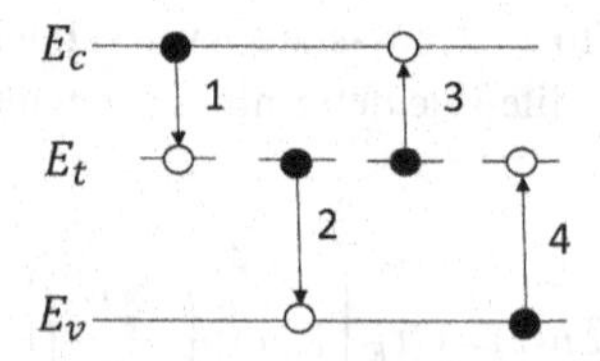

FIGURE 13.1 SRH processes. Electron (1) and hole (2) recombination. Electron (3) and hole (4) emission. Filled circles denote states occupied with electrons; arrows show electron transitions.

$$\frac{\partial(\delta n)}{\partial t} = G_{tn} + G - R_n$$

$$\frac{\partial(\delta p)}{\partial t} = G_{tp} + G - R_p, \tag{13.11}$$

where G is the external source of electron-hole excitation. Subscripts n and p denote phenomenological constants for electrons and holes, respectively.

Electron band-to-impurity recombination (capture) rate is proportional to the density of empty traps and electron concentration in the conduction band. Hole capture rate is proportional to the density of traps filled with electrons and hole concentration in the valence band:

$$R_n = c_n n N_t \left(1 - f_t\right),$$

$$R_p = c_p p N_t f_t, \tag{13.12}$$

where N_t and f_t are the trap concentration and electron filling factor, respectively. In equilibrium, f_{t0} is the Fermi function of an impurity level (see Chapter 4). As capture implies scattering of electrons and holes against impurity, the capture rate $c_{n(p)}\left[\mathrm{m^3/s}\right]$ can be presented naturally as scattering cross-section multiplied by electron (hole) velocity, averaged over Boltzmann distribution for nondegenerate carriers: $c_{n(p)} = \sigma_{n(p)}\left\langle v_{n(p)}\right\rangle, \left\langle v_{n(p)}\right\rangle = \sqrt{3k_BT/m} \approx 10^5$ m / s.

The reverse process (emission) can be written similarly: electron emission goes from filled traps while hole emission requires empty levels:

$$G_{tn} = e_n N_t f_t,$$

$$G_{tp} = e_p N_t \left(1 - f_t\right), \tag{13.13}$$

where $e_{n(p)}$ are the emission coefficients. In equilibrium, thermal emission and recombination compensate one another:

$$e_n f_{t0} = c_n n_0 \left(1 - f_{t0}\right),$$

$$e_p \left(1 - f_{t0}\right) = c_p p_0 f_{t0},$$

and then

$$e_n = c_n n_0 \frac{1-f_{t0}}{f_{t0}} \equiv c_n n_1,$$

$$e_p = c_p p_0 \frac{f_{t0}}{1-f_{t0}} \equiv c_p p_1. \tag{13.14}$$

In semiconductors doped with g-fold degenerate traps (see Eq. (4.32)),

$$f_{t0} = \frac{1}{1+g^{-1}\exp\left(\frac{E_t-\mu_0}{k_B T}\right)},$$

$$\left(1-f_{t0}\right) f_{t0}^{-1} = g^{-1}\exp\left(\frac{E_t-\mu_0}{k_B T}\right). \tag{13.15}$$

Substituting n_0 and p_0 from (4.17) and (4.20), we calculate n_1 and p_1 defined in (13.14):

$$n_1 = n_0 \frac{1-f_{t0}}{f_{t0}} = n_0\, g^{-1}\exp\left(\frac{E_t-\mu_0}{k_B T}\right) = N_c\, g^{-1}\exp\left(\frac{E_t-E_c}{k_B T}\right),$$

$$p_1 = p_0 \frac{f_{t0}}{1-f_{t0}} = N_v\, g\exp\left(\frac{E_v-E_t}{k_B T}\right) \tag{13.16}$$

Using (13.12), (13.13), and (13.14) one rewrites kinetic equation (13.11) as follows:

$$\frac{\partial(\delta n)}{\partial t} = c_n N_t\left[n_1 f_t - n\left(1-f_t\right)\right] + G,$$

$$\frac{\partial(\delta p)}{\partial t} = c_p N_t\left[p_1\left(1-f_t\right) - p f_t\right] + G. \tag{13.17}$$

Under the nonequilibrium steady-state condition, $\partial(\delta n)/\partial t = \partial(\delta p)/\partial t = 0$, right-hand sides in (13.17) are equal. As effective recombination rates for electrons and holes are equal, one can introduce electron–hole pair recombination $R_n - G_{tn} = R_p - G_{tp} \equiv R$.

Equating the right-hand sides in (13.17), one obtains occupation probability of traps:

$$f_t = \frac{c_n n + c_p p_1}{c_n\left(n_1+n\right) + c_p\left(p_1+p\right)}. \tag{13.18}$$

Using f_t and n_1, p_1 from (13.16) we obtain an effective recombination rate R as

$$R = c_n c_p N_t\left[\frac{np-n_i^2}{c_n\left(n_1+n\right)+c_p\left(p_1+p\right)}\right] \equiv \frac{np-n_i^2}{\tau_p\left(n_1+n\right)+\tau_n\left(p_1+p\right)},$$

$$n_i^2 = n_0 p_0 = N_c N_v \exp\left(-\frac{E_g}{k_B T}\right), \quad \tau_{n(p)} = \frac{1}{N_t c_{n(p)}}, \tag{13.19}$$

where capture constants $c_{n(p)}$ are being replaced with lifetimes for electrons and holes $\tau_{n(p)}$.

At high excitation level, $\delta n \approx \delta p \gg n_0, p_0;\ np \gg n_i^2$, it follows from (13.19):

$$R \approx \frac{\delta n}{\tau_n + \tau_p}, \tag{13.20}$$

where $\tau_R = \tau_n + \tau_p$ is the e–h recombination time which is controlled by carriers with lower capture rate.

At low excitation level, $\delta n \approx \delta p \ll n_0$, recombination rate (13.19) tends to SRH expression,

$$R = \frac{\delta n}{\tau_R}; \quad \tau_R^{-1} \approx \frac{n_0 + p_0}{\tau_p (n_1 + n_0) + \tau_n (p_1 + p_0)}, \tag{13.21}$$

where the lifetime depends on equilibrium carrier densities and traps filling factors.

To this point, we have considered SHR recombination on deep traps in nondegenerate semiconductors. Traps do not affect equilibrium carrier densities. However, besides traps, doped semiconductors contain shallow donors and (or) acceptors, making a particular type of equilibrium carriers dominant. In n-type semiconductors, $n_0 \gg p_0$, at low excitation level $\delta n \approx \delta p \ll n_0$. As chemical potential lies above trap levels, almost all traps are occupied with electrons and $n_0 \gg n_1, p_1$. This limit in (13.19) gives $R \approx \delta p/\tau_p$ meaning that it is the lifetime of the holes that determine the e–h pair recombination in n-type semiconductors. Similarly, in p-type semiconductors, the electron lifetime controls e–h recombination.

If surface termination and surface defects create energy levels in the bandgap, they act as traps. Recombination rate due to the surface traps – the number of recombination acts per unit area per second is $R_S = S\,\delta n$, where S[m/s] is the *surface recombination velocity.* This expression is the low-excitation limit of (13.19), where sheet trap density is used instead of volume density. Rate R_S multiplied by the surface area A gives the lifetime relative to surface recombination: $\tau_S^{-1} = AR_S$.

13.1.3 Auger Recombination

At a high excitation level, the density of nonequilibrium carriers increases to a point when three-particle interaction acts become probable. Thus the nonradiative interband e–h recombination includes a third particle (electron or hole), which absorbs the momentum and energy transfers. Auger processes are rendered in Figure 13.2. Abbreviations indicate bands involving in the process: conduction band (C), heavy hole band (H), light hole band (L), and spin-split band (S). Various Auger processes involving electron transitions between C, H, L, and S bands, are denoted in Fig. 13.2.

At low excitation levels, the CHCC Auger processes occur in n-type semiconductors, while in p-type materials, the CHHL and CHHS ones are more effective. However, if the injection level is high ($\delta n \approx \delta p \gg n_0, p_0$) like in semiconductor lasers, all Auger processes mentioned above take place irrespectively to the type of material. In III-V binary semiconductors and ternary alloys, dominant channels of the Auger recombination are CHCC and CHHS. In narrow-gap semiconductors such as GaSb and InAs, CHHS processes dominate due to resonance at $E_g \approx \Delta$. As three particles are involved, the recombination rate is proportional to $n^2 p$ or $p^2 n$, depending on which channel we deal with:

$$R_{CHCC} = \gamma_{CHCC}\, n^2 p; \quad R_{CHHS} = \gamma_{CHHS}\, p^2 n; \quad R_{CHHL} = \gamma_{CHHL}\, p^2 n, \tag{13.22}$$

where recombination coefficients $\gamma\,[\mathrm{m}^6/\mathrm{s}]$ are expressed below through the recombination rates in equilibrium R^0. Taking the CHCC process as an example, one gets

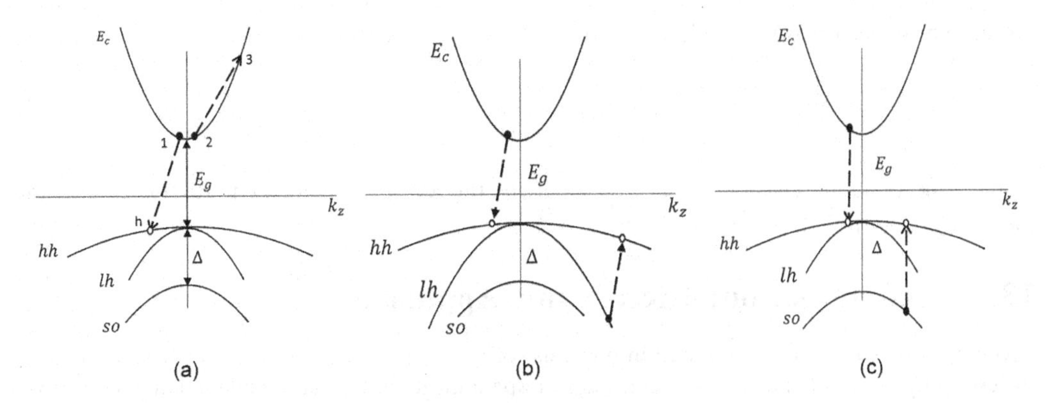

FIGURE 13.2 Auger recombination processes. Arrows denote electron transitions from the filled state (filled circle) to an empty one (open circle). (a) CHCC: c – hh recombination energy goes to accelerate an electron in the c -band. (b) CHHL: recombination excites a hole from $hh_{\_}$ to the lh-band. (c) CHHS: recombination excites a hole from hh- to so-band.

$$R^0_{CHCC} = \gamma_{CHCC} n_0^2 p_0, \quad \gamma_{CHCC} = \frac{R^0_{CHCC}}{n_0^2 p_0}, \tag{13.23}$$

where R^0_{CHCC} follows from microscopic calculations discussed later in the text. Equilibrium carrier concentrations depend on the type of material. In nondegenerate semiconductors, $n_0 p_0 = n_i^2$.

Processes, inverse to those shown in Figure 13.2, correspond to *impact ionization* – electron-electron collision in which a particle loses energy exciting an e–h pair. Ionization rate per unit volume due to energetic electrons (holes) is $\beta_n n$ ($\beta_p p$). After neglecting other generation and recombination processes, the population balance equation between impact ionization and Auger recombination has the form,

$$\frac{\partial(\delta n)}{\partial t} = \beta_n n + \beta_p p - \gamma_n n^2 p - \gamma_p p^2 n, \tag{13.24}$$

In equilibrium, the right hand side in (13.24) equals zero: $\beta_n = \gamma_n n_0 p_0$, $\beta_p = \gamma_p n_0 p_0$, where n_0, p_0 are the equilibrium densities. In intrinsic semiconductors, $n_0 = p_0 = n_i$. Under weak deviation from equilibrium, $\delta n = \delta p \ll n_0, p_0$,

$$\frac{\partial(\delta n)}{\partial t} = -(n_0 + p_0)(\gamma_n n_0 + \gamma_h p_0)\delta n = -\frac{\delta n}{\tau}. \tag{13.25}$$

Limits in (13.25) correspond to various materials:

intrinsic semiconductors,

$$\tau_i = \frac{1}{2n_i^2(\gamma_n + \gamma_p)}, \tag{13.26}$$

n-type semiconductors, $n_0 \gg p_0$,

$$\tau_n = \frac{1}{\gamma_n n_0^2},$$

p-type semiconductors, $p_0 \gg n_0$,

$$\tau_p = \frac{1}{\gamma_p p_0^2}. \tag{13.27}$$

At high injection rate typical for lasers, $n = p \gg n_0, p_0$, the Auger recombination rate (13.22) takes the form $R = (\gamma_n + \gamma_p)n^3$.

13.2 Recombination: Microscopic Approach

Recombination coefficients introduced in previous sections as phenomenological constants depend on material properties and interaction parameters corresponding to the type of recombination. For example, the radiative recombination, spontaneous and stimulated, is associated with the interaction between electrons and an electromagnetic field. Nonradiative trap capture is related to electron-impurity scattering, while Auger recombination relies on Coulomb interaction between electrons or holes. To calculate phenomenological constants, we have to use a microscopic theory. The theory requires handling electrons and the electromagnetic field based on quantum mechanics and quantum electrodynamics, respectively.

13.2.1 Radiative Recombination

Optical absorption discussed in Chapter 12 is an example of how a quantum mechanical electron system interacts with an external electromagnetic field represented by a classical plane wave propagating in a semiconductor. Within the classical approach to light-matter interaction, the interaction is absent after the electromagnetic field is off. If no electromagnetic field in a semiconductor, it is not clear why an electron injected into the upper energy level (conduction band) would recombine with a hole, emitting a photon (*spontaneous emission*) and thus showing the presence of an interaction that has ceased to exist. In other words, an excited electron would live the c-band indefinitely unless some inelastic interaction came into effect. This conclusion contradicts experiments that indicate a finite lifetime of the excited state in atoms and pure semiconductors associated with direct radiative recombination. The situation clarifies in quantum electrodynamics, which is an application of quantization to the electromagnetic field. The quantum approach treats a free electromagnetic field as the discrete number of photon modes (see Appendix 3, https://www.routledge.com/Modern-Semiconductor-Physics-and-Device-Applications/Dugaev-Litvinov/p/book/9780367250829# for details). In the quantum description, vector-potential (A3.11) looks as follows:

$$A(r,t) = \frac{1}{2}\sqrt{\frac{2\hbar}{\Omega\varepsilon_0}}\sum_{q,\alpha}\frac{1}{\sqrt{\omega_q}}\left[e_{q\alpha}c_{q\alpha}e^{(iq\cdot r - i\omega t)} + e_{q\alpha}^{*}c_{q\alpha}^{+}e^{(-iq\cdot r + i\omega t)}\right], \tag{13.28}$$

Making use (13.28) one obtains electron–photon interaction in the form similar to (12.53) (Chapter 12):

$$H_1 = \frac{i\hbar e}{m}A\cdot\nabla = F + F^{+},$$

$$F = i\sqrt{\frac{e^2\hbar^3}{2m^2\Omega\varepsilon_0}}\sum_{q,\alpha}\frac{1}{\sqrt{\omega_q}}\exp(iq\cdot r - i\omega_q t)(e_{q\alpha}\cdot\nabla)c_{q\alpha} \tag{13.29}$$

Note that gradient in (13.29) does not act on $\exp(iq\cdot r)$ as $e_{q\alpha}\cdot\nabla$ contains only gradient components perpendicular to r.

Perturbation (13.29) induces electron transitions with simultaneous photon emission and absorption. In Chapter 12, we discussed transitions due to electron interaction with a classical electromagnetic field. There are some features of these transitions that find explanation only within a quantum description.

Let us consider upward and downward transitions with the rates given by expressions (12.56) and (12.57), respectively. In the spirit of the perturbation approach, the matrix element calculates on unperturbed Bloch functions. As we deal with a quantized electromagnetic field (see Appendix 3), https://www.routledge.com/Modern-Semiconductor-Physics-and-Device-Applications/Dugaev-Litvinov/p/book/9780367250829# the photon wave functions – Fock states – come into play. The ground state of the system "electrons+photons" is the product of Bloch and Fock states: $\Psi = \varphi_{jk}\left|\ldots n_{q\alpha}\ldots\right\rangle$. Accordingly, one has to replace the matrix element as

$\left\langle \varphi_{ck}\left|\boldsymbol{A}_0\cdot\nabla\right|\varphi_{vk}\right\rangle \rightarrow \left\langle \Psi_c\left|\boldsymbol{A}_0\cdot\nabla\right|\Psi_v\right\rangle$, and examining photon emission, account for an additional photon in the final state $\Psi_v = \varphi_{vk}\left|n_{q\alpha}+1\right\rangle$ as compared to initial state $\Psi_c = \varphi_{ck}\left|n_{q\alpha}\right\rangle$. For downward transition, the first term in H_1 is the resonant one, so given relation $\left\langle n_{q\alpha}\left|c_{q\alpha}\right|n_{q\alpha}+1\right\rangle = n_{q\alpha}+1$, we obtain emission rate per unit volume as

$$R_{cv} = \frac{2\pi\hbar^2 e^2}{\Omega^2 m^2 \varepsilon_0}\sum_{k,k',q,\alpha}\frac{n_{q\alpha}+1}{\omega_q}\left|\left\langle \varphi_{ck}\left|e^{i(q\cdot r)}\boldsymbol{e}_{q\alpha}\cdot\nabla\right|\varphi_{vk'}\right\rangle\right|^2 \delta\left(E_v(\boldsymbol{k}')-E_c(\boldsymbol{k})+\hbar\omega_q\right)f_c(\boldsymbol{k})\left(1-f_v(\boldsymbol{k}')\right). \tag{13.30}$$

Momentum conservation (see Chapter 12) reduces (13.30) to

$$R_{cv} = \frac{2\pi e^2}{\Omega^2 m^2 \varepsilon_0}\sum_{k,q,\alpha}\frac{n_{q\alpha}+1}{\omega_q}\left|\boldsymbol{e}_{q\alpha}\cdot\boldsymbol{p}_{cv}\right|^2 \delta\left(E_v(\boldsymbol{k})-E_c(\boldsymbol{k}+\boldsymbol{q})+\hbar\omega_q\right)f_c(\boldsymbol{k}+\boldsymbol{q})\left(1-f_v(\boldsymbol{k})\right). \tag{13.31}$$

Similarly, in the upward transition, we account for the presence of additional photon in initial state $\Psi_v = \varphi_{vk}\left|n_{q\alpha}+1\right\rangle$ as compared to the final one, $\Psi_c = \varphi_{ck}\left|n_{q\alpha}\right\rangle$. This time the resonance term is the second term in H_1, and thus transition rate contains matrix element $\left\langle n_{q\alpha}+1\left|c^+_{q\alpha}\right|n_{q\alpha}\right\rangle = n_{q\alpha}$:

$$R_{vc} = \frac{2\pi e^2}{\Omega^2 m^2 \varepsilon_0}\sum_{k,q,\alpha}\frac{n_{q\alpha}}{\omega_q}\left|\boldsymbol{e}_{q\alpha}\cdot\boldsymbol{p}_{cv}\right|^2 \delta\left(E_v(\boldsymbol{k})-E_c(\boldsymbol{k}+\boldsymbol{q})+\hbar\omega_q\right)f_v(\boldsymbol{k})\left(1-f_c(\boldsymbol{k}+\boldsymbol{q})\right). \tag{13.32}$$

Expressions (13.31) and (13.32) imply field is in the free space. In a semiconductor, Eq. (13.28) modifies such as $\varepsilon_0 \rightarrow \varepsilon\varepsilon_0$, $\omega_q \rightarrow vq$, $v = c/n_r$, n_r is the refractive index. If then we assume a single mode in the cavity is in classical regime $n_{q\alpha} \gg 1$, the difference $\left(R_{vc}-R_{cvv}\right)/N_{q\alpha}v$, $N_{q\alpha} = n_{q\alpha}/\Omega$, coincides with the absorption coefficient (12.70) if vertical transitions are assumed.

There are two contributions to the rate of downward transitions (13.31) corresponding to terms $n_{q\alpha}+1$. The first term, proportional to $n_{q\alpha}$, relies on photons in the cavity describing *stimulated emission*. The sum over the photon modes $q\alpha$ contains the terms $n_{q\alpha} \neq 0$, in which the emitted photons are in the $q\alpha$ -state, they are coherent with those already present in the cavity. The second term, not depending on $n_{q\alpha}$, does not require photons in the cavity and describes *spontaneous emission* – the quantum-in-nature effect originating from vacuum fluctuations, as discussed in Appendix 3. The sum is not restricted in this case, meaning that a broad spectrum of photons constitutes spontaneous emission.

Assuming Ω large enough and using dispersion relation $\omega = vq$, one could replace discrete sum over photon modes $q\alpha$ by the frequency integral. Let's consider the total density of photons of two polarizations in the cavity of volume Ω:

$$N = \frac{1}{\Omega}\sum_{q\alpha} n_{q\alpha} = \frac{2}{(2\pi)^3}\int_0^\infty q^2\,dq\int_0^\pi \sin\theta\,d\theta\int_0^{2\pi} n_q\,d\varphi = \int_0^\infty n(\omega)\rho_{ph}(\omega)\,d\omega, \tag{13.33}$$

where $\rho_{ph}(\omega) = \omega^2/v^3\pi^2$ is the *photon density of states* per frequency interval. In thermal equilibrium, $n(\omega) = \left[\exp(\hbar\omega/k_BT)-1\right]^{-1}$. The volume density of photons of frequency ω per frequency interval has the form

$$N(\omega) \equiv n(\omega)\rho_{ph}(\omega) = \frac{\omega^2}{\mathrm{v}^3\pi^2\left[\exp(\hbar\omega/k_BT)-1\right]}. \tag{13.34}$$

Going back to (13.31) and (13.32), we have

$$\frac{1}{\Omega}\sum_{q\alpha}(\ldots)\left|\boldsymbol{e}_{q\alpha}\cdot\boldsymbol{p}_{cv}\right|^2 \to \frac{2}{3}\left|\boldsymbol{p}_{cv}\right|^2\int_0^\infty(\ldots)\rho_{ph}(\omega)d\omega, \tag{13.35}$$

Integral accounts for two polarization directions, $\boldsymbol{e}_1$ and $\boldsymbol{e}_2$, and calculated by choosing the z-axis along $\boldsymbol{p}_{cv}$, as shown in Figure 13.3. Factor 2 / 3 in (13.35) stems from dot product $\left|\boldsymbol{e}_{q\alpha}\cdot\boldsymbol{p}_{cv}\right|^2$, which gives additional factor $\sin^2\theta$ in integral on angles.

Integration on ω in (13.31) and (13.32) gives the rates of spontaneous and up-(down) stimulated transitions per unit volume expressed as

$$R = C\sum_{\boldsymbol{k}} r(\boldsymbol{k}), \quad C = \frac{4e^2}{3\pi\hbar^2\Omega m^2\varepsilon_0\mathrm{v}^3},$$

$$\hbar\omega_{\boldsymbol{k}} = E_c(\boldsymbol{k}) - E_{\mathrm{v}}(\boldsymbol{k}),$$

$$r_{sp}(\boldsymbol{k}) = \hbar\omega_{\boldsymbol{k}}\left|\boldsymbol{p}_{cv}\right|^2 f_c(\boldsymbol{k})\left(1-f_{\mathrm{v}}(\boldsymbol{k})\right),$$

$$r_{st}^{dn}(\boldsymbol{k}) = n_{ph}(\omega_{\boldsymbol{k}})\hbar\omega_{\boldsymbol{k}}\left|\boldsymbol{p}_{cv}\right|^2 f_c(\boldsymbol{k})\left(1-f_{\mathrm{v}}(\boldsymbol{k})\right),$$

$$r_{st}^{up}(\boldsymbol{k}) = n_{ph}(\omega_{\boldsymbol{k}})\hbar\omega_{\boldsymbol{k}}\left|\boldsymbol{p}_{cv}\right|^2 f_{\mathrm{v}}(\boldsymbol{k})\left(1-f_c(\boldsymbol{k})\right), \tag{13.36}$$

where $n_{ph}(\omega_{\boldsymbol{k}})$ is the photon occupation number of state $\hbar\omega_{\boldsymbol{k}}$.

If one supposes detailed thermal equilibrium between photons and electrons (at each photon energy $\hbar\omega_{\boldsymbol{k}}$), the rates of downward transitions emitting photons and upward ones absorbing photons are equal. As the downward transition rate comprises two parts - stimulated and spontaneous - the equilibrium condition is given by

$$r_{sp}(\boldsymbol{k}) + r_{st}^{dn}(\boldsymbol{k}) = r_{st}^{up}(\boldsymbol{k}). \tag{13.37}$$

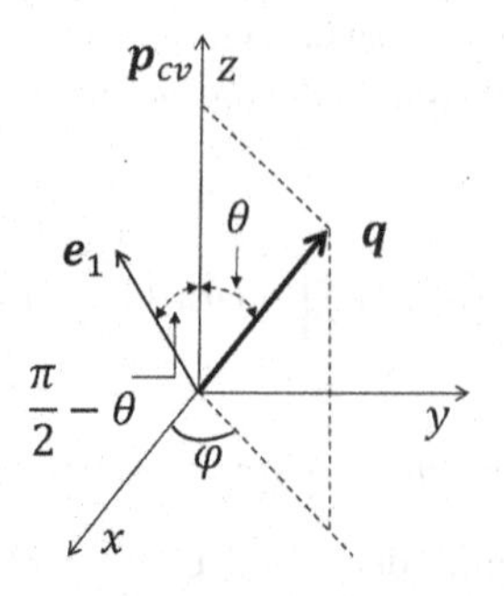

FIGURE 13.3 The coordinate system used in integral (13.35).

Substituting (13.36) into (13.37) and solving for n_{ph}, one obtains the equilibrium number of photons in mode ω_k: $n_{ph}(\omega_k) = \left[\exp(\omega_k/k_BT) - 1\right]^{-1}$, which is the Bose–Einstein occupation factor.

For absorption coefficient $\alpha = C\sum_k \beta(k)$, from (13.36), one gets the relation to spontaneous recombination,

$$\beta(k) \equiv \Omega \frac{r_{st}^{up}(k) - r_{st}^{dn}(k)}{v n_{ph}(\omega_k)} = \frac{\Omega}{\mathrm{v}} \hbar\omega_k |p_{cv}|^2 \left[f_v(k) - f_c(k)\right],$$

$$r_{sp}(k) = \frac{\mathrm{v}}{\Omega} \frac{f_c(k)(1 - f_v(k))}{f_v(k) - f_c(k)} \beta(k), \tag{13.38}$$

If $f_{c,v}(k)$ are equilibrium Fermi functions,

$$\frac{f_c(k)(1 - f_v(k))}{f_v(k) - f_c(k)} = n_{ph}(\omega_k)$$

$$r_{sp}(k) = \frac{\mathrm{V}}{\Omega} n_{ph}(\omega_k)\beta(k). \tag{13.39}$$

The total spontaneous recombination rate is

$$R_{sp} = C\sum_k r_{sp}(k) = \frac{\mathrm{V}}{\Omega}\sum_k n_{ph}(\omega_k)\alpha(k) = \frac{1}{2\pi^2 \mathrm{v}^2} \int_{E_g/\hbar}^{\infty} \frac{\omega^2 \alpha(\omega)\, d\omega}{\exp(\hbar\omega/k_BT) - 1}. \tag{13.40}$$

where $\alpha(\omega) = C\beta(\omega)$ is the absorption coefficient.

It is instructive to consider a steady-state nonequilibrium electron system where carrier injection creates quasi-Fermi levels for electrons and holes $F_{c,v}$. That is what happens in semiconductor lasers and will be discussed in Chapter 15. These conditions imply electron occupation factors $f_{c,v} = \left\{\exp\left[(E_{c,v} - F_{c,v})/k_BT\right] + 1\right\}^{-1}$. Using nonequilibrium $f_{c,v}$ in (13.36) and taking into account that stimulated recombination is the net number of downward stimulated transitions at photon energy $\hbar\omega_k$, we find the ratio of stimulated and spontaneous recombination rates:

$$\frac{r_{st}}{r_{sp}} = \frac{r_{st}^{dn}(k) - r_{st}^{up}(k)}{r_{sp}(k)} = n_{ph}(\omega_k)\left[1 - \exp\frac{\hbar\omega_k - F_c + F_v}{k_BT}\right]. \tag{13.41}$$

As follows from (13.41), the stimulated emission may prevail over spontaneous recombination which is a necessary condition for lasing in semiconductors to be discussed in Chapter 16.

Calculations of radiative recombination rates in III–V and II–VI semiconductors with spectrum shown in Figure 13.2 give recombination coefficients:

in bulk materials [1],

$$\gamma_R^{Bulk} \approx \frac{e^2 E_g^2 \hbar}{\varepsilon_0 (k_BT)^{3/2} \mathrm{v}^3 m_c^{5/2}} \frac{E_g + \Delta}{3E_g + 2\Delta} \frac{\mu_h^{3/2} + \mu_l^{3/2}}{m_h^{3/2} + m_l^{3/2}}, \quad \left[\mathrm{m^3/s}\right]$$

$$\mu_{h,l} = \frac{m_{h,l} m_c}{m_{h,l} + m_c}, \tag{13.42}$$

and in quantum wells [2]:

$$\gamma_R^{QW} \approx \frac{e^2 E_g^2}{\varepsilon_0 m_c v^3 (m_c + m_h) k_B T}, \quad \left[\frac{m^2}{s}\right]. \tag{13.43}$$

The radiative lifetime related to spontaneous optical transitions depends on optical matrix element which, being proportional to the overlap between Bloch amplitudes in conduction and valence bands, is sensitive to relative positions of band edges in $\boldsymbol{k}$-space. Optical transitions in indirect-gap materials require phonons to absorb momentum transfer, thus making transitions less probable than those in direct-gap semiconductors. For moderately doped (10^{17} cm^{-3}) indirect-gap materials such as Si, Ge, and GaP, the lifetimes are 2.5×10^{-3} s, 1.5×10^{-4} s, and 3×10^{-3} s, respectively. In III–V direct-gap semiconductors, the lifetime is much shorter: $(0.04-0.24)\times10^{-6}$ s [3].

13.2.2 Auger Recombination: Bulk Semiconductors

The Auger recombination (and impact ionization) mechanism originates from the interaction between charged particles, so the recombination rate depends on the Coulomb interaction matrix element taken between the initial and final states of the colliding particles. Since the energy of particles depends on their momenta, the simultaneous conservation of energy and momentum in a collision leads to the existence of threshold energy ε_{th} – the minimum total energy of colliding particles necessary for the process to occur. The threshold determines the exponential temperature dependence of the recombination rate, $\gamma \sim \exp(-\varepsilon_{th}/k_B T)$. In narrow-gap semiconductor alloys $(\text{In},\text{Ga})\text{Sb}$, the threshold is low enough for Auger processes to become the main channel of recombination that has critical importance for the room-temperature performance of mid-infrared semiconductor lasers (see Chapter 15 for details).

Let us consider the threshold energy in the example CHCC process rendered in Figure 13.2a. We assume parabolic dispersion in c- and v-bands, take reference point at the valence band maximum, and write the energy and momentum conservation conditions as

$$E_g + \varepsilon_c(\boldsymbol{k}_1) + \varepsilon_c(\boldsymbol{k}_2) + \varepsilon_h(\boldsymbol{k}_h) = \varepsilon_c(\boldsymbol{k}_3),$$

$$\varepsilon_{c,h}(\boldsymbol{k}) = \frac{\hbar^2 k^2}{2m_{c,h}}. \tag{13.44}$$

$$\boldsymbol{k}_1 + \boldsymbol{k}_2 + \boldsymbol{k}_h = \boldsymbol{k}_3, \tag{13.45}$$

where $\boldsymbol{k}_{1,2,3}$ are the electron momenta in states 1,2,3; $\boldsymbol{k}_h$ is the hole momentum. The total kinetic energy of two electrons and a hole before a collision is

$$\varepsilon \equiv \varepsilon_c(\boldsymbol{k}_1) + \varepsilon_c(\boldsymbol{k}_2) + \varepsilon_h(\boldsymbol{k}_h). \tag{13.46}$$

The threshold energy for the Auger process is the minimum ε under conditions (13.43) and (13.44). In virtue of Eq. (13.44), the sum of vectors $\boldsymbol{k}_1, \boldsymbol{k}_2, \boldsymbol{k}_h$ must provide for the threshold value of $\boldsymbol{k}_3$. At the same time, minimum ε is reached with the smallest modulo momenta k_1, k_2, k_h. These two conditions can be fulfilled if, at the threshold, vectors $\boldsymbol{k}_1, \boldsymbol{k}_2, \boldsymbol{k}_h$ are parallel to each other and thus parallel to $\boldsymbol{k}_3$.

To estimate ε_{th}, one may apply simple physical arguments assuming initial momenta $\boldsymbol{k}_{1,2}$ are small. Corresponding kinetic energies are of the order of $k_B T$ or the Fermi energy in degenerate semiconductors, so they are much less than E_g and neglected in (13.44):

$$E_g + \varepsilon_h(\boldsymbol{k}_h) \approx \varepsilon_c(\boldsymbol{k}_3), \quad \boldsymbol{k}_h \approx \boldsymbol{k}_3. \tag{13.47}$$

As $m_h \gg m_c$, the threshold (minimum) momentum of the excited electron and threshold energy follows from (13.46) and (13.47):

$$\varepsilon_c(\mathbf{k}_3) \approx E_g, \quad \varepsilon_{th}^{CHCC} \approx \varepsilon_h(\mathbf{k}_3). \tag{13.48}$$

In parabolic spectrum, $\varepsilon_{th}^{CHCC} \approx E_g m_c/m_h$. Within the Kane model, the electron spectrum of practical semiconductors comprises parabolic *hh*-band and nonparabolic *c*-, *lh*-, and *so*-bands.

Applying (13.48) to a nonparabolic conduction band, one gets

$$\varepsilon_{th}^{CHCC} \approx \frac{m_c E_g(\Delta + 2E_g)(2\Delta + 3E_g)}{m_h(\Delta + 3E_g)(\Delta + E_g)}, \tag{13.49}$$

where Δ is the valence band spin-split energy. In both limits $\Delta \gg E_g$ and $\Delta \ll E_g$, Eq. (13.49) tends to $\varepsilon_{th}^{CHCC} \approx 2E_g m_c/m_h$.

To obtain the recombination coefficient for a particular Auger process, one needs to calculate the matrix element of Coulomb interaction between initial and final states in conduction and valence bands. First calculations assumed the matrix element independent of electron and hole momenta and resulted in an overestimated recombination rate [4,5]. The reason is that the matrix element turns zero at the threshold when all momenta of colliding particles are parallel. This fact is taken into account in a more rigorous approach which gives for CHCC process [6] (see Ref. [7] for a review):

$$\gamma_{CHCC} = a\left(\frac{m_c}{m_h}\right)^{3/2} \frac{e^4\hbar^3\langle\varepsilon_c\rangle}{(4\pi\varepsilon_0)^2(k_BT)^{3/2}E_g^{5/2}m_c^2}\exp\left(-\frac{\varepsilon_{th}^{CHCC}}{k_BT}\right),$$

$$a = 16\pi^{5/2}\left(\frac{E_g + \Delta}{3E_g + 2\Delta}\right)^{3/2}\left(\frac{3E_g + \Delta}{2E_g + \Delta}\right)^{1/2}, \tag{13.50}$$

where $\langle\varepsilon_c\rangle$ is the average electron energy, ε_F is the Fermi energy. Due to the large effective mass of holes, they are assumed nondegenerate. The pre-exponential coefficient in (13.50) contains factor ε_c, which is sensitive to conduction electron degeneracy: $\langle\varepsilon_c\rangle$ equals $3k_BT/2$ for nondegenerate or $3\varepsilon_F/5$ for degenerate electrons (see Eq. (4.14), Chapter 4).

In nondegenerate *p*-type semiconductors, the recombination coefficient for CHHS process (Figure 13.2c) takes the form [8]:

$$\gamma_{CHHS} \approx \begin{cases} \dfrac{27\pi^4 e^4\hbar^3 m_s^{5/2}(\Delta - E_g)}{5(4\pi\varepsilon_0)^2 m_c^{3/2} m_h^3 \Delta^2(E_g + \Delta)k_BT}\exp\left(-\dfrac{\Delta - E_g}{k_BT}\right), & \Delta - E_g \gg k_BT, \\ \dfrac{216\pi^{5/2}e^4\hbar^3 m_c(\Delta + E_g)\sqrt{k_BT}}{(4\pi\varepsilon_0)^2 m_s^2 m_h E_g^5\sqrt{\varepsilon_{th}^{CHHS}}}\exp\left(-\dfrac{\varepsilon_{th}^{CHHS}}{k_BT}\right), & E_g - \Delta \gg k_BT,\ \varepsilon_{th}^{CHHS} \gg T, \\ \dfrac{e^4\hbar^3 m_s}{(4\pi\varepsilon_0)^2 m_h^3 E_g^3} f\left(\dfrac{E_g - \Delta}{k_BT}\right), & |E_g - \Delta| \ll E_g \end{cases} \tag{13.51}$$

where m_s is the effective mass in the spin-split valence band,

$$\varepsilon_{th}^{CHHS} = \frac{m_s E_g^2(E_g - \Delta)}{m_h(E_g + \Delta)(3E_g - \Delta)}, \tag{13.52}$$

TABLE 13.1 CHCC Recombination Coefficient (13.49)

	GaAs	InP	GaSb	InAs	InSb
E_g, [eV] (300 K)	1.42	1.34	0.7	0.35	0.18
$\gamma_{CHCC}\left[\mathrm{m^6/s}\right]$	1.3×10^{-42}	1.7×10^{-39}	6×10^{-37}	3×10^{-33}	1.1×10^{-31}

and $f(x)$ is determined by numerical calculations in Ref. [8].

At a fixed temperature, the Auger recombination coefficient increases with a decreasing threshold. In narrow-gap alloys based on InAs and GaSb near the resonance $E_g \approx \Delta$, the CHHS channel of recombination becomes dominant as it has a low or no threshold at all.

For the CHCC process, the threshold is proportional to E_g, so the trend is illustrated in Table 13.1 as increasing γ_{CHCC} in a row of III–V binary compounds with decreasing bandgap.

A relatively large threshold in GaAs and InP suppresses the rates of direct Auger processes. However, the recombination may occur indirectly involving phonons or impurities, and in this case, becomes thresholdless. For more details on indirect Auger channels, see Ref. [7].

13.2.3 Auger Recombination. Quantum Wells

Auger currents in bulk materials include exponential temperature factor because of the threshold nature of the Auger transition. Both momentum and energy conservation conditions determine the threshold energy required to excite the third particle. The same theoretical consideration applies to QW structures with infinite barriers where the conservation laws involve in-plane momenta and thus result in Auger rates similar to those calculated for bulk materials. In practical heterostructures, finite-height barriers allow electrons and holes to penetrate inside the barrier material. In this case, electron momentum in the growth direction is ill-defined and thus cannot conserve in the course of tunneling in the normal to QW-plane direction. In other words, for part of electrons and holes which penetrate the barriers, momentum conservation is not required, so the recombination channel is thresholdless. Thus in a QW, there exist two channels of Auger recombination, one with bulk-like characteristics (with strict momentum conservation) and another one with no threshold. The thresholdless channel can be the dominant Auger mechanism below room temperature in thin QW of 1–3 nm width. Calculations performed in Ref. [2] for CHCC and CHHS processes give

$$\gamma_{CHCC}^{QW} \approx \frac{A}{k_B T}, \quad \left[\frac{\mathrm{m}^4}{\mathrm{s}}\right].$$

$$\gamma_{CHHS}^{QW} \approx \frac{A}{E_{0c}}\left(\frac{m_c}{m_s}\right)^{3/2}\left(\frac{E_g}{E_g-\Delta}\right)^{5/2}, \tag{13.53}$$

where A is the function of QW width as well as conduction and valence band offsets, E_{0c} is the first sublevel in the conduction band. The temperature dependence of Auger rates is used in Chapter 15 to discuss the semiconductor laser threshold.

References

1. B.L. Gel'mont and G.G. Zegrya, "Temperature dependence of threshold current density for an injection laser", *Sov. Phys. Semicond.* **25**, 1216 (1991).
2. G.G. Zegrya, A.D. Andreev, N.A. Gun'ko, and E.V. Frolushkina, "Calculation of QW laser threshold currents in terms of new channels of nonradiative Auger recombination", *Proc. SPIE.* **2399**, 307 (1995).

3. P. Yu and M. Cardona, *Fundamentals of Semiconductors* (Springer, Berlin, 2001).
4. A.R. Beattie and P.T. Landsberg, "Auger effect in semiconductors", *Proc. Royal Sci. London.* A249, 16 (1959)
5. N.K. Dutta and R.J. Nelson, "The case for Auger recombination in $In_{1-x}Ga_xAs_yP_{1-y}$", *J. Appl. Phys.* **53**, 74 (1982)
6. B. Gelmont and Z. Sokolova, "Auger recombination in direct-gap semiconductors", *Sov. Phys. Semicond.* **16**, 1067 (1982).
7. V.N. Abakumov, V.I. Perel, and I.N. Yassievich, *Nonradiative Recombination in Semiconductors* (North-Holland, Amsterdam, 1991).
8. B.L. Gelmont, Z.N. Sokolova, and I.N. Yassievich, "Auger recombination in direct-gap *p*-type semiconductors", *Sov. Phys. Semicond.* **16**, 592 (1982).

14

Schottky Contacts and p–n Junctions

Metals-semiconductors contacts, as well as contacts between p- and n-type semiconductors, are the key elements of various devices such as diodes, solar cells, transistors, photodetectors, chemical sensors, light-emitting diodes, lasers, and microwave devices.

Contacts between metals and semiconductors can be either low-resistance (ohmic) or Schottky-type, in which case they work as diodes providing for ac-rectification. The same material but doped with n- and p-type impurities form p–n-junctions, while contacts between different semiconductors represent heterojunctions - structural elements of quantum wells and remotely doped layers of high electron mobility.

When two samples, each in thermal equilibrium, are brought into contact, they exchange electrons and holes due to the difference of chemical potentials on both sides of the junction. Diffusion currents create the built-in electric field across the contact, which, once the external voltage is applied, results in nonlinear current-voltage characteristics, which enable device applications of metal (semiconductor)/semiconductor junctions.

Electron and hole dynamics in inhomogeneous semiconductor structures include diffusion, drift in an electric field, excess carrier generation, and recombination. Electric fields, arising self-consistently during this process, change on a scale significantly exceeding both the lattice constant and the electron wavelength. It allows us to treat carrier dynamics within the framework of macroscopic electrodynamics and continuity equations. Drift in an electric field, diffusion due to inhomogeneous carrier distribution, carrier generation, and recombination cast into the following system of nonlinear differential equations:

$$\frac{\partial n}{\partial t} = G_n - \frac{n - n_0}{\tau_n} + \frac{1}{q}(\nabla \cdot \boldsymbol{J}_n),$$

$$\frac{\partial p}{\partial t} = G_p - \frac{p - p_0}{\tau_p} - \frac{1}{q}(\nabla \cdot \boldsymbol{J}_p),$$

$$\boldsymbol{J}_n = q\mu_n n\boldsymbol{E} + qD_n(\nabla n),$$

$$\boldsymbol{J}_p = q\mu_p p\boldsymbol{E} - qD_p(\nabla p),$$

$$\nabla \cdot \boldsymbol{E} = \frac{\rho}{\varepsilon_0 \varepsilon}, \tag{14.1}$$

where q is the elementary charge, ε_0, ρ, (n, p), $\mu_{n,p}\left[\mathrm{m^2/Vs}\right]$, $G_{n,p}\left[1/\mathrm{m^3\,s}\right]$, and $\boldsymbol{E}$ are the elementary charge, vacuum permittivity, charge density, carrier densities, mobilities, generation rates, and electric field, respectively. Lifetimes of nonequilibrium carriers $\tau_{n,p}$ describe recombination processes

DOI: 10.1201/9780429285929-14

(see Chapter 13). Current densities $J_{n,p}\left[\mathrm{A/m^2}\right]$ comprise drift (first term) and diffusion (second term) components, ε is the medium permittivity, $D_{n,p}$ are the diffusion coefficients.

Electric charge conservation law $q\partial n/\partial t=(\nabla\cdot J_n)$ is violated in (14.1) by the presence of generation and recombination terms, which cause particle number nonconservation.

Equations (14.1) determine variables n, p, J_n, J_p, E providing other parameters are known. The system implies the medium is isotropic and, in its most general three-dimensional form, can be solved with numerical methods only. Fortunately, one-dimensional models allow analytical solutions and describe well the basic properties of metal/semiconductor and semiconductor/semiconductor junctions. That is what to discuss in the next sections.

14.1 Contacts Metal-Semiconductor

14.1.1 Energy Band Diagram

Various solids have different energies needed to tear off an electron and move it to energy level E_{vac} (to place an electron in a vacuum near the surface). Also, solids have different *work functions* W illustrated in Figures 14.1 and 14.2 – chemical potentials counted from the energy level E_{vac}. If the metal is in contact with a semiconductor, the difference in chemical potentials works as a thermodynamic force and induces the flow of particles. Flow direction depends on the relation between work functions of constituent materials.

If $W_m > W_s$ (see Figure 14.1), the net electron flow goes from semiconductor to metal, thus charging the contact positively from the semiconductor side and negatively from the metal side. In other words, an electric field appears to stop the flow.

In metals, nonequilibrium electrons exist for a short time, Maxwell relaxation time, $\tau_M=\sigma/\varepsilon\varepsilon_0$, σ is the conductivity. In good metals, $\tau_M\approx10^{-16}$ s. A large density of free charges effectively screens the electric field, so the spatial distribution of the electric field is essential only on a semiconductor side of the junction. Short Maxwell time prevents the creation of space charge, so electron concentration does not depend on distance from the contact, and thus diffusion current in metals is absent ($\nabla n=0$).

In equilibrium (no currents), the energy–distance diagram with the space charge region and barrier height from the semiconductor side, $V_b=W_m-W_s>0$, is shown in Figure 14.1b.

An inverse relation between W_m and W_s results in the diagram shown in Figure 14.2 where the electric field inverses its sign.

Nonrectifying (accumulation) contact shown in Figure 14.2 can serve as an ohmic contact that is the low-resistance contact with linear current-voltage characteristics. Unfortunately, for many technologically important metal/semiconductor pairs, the relation between work functions does not satisfy condition $W_m < W_s$. That is why ohmic contacts are the rectifying Schottky contacts but thin enough to

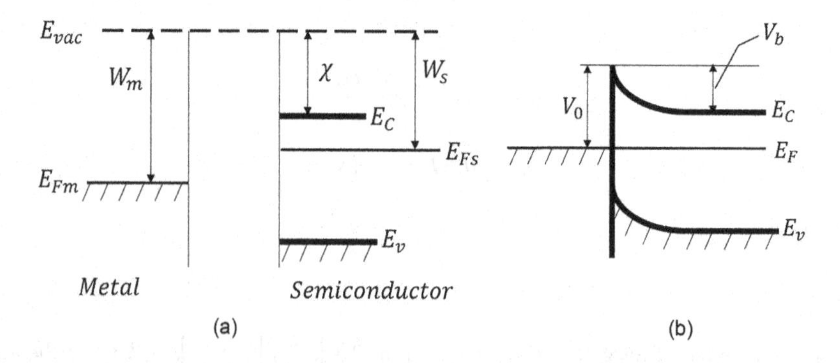

FIGURE 14.1 Metal-semiconductor rectifying contact, $W_m > W_s$. (a) Before contact. (b) After contact. V_b is the barrier height. V_0 is the Schottky barrier.

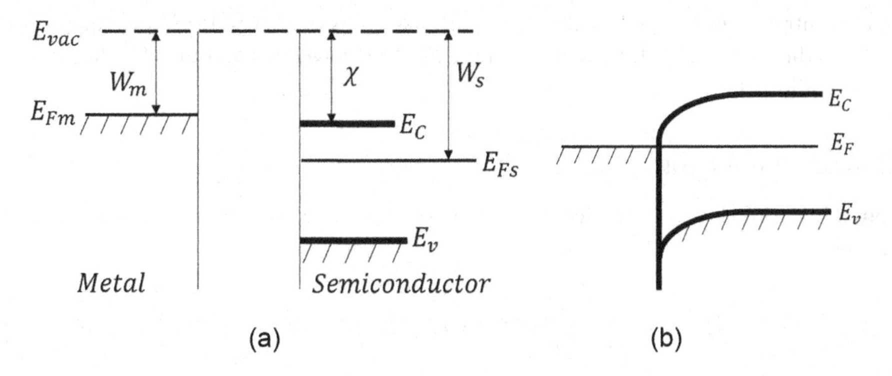

FIGURE 14.2 Metal-semiconductor accumulation contact, $W_m < W_s$. (a) Before contact. (b) After contact.

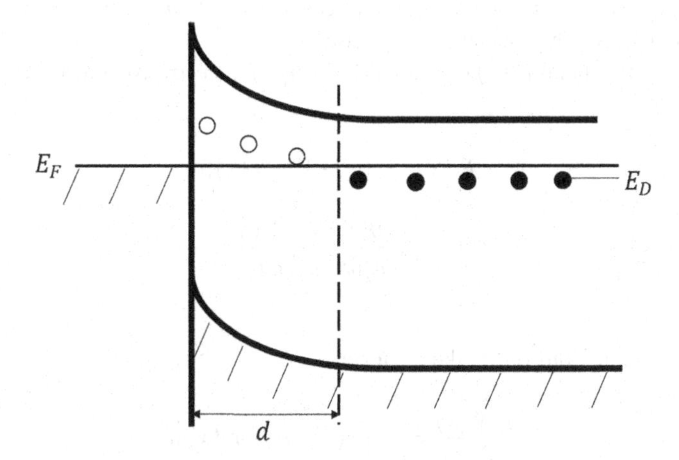

FIGURE 14.3 Space charge region in rectifying metal/*n*-semiconductor contact. E_D is the donor energy level. Hollow circles depict positively charged donors.

provide for electron tunneling and thus low resistance. Later we discuss minimum resistance of ohmic contact.

In non-degenerate semiconductors, the chemical potential does not coincide with the real electron energy and, along with the work function, depends on external fields, doping, and temperature. The parameter which is unambiguously related to a particular semiconductor is the *electron affinity* χ – the energy difference between the edge of the conduction band and the vacuum level.

After electrons leave for metal, ionized donors create the positively charged region called *depletion* or *space charge region*, as illustrated in Figures 14.2b and 14.3

The resulting electric field makes the band edges coordinate dependent, and V_b/q is the voltage drop on a depletion region. Barrier V_0 (see Figure 14.1b) for electrons flowing from metal to semiconductor, the *Schottky barrier*, is:

$$V_0 = W_m - \chi = V_b + E_c(\infty) - E_F, \tag{14.2}$$

where $E_c(\infty)$ is the edge of the conduction band outside the space charge region.

The picture described above corresponds to ideal metal-semiconductor contact and does not account for the quality of the interface. The relation between work functions on both sides of the junction as well as the electron affinity in semiconductors determines the Schottky barrier height as long as Fermi levels move freely when metal and semiconductor exchange electrons having brought in contact. Often, a large enough concentration of charged defects at the interface pins the Fermi level, and thus the interface charge affects the barrier height. For details on the *Bardeen model* that accounts for the pinning of the Fermi level, see Ref. [1].

14.1.2 Rectifying Contacts

In the one-dimensional model, the electric field distribution in the depletion region obeys the Poisson equation (14.1):

$$\frac{\partial E}{\partial x}=\frac{\rho}{\varepsilon_0\varepsilon}, \quad \rho=q\left[n_d^+(x)-n(x)-n_a\right], \tag{14.3}$$

where $n(x)$, $n_d^+(x)$, and n_a are the density of electrons, ionized (positively charged) donors, and acceptors, respectively. In an n-type semiconductor, the acceptor density n_a is small $(n_a \ll n_d)$, and all acceptors are negatively charged. The contact is placed at $x=0$.

The edge of conduction band $E_c(x)$ (the potential energy for electrons) and electric field E are given as

$$E_c(x)=E_c(\infty)+V_b-V(x),$$

$$E=-\frac{\partial E_c}{(-q)\partial x}=-\frac{\partial V(x)}{q\,\partial x}. \tag{14.4}$$

With these notations, equations (14.1) take the form:

$$\frac{\partial^2 V(x)}{\partial x^2}=\frac{q^2}{\varepsilon_0\varepsilon}\left[n(x)+n_a-n_d^+(x)\right],$$

$$J_n=-\mu_n n(x)\frac{\partial V(x)}{\partial x}+qD_n\frac{\partial n(x)}{\partial x}. \tag{14.5}$$

Equilibrium electron density in a nondegenerate semiconductor is expressed in Eq. (4.1) (Chapter 4):

$$n(x)=n_0\exp\left(-\frac{V_b-V(x)}{k_BT}\right), \quad n_0\equiv N_c\exp\left(\frac{E_F-E_c(\infty)}{k_BT}\right). \tag{14.6}$$

Concentration $n(x)$ and potential profile $V(x)$ describe depletion and band bending, respectively, and obey boundary conditions:

$$V(x)=\begin{cases} 0, & x=0, \\ V_b, & x\to\infty, \end{cases}$$

$$n(x)=\begin{cases} n_0\exp\left(-\frac{V_b}{k_BT}\right), & x=0, \\ n_0, & x\to\infty. \end{cases} \tag{14.7}$$

Schematically the potential profile $V(x)$ is shown in Figure 14.4. Exact coordinate dependence is a solution to the Poisson equation to be discussed below.

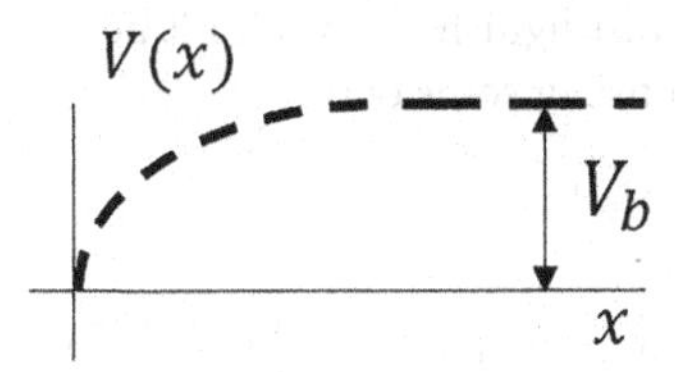

FIGURE 14.4 Potential profile in the space charge region.

Under external voltage, V_b shifts to become $V_b - qU$. A positive voltage implies plus of the dc source applied to metal.

Thermal equilibrium means that currents flowing through the contact compensate each other, and thus the net current is zero, in other words, there is a balance between drift and diffusion terms in (14.5):

$$\mu_n n(x)\frac{\partial V(x)}{\partial x} = qD_n\frac{\partial n(x)}{\partial x}, \tag{14.8}$$

Calculating the gradient of concentration (14.6), one obtains from (14.8)

$$\mu_n = \frac{qD_n}{k_BT}, \tag{14.9}$$

that is the *Einstein relation* between mobility and diffusion coefficients.

To calculate the charge on the right-hand side of Eq. (14.5), we use Boltzmann distribution (14.6) for band electrons and Fermi distributions for occupied, n_d^0, and ionized, n_d^+, donors (4.2) (Chapter 4):

$$n_d^+ = \frac{n_d}{1 + 2\exp\left(\dfrac{E_F - E_D}{k_BT}\right)}, \quad n_d^0 = \frac{n_d}{1 + \dfrac{1}{2}\exp\left(\dfrac{E_D - E_F}{k_BT}\right)}, \quad n_d^+ + n_d^0 = n_d. \tag{14.10}$$

It is convenient to represent (14.10) by finding $\exp(E_F/k_BT)$ from (14.6) and substituting in (14.10):

$$\frac{n_d^+}{n_d} = \frac{n_1}{n_1 + n(x)}, \quad \frac{n_d^0}{n_d} = \frac{n(x)}{n_1 + n(x)},$$

$$n_1 \equiv \frac{1}{2}N_c\exp\left(\frac{E_D - E_c}{k_BT}\right). \tag{14.11}$$

Note that donor ionization energy $E_D - E_c$ does not depend on x. From (14.11), one obtains

$$n(x) - n_d^+(x) = n(x) + n_a - \frac{n_1 n_d}{n_1 + n(x)}. \tag{14.12}$$

Expression (14.12) enters the r-h side of the Poisson equation and is valid for an arbitrary degree of donors ionization. Outside depletion region ($x \to \infty$), the space charge is zero, $n_d^+(x) \to n(x) = n_0$, and as follows from (14.7) and (14.12),

$$\frac{n_1 n_d}{n_1 + n_0} = n_0 + n_a. \tag{14.13}$$

Starting from this point, one may distinguish various limiting cases corresponding to different semiconductor parameters. Below we consider some of them.

14.1.3 Weak Band Bending

If $V_d \ll k_BT$, expanding (14.12) up to the first order on $\left[V_b - V(x)\right]/k_BT$ and using (14.13), one obtains

$$n(x) - n_d^+(x) \approx n_0 \frac{n_1 + n_a + 2n_0}{n_1 + n_0} \frac{V(x) - V_b}{k_BT}. \tag{14.14}$$

Finally, the Poisson equation reads as

$$\frac{\partial^2 V(x)}{\partial x^2} = \frac{V(x) - V_b}{L^2}, \quad L^2 \equiv \frac{\varepsilon_0 \varepsilon k_B T}{q^2 n_0} \frac{n_1 + n_0}{n_1 + n_a + 2n_0}, \quad \left[m^2\right]. \tag{14.15}$$

The solution to (14.15) with boundary conditions (14.7) has the form,

$$V(x) = V_b \left[1 - \exp\left(-\frac{x}{L}\right)\right]. \tag{14.16}$$

and is illustrated in Figure 14.4.

Length of the depletion region L depends on relative positions of chemical potential and donor level. If $E_F < E_D$, all donors are fully ionized, $n_1 \gg n_0$, and thus

$$L \approx \sqrt{\frac{\varepsilon_0 \varepsilon k_B T n_1}{q^2 n_0 (n_1 + n_a)}}. \tag{14.17}$$

In the limit $n_1 \gg n_a$, L becomes the Debye screening length $L_D = \sqrt{\varepsilon_0 \varepsilon k_B T / q^2 n_0}$. In this regime, depletion region length is determined by the equilibrium density of electrons far from the contact ($n_0 = n_d$). With decreasing n_0, the Debye length goes up until n_0 becomes lower than the charged impurity concentration, and then screening is determined by impurities. For example, when in compensated n-semiconductors the acceptor density becomes high enough ($n_0 < n_a$), the chemical potential at low temperatures is close to E_D, thus $n_1 \approx n_0$, and

$$L \approx \sqrt{\frac{2\varepsilon_0 \varepsilon k_B T}{q^2 (n_a + 3n_0)}}. \tag{14.18}$$

A further decrease in n_0 does not increase the screening length, which is determined now by the density of compensating impurities n_a.

14.1.4 Strong Band Bending

The regime of fully ionized donors was studied in (14.17) while assuming weak band bending. However, a more relevant case in real contacts is full donor ionization along with a strong near-interface electric field.

Donors are fully ionized if $E_D \geq E_F + 2k_BT$. The case corresponds to general expressions (14.12), (14.13) in the limit $n_1 \gg n(x)$:

$$\rho = q\left[n(x) - n_d + n_a\right],$$

$$n_d = n_0 + n_a,$$

or

$$\rho = q\left[n(x) - n_0\right]. \tag{14.19}$$

Full depletion means $n(x) \approx 0$ in the region $0 \le x \le L$, where L follows from the boundary conditions to Poisson equation,

$$\frac{\partial^2 V}{\partial x^2} = -\frac{q^2}{\varepsilon_0 \varepsilon} n_0,$$

$$x = 0, \quad V(x) = 0$$

$$x = L, \quad V(x) = V_b - qU. \tag{14.20}$$

The solution to (14.20) has the form:

$$V(x) = V_b - qU - \frac{q^2 n_0}{2\varepsilon\varepsilon_0}(L - x)^2, \tag{14.21}$$

and from $V(0) = 0$ we have

$$L = \sqrt{\frac{2\varepsilon\varepsilon_0(V_b - qU)}{q^2(n_d - n_a)}} \approx \sqrt{\frac{2\varepsilon\varepsilon_0(U_b - U)}{qn_d}}, \tag{14.22}$$

where we used n_0 from (14.19), neglected the density of compensating impurities, $n_a \ll n_d$, and replaced energy parameter V_b with voltage drop $U_b = V_b/q$. Forward bias corresponds to $U > 0$.

For example, for barrier height $V_b = 0.5\,\text{eV}$, $n_d = 10^{15}\ \text{cm}^{-3}$, and $\varepsilon \approx 10$, $L \approx 7 \times 10^{-5}$ cm, that is macroscopic length more than 100 times the lattice constant in most semiconductors. As seen from what follows, an important parameter is L as compared to the mean free path l in semiconductors. Both lengths are of the same order of magnitude, and the relation between them discerns two different mechanisms of current transmission through the junction.

14.1.5 Inverse Contact

Electrical current through a Schottky contact is carried out by majority carriers – electrons in metal/n-semiconductor contacts. Because of band bending, in the space charge region, there are holes of higher density as compared to the bulk (see Figure 14.5)

In an *inverse contact*, the hole density may exceed the electron density in the near-contact region, thus forming the *p–n* junction. It happens when $V_0 \ge E_g/2$ or $V_b + E_c - E_F \ge E_g/2$.

Assuming all donors are ionized, one obtains $V_b + E_c - E_F = V_b + k_B T \log(N_c/n_d)$. Expressing E_g through intrinsic carrier density n_i (see Chapter 4), one finds a condition preventing the formation of the inverse layer:

$$V_b \le k_B T \log\frac{\sqrt{N_v}\, n_d}{\sqrt{N_c}\, n_i}. \tag{14.23}$$

The restriction to V_b better holds in wide bandgap semiconductors where intrinsic carrier concentration is low. For example, assuming $n_d = 0.075 N_c$, $N_c \approx N_v$, at room temperature, we obtain $V_b < 0.26\,\text{eV}(\text{Ge})$, $V_b < 0.45\,\text{eV}(\text{Si})$, $V_b < 0.57\,\text{eV}(\text{GaAs})$.

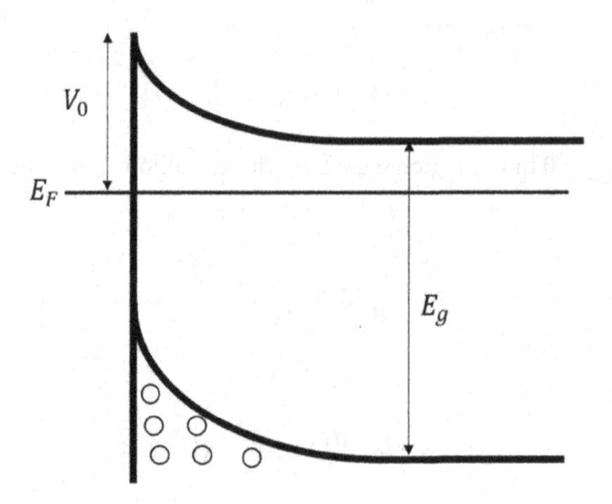

FIGURE 14.5 Minority carriers (holes) accumulated in the space charge region.

14.1.6 Current–Voltage Characteristics

The forward bias applied to the contact always forces majority carriers to flow from semiconductor to metal. For the band diagram, shown in Figure 14.6, the majority carriers are electrons (n-type semiconductor), so under the forward bias, electrons flow mostly from semiconductor to metal. In a p-type semiconductor, the forward bias would correspond to the opposite polarity of an external source (signs +,− in the picture).

The electron transport through the contact means that electrons overcome barriers of different heights: $V_b \pm qU$ from the semiconductor side, and V_0 from the metal side. Also, the barrier height depends on an applied voltage. All this makes the current-voltage characteristics non-linear and imparts rectifying properties to Schottky contacts.

Negative bias applied to a semiconductor decreases the barrier for electrons flowing to metal, so the forward current J_{sm} rises sharply, exceeds J_{ms}, and the difference $J = J_{sm} - J_{ms}$ forms the positive branch

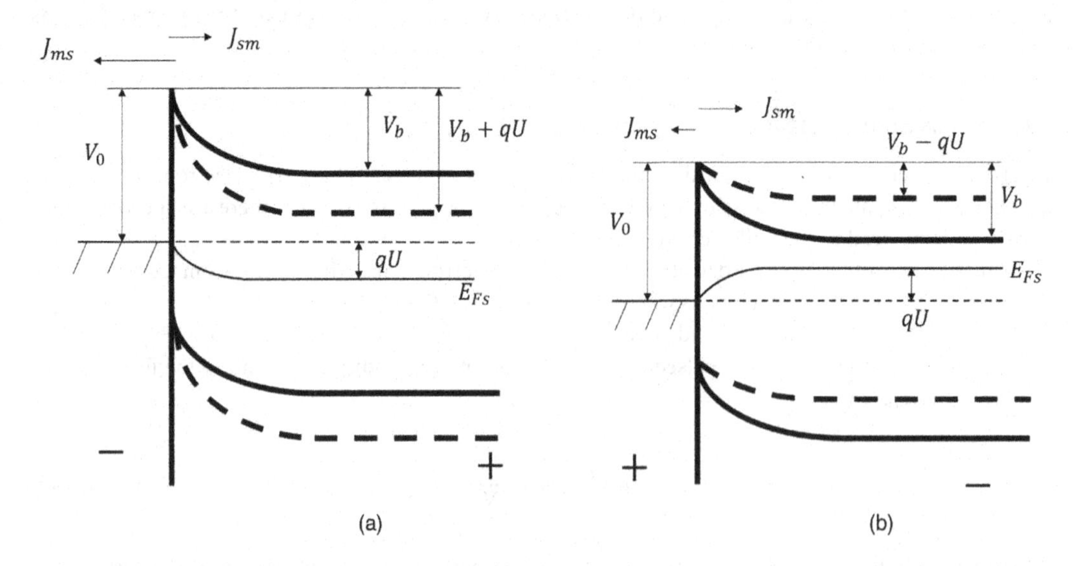

FIGURE 14.6 Metal/n-semiconductor contact under an applied voltage. (a) Reverse-biased. (b) Forward-biased. Arrows show currents direction, where J_{ms}, J_{sm} correspond to electrons flowing from metal to semiconductor and from semiconductor to metal, respectively.

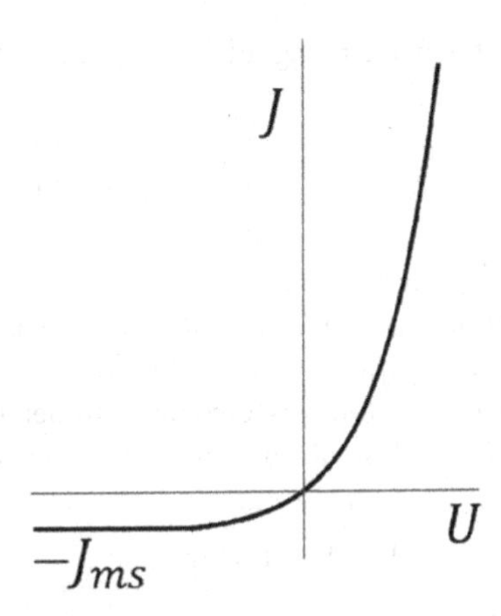

FIGURE 14.7 Current–voltage characteristics of rectifying Schottky contact.

of current-voltage characteristics. Reverse bias prevents electron flow from semiconductor, so the total current turns into a negative branch $J = -J_{ms}$, which is determined by V_0 and does not depend on an applied voltage. A qualitatively rendered *I-V* curve is given in Figure 14.7

To treat the *I-V* curve formally, we have to specify mechanisms of current transmission through the depletion region. There are two main regimes: one includes both drift and diffusion currents, the other operates with thermionic emission currents. To discern the two, it is instructive to remind that the very concept of diffusion implies that the characteristic scale of the concentration gradient is larger than the mean free path so that multiple collisions occur on a length of concentration drop. Formally, it reads as a small concentration change on the mean free path length l:

$$l\left|\nabla n\right| \ll n, \tag{14.24}$$

Taking n from (14.6), one obtains

$$\left|\frac{lqE}{k_B T}\right| \ll 1, \tag{14.25}$$

and substituting the average electric field in the depletion region as $E \approx V_b/L$, we obtain the condition under which we apply the drift-diffusion approach:

$$\left|\frac{qV_b}{k_B T}\right|\frac{l}{L} \ll 1. \tag{14.26}$$

Below we obtain the current-voltage (*I-V*) characteristics in both thermionic and drift-diffusion models.

14.1.7 Thermionic Emission Model

In high electron mobility Group-IV and III–V semiconductors, the mean free path exceeds the length of the depletion region ($l \gg L$), and diffusion can be neglected. Electrons reach the contact almost without collisions, and those have thermal energy enough to overcome the barrier, cross the interface. That is the *Bethe thermionic emission model.* The model implies that the barrier thickness is much larger than the electron wavelength ($L > \lambda$). This condition excludes tunneling, thus making over-barrier thermionic current transmission the only possible. The reverse condition corresponds to the barrier thin enough for tunneling and turns the contact ohmic that is discussed later in this Chapter.

Electrons in metal, overcoming the barrier height V_0, form the thermionic current density J_{ms}:

$$J_{ms} = RT^2 \exp\left(-\frac{V_0}{k_B T}\right), \quad R = \frac{4q\pi m k_B^2}{(2\pi\hbar)^3}, \tag{14.27}$$

where m is the electron effective mass, R is the effective Richardson constant $\left[\mathrm{Am^{-2}\,K^{-2}}\right]$, which origin becomes clear in the derivation of Eq. (14.42). In equilibrium, J_{ms} is balanced by the current flowing from semiconductor to metal $J_{sm} = qn'v_T$, $v_T = \sqrt{8k_BT/\pi m}$ is the thermal electron velocity, n' is the density of electrons taking part in conductivity in an unbiased contact. To determine n', we use the equilibrium condition $J_{sm} = J_{ms}$ and express V_0 through semiconductor parameters (14.2):

$$V_0 = V_b + E_c(\infty) - E_F,$$

$$n' = \frac{J_{ms}}{qv_T} = \frac{RT^2}{qv_T}\exp\left(-\frac{V_b + E_c(\infty) - E_F}{k_BT}\right) = \frac{(mk_BT)^{3/2}}{2q(2\pi\hbar^2)^{3/2}}\exp\left(-\frac{E_c(\infty)-E_F}{k_BT}\right)\exp\left(-\frac{V_b}{k_BT}\right)$$

$$= \frac{1}{4}N_c\exp\left(-\frac{E_c(\infty)-E_F}{k_BT}\right)\exp\left(-\frac{V_b}{k_BT}\right) = \frac{1}{4}n_0\exp\left(-\frac{V_b}{k_BT}\right) = \frac{1}{4}n(0) \tag{14.28}$$

where $n(0)$ is electron density in a semiconductor near the interface at zero bias (14.6):

$n(0) = n_0\exp(-V_b/k_BT)$. Replacing $V_b \to V_b - qU$ in (14.27), one obtains interface density at $U \neq 0$, $n'(U) = n(0)\exp(qU/k_BT)/4$, and current $J_{sm} = qn(0)v_T\exp(qU/k_BT)/4$. Current in opposite direction follows from (14.27): $J_{ms} = qv_Tn(0)/4$. Then, the current–voltage relation of rectifying contact takes the form:

$$J = J_{sm} - J_{ms} = J_s\left[\exp\left(\frac{qU}{k_BT}\right) - 1\right],$$

$$J_S = \frac{1}{4}qv_Tn(0). \tag{14.29}$$

The current density increases if the forward bias is applied, as illustrated in Figure 14.7. Relation (14.28) has sense if electrons in a semiconductor feel the barrier while flowing to metal, in other words, until the forward bias does not exceed V_b/q. Under further increased U, the voltage drop would occur on semiconductor series resistance, and the *I-V* relation becomes Ohm's law.

Voltage U in (14.28) is the voltage drop in the depletion region. In a diode as a device, one has to account for bulk semiconductor resistance R in series with depletion region, so U becomes $U = V - IR$, I is the current through the diode, V is the bias applied to the terminals.

The saturation current density at reverse bias, $J_S = qv_Tn(0)/4$, is determined by the thermal velocity as well as density of the majority carriers $n(0)$, which is assumed independent of an applied voltage.

14.1.8 Drift-Diffusion Model

In low-mobility semiconductors such as Se or Cu_2O, the mean free path is less than depletion length, so electrons flowing to metal (direct current) experience multiple collisions while traversing the space charge region. Forward bias induces electron drift in an applied electric field, and also the increase in Fermi level on a semiconductor side relative to metal creates diffusion flux toward metal. In low-mobility semiconductors such as Se or Cu_2O, the mean free path is less than depletion length, so electrons flowing to metal (direct current) experience multiple collisions while traversing the space charge region. Forward bias induces electron drift in an applied electric field, and also the increase in Fermi level on a semiconductor side relative to metal creates diffusion flux toward metal. Two equations (14.5) self-consistently regulate current and electric field in the space charge region. We dwell on a balance between drift and diffusion currents leaving for a while the potential profile as a function of distance to be specified. Taking into account Einstein relation (14.9), from (14.5) one obtains the equation for concentration profile $n(x)$,

$$\frac{J}{\mu k_B T}+\frac{1}{k_B T}n(x)\frac{\partial V(x)}{\partial x}-\frac{\partial n(x)}{\partial x}=0,$$

$$x=0: n=n(0). \tag{14.30}$$

As $V(0)=0$, the solution can be verified by direct substitution in (14.30) and is given below:

$$n(x)=exp\left(\frac{V(x)}{k_B T}\right)\left[n(0)+\frac{J}{\mu k_B T}\int_0^x exp\left(-\frac{V(z)}{k_B T}\right)dz\right], \tag{14.31}$$

where $n(0)$ is given in (14.7).

Let us determine the current crossing the plane outside the depletion region at $x=x_1>L$. In this region, $n(x)=n_0$, $V(x)=V_b-qU$, so the concentration profile $n(x)$ takes the form,

$$n_0=\exp\left(\frac{V_b+qU}{k_B T}\right)\left[n_0\exp\left(-\frac{V_b}{k_B T}\right)+\frac{J}{\mu k_B T}\int_0^{x_1}\exp\left(-\frac{V(z)}{k_B T}\right)dz\right],$$

or

$$J=J_s\left[\exp\left(\frac{qU}{k_B T}\right)-1\right],$$

$$J_s\equiv\mu k_B T n(0)\left[\int_0^{x_1}\exp\left(-\frac{V(z)}{k_B T}\right)dz\right]^{-1}\approx\mu n(0)\left.\frac{\partial V}{\partial x}\right|_{x=0} \tag{14.32}$$

Linear approximation to $V(x)$ in the last row of (14.31) gives the saturation current proportional to the built-in electric field at the contact plane. The only difference between thermionic (14.29) and drift-diffusion (14.32) I-V characteristics is the value of saturation currents.

In practical diodes, the barrier height decreases with an applied voltage that requires introducing the ideality factor [2] into current–voltage characteristics,

$$J=J_s\left[\exp\left(\frac{qU}{mk_B T}\right)-1\right]. \tag{14.33}$$

In high-quality Schottky diodes, the typical value of the ideality factor m is between 1 and 2. Similar to the previously discussed thermionic model, $n(0)$ and thus saturation current is assumed independent of an applied voltage. In real contacts, $n(0)$ depends on the applied voltage. Under forward bias, this dependence is negligible as the electric field is weak. However, high reverse bias increases J_s due to the barrier lowering effect, which is to discuss in the next section.

14.1.9 Barrier Height Lowering

An electron removed from metal feels an attraction force toward the surface. The force is electrostatic and equal to the Coulomb attraction force between the electron and the "image" positive charge, which the electron "leaves behind" in the metal, as illustrated in Figure 14.8.

To account for image force, one needs to update the electron potential energy $E_c(x)$ from (14.4) as follows:

$$E_c(x) = E_c(\infty) + V_b - V(x) - \frac{q^2}{16\pi\varepsilon\varepsilon_0 x}, \quad (14.34)$$

Barrier lowering that follows from (14.34) is illustrated in Figure 14.9. Calculating the maximum of $E_c(x)$, one finds

$$x_m = \frac{1}{4}\sqrt{\frac{q}{\pi\varepsilon\varepsilon_0 |E|}}, \quad q|E| \equiv \left.\frac{\partial V(x)}{\partial x}\right|_{x=0},$$

$$\Delta\Phi \equiv E_c(\infty) + V_b - E_c(x_m) = \frac{1}{2}\sqrt{\frac{q^3|E|}{\pi\varepsilon\varepsilon_0}}. \quad (14.35)$$

shown in Figure 14.1 Solid line-profile corrected to account for image force.

The resulting barrier height becomes $V_0 - \Delta\Phi$ from the metal side and $V_b - \Delta\Phi$ from the semiconductor. Saturation currents in current–voltage characteristics (14.29) and (14.32) are sensitive to barrier lowering through $n(0)$.

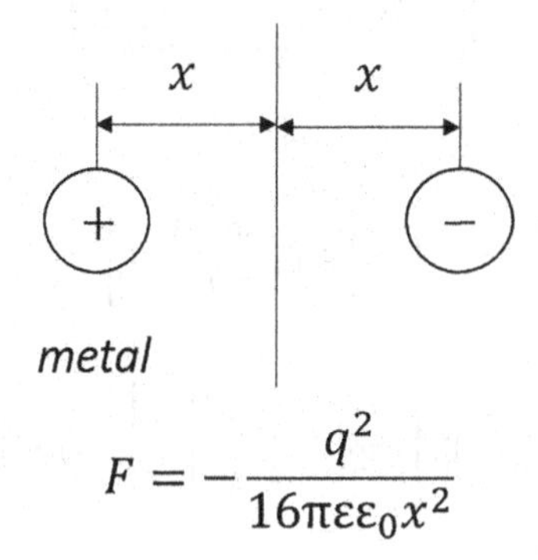

FIGURE 14.8 Attraction by image force, ε is the dielectric constant of the medium interfaced with metal.

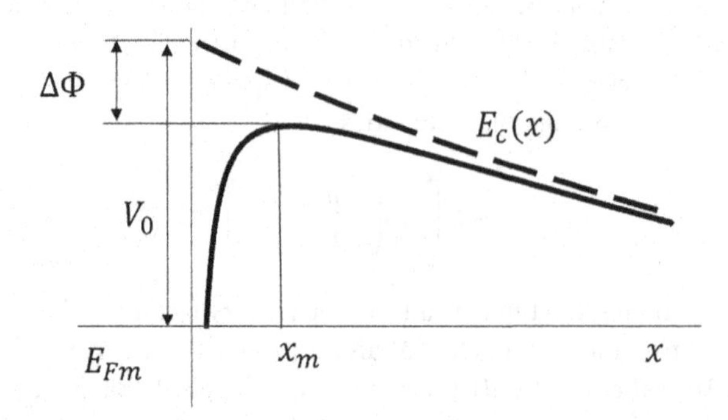

FIGURE 14.9 Potential profile for conduction electrons. Dashed line – band bending.

Graphs in Figure 14.9 are shown to scale for clarity. Maximum $E_c(x)$ is very close to the interface $(x_m \ll L)$. That allows us to use the electric field at the interface in (14.35). The field follows from (14.21) as

$$|E| = \frac{1}{q}\frac{\partial V}{\partial x}\bigg|_{x=0} = \sqrt{\frac{2n_0 q(V_b - qU)}{\varepsilon\varepsilon_0}}. \tag{14.36}$$

If reverse bias is large enough, $|E| \sim \sqrt{|U|}$, $\Delta\Phi \sim |U|^{1/4}$, and thus the reverse current depends on an applied voltage as $n(0) \sim \exp\left(|U|^{1/4}\right)$.

14.1.10 Transmission Modes and Ohmic Contacts

When the doping level in the semiconductor increases, the barrier shape changes, and one may distinguish three transmission modes. If the width of the barrier is much larger than the electron wavelength at fixed energy, tunneling is not possible. The main channel of current transmission is the previously discussed thermionic emission (arrow 1 in Figure 14.10). Since a strong electric field in the near-barrier region makes the top of the barrier thin enough for electrons to tunnel through, that forms *thermionic field* emission mode that combines over-barrier and tunnel electrons as depicted by arrow 2. In a heavily doped semiconductor, the barrier becomes thin enough, and thus tunneling becomes the dominant mechanism of current transmission called *field emission* (arrow 3). In examples of Figure 14.10, electron concentration n_{02} corresponds to degenerate semiconductor, where tunneling is possible at the Fermi level.

The higher n_{02}, the thinner the barrier, and at some point, the tunneling probability reaches a maximum, the current becomes independent of the height of the barrier and obeying Ohm's law. Ohm's law and low contact resistance are the factors that characterize an ideal ohmic contact. Below we estimate the lower limit of contact resistance in ohmic contact.

The total current density in the x-direction is the difference between counter-propagating flows: $J = J_{sm} - J_{ms}$. As far as tunneling is the mechanism of transmission, the electron flow from semiconductor to metal takes the form,

$$J_{sm} = \frac{2q}{\Omega}\sum_{k} \mathrm{v}_x(\boldsymbol{k}) f_s(E) T(E_x)\left[1 - f_m(E)\right],$$

$$\hbar \mathrm{v}_x = \partial E_x / \partial k_x,$$

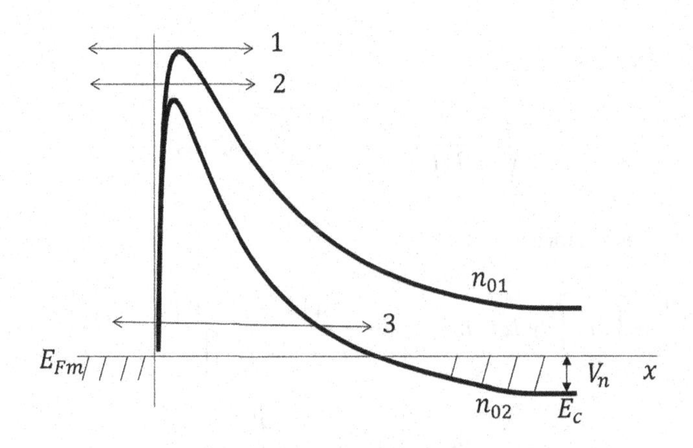

FIGURE 14.10 Transmission modes. 1 – thermionic, 2 – thermionic-field, 3 – field. $n_{02} \gg n_{01}$.

$$E = \hbar^2 k^2 / 2m^* \equiv E_x + E_y + E_z,$$

$$f_m = \frac{1}{1+\exp\left[(E-E_F)/k_B T\right]}, \quad f_s = \frac{1}{1+\exp\left[(E-qU-E_F)/k_B T\right]}, \tag{14.37}$$

where Ω is the crystal volume, coefficient 2 is the spin-factor, $\boldsymbol{k}$ is the three-dimensional wave vector, T, $f_s\left(f_m\right)$ are the tunneling probability and electron fill-factors in semiconductor (metal), respectively, U is the applied voltage. Filled states in metal with occupation probability f_m reflect electrons to the semiconductor, so factor $1-f_m$ accounts for the unoccupied states in metal. As the tunnel transmission occurs in the x-direction, the tunneling probability depends on energy component E_x only.

Current density J_{ms} is given by (14.37) with replacement $f_m \leftrightarrow f_s$, so the total current,

$$J = J_{sm} - J_{ms} = \frac{2q}{\Omega}\sum_{\boldsymbol{k}} \mathrm{v}_x(\boldsymbol{k}) T(E_x)\left[f_s(E) - f_m(E)\right]. \tag{14.38}$$

To calculate the current density (14.38), we use (14.37) for $\mathrm{v}_x(\boldsymbol{k})$ and replace $\boldsymbol{k}$-sums with integrals:

$$\sum_{\boldsymbol{k}}(\ldots) \to \frac{\Omega}{(2\pi)^3}\int d^3k,$$

$$J_{sm} = \frac{2q}{(2\pi)^3\hbar}\int_{-\infty}^{\infty} dk_y\, dk_z \int_{qU}^{\infty} dE_x\, T(E_x)\left[f_s(E) - f_m(E)\right]. \tag{14.39}$$

Notations (14.37) imply that the energy reference point is the edge of the conduction band E_c far from the contact (Figure 14.10). The minimum energy level from which electrons could tunnel to metal and back is $E_c = 0$, and under bias is qU. Substituting difference $f_s - f_m$ in (14.39), one obtains,

$$J = \frac{q}{4\hbar\pi^3}\int_{qU}^{\infty} dE_x \int_0^{\infty} dk_y\, dk_z\, T(E_x) \frac{\exp\left[\frac{(E-E_F)}{k_B T}\right]\left(1-\exp\left[\frac{-qU}{k_B T}\right]\right)}{\left[1+\exp\left[\frac{(E-E_F)}{k_B T}\right]\right]\left[1+\exp\left[\frac{(E-qU-E_F)}{k_B T}\right]\right]} \tag{14.40}$$

After replacing variables

$$\sqrt{\frac{\hbar^2}{2m^* k_B T}}k_y = y, \quad \sqrt{\frac{\hbar^2}{2m^* k_B T}}k_z = z,$$

the current density takes the form

$$J = \frac{qm^* k_B T}{2\hbar^3\pi^3}\int_{qU}^{\infty} dE_x \int_0^{\infty} dy\, dz\, T(E_x) \frac{\exp\left(\tilde{E}_x + y^2 + z^2\right)\left(1-\exp[-u]\right)}{\left[1+\exp\left(\tilde{E}_x + y^2 + z^2\right)\right]\left[1+\exp\left(\tilde{E}_x + y^2 + z^2 - u\right)\right]}$$

$$\tilde{E}_x \equiv \frac{E_x - E_F}{k_B T}, \quad u \equiv \frac{qU}{k_B T}. \tag{14.41}$$

Introducing polar coordinates,

$$\int_{-\infty}^{\infty} dy\,dz = \int_{0}^{\infty} r\,dr \int_{0}^{2\pi} d\varphi = 2\pi \int_{0}^{\infty} r\,dr,$$

$$J = \frac{RT}{k_B} \int_{qU}^{\infty} dE_x \int_{0}^{\infty} dr^2 T(E_x) \frac{\exp\left(\tilde{E}_x + r^2\right)\left(1 - \exp[-u]\right)}{\left[1 + \exp\left(\tilde{E}_x + r^2\right)\right]\left[1 + \exp\left(\tilde{E}_x + r^2 - u\right)\right]}. \tag{14.42}$$

where $R = 4q\pi m k_B^2/(2\pi\hbar)^3$ is the effective Richardson constant. Assuming $m = m_0$, the free-electron mass, $R = 120\ \mathrm{A\,cm^{-2}\,K^{-2}}$. In n-type Si, the constant is $R = 110\ \mathrm{A\,cm^{-2}\,K^{-2}}$.

In the limit $T(E_x) \to 1$, one obtains maximal tunneling current,

$$J_{\max} = \frac{RT}{k_B} \int_{qU}^{\infty} dE_x \log\left[\frac{1 + \exp\left(\dfrac{-E_x + E_F + qU}{k_B T}\right)}{1 + \exp\left(\dfrac{-E_x + E_F}{k_B T}\right)}\right],$$

and minimum *specific contact resistance*,

$$\rho_{\min}^{-1} = \left.\frac{\partial J_{\max}}{\partial U}\right|_{U=0} = \frac{Rq}{k_B^2}\left[\int_{-E_F}^{\infty} \frac{d(E_x - E_F)}{1 + \exp\left(\dfrac{E_x - E_F}{k_B T}\right)}\right] = \frac{RTq}{k_B} \log\left[1 + \exp\left(\frac{E_F}{k_B T}\right)\right]. \tag{14.43}$$

Relative to energy reference point $E_c = 0$, we express the Fermi energy as $E_F = V_n$ (see Figure 14.10) and finally,

$$\rho_{\min} = \frac{k_B}{qRT \log\left[1 + \exp\left(\dfrac{V_n}{k_B T}\right)\right]}, \quad \left[\Omega \cdot \mathrm{m}^2\right]. \tag{14.44}$$

Result (14.44) presents theoretically minimal contact resistance. The high doping level near the interface makes the tunnel barrier thinner and increases V_n. That is a practical way to suppress the contact resistance if non-rectifying contact (see Figure 14.2) is not possible to form. In strongly degenerate samples ($V_n \gg k_B T$),

$$\rho_{\min} \approx \frac{k_B^2}{Rq V_n}. \tag{14.45}$$

In a nondegenerate semiconductor, $V_n < 0$, thus only thermal electrons could take part in tunneling,

$$\rho_{\min} = \frac{k_B}{qRT \log\left[1 + \exp\left(-\dfrac{|V_n|}{k_B T}\right)\right]} \approx \frac{k_B}{RTq} \exp\left(\frac{|V_n|}{k_B T}\right) \approx \frac{k_B}{RTq} \frac{N_c}{n_d}. \tag{14.46}$$

Last equality in (14.46) follows from (4.28) (Chapter 4), providing all donors are ionized.

Known from experiments trend $\rho \sim n_d^{-1}$ was obtained theoretically in Ref. [3].

14.1.11 Equivalent Circuit and Frequency Response

When an alternating voltage applies to the contact, the current cannot immediately follow the bias and the delay limits the high-frequency operation of the diode. The delay occurs due to the finite time of accumulation and dissipation of the nonequilibrium charge during the operation of the device (*time constant*). There are two main contributions to the time constant: one is due to conduction mechanisms, and another one originates from the device structure and parasitic elements inherent to it. Time constants due to conduction mechanisms are determined by the diffusion of minority carriers (see Figure 14.5) and cooling down of hot electrons injected into metal (forward bias) or into semiconductor (reversed bias). These two processes have time constants in the range of $\left(10^{-11}-10^{-12}\right)$ s, thus do not hamper operation up to the submillimeter frequency range. The essential restriction comes from the recharging of the depletion region, which capacitance depends on the device structure. Providing all donors are ionized, the charge accumulated in the L-long depletion region of the area S is $Q=Sqn_dL$. Then, using (14.22), one obtains the differential capacitance at bias U_0:

$$C_d=\left|\frac{\partial Q}{\partial U}\right|=S\sqrt{\frac{\varepsilon\varepsilon_0 qn_d}{2(U_b-U_0)}}. \tag{14.47}$$

The equivalent circuit of the diode includes depletion capacitance C_d, differential resistance $R_d=\partial V/\partial I$, and series resistance of neutral semiconductor region between the end of depletion region and the ohmic contact from semiconductor side (R_S) as illustrated in Figure 14.11.

Assuming l is the length of the neutral region,

$$R_S=\frac{l}{S\sigma_S}=\frac{l}{Sqn_d\mu_n}, \tag{14.48}$$

where μ_n is the mobility of majority carriers.

Let us estimate the frequency above which the diode loses its rectifying properties. Rectification takes place when reverse and forward impedance between terminals a,b (see Figure 14.11) obey the relation: $|Z_r/Z_f|>1$. Calculating total impedance between the terminals, one obtains

$$Z_{ab}=R_S+\frac{R_d}{i\omega C_d R_d+1}, \tag{14.49}$$

Assuming the diode is ideal, in other words, $R_d=0$ under the forward bias and $R_d=\infty$ under reversed bias, from (14.49) one obtains

$$\left|\frac{Z_r}{Z_f}\right|=\frac{1}{\omega C_d R_S}. \tag{14.50}$$

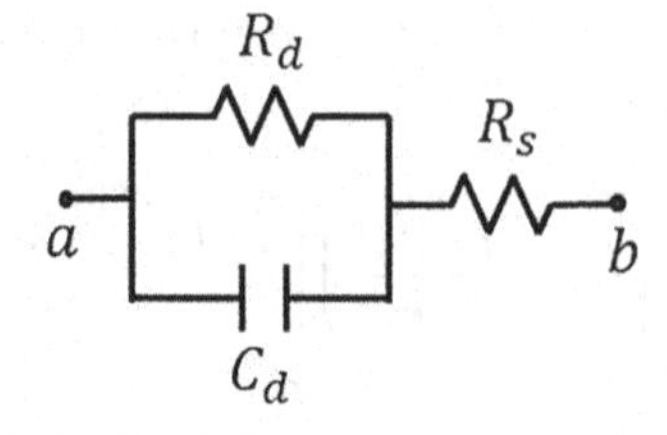

FIGURE 14.11 Small-signal equivalent circuit of the Schottky diode.

Condition $|Z_r/Z_f|>1$ gives $\omega<(C_d R_S)^{-1}$. Time constant

$$C_d R_S = \frac{l}{q\mu_n}\sqrt{\frac{\varepsilon\varepsilon_0 q}{2n_d(U_b-U_0)}} \tag{14.51}$$

determines the frequency limit for rectifying operation of the diode. When periodic bias $u_0\exp(i\omega t)$ is applied to the terminals a,b, the rectified signal is proportional to the cycle averaged power $P_{av}=\mathrm{Re}\left[U_d I_d^*/2\right]$, where U_d and I_d are the complex amplitudes of voltage drop and current through the diode (resistive element R_d in Figure 14.11):

$$P_{av} = \frac{u_0^2}{2R_d}\left[\left(1+\frac{R_s}{R_d}\right)^2+\omega^2 C_d^2 R_s^2\right]^{-1}, \tag{14.52}$$

where $R_d=(1/S)\partial U/\partial J$ is the differential resistance at some point on the *I-V* curve chosen experimentally to maximize P_{av}. It follows from (14.51) that short diode length l and high electron mobility μ_n both favor higher operation frequency.

14.2 P–n Junctions

Contact between two identical semiconductor crystals but one doped with n-type and the other one with p-type impurities forms *p–n* homojunction. Since chemical potentials on both sides of the junction are different, similarly to contact metal-semiconductor, the diffusion creates electron flow from n- to a p-region and hole flow from p- to an n-region. Diffusion takes place until the built-in electric field forms to stop the process. The built-in electric field generates the drift current so that drift and diffusion components in J_n and J_p (14.1) cancel each other. Zero total currents manifest the onset of thermal equilibrium in which chemical potentials on both sides of the contact become equal. The resulting band diagram is shown in Figure 14.12a.

Positive U (forward bias) means the negative polarity of the source applied to an n-type semiconductor. Planes x_1 and $-x_2$ delimit the depletion region where an electric field exists. In the next section, we discuss how the length of the depletion region $L=x_1+x_2$ depends on an applied voltage. The potential

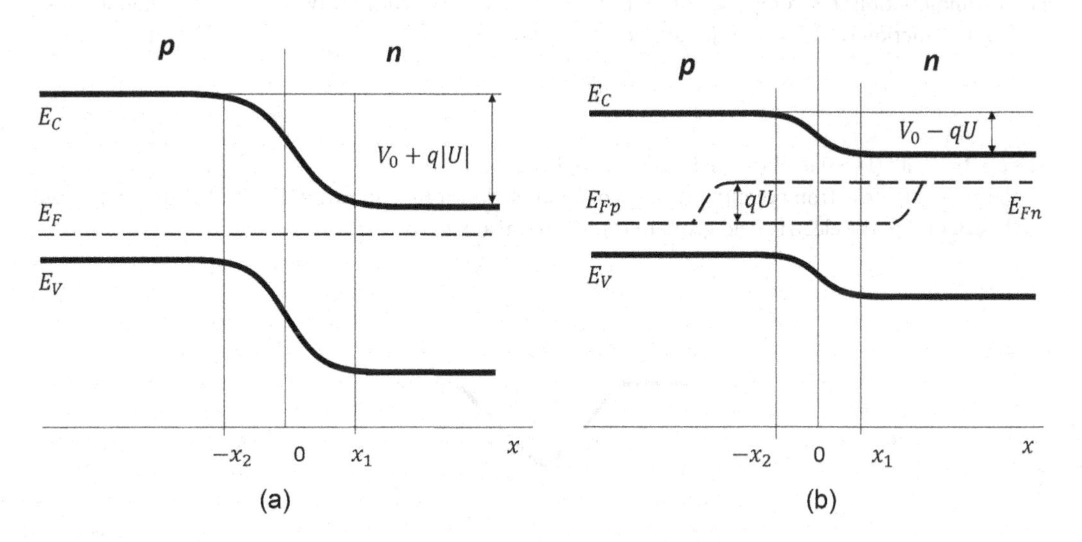

FIGURE 14.12 Band diagram of the *p–n* junction. (a) Thermal equilibrium. (b) Under the forward bias.

step in the depletion region is V_0 in equilibrium and $V_0 - qU$ under external voltage. As follows from Figure 14.12a, the potential step equals

$$V_0 = E_g - (E_c - E_F) - (E_F - E_v) = k_B T \log \frac{n_d n_a}{n_i^2}, \tag{14.53}$$

where n_d, n_a are densities of fully ionized donors in n-region and acceptors in p-region, respectively. In (14.53), we assumed Boltzmann statistics and excluded E_g and E_F using equations (4.24), (4.28), and (4.31) (Chapter 4). The value of V_0 cannot exceed the bandgap energy which is $(0.7, 1.1, 1.4, 3.4)$ eV for (Ge, Si, GaAs, GaN), respectively.

14.2.1 Depletion Region

We are going to estimate the width of the space charge region assuming all donors in the range $0 < x < x_1$ and all acceptors in $-x_2 < x < 0$ are ionized, and no free electrons and holes present in these regions (full depletion). From the Gauss law in (14.1), we have

$$\frac{\partial E}{\partial x} = \frac{\rho}{\varepsilon_0 \varepsilon},$$

$$\rho = \begin{cases} -qn_a, & -x_2 < x < 0 \\ qn_d, & 0 < x < x_1, \end{cases} \tag{14.54}$$

where n_d and n_a are the concentrations of ionized donors and acceptors, respectively. Integration in (14.54) with boundary conditions $E(x_1) = E(-x_2) = 0$ gives

$$E = \begin{cases} -\dfrac{qn_a}{\varepsilon_0 \varepsilon}(x + x_2), & -x_2 < x < 0 \\ \dfrac{qn_d}{\varepsilon_0 \varepsilon}(x - x_1), & 0 < x < x_1, \end{cases} \tag{14.55}$$

For a homojunction that comprises similar semiconductors, the continuity of electrical induction normal to the junction reads as equality of electric fields at $x = 0$:

$$n_d x_1 = n_a x_2. \tag{14.56}$$

The electric field (14.55) is illustrated in Figure 14.13

The step of the electron potential energy in the depletion region shown in Figure 14.12b is expressed by the integral of the electric field given in (14.55) (see also Figure 14.13):

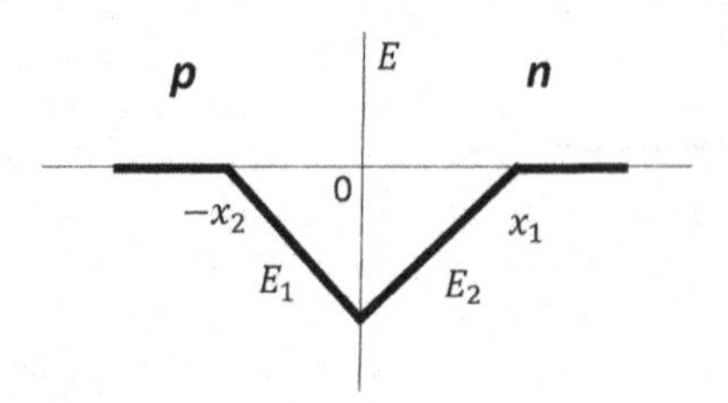

FIGURE 14.13 Electric field distribution across the depletion region.

$$V_0 - qU = -q\left(\int_{-x_2}^{0} E_1\,dx + \int_{0}^{x_1} E_2\,dx\right) = \frac{q^2}{\varepsilon_0\varepsilon}\left[n_a\int_{-x_2}^{0}(x+x_2)dx - n_d\int_{0}^{x_1}(x-x_1)dx\right] = \frac{q^2}{2\varepsilon_0\varepsilon}\left(n_a x_2^2 + n_d x_1^2\right). \tag{14.57}$$

Using (14.56) and (14.57), one obtains the size of the depletion region W_d as related to the applied voltage.

$$x_1 = \sqrt{\frac{2\varepsilon_0\varepsilon n_a\left(V_0 - qU\right)}{q^2 n_d\left(n_d + n_a\right)}},$$

$$x_2 = \sqrt{\frac{2\varepsilon_0\varepsilon n_d\left(V_0 - qU\right)}{q^2 n_a\left(n_d + n_a\right)}},$$

$$W_d = x_1 + x_2 = \sqrt{\frac{2\varepsilon_0\varepsilon\left(n_d + n_a\right)\left(V_0 - qU\right)}{q^2 n_d n_a}}. \tag{14.58}$$

As it follows from (14.58), the lower the doping, the larger the depletion region. For example, at $n_d \gg n_a$ the most of the depletion region is on the *p*-side of the junction.

A similar expression for W_d is valid for *p–n* heterojunctions in which dielectric constants of constituent semiconductors are different and thus the continuity of electrical induction reads as $\varepsilon_p E_1 = \varepsilon_n E_2$:

$$W_d = \sqrt{\frac{2\varepsilon_0\varepsilon_p\varepsilon_n\left(n_d + n_a\right)^2\left(V_0 - qU\right)}{q^2 n_d n_a\left(\varepsilon_n n_d + \varepsilon_p n_a\right)}}. \tag{14.59}$$

If $\varepsilon_p = \varepsilon_n = \varepsilon$, (14.59) coincides with the expression for homojunction (14.58).

14.2.2 Carrier Distributions

Diffusion creates minority carriers on both sides of the junction (electrons in *p*-region and holes in *n*-region), as illustrated in Figure 14.14a and b.

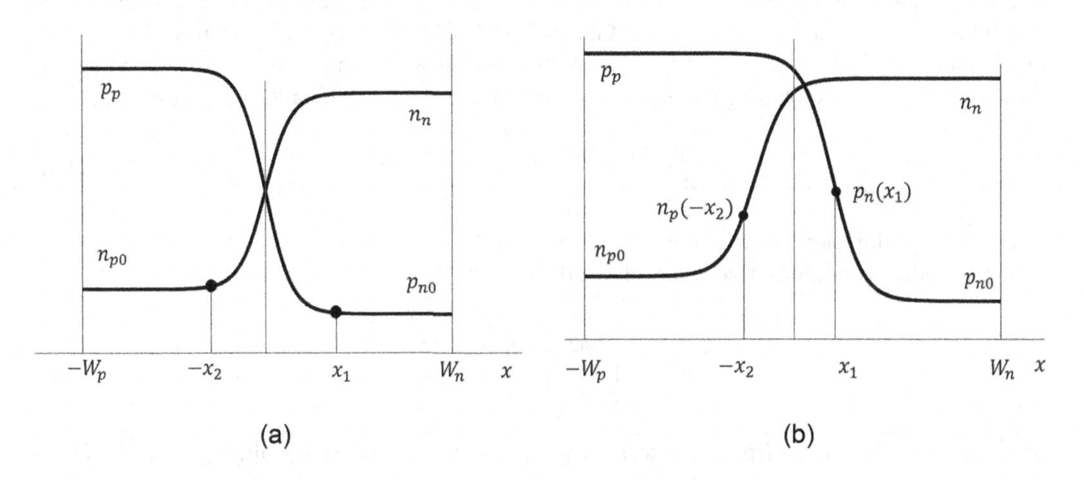

FIGURE 14.14 (a) Carrier distribution across equilibrium *p-n* junction. (b) Forward voltage. p_{n0} and n_{p0} are equilibrium minority carrier density outside the space charge region, $W_n, -W_p$ are the ends of the diode (positions of ohmic contacts).

The energy of electrons (holes) on both sides of the junction differs by V_0 (see Figure 14.12a). Therefore, providing electrons and holes obey Boltzmann statistics, the relation between concentrations on both sides of the junction in equilibrium takes the form:

$$n_{p0} = n_n \exp\left(-\frac{V_0}{k_B T}\right), \quad p_{n0} = p_p \exp\left(-\frac{V_0}{k_B T}\right), \tag{14.60}$$

where $n_n\left(p_{n0}\right)$ and $p_p\left(n_{p0}\right)$ are equilibrium majority (minority) carrier density on n- and p-sides, respectively, as they depicted in Figure 14.14. We do not use subscript 0 for equilibrium majority carriers as in what follows we consider small currents, which do not affect n_n and p_p.

14.2.3 Forward Bias

Forward bias injects electrons (holes) from n-(p-) into the depletion region where majority carriers coming from opposite sides recombine. So, the total flow by majority carriers is the recombination current $J_R = J_n^n + J_p^p$. Part of the majority electrons (holes), which do not recombine, go through the depletion region, penetrate the p-(n-) region where they become minority carriers. Thus, if no source of generation, the total current through the junction comprises recombination current inside the depletion region, and minority carrier flows outside the depletion region: $J = J_R + J_n^p(-x_2) + J_p^n(x_1)$. How many carriers could go through the junction depends on the relation between their recombination lifetimes $\tau_{n,p}$ and the time of diffusion through the depletion region. At this point, we assume lifetimes long enough to make diffusion length larger than thickness $d = x_1 + x_2$. In this approximation, one can neglect the recombination current: $J \approx J_n^p(-x_2) + J_p^n(x_1)$.

To get details on concentration profile $\delta p_n(x) = p_n(x) - p_{n0}$ outside the depletion region ($x \geq x_1$ in Figure 14.14), we consider the continuity equation for the hole current (14.1):

$$\frac{\delta p_n(x)}{\tau_{pn}} + \frac{1}{q}\frac{\partial J_p^n(x)}{\partial x} = 0,$$

$$J_p^n(x) = -qD_{pn}\frac{\partial p_n(x)}{\partial x}, \tag{14.61}$$

where $\delta p_n(x)$ is the excess minority holes density in n-region, τ_{pn} and D_{pn} are the recombination lifetime and diffusion coefficient for minority holes. Equation (14.61) implies we consider the steady-state carrier distribution with no external source and zero electric field, as we are interested in the solution outside the depletion region. Combining the first and second line in (14.61), one obtains diffusion equation

$$\frac{\partial^2 \delta p_n(x)}{\partial x^2} = \frac{\delta p_n(x)}{L_{pn}^2}, \quad L_{pn} = \sqrt{\tau_{pn} D_{pn}}. \tag{14.62}$$

where L_{pn} is the diffusion length of minority holes.

Two boundary conditions needed for solving (14.62) are

$$\delta p_n(x) = \begin{cases} \delta p_n(x_1), & x = x_1, \\ 0, & x = W_n, \end{cases} \tag{14.63}$$

where the boundary value $\delta p_n(x_1)$ is to be determined. The boundary condition at $x = W_n$ implies no excess carriers exist at the end of the diode that is true either if $W_n \gg L_{pn}$ or if there is infinite surface recombination at the electrode. Boundary conditions applied to the general solution $\delta p_n(x) = A\exp\left(-x/L_{pn}\right) + B\exp\left(x/L_{pn}\right)$ give two equations for A, B and thus the solution

$$\delta p_n(x)=\delta p_n(x_1)\frac{\sinh\left(\dfrac{W_n-x}{L_{pn}}\right)}{\sinh\left(\dfrac{W_n-x_1}{L_{pn}}\right)}. \tag{14.64}$$

Substituting (14.64) into (14.61) one obtains the hole minority current,

$$J_p^n(x)=\frac{qD_{pn}\delta p_n(x_1)}{L_{pn}}\frac{\cosh\left(\dfrac{W_n-x}{L_{pn}}\right)}{\sinh\left(\dfrac{W_n-x_1}{L_{pn}}\right)}, \tag{14.65}$$

The total current becomes

$$J_p^n(x_1)=\frac{qD_{pn}\delta p_n(x_1)}{L_{pn}}\coth\left(\frac{W_n-x_1}{L_{pn}}\right). \tag{14.66}$$

To complete the calculation of minority holes current, one has to determine $\delta p_n(x_1)=p_n(x_1)-p_{n0}$ (see Figure 14.14b). For that, we need to dwell on carrier statistics inside the depletion region. Under the forward voltage, nonequilibrium carriers injected from both sides of the junction into the depletion region quickly thermalize due to fast processes of intraband electron-phonon energy relaxation. As the recombination lifetime $(10^{-9}-10^{-6})$s is much longer than thermalization processes $(\sim 10^{-12}\text{ s})$, electrons and holes come to thermal equilibrium while remaining in conduction and valence bands, respectively. As a result, carriers reach equilibrium in c- and v-bands with quasi-Fermi levels $E_{Fn}-E_{Fp}=qU$, as illustrated in Figure 14.12b. Next, we assume Boltzmann statistics and take into account that the density of injected minority holes in point x_1 is determined by quasi-Fermi level E_{Fp} as illustrated in Figure 14.12b:

$$n_n=N_{cn}\exp\left(\frac{E_{Fn}-E_c}{k_BT}\right),\quad p_n(x_1)=N_{vn}\exp\left(\frac{E_v-E_{Fp}}{k_BT}\right)$$

$$n_np_n(x_1)=N_{cn}N_{vn}\exp\left(\frac{qU-E_g}{k_BT}\right). \tag{14.67}$$

Excluding the bandgap in (14.67) with (4.24) (Chapter 4), $n_np_{n0}=N_{cn}N_{vn}\exp\left(-E_g/k_BT\right)$, we finally obtain

$$p_n(x_1)=p_{n0}\exp\left(\frac{qU}{k_BT}\right),\quad \delta p_n(x_1)=p_{n0}\left[\exp\left(\frac{qU}{k_BT}\right)-1\right]. \tag{14.68}$$

Substituting $\delta p_n(x_1)$ into (14.66), one obtains the minority hole current,

$$J_p^n(x_1)=\frac{qD_{pn}p_{n0}}{L_{pn}}\left[\exp\left(\frac{qU}{k_BT}\right)-1\right]\coth\left(\frac{W_n-x_1}{L_{pn}}\right), \tag{14.69}$$

A similar calculation of $J_n^p(-x_2)$ gives one more contribution to the current–voltage characteristics of the junction:

$$J_n^p(-x_2) = \frac{qD_{np}n_{p0}}{L_{np}}\left[\exp\left(\frac{qU}{k_BT}\right)-1\right]\left[\coth\left(\frac{W_p - x_2}{L_{np}}\right)\right]. \tag{14.70}$$

For long diodes, $W \gg L$, the *I-V* characteristics becomes

$$J \approx J_n^p(-x_2) + J_p^n(x_1) = \left(\frac{qD_{pn}p_{n0}}{L_{pn}} + \frac{qD_{np}n_{p0}}{L_{np}}\right)\left[\exp\left(\frac{qU}{k_BT}\right)-1\right], \tag{14.71}$$

that is the ideal *diode equation* as we neglected recombination on traps in the depletion region, which enters J_R . When taking J_R into account, one obtains *I-V* relation modified by non-ideality factor m [1,4]:

$$J = J_S\left[\exp\left(\frac{qU}{mk_BT}\right)-1\right] \tag{14.72}$$

The nonideality factor follows from experimental data. The shape of the *I-V* characteristic is similar to that in metal-semiconductor contact (see Figure 14.7). Different mechanisms manifest themselves in different saturation currents.

14.2.4 Reverse Bias

Reverse bias ($U < 0$) increases the energy barriers for majority carriers preventing their flows in a forward direction. On the contrary, the barrier is absent for minority carriers, which easily penetrate the depletion region to recombine there (see Figure 14.15).

The current density of thermally generated minority carriers can be expressed through the generation rate $G_t\left[\mathrm{m^{-3}\,s^{-1}}\right]$ by bearing in mind that carriers contribute to the current during their lifetime or, the same, those generated on the length of diffusion:

$$J_S = q\left(G_{tn}L_{np} + G_{tp}L_{pn}\right) \tag{14.73}$$

In the steady-state, G_t can be replaced with (see Eq. (13.6), Chapter 13),

$$n_{p0} = G_{tn}\tau_{np},\ p_{n0} = G_{tp}\tau_{pn}\ , \tag{14.74}$$

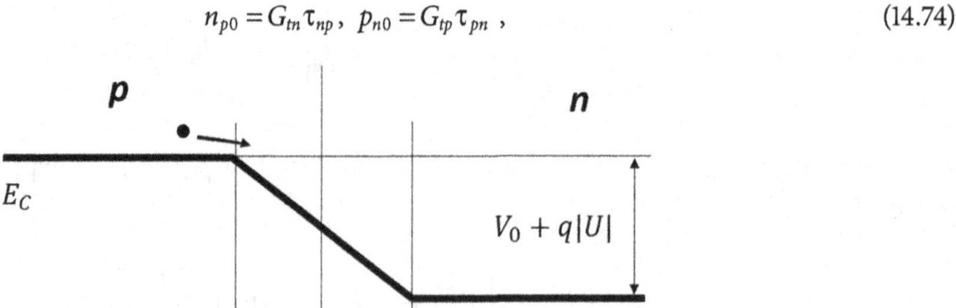

FIGURE 14.15 Reverse current in a schematic *p–n* junction as diffusion of minority carriers.

that results in

$$J_S = q\left(\frac{n_{p0}L_{np}}{\tau_{np}} + \frac{p_{n0}L_{pn}}{\tau_{pn}}\right) = q\left(\frac{n_{p0}D_{np}}{L_{np}} + \frac{p_{n0}D_{pn}}{L_{pn}}\right), \tag{14.75}$$

the expression which coincides with saturation current (14.71). This equivalence reveals the physical meaning of saturation current as the one created by thermally generated minority carriers on both sides of the *p-n* junction.

14.2.4.1 Diode Breakdown

Zener breakdown. Under high enough reverse voltage, $q|U| + V_0 > E_g$, the relative positions of conduction and valence bands make possible interband electron tunneling in an electric field (*Zener effect*), which may lead to diode tunnel breakdown (Figure 14.16a). Tunneling requires a thin enough depletion region of the order of electron wavelength (hundreds of Angstroms) and thus occurs in heavily doped or even degenerate semiconductors.

Electron tunneling can take place under the forward bias as well. In this case, the junction presents a tunnel diode where the Zener effect is the main mechanism of the forward and reverse conductivities. The main feature of the tunnel diode is the negative differential resistance under the forward bias.

Avalanche breakdown. In low-doped semiconductors, the Zener effect is not possible because the space charge region is wide enough to suppress tunneling. The main mechanism that limits reverse voltage becomes the avalanche breakdown. Carriers in the depletion region move in a strong electric field and gain energy high enough to excite an electron–hole pair by impact ionization as illustrated in Figure 14.16b. After voltage reaches the threshold value, the process develops as a multiplication of electrons and holes and then an abrupt increase in reverse current.

The empirical rule for Ge and Si diodes is that once the breakdown happens at $q|U| < 4E_g$, it is due to tunneling effects. Both tunneling and avalanche mechanisms occur at $6E_g > q|U| > 4E_g$. At $q|U| > 6E_g$, the avalanche mechanism prevails.

Thermal breakdown. At high reverse current, the power dissipated in the junction increases device temperature and thus the rate of electron-hole pair generation. The process develops avalanche-like and ends with the diode breakdown. The thermal breakdown mostly occurs in dc mode of operation and can be avoided in pulse regimes with duty cycles appropriate for effective heat dissipation.

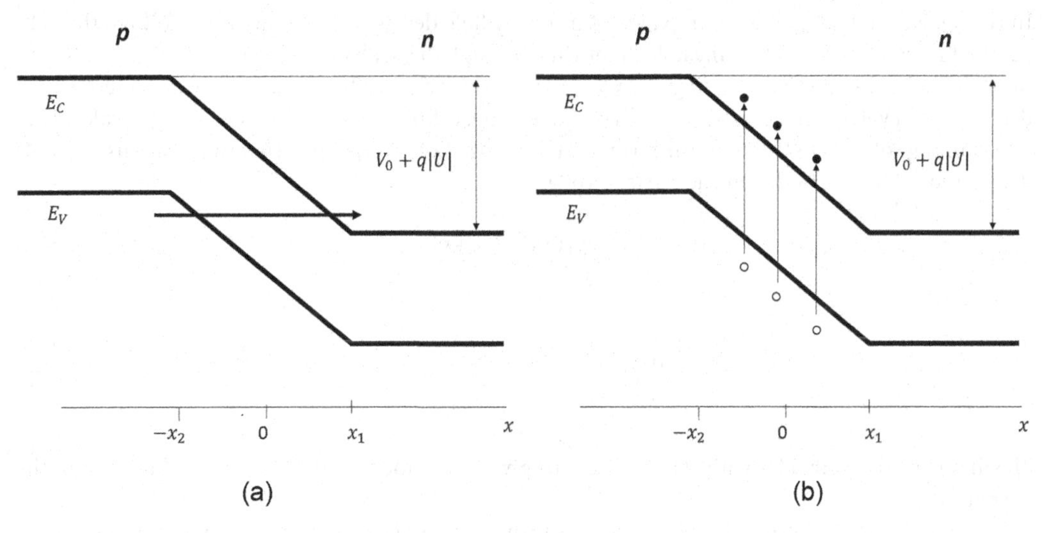

FIGURE 14.16 Diode breakdown. (a) Zener effect. (b) Avalanche breakdown (impact ionization). Zener and avalanche processes can occur simultaneously.

14.2.5 Frequency Response

The small-signal equivalent circuit for the Schottky diode in Figure 14.11 is applicable to the *p–n* junction as well. However, Schottky and *p–n* junctions are different in mechanisms of current transmission that modifies the physical meaning of capacitance in the *p–n* junction. The point is that the Schottky diode is a majority carrier device. Under the forward bias, the nonequilibrium charge of majority carriers in metal dissipates fast during Maxwell relaxation time. That is why a slower process of recharging the depletion region restricts the frequency response, $\omega C_d R < 1$. Under the forward voltage in the *p–n* junction, apart from recharging the depletion region, there are also minority carriers injected from both sides of the junction. Recharging of nonequilibrium minority carriers should occur during their lifetimes, so we anticipate the frequency limit originates from the slow diffusion process. In what follows, we are going to confirm by calculations that the time constant, which limits rectification at high frequency, is determined by the minority carrier lifetime. From the view of circuit parameters, it corresponds to introducing *diffusion capacitance*, which origin we consider below.

Actually, we need to derive the *I-V* characteristic (14.71) under periodic in time external voltage: $U(t) = U + u\exp(i\omega t)$. Again, we neglect the recombination current, assume the small-signal approximation, $qu < k_B T$, and low-injection level $p_n(x_2) \ll n_n$. The last condition guarantees the total voltage drops on a depletion region only.

Diffusion equation for minority holes in the n -side of the junction now contains the time derivative, which was neglected in (14.61):

$$\frac{\partial \delta p_n(x)}{\partial t} = D_{pn}\frac{\partial^2 \delta p_n(x)}{\partial x^2} - \frac{\delta p_n(x)}{\tau_{pn}}. \tag{14.76}$$

Small-signal boundary condition (14.68) now takes the form

$$\delta p_n(t, x_1) \approx p_{n0}\exp\left(\frac{qU}{k_B T}\right)\left(1 + \frac{qu}{k_B T}\exp(i\omega t)\right) - p_{n0} = \delta p_n(x_1) + \delta p_{1n}(x_1)\exp(i\omega t),$$

$$\delta p_{1n}(x_1) \equiv \frac{qup_{n0}}{k_B T}\exp\left(\frac{qU}{k_B T}\right). \tag{14.77}$$

In (14.77), we implied the boundary value of minority hole density at x_1 immediately follows the voltage, that is true if $\omega < 1/t_f$, t_f is the hole flight-time through the depletion region (typically, $t_f \approx 10^{-11}$ s). Minority holes distribution in n -region $(x > x_1)$ sets up by a slow diffusion process, which creates time delay between voltage and current. The delay is due to the diffusion capacity we are going to calculate.

Assuming solution to (14.76) in the form $\delta p_n(x,t) = \delta p_n(x) + \delta p_{1n}(x)\exp(i\omega t)$, one can discern hole distributions induced by dc- and ac-parts of voltage:

$$\frac{\partial^2 \delta p_n(x)}{\partial x^2} = \frac{\delta p_n(x)}{L_{pn}^2}.$$

$$\frac{\partial^2 \delta p_{n1}(x)}{\partial x^2} = \frac{\delta p_{1n}(x)(1 + i\omega\tau_{pn})}{L_{pn}^2}, \tag{14.78}$$

The first equation coincides with (14.62) and thus gives the stationary distribution and dc-part of the current (14.69).

The solution to the second equation which satisfies boundary condition $x \to \infty$, $\delta p_{n1}(x) \to 0$, and $x = x_1$, $\delta p_{1n}(x) = \delta p_{1n}(x_1)$, has the form

$$\delta p_{1n}(x,\omega)=\delta p_{1n}(x_1)\exp\left(-\frac{\sqrt{1+i\omega\tau_{pn}}}{L_{pn}}(x-x_1)\right) \tag{14.79}$$

Using (14.79) in (14.61) one obtains frequency dependent contribution to the minority holes current in the n-side of the diode at $x=x_1$:

$$J_p^n(\omega,x_1)=-qD_{pn}\left.\frac{\partial\omega p_{1n}(x,\omega)}{\partial x}\right|_{x_1}=\frac{q^2p_{n0}D_{pn}u}{k_BTL_{pn}}exp\left(\frac{qU}{k_BT}\right)\sqrt{1+i\omega\tau_{pn}}\,. \tag{14.80}$$

A similar contribution comes from minority electrons on the p-side of the structure. The total current density amplitude has the form:

$$J(u,\omega)=J_p^n(\omega,x_1)+J_n^p(\omega,-x_2)=\frac{q^2p_{n0}D_{pn}u}{k_BTL_{pn}}\exp\left(\frac{qU}{k_BT}\right)\sqrt{1+i\omega\tau_{pn}}+\frac{q^2n_{p0}D_{np}u}{k_BTL_{np}}\exp\left(\frac{qU}{k_BT}\right)\sqrt{1+i\omega\tau_{np}}\,. \tag{14.81}$$

In an asymmetric junction with p-side heavily doped ($n_n \ll p_p$), expression (14.80) presents the ac-part of the total *I-V* characteristics $J(u,\omega)=J_p^n(\omega,x_1)$, from which one can get the ac-conductivity as

$$G=S\left(\frac{\partial J(u,\omega)}{\partial u}\right)=\frac{Sq^2p_{n0}D_{pn}}{k_BTL_{pn}}\exp\left(\frac{qU}{k_BT}\right)\sqrt{1+i\omega\tau_{pn}}\equiv g_{diff}+i\omega C_{diff}\,, \tag{14.82}$$

Presenting real and imaginary parts as $\sqrt{1+ix}=\left(\sqrt{\sqrt{1+x^2}+1}+i\sqrt{\sqrt{1+x^2}-1}\right)\Big/\sqrt{2}$, one obtains active diffusion conductivity and diffusion capacitance as follows:

$$g_{diff}=R_{diff}^{-1}=\frac{Sq^2p_{n0}D_{pn}}{k_BTL_{pn}\sqrt{2}}exp\left(\frac{qU}{k_BT}\right)\sqrt{\sqrt{1+\omega^2\tau_{pn}^2}+1}\,,$$

$$C_{diff}=\frac{Sq^2p_{n0}D_{pn}}{k_BTL_{pn}\omega\sqrt{2}}exp\left(\frac{qU}{k_BT}\right)\sqrt{\sqrt{1+\omega^2\tau_{pn}^2}-1}\,. \tag{14.83}$$

The same arguments used in (14.50) for Schottky diodes confirm that the frequency limit for rectification is determined by the time constant

$$C_{diff}R_{diff}=\frac{\tau_{pn}}{\sqrt{1+\omega^2\tau_{pn}^2}+1}, \tag{14.84}$$

which gives $\omega<1/C_{diff}R_{diff}$. This frequency is determined by τ_{pn} and thus much lower than in Schottky diodes where it relies on the depletion capacitance only. At a frequency higher than $1/C_{diff}R_{diff}$, the depletion capacitance comes into play. The charge stored in the depletion region depends on an applied voltage:

$$Q=qS\int_0^{x_1}n_d\,dx=qSn_dx_1, \tag{14.85}$$

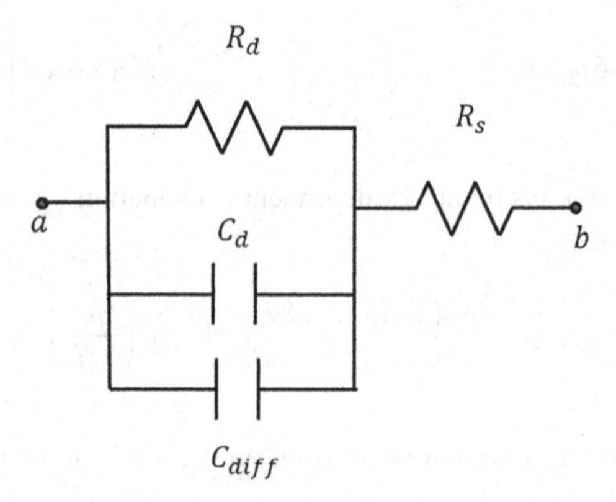

FIGURE 14.17 Small-signal equivalent circuit of a *p-n* junction.

Using (14.58), one obtains depletion capacitance as

$$C_d = \left|\frac{\partial Q}{\partial U}\right| = S\sqrt{\frac{\varepsilon_0 \varepsilon q^2 n_d n_a}{2(n_d + n_a)(V_0 - qU)}}. \tag{14.86}$$

The small-signal equivalent circuit of the *p–n* junction is similar to that in Schottky diodes but contains two capacitances as shown in Figure 14.17.

Among the two, Schottky contacts are more favorable for high-frequency applications than *p–n* junctions, which have two contributions to capacitance and thus slower speed of operation.

References

1. M. Shur, *Physics of Semiconductor Devices* (Prentice Hall, Englewood Cliffs, NJ, 1990).
2. E.H. Rhoderick, *Metal-Semiconductor Contacts* (2nd ed., Oxford University Press, 1988).
3. R.K. Kupka and W.A. Anderson, "Minimal ohmic contact resistance limits to *n*-type semiconductors", *J. Appl. Phys.* **69**, 3623 (1991).
4. C.M. Wolf, N. Holonyak Jr, and G.E. Stillman, *Physical Properties of Semiconductors* (Prentice-Hall, Inc., Englewood Cliffs, NJ, 1989).

15

Field-Effect Transistors

Semiconductor *p–n* junctions are the main elements of three-terminal devices – bipolar *p–n–p* and *n–p–n* transistors, which serve as dc- and high-frequency amplifiers. As we discussed in Chapter 14, injection and diffusion of minority carriers in *p–n* junctions makes their high-frequency operation slower if compared to metal-semiconductor diodes, where only majority carriers take part in current transmission.

The development of semiconductor electronics had called for three-terminal devices that control the current by majority carriers and thus provide for speed higher than in bipolar transistors. Devices called *field-effect transistors* (FETs) operate with majority carriers accumulated near the metal-insulator-semiconductor (MIS) or semiconductor-semiconductor heterointerface. Constituent elements of FETs, common to all their modifications, are channel, source, drain, and gate. The drain current flows through the channel between the source and drain, being modulated by gate bias. A specific implementation of elements defines the FET type and variety of corresponding abbreviated names. Metal-semiconductor contacts, *p–n*-junctions, high-electron-mobility semiconductor heterojunctions, and MIS structures – all can serve as elements of FETs: MESFET, JFET, HEMT, MISFET (MOSFET), respectively. Below we discuss typical example structures.

15.1 Energy Band Diagrams

Most common are metal-insulator-semiconductor devices (MISFET), where gate voltage affects the drain current through a capacitive coupling between the gate and the channel. Capacitive coupling implies the presence of an insulating layer under the gate electrode. In a frequently used abbreviation MOSFET, "O" stands for "Oxide" as an insulator. The use of a native semiconductor oxide layer as an insulator, by itself, does not affect the principles of device operation while still being important from a fabrication technology standpoint.

The energy band diagram of the MIS structure is shown in Figure 15.1.

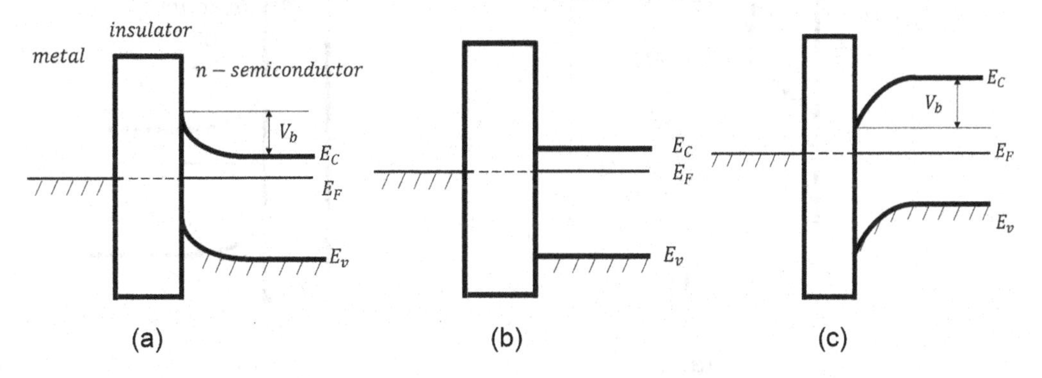

FIGURE 15.1 MIS structures in equilibrium. (a) Rectifying contact. (b) Flat-band condition. (c) Non-rectifying contact. Flat-band voltage $U_{FB} = V_b/q$.

DOI: 10.1201/9780429285929-15

If there is no difference between work functions in metal and semiconductor, the equilibrium flat-band picture in Figure 15.1b is exact. It is not the case in most structures. Typically, to flatten the bands, one has to apply positive or negative flat-band voltage $U_{FB} = V_b/q$ to the metal in rectifying (Figure 15.1a) or accumulation (Figure 15.1c) contact, respectively. However, since metal serves as a gate electrode usually biased in the range of $(1-10)$ eV, the equilibrium band bending $V_b \approx (0.1-1)$ eV can be neglected, and thus study FET operation principles under the flat-band condition is a good approximation.

The gate creates either accumulation or depletion layer depending on the sign of gate voltage relative to the semiconductor, as shown in Figure 15.2. As the exchange of electrons between metal and semiconductor through the insulator is absent, the energy band diagram shows the jump in Fermi level.

A negatively biased gate puts MIS into a depletion regime, as shown in Figure 15.2b. Further increased gate voltage forms the *induced channel regime* that is an inversion $p(n)$-channel in $n(p)$-type semiconductor (see Figure 15.3).

Depletion, accumulation, and inversed MIS structures, shown in Figures 15.2 and 15.3 for n-type semiconductor base, can also be made of p-semiconductors with corresponding gate voltage sign change. The FET structure can employ n- or p-semiconductor or any of the three MIS options mentioned above. Despite this creates a variety of devices, the principles of their operation have much in common. Below we consider current-voltage characteristics in accumulation and depletion regimes.

At zero gate bias in the flat-band approximation, the equilibrium electron density in semiconductor, n_0, does not depend on the distance from the contact. The gate bias allows manipulating charges stored in the channel by charging series-connected capacitances: d – thick insulator C_I and L – thick semiconductor space charge layer C_S (see Figure 15.2b),

$$C = \frac{C_S C_I}{C_I + C_S},$$

$$C_S = \frac{S\varepsilon_0\varepsilon_S}{L}, \quad C_I = \frac{S\varepsilon_0\varepsilon_I}{d}, \tag{15.1}$$

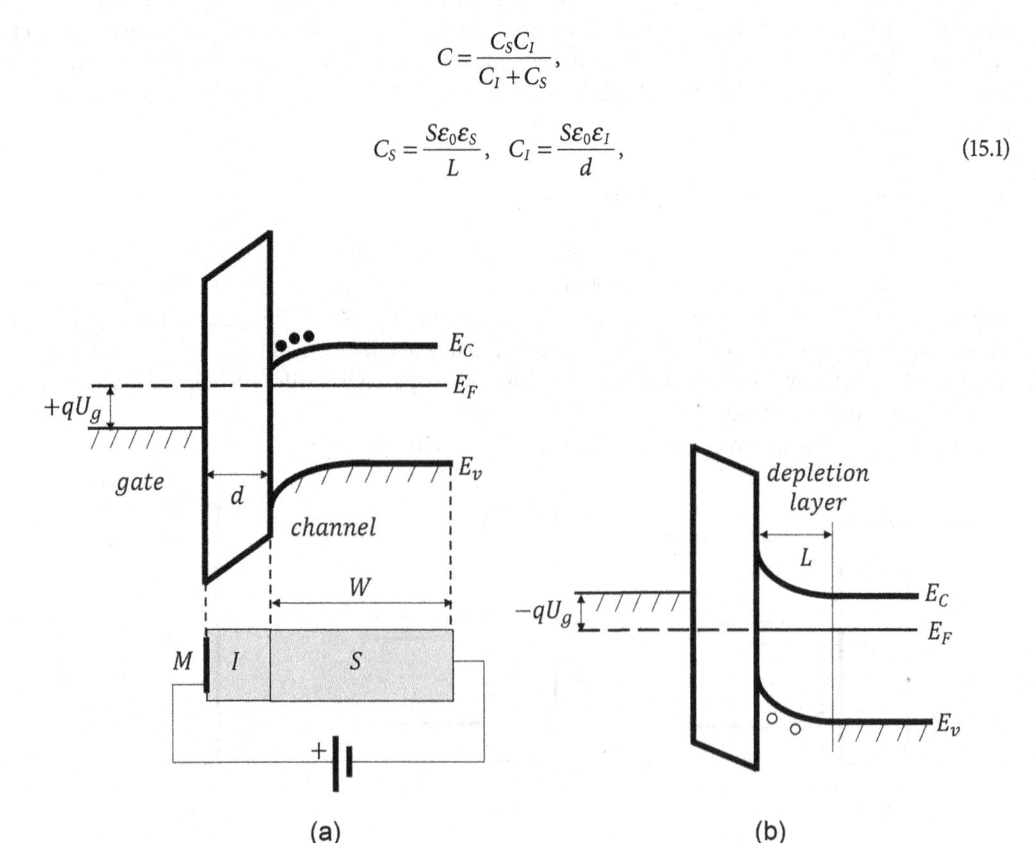

FIGURE 15.2 MIS structure in accumulation (a) and depletion (b) regime.

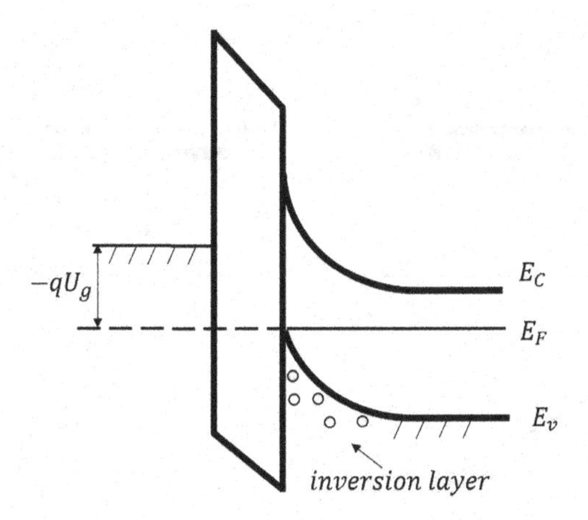

FIGURE 15.3 Inversion-induced p-channel.

where S is the gate area and ε_S (ε_I) is the dielectric constant in semiconductor (insulator). Nonequilibrium charge created by the gate voltage modulates conductivity on a semiconductor side of the MIS structure (channel).

15.2 Accumulation Regime

In the MIS structure in Figure 15.2, a high electron concentration near the interface makes the gate voltage, U_g, mostly drop on the insulator. Nonequilibrium charge accumulated due to applied gate voltage equals $Q_{ne} = C_I U_g$. The equilibrium charge $(U_g = 0)$ is $Q_e = qn_0WS$, where n_0 is the ionized donor density. The modulation becomes significant when an additional charge equals the equilibrium one: the electron concentration doubles, so the conductivity. It occurs when the gate voltage reaches the threshold:

$$U_{th} = \frac{Q_e}{C_I} = \frac{qn_0Wd}{\varepsilon_0\varepsilon_I}. \tag{15.2}$$

For typical Si/SiO$_2$ FETs the U_{th}-magnitude varies in the range $(0.1 \div 1)\,\text{V}$ for $n_0 = (10^{21} \div 10^{22})\,\text{m}^{-3}$, respectively. The total charge in the semiconductor side of MIS (channel):

$$Q_t = Q_e + Q_{ne} = C_I\left(U_{th} + U_g\right) \tag{15.3}$$

The device layout that corresponds to the MIS structure in Figure 15.2a is shown in Figure 15.4.

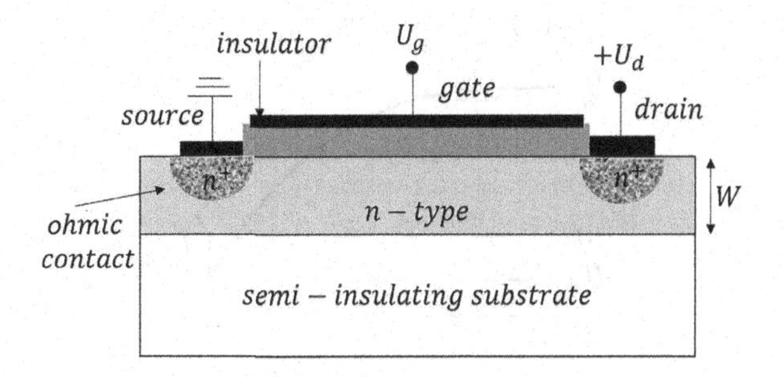

FIGURE 15.4 Schematic n-channel MISFET in accumulation regime, gate bias $U_g > 0$.

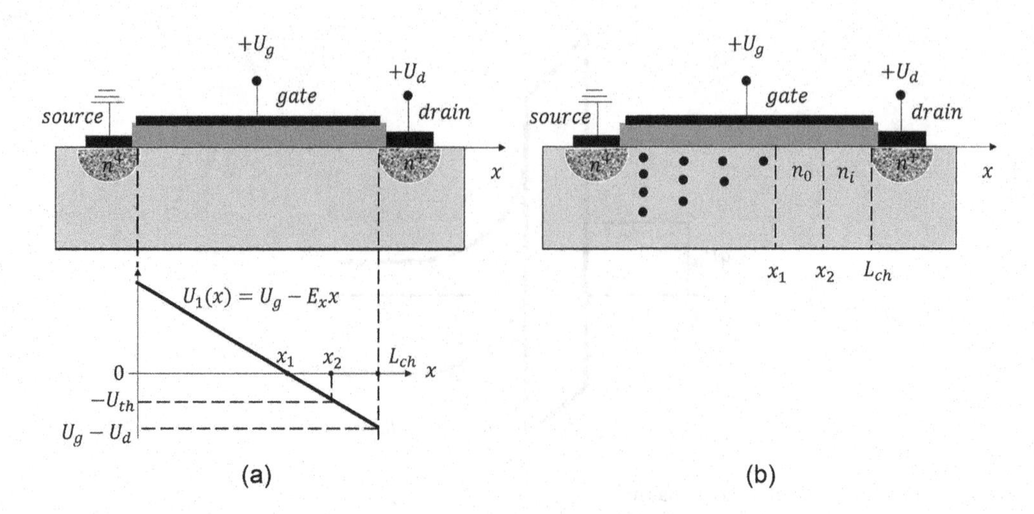

FIGURE 15.5 FET structure under the drain and gate voltages. (a) Semiconductor-gate voltage drop along the channel. (b) Carrier distribution. Black circle – nonequilibrium carriers induced by a gate voltage.

When the drain is biased with driving voltage, U_d, the current flows in the channel (accumulation layer in Figure 15.2a) parallel to the insulator-semiconductor interface. Gate voltage modulates this current, and this is the operation of the transistor. If the drain is unbiased $(U_d = 0)$, the gate-semiconductor voltage difference equals to gate bias U_g. Positive drain bias U_d creates the driving electric field along the channel so that the voltage difference between the gate and semiconductor depends on a distance, as illustrated in Figure 15.5a.

The potential of the semiconductor-insulator interface $U_1(x) = U_g - E_x x$ varies along the channel, as shown in Figure 15.5a. Linear distance dependence implies constant electric field, $E_x = U_d / L_{ch}$, – an approximation we use, assuming that device operation basics are not sensitive to exact field distribution. In point $x_1 = U_g L_{ch} / U_d$, the nonequilibrium charge induced by gate voltage becomes zero (flat band situation), and carrier density in the channel is the equilibrium density n_0 (see Figure 15.5b) determined by ionized donors. In the region $x \geq x_1$, band bending changes sign, causing depletion of electrons. In point $x_2 = \left(U_g + U_{th}\right) L_{ch} / U_d$, all free equilibrium charges repel from the depletion region, meaning the channel being *pinched-off*. In the region $x_2 < x < L_{ch}$, the semiconductor is in a highly resistive state with intrinsic carrier concentration determined by temperature and the bandgap. The pinch-off condition could read as $x_2 \leq L_{ch}$, or $U_d \geq U_g + U_{th}$. In reality, pinch-off does not occur. The point is the electric field is not constant along the channel: it is low near the source where the carrier density is high (see Figure 15.5b), then gradually increases, reaching the maximum in the drain region, where depletion works, and thus the carrier density is low. So, in a more realistic picture, almost total drain voltage drops on the near-the-drain region. Unlike Figure 15.5a, the voltage distribution looks like that shown in Figure 15.6.

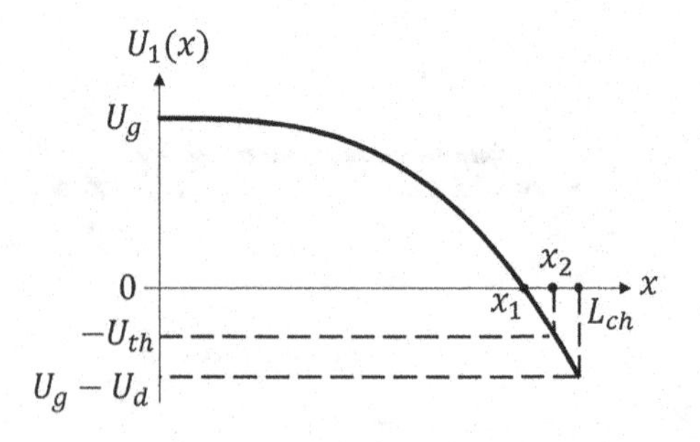

FIGURE 15.6 Channel potential distribution along the channel of length L_{ch}.

According to the distribution in Figure 15.6, the accumulation $(U_g > 0)$ spreads on almost the whole channel, whereas neutral region $x_1 < x < x_2$ and depletion region $x_2 < x < L_{ch}$ both located close to the drain, together with n^+ Ohmic contact, form n–i–n^+ junction. The n–i junction is forward-biased (plus on neutral i side) and emits electrons. The i–n^+ contact is reverse-biased and thus saturates the current (see Chapter 14). So, source-drain current saturates when the drain voltage reaches the value $U_d = U_g + U_{th}$ at which depletion occurs at the drain end of the channel.

15.2.1 Current–Voltage Characteristics

Modeling of dc- and ac- characteristics of real devices requires knowledge of the field and carrier density distributions specific to the choice of materials and structure geometry. It can be done numerically [1,2]. However, one can demonstrate the principles of operation analytically in a simple linear model that assumes field distribution shown in Figure 15.5a.

Electron transport in the channel is monopolar, so the drain current density is

$$j_d = \frac{\bar{Q}}{\Omega}\text{v}, \tag{15.4}$$

where $\bar{Q}$ is the average charge in the channel, Ω is the channel volume, v is the electron drift velocity. Introducing mobility, $\text{v} = \mu E$, $E = U_d / L_{ch}$, one obtains current,

$$I_d = \frac{\bar{Q}\mu U_d}{L_{ch}^2}, \quad [A]. \tag{15.5}$$

The average charge is the integral over channel length,

$$\bar{Q} = \frac{1}{L_{ch}} \int_0^{L_{ch}} Q(x)\,dx, \tag{15.6}$$

where $Q(x)$ is the charge (Figure 15.3) where U_g is replaced by $U_1(x)$, the actual voltage between semiconductor and insulator at $U_d > 0$:

$$Q(x) = C_I\left(U_{th} + U_1(x)\right) = C_I\left(U_{th} + U_g - \frac{U_d}{L_{ch}}x\right). \tag{15.7}$$

Integration in Eq.(15.6) gives the current–voltage characteristics $I_d(U_d)$:

$$I_d = \begin{cases} \dfrac{\mu C_I U_d}{L_{ch}^2}\left(U_{th} + U_g - \dfrac{U_d}{2}\right), & U_d \le U_{th} + U_g \\ \dfrac{\mu C_I}{2L_{ch}^2}\left(U_{th} + U_g\right)^2, & U_d > U_{th} + U_g, \end{cases} \tag{15.8}$$

Parameters which characterize FET follow from (15.8): *conductance* g_d that is the inverse differential resistance of the channel, and *transconductance* g_m which is derivative of *transfer characteristic* $I_d(U_g)$ and shows how input voltage U_g affects output current I_d:

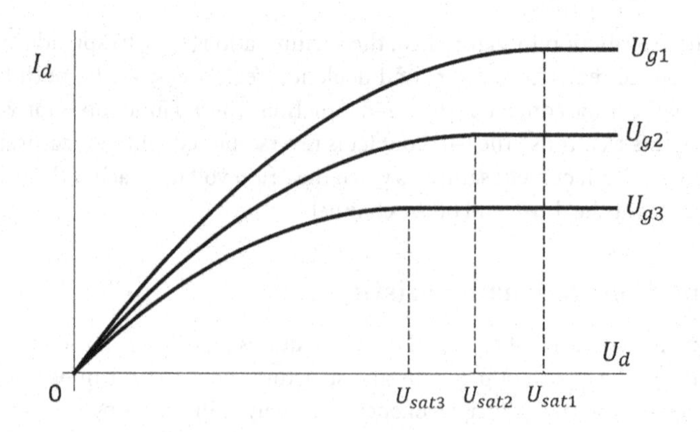

FIGURE 15.7 Drain current, $U_{g1} = 2.5\text{V}, U_{g2} = 2\text{V}, U_{g3} = 1.5\text{V}$. Saturation voltage: $U_{sat} = U_g + U_{th}$.

$$g_d = \frac{\partial I_d}{\partial U_d} = \begin{cases} \frac{\mu C_I}{L_{ch}^2}\left(U_{th} + U_g - U_d\right), & U_d \le U_{th} + U_g \\ 0, & U_d > U_{th} + U_g, \end{cases}$$

$$g_m = \frac{\partial I_d}{\partial U_g} = \begin{cases} \frac{\mu C_I U_d}{L_{ch}^2}, & U_d \le U_{th} + U_g, \\ \frac{\mu C_I \left(U_{th} + U_g\right)}{L_{ch}^2}, & U_d > U_{th} + U_g. \end{cases} \tag{15.9}$$

The first row in (15.8) follows from (15.6) and (15.7) and, as mentioned in the preceding section, is valid until $U_d = U_{th} + U_g$, then, current saturates, and conductance in the channel becomes zero. The second rows in Eqs. (15.8) and (15.9) give saturation values of conductance and transconductance, respectively. Series of graphs in Figure 15.7 describe *I-V* characteristics (15.8) depending on U_g as a parameter.

So, by increasing positive gate voltage, one increases the drain current in FET working in the accumulation mode. Expression (15.8) is the total current flowing through the channel of volume $\Omega = L_{ch}WY$, where Y is the channel width in the direction perpendicular to the plane of Figure 15.5. Substituting $C_I = L_{ch}Y\varepsilon_0\varepsilon_I/d$ into (15.8) one obtains current per unit length in the Y-direction:

$$j_d = \frac{I_d}{Y} = \frac{\varepsilon_0\varepsilon_I\mu U_d}{L_{ch}d}\left(U_{th} + U_g - \frac{U_d}{2}\right), \quad U_d \le U_{th} + U_g, \quad [\text{A/m}]. \tag{15.10}$$

This current density and corresponding transconductance, $g_m = \partial j_d/\partial U_g$, [Siemens/m], are those usually used for FET characterization.

Analyzing the current–voltage characteristics in saturation region $U_d > U_{th} + U_g$, one has to account that length of the depletion region l_D at the drain side of the channel grows with U_d, thus making effective channel shorter: $L_{ch}^{eff} = L_{ch} - l_D$. As drain current is inversely proportional to the channel length, the channel shortage makes drain current increase in the saturation region as shown in Figure 15.8 (*Early effect*):

15.3 Depletion Regime

In MIS structures with the depletion region shown in Figure 15.2, there are two regimes of operation. One is for the depletion region less than semiconductor thickness ($L < W$), so at zero gate bias, the channel is open and called normally-ON. If the semiconductor layer is thin, the depletion region may exceed

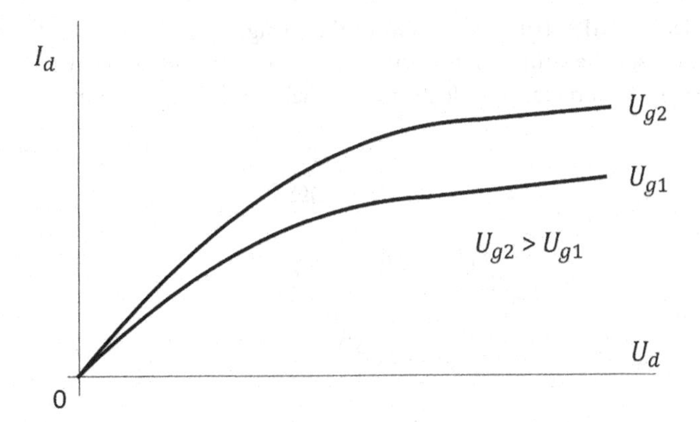

FIGURE 15.8 Weak increase in drain current at $U_d > U_{th} + U_g$.

the semiconductor thickness and close the channel ($L > W$). That is normally-OFF FET, which requires a finite positive gate bias to get the channel open.

Device operating in the depletion regime employs equilibrium MIS structure shown in Figure 15.2b and has the same layout as in Figure 15.4 but at negative gate voltage. In a flat-band state, ionized donors determine the equilibrium electron density in semiconductor, n_0. Under gate voltage, U_g, electron transfer from semiconductor to the metal gate creates a depletion region. The charge transferred depends on MIS capacitance, which, in this case, is represented by the total capacitance (15.1) comprising the insulator and the depletion region, as shown in Figure 15.9.

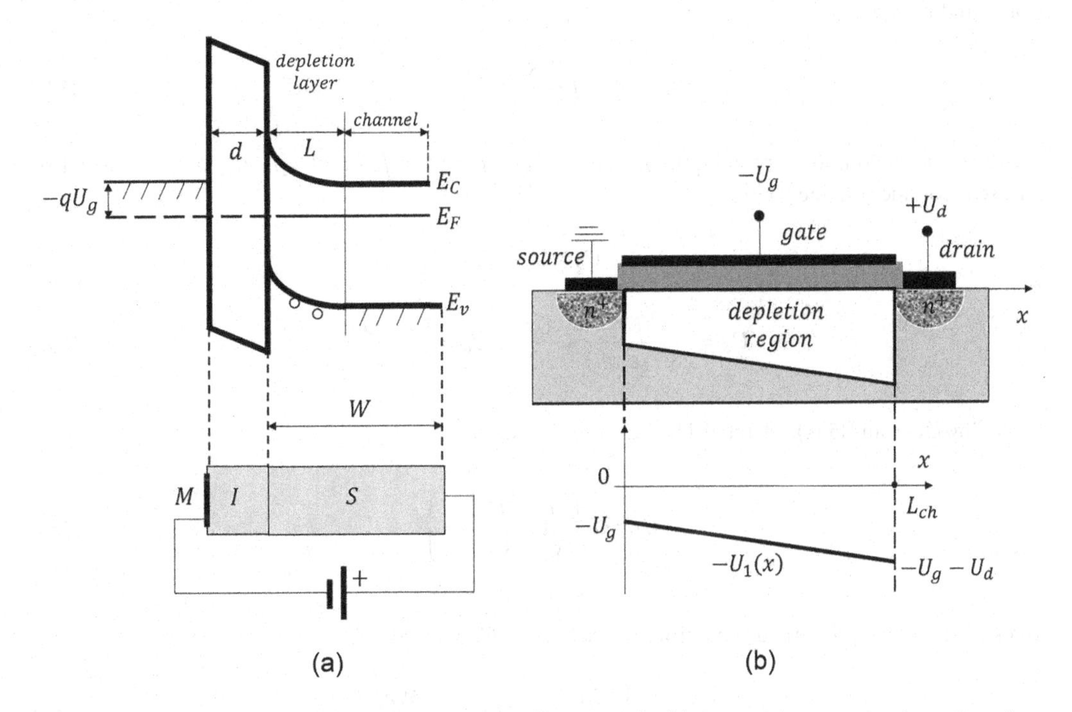

FIGURE 15.9 (a) MIS structure in a depletion normally-ON mode. (b) Semiconductor-gate voltage drop along the channel, $U_1(x) = U_g + xE_x$, $E_x = U_d/L_{ch}$.

First, we consider the MIS structure at zero drain voltage. Gate bias forces the transfer of charge $Q = CU_g$ and thus causes the drop of electron density in a semiconductor by CU_g/qSW, S being the channel area. The density of remaining electrons and their total charge become

$$n_r = n_0 - \frac{CU_g}{q\Omega},$$

$$Q_t' = qn_r\Omega = C\left(U_{th}' - U_g\right),$$

$$U_{th}' \equiv \frac{qn_0SW}{C} = U_{th}\frac{C_I}{C}, \tag{15.11}$$

where $\Omega = WS$ is the semiconductor volume, U_{th} is given in (15.2). At the threshold gate voltage, the channel is pinched-off: $n_r = 0$, meaning that the depletion region extends to the entire thickness of the semiconductor layer, completely blocks the channel, and should the voltage applied to drain, prevents current from flowing.

Insulator and depletion region present two capacitance in series with equal stored charges, so $CU_g = C_IU_I$, where U_I is the voltage drop on the insulator. Then, the total charge (15.11) expressed through U_I has the form,

$$Q_t' = C_I\left(U_{th} - U_I\right), \tag{15.12}$$

Expression (15.12) differs from (15.3) in that U_g is replaced with $-U_I$.

To analyze (15.12), it is instructive to find the relation between voltage drop on dielectric U_I and applied gate bias U_g. Nonequilibrium charge, accumulated in L-long depletion region, equals the one stored in dielectric:

$$qn_0SL = \frac{S\varepsilon_0\varepsilon_IU_I}{d}, \tag{15.13}$$

where geometry notations are in Figure 15.9. Bearing in mind $L = \sqrt{2\varepsilon_0\varepsilon_SU_S/qn_0}$ (U_S is the voltage drop on a semiconductor), one obtains

$$U_g = U_I + U_S$$

$$U_S = \frac{C_I^2U_I^2}{2S^2\varepsilon_0\varepsilon_Sqn_0} \equiv \frac{U_I^2}{2U_0}, \quad U_0 = \frac{S^2\varepsilon_0\varepsilon_Sqn_0}{C_I^2}. \tag{15.14}$$

Excluding U_S from (15.14), we relate U_I and U_g as follows:

$$U_I = U_0\left(\sqrt{1 + \frac{2U_g}{U_0}} - 1\right). \tag{15.15}$$

To estimate ratio U_g/U_0, we use maximal value $U_{g\max} = U_{th}' = qn_0SW/C$:

$$\frac{U_{g\max}}{U_0} = \left.\frac{WC_I\left(C_I + C_S\right)}{S\varepsilon_0\varepsilon_SC_S}\right|_{C_S \gg C_I} \approx \frac{W\varepsilon_I}{\varepsilon_Sd}. \tag{15.16}$$

Large C_S (see Eq. (14.47) Chapter 14) corresponds to the high conductivity and short depletion region in the semiconductor layer. Following (15.16), $U_g/U_0 < 1$ if $W\varepsilon_I < \varepsilon_S d$. That condition defines the *thin-film transistor* (TFT), meaning that the semiconductor layer is thinner than the insulator. As follows from (15.15) $U_I \approx U_g$, the entire gate voltage drops on the insulator. So, following (15.12), the total charge taking part in transport is

$$Q_t' \approx C_I\left(U_{th} - U_g\right). \tag{15.17}$$

When the drain is unbiased, the gate voltage modulates the charge in the channel, according to (15.17).

15.3.1 Current–Voltage Characteristics

Once positive voltage applied to drain, U_g, in (15.17) should be replaced by $U_1(x) = U_g + E_x x$, $E_x = U_d/L_{ch}$, and

$$Q_t'(x) = C_I\left(U_{th} - U_1(x)\right), \tag{15.18}$$

Semiconductor-gate voltage $U_1(x)$ is illustrated in Figure 15.9b. The next steps toward the calculation of *I-V* characteristics are the same as those done in Section 15.2.1. The result is

$$I_d = \frac{\mu C_I U_d}{L_{ch}^2}\left(U_{th} - U_g - \frac{U_d}{2}\right), \quad U_d \le U_{th} - U_g, \tag{15.19}$$

where $U_{th} - U_g$ is the saturation drain current. The only difference with (15.8) is the sign before U_g.

The *I-V* relation (15.19) is valid if $U_1(x) < U_0 < U_{th}'$. As shown in Figure 15.9b, at some point x, voltage $U_1(x)$ may exceed U_0 remaining less than pinch-off value U_{th}'. Approximation (15.17), which neglects the voltage drop on the semiconductor layer, does not hold, so the voltage drop in the depletion region, U_S, should be taken into account. If U_S is large enough (see Eq. (14.23), Chapter 14), the device switches to the inversion regime. Switching creates an *induced channel* at the near-insulator surface instead of the former neutral semiconductor region, as illustrated in Figure 15.3. An induced channel cannot form in thin-film transistors as upon increasing U_g, pinch-off happens before inversion.

The channel potential depends on U_g and varies along the channel, as shown in Figure 15.9b. Inversion happens first in the region where channel potential, $U_1(x)$, is maximal (near the drain), and then, upon increasing gate voltage, extends everywhere in the channel. So, induced channel FET operates at gate *threshold voltage* $U_g > U_T$, which provides for inversion channel opening. Example device layout is in Figure 15.10.

Induced channel FETs rely on a layer with inverted conduction and require large threshold gate bias to open the channel. In some flavors of FET, additional doping helps to create an inverted channel, and thus, to decrease the threshold. For example, for the option shown in Figure 15.10, additional p-doping

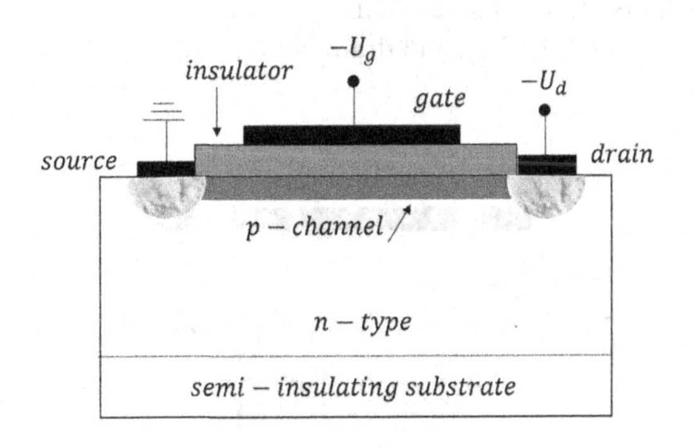

FIGURE 15.10 p-channel induced by a negatively biased gate (inverse layer).

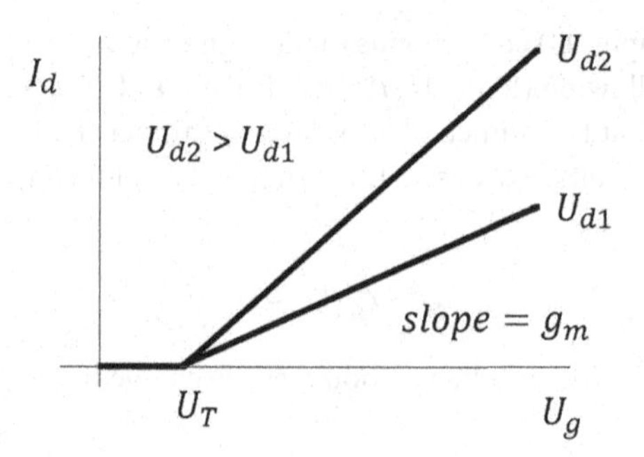

FIGURE 15.11 Transfer characteristics, U_T, is the threshold voltage.

in the near-interface region by acceptor ion implantation enhances carrier density and thus decreases threshold gate voltage. This type of n- and p-channels operate in so-called *enhancement mode*.

The drain current is zero at $0 < U_g < U_T$ (normally-off FET), while above the threshold $(U_g \geq U_T)$, the current–voltage characteristic is similar to (15.8) and (15.19):

$$I_d = \frac{\mu C_I U_d}{L_{ch}^2}\left(U_g - U_T - \frac{U_d}{2}\right), \quad U_d \leq U_g - U_T, \tag{15.20}$$

The current reaches the maximum at $U_{dsat} = U_g - U_T$: $I_{d\max} = \mu C_I \left(U_g - U_T\right)^2 / 2L_{ch}^2$. The *I-V* relation (15.20) is qualitatively the same as shown in Figure 15.7, while transfer characteristic $I_d(U_g)$ is illustrated in Figure 15.11.

The n-channel has higher electron mobility. Similar to p-channel FET shown in (15.10), one can arrange the induced n-channel in p-semiconductor by applying the positive above-threshold gate voltage.

Bearing in mind signs of U_g in (15.8), (15.19), and (15.20), one concludes that transconductance is positive in accumulation and inversion regimes and negative in the depletion mode.

When explaining the fundamentals of the device operation, Eqs. (15.8)–(15.10) and (15.19) were obtained within the linear model in a flat-band approximation. To create the flat-band alignment, one needs to apply additional "flat-band voltage". To account for the flat-band voltage (it equals the built-in potential in a space charge region), one has to shift U_g in (15.8) and (15.19): $U_g \rightarrow U_g \pm U_{FB}$, where the sign depends on operation regime. For more details on FET regimes and modes of the operation, see Ref. [3].

15.4 FET Extrinsic Parameters

Source and drain voltages used in (15.8)–(15.10) and (15.19) do not account for finite resistance of metal/n^+-semiconductor contacts, as shown in Figure 15.12.

Measurable (extrinsic) gate-source, U_{gs}, and drain-source, U_{ds}, voltages account for voltage drop on parasitic resistances:

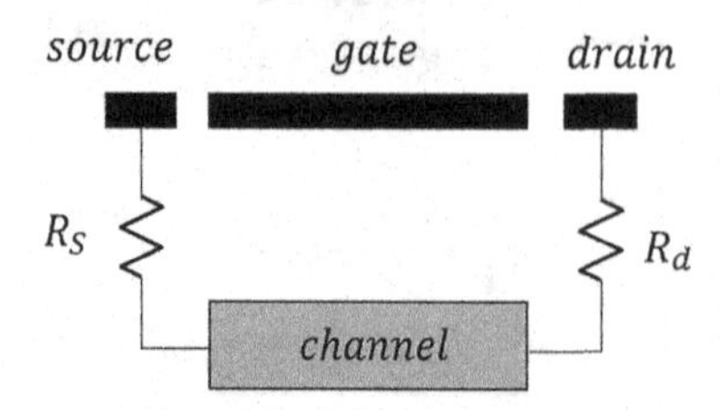

FIGURE 15.12 Parasitic resistances in source and drain contacts.

$$U_{ds} = U_d + I_d\left(R_d + R_S\right),$$

$$U_{gs} = U_g + I_d R_S. \tag{15.21}$$

Extrinsic conductance g'_d and transconductance g'_m can be expressed through their intrinsic values (15.9) by calculating derivatives of drain current (15.8) as an implicit function $F\left(U_{gs}, U_{ds}\right) = I_d - \mu C_I U_d \left(U_{th} + U_g - \frac{U_d}{2}\right) \Big/ L_{ch}^2$ where U_g and U_d are given in (15.21):

$$\frac{g_d}{g'_d} = \frac{g_m}{g'_m} = 1 + g_d\left(R_d + R_S\right) + g_m R_S. \tag{15.22}$$

Values of g'_d and g'_m follow from experimental dc *I-V* characteristics.

15.5 High-Electron Mobility Transistors

Conductivity in the channel depends on carrier density and mobility. To increase carrier concentration, one needs a heavily doped semiconductor, which suppresses carrier mobility because of impurity scattering. The issue resolves in FET flavor called HEMT (high electron mobility transistor). Unlike MISFET that uses the MIS structure, the HEMT is a Metal-Semiconductor-Semiconductor structure, where two semiconductors of different bandgaps form a heterostructure.

15.5.1 Remote Doping

Example heterostructure, shown in Figure 15.14, comprises n-doped $Al_xGa_{1-x}As$ and not intentionally doped GaAs. Electron exchange across the interface creates the depletion region in the barrier and accumulates 2D electrons in the well. The built-in electric field resulting from the redistribution of electrons reaches 10^7 V/m and manifests itself as band bending in Figure 15.13.

Filling up the GaAs quantum well by electrons from ionized donor levels in the AlGaAs barrier separates electrons from scattering centers, as illustrated in Figure 15.13a. That is *remote doping* [4]. The spatial separation of ionized impurities from channel electrons suppresses scattering, making the mobility of 2D electrons much higher than in bulk GaAs. Superlattices doped selectively and periodically (see Figure 15.13b) present *modulation doping.* Modulation-doped superlattice allows achieving an electron density of 5×10^{16} cm^{-3} and electron mobility that exceeds 10^6 cm^2/Vs at cryogenic temperatures.

High mobility implies that the interface is sharp and free of defects, so the scattering on interface potential is suppressed. First high-quality interfaces were grown with $Al_xGa_{1-x}As$/GaAs material system as the interface is lattice-matched at any Al content, which prevents the formation of mismatch

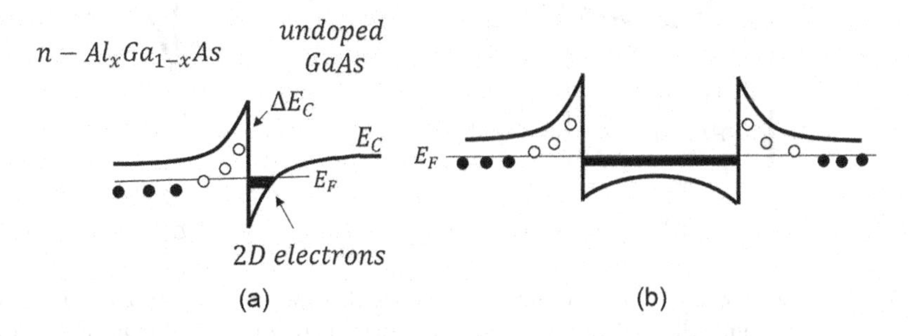

FIGURE 15.13 (a) Single interface, ΔE_C is the conduction band offset. (b) A single period in a superlattice.

dislocations. Record mobility in other III-V lattice-matched combinations grown by molecular-beam epitaxy and metal-organic chemical vapor deposition has also been reported [5].

In HEMT, the heterostructure forms the Schottky barrier to gate metal, as illustrated in Figure 15.14. If the thickness of the AlGaAs layer is not larger than the depletion region in the Schottky contact, the built-in electric field forces all donors to ionize and supply electrons to the well. If the barrier thickness is large, additional negative gate voltage does the same and thus increases the electron density in the channel (Figure 15.14a).

Triangular quantum well is $(50-100)$ $\dot{A}$ wide and $\Delta E_C(x=0.3)\approx 0.3\,\text{eV}$ deep. The wave function of a 2D electron is a solution of the Schrödinger equation that is found by the variational method. Fang-Howard trial function $\Psi_1(z)$ has the form,

$$\Psi_1(z)=\begin{cases} \dfrac{A\,z_0 b^{3/2}}{\sqrt{2}}\exp(-\kappa_b z), & z\le 0, \\ \dfrac{A(z+z_0)b^{3/2}}{\sqrt{2}}\exp(-bz/2), & z>0, \end{cases}$$

$$\kappa_b=\sqrt{2m_{zB}\Delta E_C/\hbar^2}, \tag{15.23}$$

where b is the variational parameter, m_{zB} is the electron effective mass of barrier material in the z-direction, z_0 and A follow from normalization and matching conditions for the electron flux through the interface at $z=0$:

$$z_0^{-1}=\frac{\kappa_b m_{zC}}{m_{zB}}+\frac{b}{2},\quad A^2=\frac{4z_0^{-2}}{\left(2z_0^{-1}+b\right)^2+b^2+b^3\kappa_b^{-1}}, \tag{15.24}$$

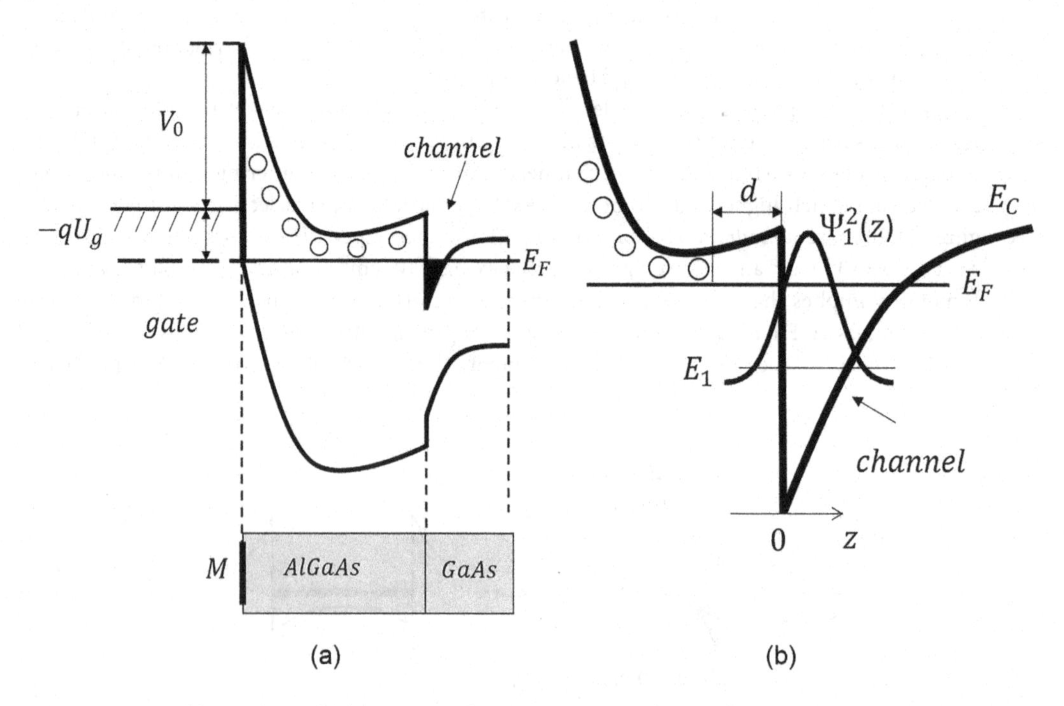

FIGURE 15.14 (a) Gated Metal/AlGaAs/GaAs heterostructure. Hollow circles depict ionized (empty) donors. The black area is a quantum well filled with electrons. V_0 is the Schottky barrier. (b) One-electron probability density in the quantum well ground state $\psi_1(z)$.

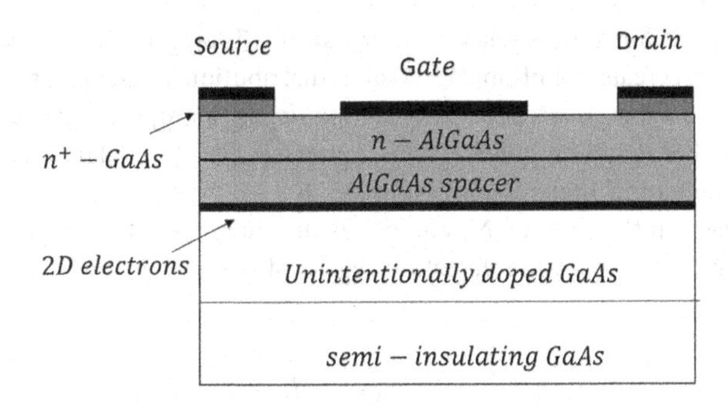

FIGURE 15.15 Cross-section of AlGaAs / GaAs HEMT.

where m_{zC} is the effective mass in the channel.

The expectation value of total average electron energy E_{av} consists of three parts: the kinetic energy, the energy in the channel electric field $F_C(z)$, and the potential energy induced by other electrons in the channel:

$$E_{av} = \langle T \rangle + \langle V_C \rangle + \frac{1}{2}\langle V_S \rangle, \quad \langle T \rangle = \left\langle -\frac{\hbar^2}{2m_{zC}} \frac{\partial^2}{\partial z^2} \right\rangle,$$

$$\langle V_C \rangle = q F_C \langle z \rangle, \quad \langle V_S \rangle = \frac{q^2 N_S}{\varepsilon_0 \varepsilon_C} \left\langle z \int_z^{\infty} \Psi_1^2(y)dy + \int_0^z y\, \Psi_1^2(y)\, dy \right\rangle,$$

$$\langle \ldots \rangle \equiv \int_{-\infty}^{\infty} \Psi_1(z)(\ldots)\Psi_1(z)dz, \tag{15.25}$$

where ε_C is the dielectric constant of channel material. For details, see Refs. [6,7].

Average electron energy in the well E_{av} is subject to minimization with respect to parameter b. Energy $E_1 = \min[E_{av}]$ is the ground state level in the channel. Parameter $b_{\min}$ determines the spatial size of the wave function in the z-direction. Numerical solution $\Psi^2(z)$ is shown in Figure 15.14b. Due to the finite barrier height, $\Psi(z)$ penetrates the barrier and causes electron scattering on ionized impurities (penetration renders as a tail of $\Psi^2(z)$ inside AlGaAs in Figure 15.14). For further suppression of scattering, usually, an undoped AlGaAs spacer is placed between the doped AlGaAs region and the interface: $-d < z < 0$ in Figure 15.14b.

The ability to manipulate the carrier density in the channel with a gate voltage along with high electron mobility makes the remotely doped structure a high electron mobility field-effect transistor (HEMT). The first demonstration of HEMT was reported in 1980 [8]. Schematic layout of AlGaAs HEMT is illustrated in Figure 15.15.

Due to high electron mobility, the device operates at a high speed that makes it suitable for various millimeter-wave applications.

15.5.2 Polarization Doping in III-Nitride HEMTs

High-mobility 2D-electron channels by remote doping also exist in wide bandgap semiconductors such as III-Nitride ternary alloys AlGaN and InGaN. Due to wide bandgap, GaN/AlGaN structures withstand breakdown voltage much higher than their narrow-gap GaAs -based counterparts and thus are suitable for high-power device applications.

When standing alone, the AlGaN layer shows spontaneous polarization. If the substrate is lattice-mismatched to AlGaN, the strain induces piezoelectric polarization. Once layer thickness exceeds the critical value, dislocations form, thus relaxing misfit strain. Total polarization in thin layers develops built-in electric fields and creates additional electron redistribution, thus increasing the electron density in the channel. So, there appears so-called *polarization doping* along with remote doping discussed in the preceding section. For detail on polarization doping, see Ref. [9]. Polarization and remote doping result in high-mobility channel illustrated in Figure 15.16.

Sheet electron density in the channel N_S comprises the charge induced by polarization and the one transferred from ionized donors in the depletion region of length l (if the barrier is doped to volume density n_D):

$$N_S = n_S + n_D l,$$

$$n_S = -\frac{P}{q} - \frac{\varepsilon_0 \varepsilon_B}{q^2 d}\left[V_0 - qU_g - \Delta E_C + E_F\right]. \tag{15.26}$$

where ε_B is the AlGaN dielectric constant and other parameters are defined in Figure 15.16. The total polarization, P, is negative in the structure under consideration for any Al content in AlGaN. To determine N_S, we use (15.26) along with condition for chemical potential μ in 2D electron gas (see Chapter 4):

$$N_S = \frac{mk_BT}{\pi\hbar^2}\log\left[1+\exp\left(\frac{\mu - E_1}{k_BT}\right)\right], \tag{15.27}$$

where E_1 is the ground state energy in the well shown in Figure 15.16, m is the in-plane electron effective mass. Remote and polarization doping result in electron density in the channel of the order of 10^{13} cm^{-2}. Electron mobility in heterostructures reaches values (1,000–2,000) cm^2/Vs at room temperature, which is several times higher than mobility in thick GaN layers (Figure 15.17).

Materials used in device fabrication may vary. For high-power HEMT, SiC and Si substrates are more preferred as they have high thermal conductivity as compared to Sapphire. Native free-standing GaN substrate might considerably decrease the density of threading dislocations, and would be the best choice

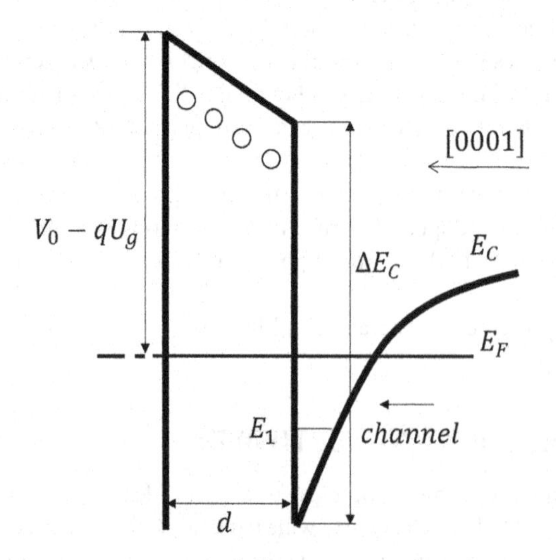

FIGURE 15.16 Conduction band profile in AlGaN/GaN heterostructure in the growth direction.

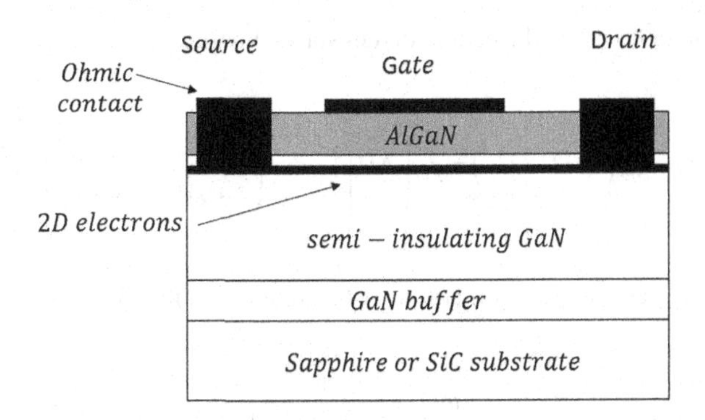

FIGURE 15.17 Schematic layout of GaN/AlGaN HEMT.

at the time the growth technology of bulk GaN becomes mature. The barrier material $Al_xGa_{1-x}N$ could be used for as high Al content as $x = 1$, keeping barrier thickness below the critical value $(30-50)\dot{A}$ that prevents dislocation formation.

Wide bandgap and high conduction band offset in GaN / AlGaN material system enhance electron confinement, provide for high electron saturation velocity of 2.5×10^7 cm/s, and breakdown electric field of 3.3 MV/cm, thus facilitating device operation at high output power.

Example of GaN/AlGaN HEMT layout is shown in Fig.15.17.

For more details on III-Nitride HEMTs, see the review [10].

15.6 Frequency Response and Power Characteristics

If the gate voltage exceeds the threshold and an ac-signal is applied, full lump element equivalent circuit of FET is given in Figure 15.19 [11]. The circuit comprises intrinsic device elements: gate-source resistance and capacitance, R_{gs}, C_{gs}, gate-drain capacitance C_{gd}, as well as drain-source resistance and capacitance R_{ds}, C_{ds}. The circuit includes terminal resistances R_d, R_s, R_g, which could be internal contact resistances or lump circuit elements. Resistance R_m is the variable element aimed at matching the input to the rest of the circuit.

Two current sources correspond to the drain current, which is the sum of dc-current in operating point (I_d, U_d) and ac-current i_d induced by a signal applied to the gate.

15.6.1 Switching Time

When the channel is closed (normally-off device), it takes time for the gate voltage to reach above-threshold value U_T. The delay is the time constant $R_g\left(C_{gs}+C_{gd}\right)$. If the channel is closed, both capacitances should be taken into account, as usually $C_{gs} \geq C_{gd}$. Should abrupt voltage step ΔU be applied to the gate, the gate voltage would follow the temporal evolution

$$U_g(t) = \Delta U\left[1-\exp\left(-\frac{t}{R_g\left(C_{gs}+C_{gd}\right)}\right)\right]. \tag{15.28}$$

The *turn-on time delay* follows from condition $U_g(t_d) = U_T$:

$$t_d = -R_g\left(C_{gs}+C_{gd}\right)\log\left(1-\frac{U_T}{\Delta U}\right). \tag{15.29}$$

Following the turn-on delay time, the drain current starts flowing:

$$I_d(t) = g_m\left(U_g(t) - U_T\right) = g_m\left(\Delta U\left[1 - \exp\left(-\frac{t}{R_g\left(C_{gs} + C_{gd}\right)}\right)\right] - U_T\right). \tag{15.30}$$

Frequency of switching between turn-on and turn-off states is limited by

$$\omega_0 = \frac{1}{R_g\left(C_{gs} + C_{gd}\right)}. \tag{15.31}$$

High-frequency switching depends on parasitic capacitances managed to some extent by varying geometry of the device and technology of its fabrication.

15.6.2 Output Power, Power Gain, and Power-Added Efficiency

As illustrated in the example in Figure 15.9b, under the drain voltage, the depletion region increases (its capacitance decreases) towards the drain, thus at $U_d \neq 0$, C_{gd} drops even more compared to C_{gs}: $C_{gs} \gg C_{gd}$ (at least, one order of magnitude). Small C_{gd} allows us to simplify the equivalent circuit in Figure 15.18. Assuming also that resistance of the insulator layer $R_{gs} \rightarrow \infty$, the small-signal equivalent circuit takes the form rendered in Figure 15.19.

Circuit elements in Figure 15.19 include the source and load impedances Z_S and Z_L, respectively. Input current i_{in}, gate voltage u_g, output (load) current i_{out}, and output voltage u_L in Figure 15.19 are:

$$i_{in} = \frac{u_{in}}{Z_S + Z_{in}}, \quad i_{out} = \frac{i_d Z_{out}}{Z_L + Z_{out}}, \quad u_g = \frac{u_{in} Z_{in}}{Z_S + Z_{in}},$$

$$u_{out} = i_{out}\, Z_L = \frac{i_d\, Z_{out} Z_L}{Z_L + Z_{out}}, \quad i_d = g_m\, u_g, \tag{15.32}$$

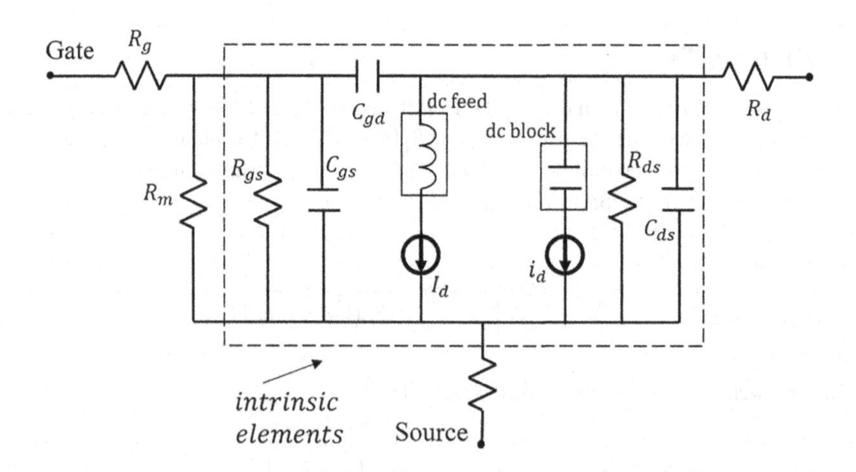

FIGURE 15.18 Lump-element circuit of the common-source amplifier.

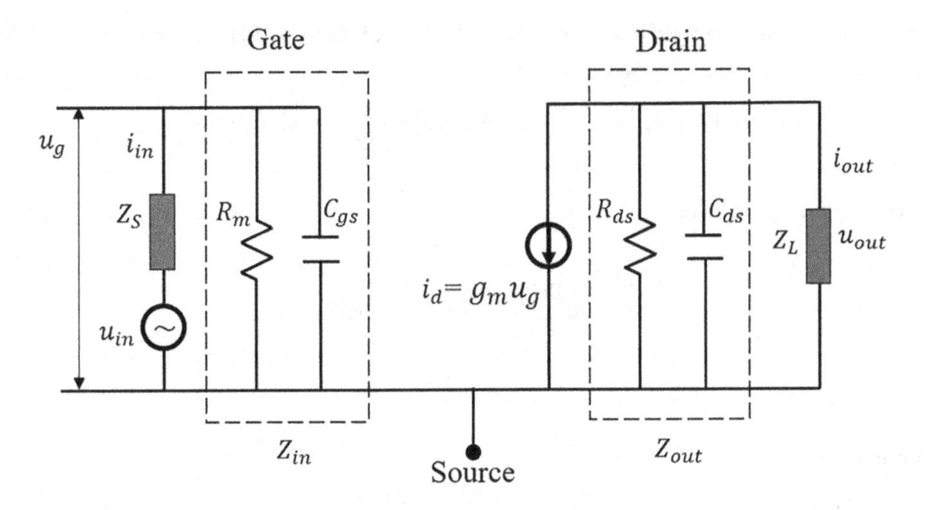

FIGURE 15.19 Small-signal equivalent circuit of common-source amplifier: u_{in}, u_g, and i_d are the complex amplitudes of signal, gate voltage, and drain ac-current, respectively.

where

$$Z_{in} = \frac{R_m}{i\omega C_{gs} R_m + 1}, \quad Z_{out} = \frac{R_{ds}}{i\omega C_{ds} R_{ds} + 1}. \tag{15.33}$$

Current and voltage gains. To characterize signal amplification, one uses parameters such as current gain, voltage gain, and power gain: the ratio of output and input values. From (15.32), one obtains current and voltage gain as follows:

$$A_i = \frac{i_{out}}{i_{in}} = \frac{g_m Z_{in} Z_{out}}{(Z_L + Z_{out})}, \quad A_u = \frac{u_{out}}{u_g} = \frac{g_m Z_{out} Z_L}{Z_L + Z_{out}}. \tag{15.34}$$

Maximum frequency. Frequency that allows amplification is restricted by the condition $|A| > 1$. For the current gain:

$$\left| \frac{g_m Z_{in} Z_{out}}{(Z_L + Z_{out})} \right| > 1,$$

or, in short circuit $(Z_L = 0)$,

$$g_m |Z_{in}| \approx \frac{g_m}{\omega C_{gs}} > 1,$$

$$\omega < \frac{g_m}{C_{gs}}. \tag{15.35}$$

We have used $\omega R_m C_{gs} \gg 1$ – realistic condition for radio-frequency and microwave range. Similarly to (15.35), the frequency limit for voltage gain is $\omega < g_m / C_{ds}$.

As follows from (15.9) and (15.35), limiting frequency depends on transfer characteristics. High carrier mobility and short channel length favor high maximum operating frequency. These parameters are the functions of device geometry and choice of materials.

Power gain. The cycle averaged power taken from the source is expressed as (* means complex conjugate)

$$P_{in} = \frac{1}{2}\mathrm{Re}\left[i_{in}u_g^*\right] = \frac{|u_S|^2}{2\,|Z_S + Z_{in}|^2}\mathrm{Re}\left[Z_{in}\right] = \frac{|u_S|^2 R_m}{2\,|Z_S + Z_{in}|^2\left(1+\omega^2 C_{gs}^2 R_m^2\right)}. \quad (15.36)$$

Power transferred to the load:

$$P_{out} = \frac{1}{2}\mathrm{Re}\left[\,i_{out}u_{out}^*\,\right] = \frac{|i_d|^2 |Z_{out}|^2 R_L}{2\,|Z_L + Z_{out}|^2},$$

$$R_L \equiv \mathrm{Re}\left[Z_L\right]. \quad (15.37)$$

and power gain,

$$A_P = \frac{P_{out}}{P_{in}} = \frac{|i_d|^2\,|Z_{out}|^2 |Z_S + Z_{in}|^2 R_L}{|u_S|^2 |Z_L + Z_{out}|^2\,\mathrm{Re}\left[Z_{in}\right]}. \quad (15.38)$$

If the source and load ideally match the input and output, respectively, $Z_{in} = Z_S^*$ and $Z_{out} = Z_L^*$, power gain can be simplified to

$$A_P = \frac{|i_d|^2 R_{ds} R_m}{|u_S|^2\left(1+\omega^2 R_m^2 C_{gs}^2\right)}. \quad (15.39)$$

Power of amplified signal as a fraction of dc-power taken from the source is called *power-added efficiency* and defined as

$$\eta = \frac{P_{out} - P_{in}}{P_{dc}},$$

$$\eta_{dB} = 10\log_{10}(\eta), \quad (15.40)$$

where η_{dB} is the efficiency in decibels, $P_{dc} = I_d U_d$ and dc-current and voltage correspond to an operating point on *I-V* characteristics.

Class A and Class B operations. Intrinsic circuit elements depend on implementations in design and materials. To extract parameters, one may use a finite-element numerical package. Estimated with the Atlas-Silvaco simulator [12] and Eq. (15.38), the small-signal power gain in $GaN/Al_{0.3}Ga_{0.7}N$ HEMT is 15.3 dB (10 GHz).

Analytical description in (15.32)–(15.39) implies using full swings of drain current and gate voltage, which is appropriate under small-signal conditions. The large-signal calculations require numerical technics and results depend on the regime of operation, which is determined by the operating point shown in Figure 15.20. Under class A operation conditions, one chooses the gate bias U_{TA} to accommodate the full signal swing in the above-threshold region (point *A* in Figure 15.20). This type of operation provides highly linear amplification.

Under Class B conditions, only half of the gate voltage cycle is above the threshold (point B in Figure 15.20). The other half keeps the channel pinched-off. Under Class B conditions, one may use signal amplitude much larger than in point A, thus supporting high power operation at high efficiency, however violating linearity typical for A-class operation. To reinstate linearity, one needs two Class B-operating complementary transistors. For details of using FETs in electronic circuits, see, for example, Ref. [13].

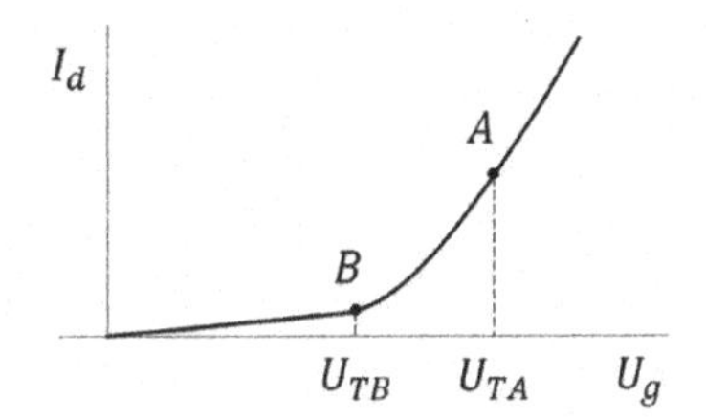

FIGURE 15.20 Operating points for class A and class B operations.

References

1. http://bsim.berkeley.edu/.
2. F. Sacconi, A. Di Carlo, P. Lugli, and H. Morkoç, "Spontaneous and piezoelectric polarization effects on the output characteristics of AlGaN/GaN heterojunction modulation doped FETs", *IEEE Trans. Electron Devices.* **48**(3), 450 (2001).
3. M. Shur, *Physics of Semiconductor Devices* (Prentice Hall, Englewood Cliffs, NJ, 1990).
4. R. Dingle, H.L. Stormer, A.C. Gossard, and W. Wiegmann, "Electron mobilities in modulation-doped semiconductor heterojunction superlattices", *Appl. Phys. Lett.* **33**, 665 (1978).
5. E.E. Mendez, "Electron mobility in semiconductor heterostructures", *IEEE J. Quant. Electr.* **QE-22**(9) (1986).
6 T. Ando, B. Fowler, and F. Stern, "Electronic properties of two-dimensional systems", *Rev. Mod. Phys.* **54**(2), (1982).
7. G. Bastard, *Wave Mechanics Applied to Semiconductor Heterostructures* (J. Wiley & Sons Inc, New York, 1988).
8. T. Mimura, S. Hiyamizu, T. Fujii, and K. Nanbu, "A new field-effect transistor with selectively doped GaAs/n-AlGaAs heterojunctions", *Jpn. J. Appl. Phys.* **19**, L225 (1980).
9. V.I. Litvinov, *Wide Bandgap Semiconductor Spintronics* (Pan Stanford, Singapore, 2016).
10. F. Roccaforte, G. Greco, P. Fiorenza, and F. Iucolano, "An overview of normally-off GaN-based high electron mobility transistors", *Materials.* **12**, 1599 (2019).
11. B. Toner and V.F. Fusco, "Large-signal modeling of frequency dispersion effects in submicron MOSFET devices", *Microw. Opt. Technol. Lett.* **34**, 429 (2002).
12. https://silvaco.com/examples/tcad/.
13. A.S. Sedra and K.C. Smith, *Microelectronic Circuits* (6th ed., Oxford University Press, Oxford, UK, 2010).

16

Semiconductor Lasers

The interband optical absorption coefficient in semiconductors (see Eq. (12.69) Chapter 12) depends on occupation numbers in conduction and valence bands, $\alpha \sim f_v - f_c$ where $f_{c,v}$ are Fermi functions with the equilibrium chemical potential same for electrons and holes. The absorption coefficient determines the exponential drop of electromagnetic power with distance (see Chapter 12): $\bar{\boldsymbol{P}} \sim \exp(-\alpha z)$. If in some range of $\boldsymbol{k}$ the difference $f_v(\boldsymbol{k}) - f_c(\boldsymbol{k})$ and thus α were negative, it would correspond to amplification at all frequencies $\hbar\omega = E_c(\boldsymbol{k}) - E_v(\boldsymbol{k})$ in that range, meaning that the medium becomes active and thus power increases with distance. Sign change is called *population inversion*. That is a nonequilibrium state, which converts the active region into a gain medium and favors the dominance of stimulated emission over recombination. So, population inversion presents the necessary condition for coherent emission generated by a laser ("light amplification by stimulated emission of radiation"). A more detailed discussion of lasing conditions follows.

16.1 Quasi-Fermi Levels and Population Inversion

Nonequilibrium electron–hole pairs appear by the valence-to-conduction band optical pumping or carrier injection into a *p-n* semiconductor diode. By the method of creating nonequilibrium carriers, one distinguishes injection and optically pumped semiconductor lasers.

In injection lasers, the active region, where light emission comes from, is sandwiched between the *p*- and *n*-type semiconductors in a *p-n* diode or between cladding layers in the *p-n* double heterostructure. The forward bias applied across the *p-n* junction injects carriers into the active region.

When steady-state *e*-*h* generation occurs at energy $\varepsilon_{0c,v}$, the shape of initial distribution functions for hot electrons and holes are close to $\delta(\varepsilon - \varepsilon_{0c,v})$ as schematically shown in Figure 16.1a. Recombination (radiative and non-radiative) determines the lifetime of excited carriers, τ_{rec}. In the transient process, during the time $\tau_T \ll \tau_{rec}$ carriers thermalize due to phonon emission caused by inelastic electron-phonon interaction as illustrated in Figure 16.1b.

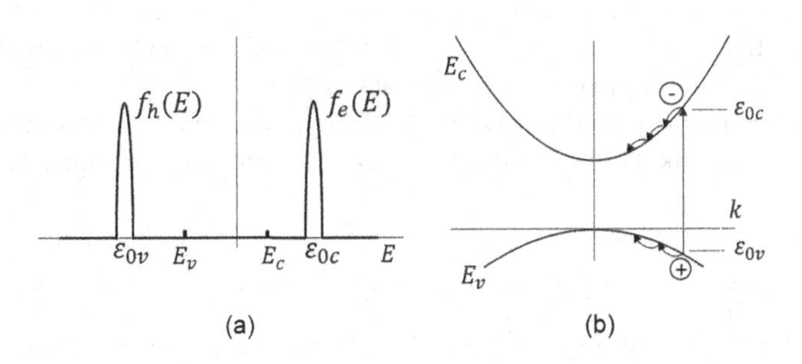

FIGURE 16.1 Dynamics of population inversion. (a) Initial carrier distributions, (b) phonon-induced thermalization of electrons and holes.

DOI: 10.1201/9780429285929-16

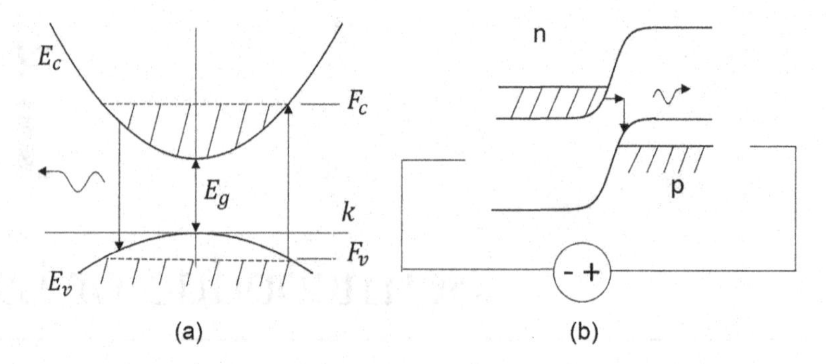

FIGURE 16.2 Population inversion. Dashing shows regions filled with electrons. Wavy arrows indicate photon emission. (a) An upward arrow depicts optical pumping. (b) Space charge region in *p-n* junctions; arrow renders tunneling-assisted transition in forward-biased diode lasers.

Thermalization means that initially hot electrons and holes cool down, so the carrier distributions tend to Fermi functions at ambient temperature but with different *quasi-Fermi levels* $F_{c,v}$ determined by occupation numbers in c- and v-bands, as shown in Figure 16.2.

In the steady-state, the nonequilibrium difference in electron occupation numbers of states related by interband vertical transition $\hbar\omega > E_g$ equals to

$$f_v(\mathbf{k}) - f_c(\mathbf{k}) = f(E_{v\mathbf{k}}) - f(E_{v\mathbf{k}} + \hbar\omega) = \frac{1}{1+\exp\left[\beta(E_{v\mathbf{k}} - F_v)\right]} - \frac{1}{1+\exp\left[\beta(E_{v\mathbf{k}} + \hbar\omega - F_c)\right]},$$

$$\beta = \frac{1}{k_B T}. \tag{16.1}$$

Negative absorption corresponds to light amplification and takes place when the difference (16.1) is negative:

$$F_c - F_v > \hbar\omega \geq E_g. \tag{16.2}$$

In other words, the amplification is possible if downward transition reduces free energy by the amount greater than photon energy.

Net stimulated emission rate r_{st}, that is, a difference between rates of photon-stimulated downward and upward transitions, is proportional to the term $f_c(\mathbf{k}) - f_v(\mathbf{k})$, which becomes positive in a nonequilibrium state. The relation between spontaneous and stimulated emission rates [1,2] (see Chapter 13):

$$\frac{r_{st}}{r_{sp}} = 1 - \exp\left[-\beta(F_c - F_v - \hbar\omega)\right]. \tag{16.3}$$

As r_{sp} is always positive, $r_{st} > 0$ if the population inversion holds for transitions with energy $\hbar\omega$. The stimulated emission is at the origin of the laser operations.

At all frequencies that satisfy condition (16.2), the absorption coefficient is negative and called *material gain*, $g(\omega) = -\alpha(\omega)$. Taking the absorption coefficient from (12.69) (Chapter 12), one can present gain as follows:

$$g(\omega) = C\rho(\hbar\omega - E_g) \times \left\{ \frac{1}{1+\exp\left[\beta\left(\frac{E_g}{2} + \frac{m_r}{m_c}(\hbar\omega - E_g) - F_c\right)\right]} - \frac{1}{1+\exp\left[\beta\left(-\frac{E_g}{2} - \frac{m_r}{m_v}(\hbar\omega - E_g) - F_v\right)\right]} \right\},$$

$$C \equiv \frac{\pi e^2 \mu Z_0}{m^2 n_r \omega} \left| \boldsymbol{e} \cdot \boldsymbol{p}_{cv}(0) \right|^2, \quad \rho(E) = \frac{(2m_r)^{3/2}}{2\pi^2 \hbar^3} \sqrt{E}. \tag{16.4}$$

Intraband scattering processes take part in carrier thermalization and relax energy conservation in interband transition that can be described by replacing δ-function with Laplacian in (12.69) (Chapter 12):

$$\delta\left(E_v(\boldsymbol{k}) - E_c(\boldsymbol{k}') + \hbar\omega\right) \to L(\omega, E) = \frac{1}{2\pi} \frac{\gamma}{\left(E_g + E - \hbar\omega\right)^2 + (\gamma/2)^2}, \tag{16.5}$$

where γ is the linewidth due to intraband relaxation caused by electron interaction with phonons, electrons, and impurities. The energy uncertainty relates to the intraband relaxation time as $\tau_i = \hbar/\gamma$. The gain takes the form

$$g(\omega) = C \int dE\, \rho(E) L(\omega, E) \left\{ \frac{1}{1 + \exp\left[\beta\left(\frac{E_g}{2} + \frac{m_r}{m_c} E - F_c\right)\right]} - \frac{1}{1 + \exp\left[\beta\left(-\frac{E_g}{2} - \frac{m_r}{m_v} E - F_v\right)\right]} \right\}, \tag{16.6}$$

Spectral gain (16.4) normalized on its maximum value is shown in Figure 16.3a.

The graph in Figure 16.3a shows that absorption turns zero in *transparency point* $\hbar\omega = F_c - F_v$, and then the gain reaches a maximum at $\hbar\omega_0$, the lasing frequency. Forward current supplies nonequilibrium electrons and holes in the active region $n = p$, increases $F_c - F_v$, thus controlling maximum gain g_m, as illustrated in Figure 16.3b.

Near transparency point, one can approximate maximum gain by the linear function of nonequilibrium carrier density, $g_m = g_d(n - n_{tr})$, where g_d is the *differential gain*, and n_{tr} is the *transparency carrier density*.

For lasing, the gain should exceed optical losses. Increasing driving current makes higher the maximum gain until it exceeds the threshold value for lasing to start. A further current increase does not affect gain as all additional electrical power goes to optical intensity of stimulated emission. After the current reaches the threshold value, the emission intensity and density of e–h pairs are self-consistent.

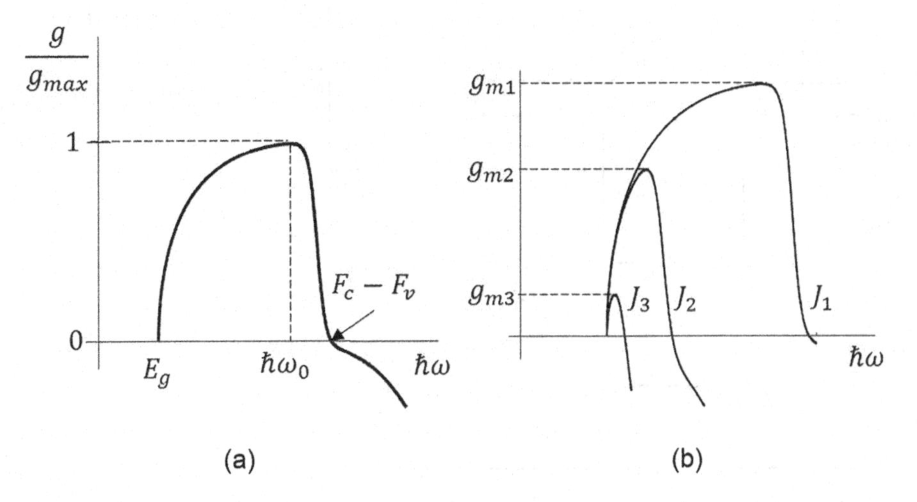

FIGURE 16.3 (a) Normalized spectral gain (16.4). Lasing frequency $\hbar\omega_0$ corresponds to maximum gain. (b) Gain spectrum at different injection levels: $J_1 > J_2 > J_3$.

As laser operation relies on stimulated emission, the device design should provide for as many photons in the active region as possible, so it requires creating a cavity where photons make roundtrips before they leave the gain region. In the next section, we discuss how the device design determines the gain, optical losses, and the threshold carrier density.

16.2 Diode Laser Design

The layout rendered in Figure 16.4 shows a typical structure of an edge-emitting semiconductor laser.

There are several options for implementing the generic active region in Figure 16.4. Historically first and most simple is a single *p-n* homojunction, which has the band diagram shown in Figure 16.2b (for a review, see [3]). Other options include double heterostructure, single or multiquantum well (MQW), and quantum dots as illustrated in Figure 16.5.

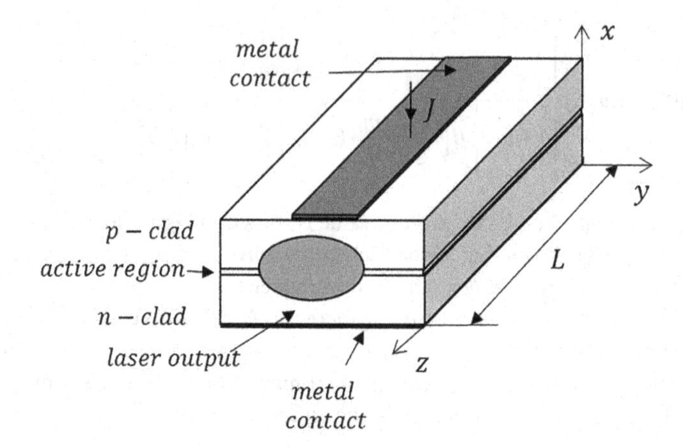

FIGURE 16.4 Schematics of edge-emitting injection laser. Single *p-n* junction (or heterojunction) gain-guided device.

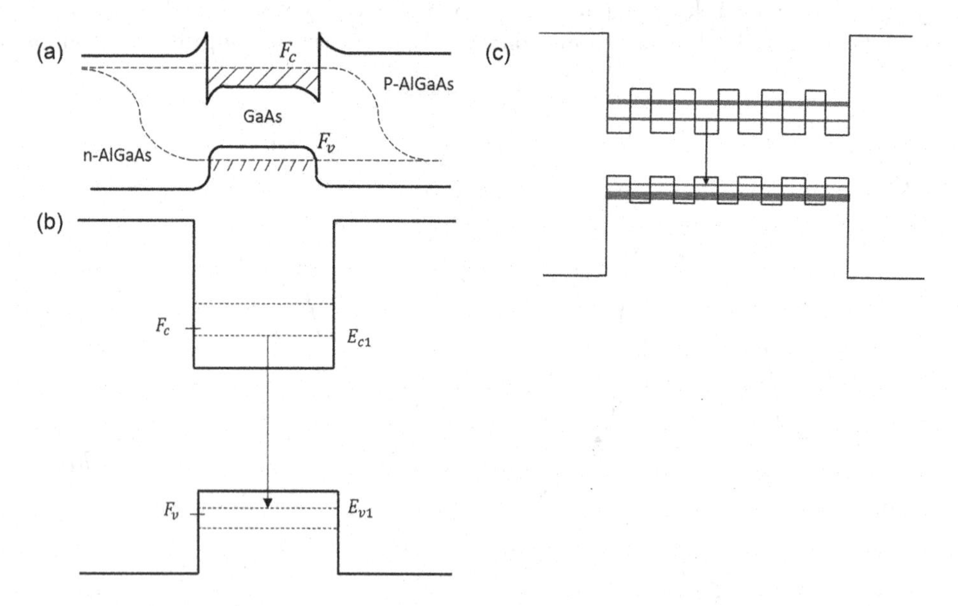

FIGURE 16.5 Band diagram in an active region. (a) Double heterostructure. (b) Single quantum well. (c) Multiquantum well. Arrows indicate laser emission that involves the lowest size-quantized energy levels.

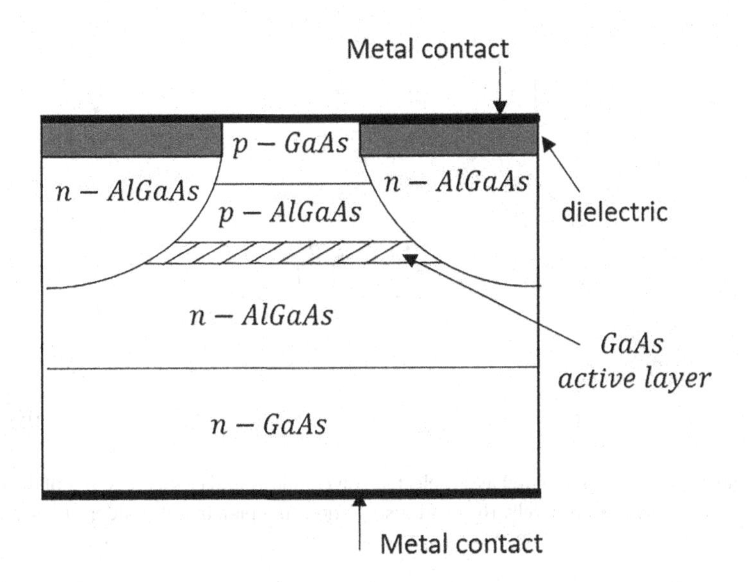

FIGURE 16.6 Example double-heterostructure laser. The band diagram in a double heterostructure active region is shown in Figure 16.5a.

Technologically, QWs are being fabricated with metal-organic chemical vapor deposition or molecular beam epitaxy. The growth mode, depending on various technological parameters, can be adapted for the growth of quantum dots – instead of the homogeneous semiconductor QW layer (see Figure 16.5b) one obtains the wetting layer with periodically self-assembled quantum dots. Such layers serve as active regions in quantum dot lasers.

In a simple design shown in Figure 16.4, the gain region is limited within a space-charge area of the *p-n* junction in the x-direction; and the y-z area under the metal strip where maximum current J_x flows. Far-from-the-contact p- and n-regions in the x-direction, and lateral area in the y-z plane, where under-threshold current flows, are lossy and inclusive for interband absorption loss along with all other absorption mechanisms. So, the gain region creates an optical waveguide where localization of radiation occurs due to amplification in the gain region and absorption all around it. This design does not imply a physical boundary that separates the gain region from the rest of the optical cavity and is called *gain-guided*. The factor which controls the optical confinement is the contrast in refraction and absorption between the active region and the rest of the structure: $\delta\varepsilon$, where ε is the complex-valued dielectric constant.

Efficient laser operation implies maximum population inversion at minimal driving current as well as maximal photon density in the active region to make stimulated emission as efficient as possible. Both requirements are satisfied in double-heterostructure (DH) design where energy barriers, originating from conduction and valence band offsets, provide for carrier confinement illustrated in Figure 16.5a [4,5]. Compared to the gain-guided lasers, the DH and QW devices manifest enhanced optical confinement as a high refractive index step between GaAs and barrier layer turns the gain region into a highly efficient waveguide. The *index-guided* design is illustrated in Figure 16.6.

Optical power concentrates in the high-index medium that decreases optical losses, which accumulate while evanescent tails of optical mode propagate through lossy AlGaAs layers.

16.3 Resonant Cavity and Longitudinal Modes

In index-guided lasers, there are two mechanisms of optical confinement: first is the same as discussed above in gain-guided devices, and second is due to the difference in the real parts of dielectric constant, $\mathrm{Re}[\delta\varepsilon]$. For example, in AlGaAs / GaAs structures, dielectric constants are $\varepsilon_{\mathrm{AlGaAs}} = 12.6$, $\varepsilon_{\mathrm{GaAs}} = 12.96$.

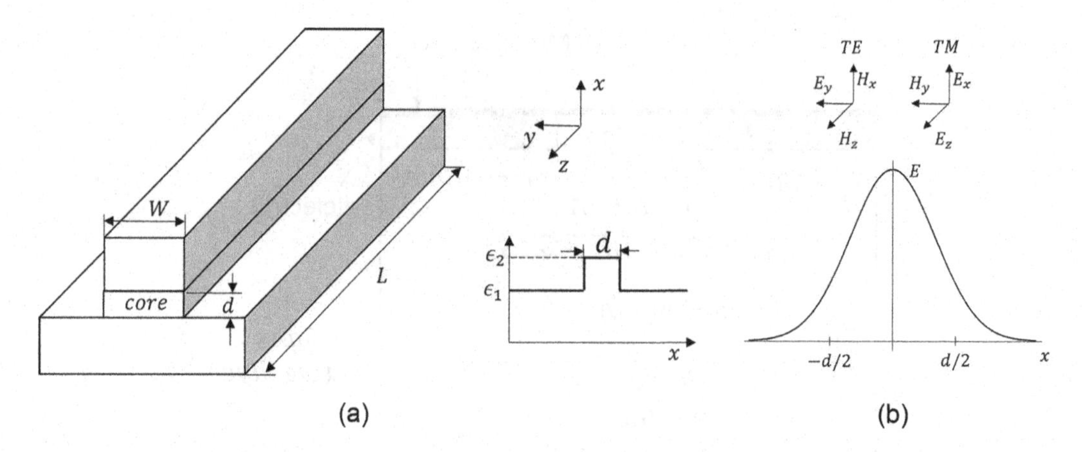

FIGURE 16.7 (a) The active region (core) as a dielectric waveguide; ϵ_1 and ϵ_2 are real parts of dielectric constants in cladding and active layers, respectively. (b) Field distribution in a planar waveguide: $f(x)=E_y$ in TE-wave or $f(x)=E_x$ in TM-wave.

Usually, it is the step that determines optical confinement in the index-guided design, so an active layer can be represented by a waveguide in the form of a multilayer dielectric slab as shown in Figures 16.6 and 16.7.

If in a planar – not limited in the y-z plane – waveguide, the refractive index profile is symmetric, the transverse eigenmodes TE and TM exist at any thickness d (no cut-off). Fundamental TE(TM) mode is even and propagates in the z-direction with amplitude distribution in the x-direction shown in Figure 16.7b. If thin enough, the waveguide supports fundamental modes only. Once thickness increases, an odd eigenmode appears at $D>\pi$, where $D=2\pi d\sqrt{\varepsilon_2-\varepsilon_1}/\lambda$ is the optical thickness of the active region, λ is the vacuum wavelength of the propagating mode.

In practical laser devices, optical modes are determined by the waveguide of finite dimensions W and L, as shown in Figure 16.7. Size limitation in the z-direction is implementing by fabrication of parallel cleaved front and rear faces-mirrors. Mirrors create the L-long Fabry-Perot resonator, designed to increase the photon lifetime in the active region: high-reflectivity coating on both ends confines photons in the cavity, forcing them to multiply through stimulated emission.

To find a mode structure, we consider TE(TM) propagation in a lossless medium of lateral dimensions L and W shown in Figure 16.8.

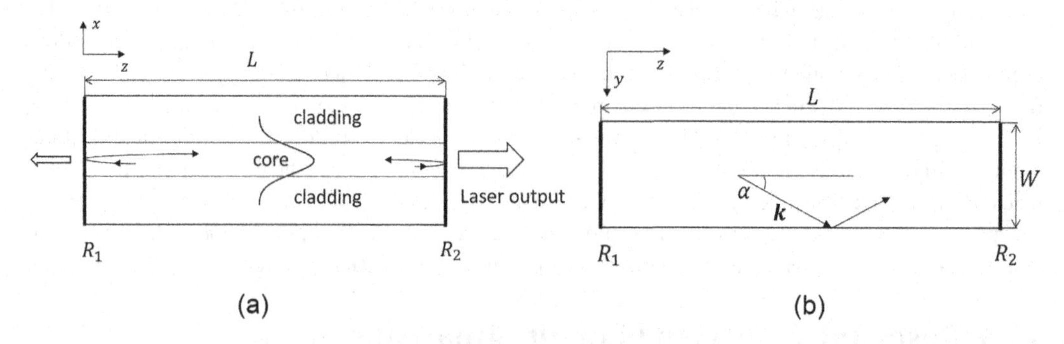

FIGURE 16.8 (a) Longitudinal Fabry-Perot resonators, $R_{1,2}$, are the reflectivity of rear and front mirrors, respectively. Core and claddings are gain and absorption regions, respectively. Field distribution $\varphi(x)$ renders optical confinement. (b) Reflection in the y-z plane creates discrete lateral modes and α equals the angle of incidence on the mirror.

If the waveguide were planar $(W \to \infty,\ L \to \infty)$ the dispersion relation would read as (12.31) (Chapter 12): $k^2c^2 = \omega^2 n_r^2$, where $\boldsymbol{k}$ is the continuous variable presenting propagation vector in the y-z plane, and $n_r^2 = \mu\varepsilon$.

Size restriction in the z-direction due to mirror reflection converts continuous $\boldsymbol{k}$ into a discrete set, which is formed by the condition that the roundtrip phase is the integer multiple of 2π. Assuming mirrors identical, one may neglect reflection phase contributions and get mode spectrum as

$$\exp(2Lk_z) = 1,\quad k_{zq} = \frac{\pi q}{L\cos\alpha},\quad q = 0, \pm 1, \ldots,$$

$$\omega_q = \frac{\pi c}{Ln_r \cos\alpha} q, \tag{16.7}$$

where α is the off-axis angle in Figure 16.8b. As long as $W \to \infty$, longitudinal Fabry-Perot modes (16.7) continuously depend on α. If W is finite, k_y becomes a discrete variable enumerated by the lateral mode index:

$$k_{yp} = \frac{\pi p}{W \sin\alpha},\quad p = 0, \pm 1, \ldots \tag{16.8}$$

Angle α becomes discrete and relates to the lateral mode index as

$$\alpha_{pq} = arctg \frac{k_{yp}}{k_{zq}} = arctg \frac{Lp}{Wq}, \tag{16.9}$$

Dispersion (16.7) takes the form

$$\omega_{qp} = \omega_{q0}\sqrt{1 + \left(\frac{Lp}{Wq}\right)^2},\quad \omega_{q0} = \frac{\pi c q}{Ln_r}. \tag{16.10}$$

Lateral index $p = 0$ corresponds to pure longitudinal modes ω_{q0} propagating along the z-direction. We deal with longitudinal modes only as far as condition $Lp \ll Wq$ holds.

Spectral resolution between adjacent longitudinal modes $(\Delta q = 1)$, can be found from (16.10):

$$\Delta\omega = \frac{\partial\omega}{\partial q}\Delta q = \frac{\partial\omega}{\partial q} = \frac{\pi c}{Ln_g}, \tag{16.11}$$

where $n_g = n_r + \omega\, \partial n_r/\partial\omega$ is the *mode group index*. In terms of free-space wavelength, the mode spectral resolution equals $\Delta\lambda = \lambda^2/2Ln_g$.

In practical III–V semiconductor lasers, the scale of wavelength spacing is 0.5–0.8 nm, while the gain spectral band in Figure 16.3 is several tens of nanometers. All modes in Figure 16.9, which fall within the gain bandwidth, amplify.

Single-mode operation implies that the peak gain for only one of the longitudinal modes reaches the lasing condition by exceeding threshold g_{th}. Otherwise, the laser operates in a multimode regime.

16.3.1 Distributed Mirrors and Mode Selectivity

For the single-mode operation, a short cavity is preferred, however often not feasible in fabrication. In this case, the cavity of an arbitrary length can be processing by fabricating distributed mirrors, as shown in Figure 16.10. Bragg reflectors are usually made as gratings in the cladding layer (Distributed Bragg Reflector Laser (DBR)) or the active layer (Distributed Feedback Laser (DFB)).

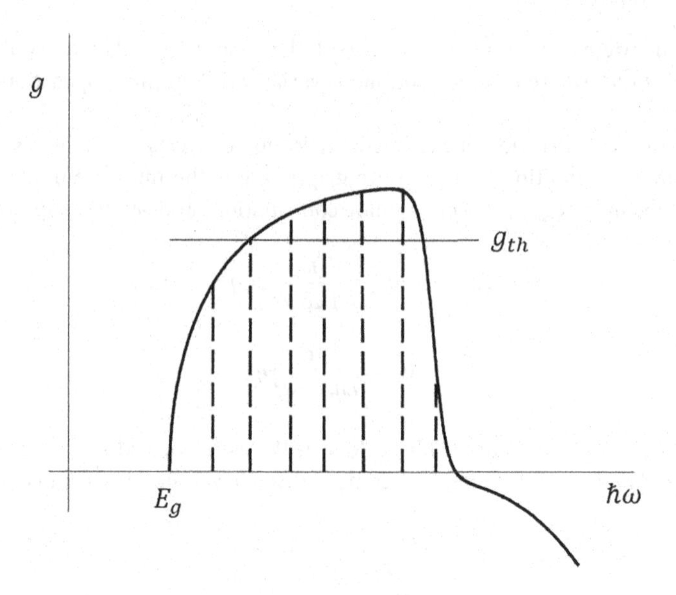

FIGURE 16.9 Longitudinal modes (dashed lines) and gain spectrum.

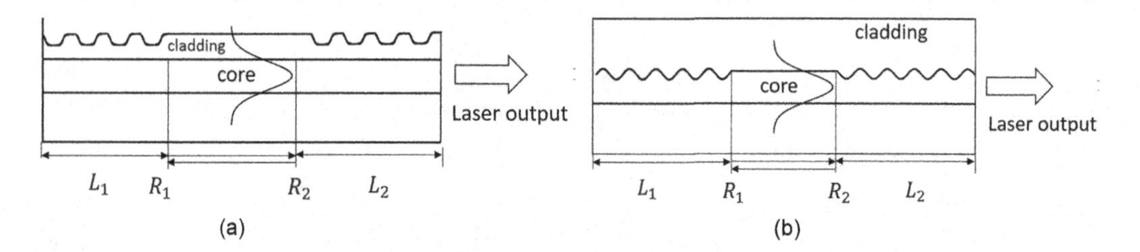

FIGURE 16.10 (a) Distributed Bragg Reflector Laser (DBR). (b) Distributed Feedback Laser (DFB).

The reflectivity of the distributed Bragg reflector depends on the length of the grating section. Mode selectivity is achieved by tuning the grating period to $\lambda / 2$ and choosing $L_{1,2}$ to provide desirable reflectivity $R_{1,2}$ at cavity boundaries.

16.4 Modal Gain and Threshold Condition

There is a range of W where the waveguide supports fundamental modes with field distributions similar to TE_0 and TM_0 in a planar waveguide, respectively. As for which of the transverse mode the laser will radiate in a single-mode regime, depends on the maximum gain as it is different for TE and TM modes.

Electric field distribution in fundamental modes, shown in Figure 16.7b, is a characteristic of laser operation that allows estimating the optical loss in the cavity. The point is that only part of the optical power propagates in the active region where amplification takes place. Another part propagates in cladding layers where it is subject to attenuation. We define the *modal confinement factor* Γ as a part of the total flux $\overline{\boldsymbol{P}}$ (12.35) (Chapter 12) confined in the active region of thickness d:

$$\Gamma = \int_{-d/2}^{d/2} \overline{\boldsymbol{P}}_z \, dx \, dy \Big/ \int_{-\infty}^{\infty} \overline{\boldsymbol{P}}_z \, dx \, dy = \int_{-d/2}^{d/2} \left| E(x,y) \right|^2 dx \, dy \Big/ \int_{-\infty}^{\infty} \left| E(x,y) \right|^2 dx \, dy, \tag{16.12}$$

Configuration of $E(x, y)$ and thus Γ depends on transverse mode under consideration.

Spectral gain and optical absorption are both defined for a plane wave. In a multimode waveguide, each mode has a particular spatial distribution of the electromagnetic field, gain, and absorption. For a given mode, absorption and gain, after being multiplied by corresponding confinement factors, are called *modal gain* $g\Gamma$ and *modal loss* $\alpha\Gamma'$, respectively. Total internal optical loss in the cavity reads as

$$\alpha_i = \alpha_g \Gamma + \alpha_u \Gamma_u + \alpha_l \Gamma_l, \tag{16.13}$$

where α_g is absorption in the gain region due to all mechanisms, but interband transitions; $\alpha_{u,d}$ and $\Gamma_{u,l}$ are the absorption coefficients and confinement factors in the upper and lower cladding layers, respectively. If the waveguide in Figure 16.7 is symmetric, $\alpha_u = \alpha_l \equiv \alpha_c$. As it follows from (16.12), $\Gamma + \Gamma_u + \Gamma_l = 1$, and thus $\alpha_u \Gamma_u + \alpha_l \Gamma_l = \alpha_c (1 - \Gamma)$. The internal loss (16.13) becomes

$$\alpha_i = \alpha_g \Gamma + \alpha_c (1 - \Gamma). \tag{16.14}$$

One more source of loss is mirror loss at the facets of the Fabry-Perot cavity. Propagating power in the cavity is proportional to $\exp\left[\left(g\Gamma - \alpha_g \Gamma - \alpha_c (1 - \Gamma)\right) z\right]$. Complete roundtrip changes the power by a factor of $R_1 R_2 \exp\left[\left(g\Gamma - \alpha_g \Gamma - \alpha_c (1 - \Gamma)\right) 2L\right]$, where $R_{1,2}$ are the mirror reflectivities. For small gain, this factor is small, meaning the power attenuates as photons, if not die due to internal absorption, leave the cavity through the mirrors. If maximum gain reaches the threshold value, the stimulated emission compensates for all losses:

$$R_1 R_2 \exp\left[\left(g_{th} \Gamma - \alpha_g \Gamma - \alpha_c (1 - \Gamma)\right) 2L\right] = 1,$$

or

$$g_{th} \Gamma = \alpha_i + \alpha_m,$$

$$\alpha_i = \alpha_g \Gamma + \alpha_c (1 - \Gamma),$$

$$\alpha_m = \frac{1}{2L} \ln \frac{1}{R_1 R_2}. \tag{16.15}$$

Terms α_i and α_m in (16.15) are the internal absorption loss and mirror loss, respectively. The mirror loss describes the escape of photons through semi-transparent mirrors and thus determines the useful output power of the laser.

At the threshold, the density of injected carriers follows from the relation $g_{th} = g_d \left(n_{th} - n_{tr}\right)$. Using (16.15), one finds the difference between *threshold carrier density* and transparency carrier density proportional to optical losses in the system:

$$n_{th} - n_{tr} = \frac{\alpha_i + \alpha_m}{g_d \Gamma}. \tag{16.16}$$

Carrier concentrations n_{th} and n_{tr} are supplied by the injection current, as discussed below.

16.5 Recombination Currents

Total injection current density relates to injected carrier concentration as

$$J = \frac{edn}{\tau(n)}, \quad \tau^{-1} = \tau_R^{-1} + \tau_{NR}^{-1}, \tag{16.17}$$

where $\tau(n)$ is the lifetime of *e-h* pairs in the active region of thickness d. Under below-threshold conditions, the total recombination rate comprises spontaneous radiative recombination, $\tau_R^{-1} \cong Bn$, and nonradiative recombination, $\tau_{NR}^{-1} \cong A + Cn^2$, where A, B, C are the Shockley-Read, radiative, and Auger recombination coefficients, respectively. Coefficient A is proportional to the density and cross-section of bulk and interface defects. Expressing threshold current density through the threshold carrier concentration, one obtains:

$$J_{th} = ed\left(An_{th} + Bn_{th}^2 + Cn_{th}^3\right) \equiv J_{SR} + J_R + J_A + J_L, \tag{16.18}$$

where J_{SR}, J_R, J_A correspond to Shockley-Read, radiative, and Auger currents, respectively. In III–V semiconductors, the Auger processes involve a spin-split valence band, so the Auger current comprises two terms corresponding to CHCC and CHHS channels. Term J_L in (16.18) is the leakage current that always exists in forward-biased *p-n* structures. Carrier confinement occurs in the active region between energy barriers shown in Figure 16.5. Leakage current consists of a fraction of injected electrons and holes with thermal energy sufficient to overcome the barriers and escape the active region where otherwise they would recombine radiatively. After leaving the active region not being recombined, electrons and holes flow to the n- and p-cladding layers, as minority carries, and form the diffusion and drift leakage currents. Leakage increases the threshold current and affects laser performance at a high injection level. For estimation of the leakage current, see Ref. [2].

16.6 Light-Current Characteristics and Efficiency

At the threshold, multiplication of the photons compensates their loss, and the steady-state laser emission starts. Injection current above the threshold level is converting to the intensity of the electromagnetic field through stimulated emission, the lifetime of electron–hole pairs decreases, and carrier density stays pinned to $n = n_{th}$. In other words, under above-threshold conditions, the current density (16.17) includes an additional term - radiative current caused by stimulated emission.

$$J - J_{th} = edR_{st} = edg_{th}\mathrm{v}_g N, \tag{16.19}$$

where R_{st} is the stimulated recombination rate $\left[\mathrm{m}^{-3}\,\mathrm{s}^{-1}\right]$, $\mathrm{v}_g = c/n_g$, N is the photon density in the cavity. Equation (16.19) implies that all injected *e-h* pairs contribute to radiative recombination. However, part of *e-h* pairs undergoes nonradiative decay; the process characterizes by the *internal quantum efficiency* – a fraction of the radiative recombination rate in the total recombination:

$$\eta_i = \frac{1/\tau_R}{1/\tau} = \frac{\tau_{NR}}{\tau_R + \tau_{NR}}. \tag{16.20}$$

Photon density follows from (16.19) and after being corrected with η_i takes the form

$$N = \frac{\Gamma\tau_{ph}}{ed}\,\eta_i\left(J - J_{th}\right),$$

$$\tau_{ph}^{-1} = \mathrm{v}_g\left(\alpha_i + \alpha_m\right), \tag{16.21}$$

where photon loss rate τ_{ph}^{-1} (inverse lifetime) comprises two terms: absorption rate $\mathrm{v}_g\alpha_i$ and escape (external emission) rate $v_g\alpha_m$. Total photon energy density $N\hbar\omega$ multiplied by the escape rate $v_g\alpha_m$ and by the optical cavity volume LWd/Γ gives the power coming out of the cavity:

$$P = \frac{\hbar\omega}{e}\eta_0\eta_i\left(I - I_{th}\right), \quad \eta_0 = \frac{\alpha_m}{\alpha_m + \alpha_i}, \tag{16.22}$$

where I is the current [A] and η_0 is the *output coupling efficiency.* Light-current characteristics (16.22) define *differential quantum efficiency*:

$$\eta_d = \frac{e}{\hbar\omega}\frac{dP}{dI} = \eta_i\eta_0. \tag{16.23}$$

If an external voltage is applied to laser terminals, the total consumed power includes Joule loss in the circuit, and thus electric to optical power conversion efficiency (often called *wall-plug efficiency*) reads as

$$\eta_{wp} = \frac{\hbar\omega\eta_d\left(I - I_{th}\right)}{e\left(I^2R_s + IU\right)}, \tag{16.24}$$

where U is the voltage at device terminals and R_s is the series resistance.

III–V DH and QW lasers operating on interband transitions became widely used once the technology of epitaxial growth had succeeded in fabricating dislocation-free layers. Dislocations are the source of non-radiative recombination that degrades the internal quantum efficiency η_i. High-quality heterojunctions can be grown almost dislocation-free either if interfaces are lattice-matched or if coherent epitaxial layers have a thickness less than critical, thus preventing dislocations from developing. In such heterojunctions and QWs, the correct choice of materials among the III–V family provides efficiency as high as $\eta_i \approx 1$. It concerns short-wavelength semiconductor lasers operating in the range of $\lambda < 2\ \mu m$. However, mid-infrared devices operating at $\lambda > 2\ \mu m$ require an active region with a smaller bandgap that enhances intrinsic non-radiative mechanism – Auger recombination (see Chapter 13). Auger recombination suppresses the internal quantum efficiency, makes the threshold current temperature-sensitive, thus preventing room-temperature operation. Quantum cascade lasers are free from this shortcoming as the long-wavelength optical transitions do not involve valence band and occur between conduction subbands, so Auger recombination is not a significant factor.

Temperature sensitivity of threshold current in interband DH and QW lasers will be discussing in the next section.

16.7 Temperature Sensitivity of the Threshold Current

The ultimate goal of semiconductor laser technology is to increase the operating temperature as much as possible. Thermal processes that increase the threshold current include leakage and non-radiative currents related to Auger recombination. High injection current heats the active region above the heat sink temperature that causes thermal damage to the device.

The leakage current and Shockley–Read recombination can be diminished by increasing carrier confinement with appropriate laser design and improving material quality. In a separate-confinement structure, two sets of cladding layers separately confine current and light. Also, contemporary technology of epitaxial growth allows fabricating heterointerfaces of high quality in the sense of suppressed recombination on traps. Progress in growth technology and design results in a practical demonstration of the devices emitting at $\lambda = 2.7\ \mu m$ in continuous wave (CW) mode up to $T = 234$ K [6].

A standard example of the DH laser is the InP / InGaAsP / InP structure operating at $\lambda = 1.3\ \mu m$ and demonstrating almost exponential temperature dependence of the threshold current: $J_{th} \approx J_0 \exp\left(T/T_0\right)$. Characteristic temperature $T_0 = \left[d\left(\log J_{th}\right)/dT\right]^{-1}$ characterizes the temperature sensitivity of the threshold current: higher T_0 corresponds to weaker temperature dependence. Usually, T_0 is not constant in the whole temperature range of operation. Nevertheless, this parameter is widely used to characterize the temperature sensitivity of lasers. At certain ambient temperatures, parameter T_0 is determined by the dominant recombination channels, as discussed later in this chapter.

In the spectral range of interest for various sensing applications (3–12) μm, the Auger recombination becomes an intrinsic limitation preventing interband injection lasers from operating at room temperature in CW mode. III–V interband lasers operating in the range of (3–5) μm are lattice-matched InAsSb / InAsSbP-based double heterostructures and QWs with the bandgap not exceeding 0.4 eV: $\lambda(\mu\text{m}) \approx 1.24/E_g\,(\text{eV})$.

To estimate the theoretical limits of T_0, we will analyze the temperature dependence of all contributions to threshold current (16.18), assuming leakage negligible $(J_L = 0)$.

16.7.1 Threshold Carrier Density

We assume that $(\alpha_i + \alpha_m)/g_d\Gamma n_{tr} \ll 1$, so following (16.16), the threshold is close to the transparency point, and thus we can estimate threshold carrier density using transparency condition. To find the transparency carrier density, one needs two equations to determine quasi-Fermi levels for electrons and holes. First is the neutrality equation at high injection level, $n = p \gg n_0$, n_0 being equilibrium concentration, and second is the transparency condition (16.2).

The energy spectrum in narrow-gap III–V and II–VI semiconductors is illustrated in Figure 16.11. Under transparency condition

$$\mu_c + \mu_v = 0. \tag{16.25}$$

Due to a large difference in effective masses, when injection increases to the transparency level, electrons become degenerate while holes are not. In the three-dimensional active region of DH lasers, carrier distributions read as

$$n = \frac{1}{3\pi^2}\left(\frac{2m_c\mu_c}{\hbar^2}\right)^{3/2},$$

$$p = 2\left(\frac{m_h k_B T}{2\pi\hbar^2}\right)^{3/2}\exp\left(\frac{\mu_v}{k_B T}\right),$$

$$m_h^{3/2} = m_{hh}^{3/2} + m_{lh}^{3/2}, \tag{16.26}$$

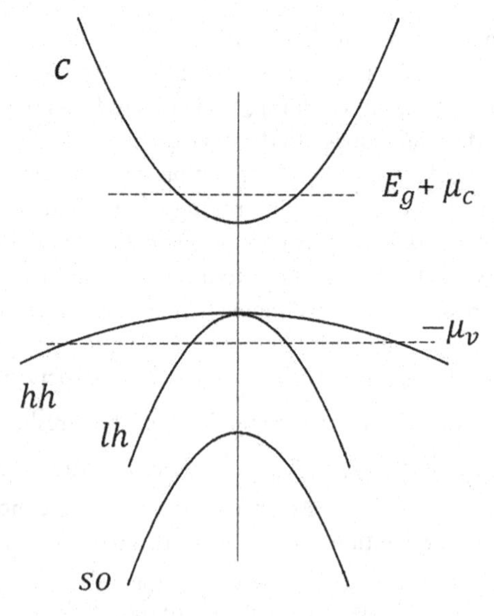

FIGURE 16.11 Energy spectrum in III–V and II–VI narrow-gap semiconductors; μ_c and μ_v are referenced to the valence band maximum quasi-Fermi levels for electrons and holes, respectively.

At the threshold, the neutrality condition $n = p$ and its solution for μ_c/k_BT become

$$\frac{4}{3\sqrt{\pi}}\left(\frac{\mu_c}{k_BT}\right)^{3/2}\exp\left(\frac{\mu_c}{k_BT}\right)=\left(\frac{m_h}{m_c}\right)^{3/2},$$

$$\mu_c = Pk_BT, \tag{16.27}$$

where P is the temperature-independent coefficient.

For two-dimensional electrons and holes in QW lasers, energy reference is the first valence subband maximum, and μ_c and μ_v describe the occupation of first conduction and valence subbands. Using (4.15) (Chapter 4), at threshold point, one obtains

$$n=\frac{m_c\mu_c}{\pi\hbar^2 d},\quad p=\frac{m_h k_BT}{\pi\hbar^2 d}\exp\left(-\frac{\mu_c}{k_BT}\right), \tag{16.28}$$

so equation $n = p$ and its solution take the form

$$\frac{\mu_c}{k_BT}\exp\left(\frac{\mu_c}{k_BT}\right)=\frac{m_h}{m_c},$$

$$\mu_c = Qk_BT, \tag{16.29}$$

where Q is the temperature-independent coefficient.

Substituting μ_c (16.27) in (16.26), one obtains volume threshold density in DH active region:

$$n_{th}=\frac{1}{3\pi^2}\left(\frac{2m_cP}{\hbar^2}\right)^{3/2}(k_BT)^{3/2}. \tag{16.30}$$

Similarly to the QW active region, substitution μ_c (16.29) in (16.28) gives the threshold sheet carrier density:

$$n_{th}^{2D}=\frac{m_cQ}{\pi\hbar^2 d}k_BT. \tag{16.31}$$

In conclusion, the threshold carrier density as a function of the temperature follows the corresponding density of states: $n_{th} \sim T^s$, where $s = 3/2$ for DH and $s = 1$ for QW lasers.

16.7.2 Radiative Recombination Current

In DH lasers, the threshold current density (16.18) includes the radiative term,

$$J_R^{DH}=edB_R^{DH}n_{th}^2=J_{R0}^{DH}\left(\frac{k_BT}{E_g}\right)^{3/2},\quad \left[\mathrm{A/m^2}\right], \tag{16.32}$$

where n_{th} follows from (16.30) while recombination coefficient scales as $B_R^{DH} \sim T^{-3/2}$ [7]. In QW lasers,

$$J_R^{QW}=eB_R^{QW}\left(n_{th}^{2D}\right)^2=J_{R0}^{QW}\frac{k_BT}{E_g},\quad \left[\mathrm{A/m^2}\right] \tag{16.33}$$

where recombination coefficient $B_R^{QW} \sim T^{-1}$ [8] and n_{th}^{2D} is given in (16.31). Factors J_{R0}^{DH} and J_{R0}^{QW} do not depend on temperature explicitly. Recombination coefficients are defined so that Bn^2 gives the number of radiative transitions per second per unit volume (area) in DH (QW) active region.

Similarly to carrier density (16.30) and (16.31), the threshold current as a function of the temperature follows the corresponding 3D and 2D density of states.

16.7.3 Auger Recombination Currents

The lower the bandgap, the smaller the Auger threshold energy. Under weak excitation, the dominant processes are CHCC for n-type and both CHHL and CHHS for p-type active region. At a high injection level, the densities of electrons and holes exceed their equilibrium values. So, near the threshold carrier density, Auger processes involving two holes become important irrespectively to the type of semiconductor under consideration. Besides, the CHHS process may become resonant in narrow-gap semiconductors if the bandgap is close to the spin-split energy. Among all possible Auger processes, CHCC and CHHS channels are considered dominant in mid-infrared semiconductor lasers with n-type active regions [2].

In bulk materials with degenerate electrons and nondegenerate holes, the CHCC recombination coefficient $\left[\mathrm{m}^6/\mathrm{s}\right]$ has the form [9,10]:

$$\gamma_{CHCC} = \gamma_{CHCC}^0 \left(\frac{E_g}{k_B T}\right)^{3/2} \exp\left(-\frac{\Delta_{CHCC}}{k_B T}\right), \tag{16.34}$$

where $\Delta_{CHCC} \approx 2E_g m_c / m_{hh}$ is the threshold energy and γ_{CHCC}^0 is the factor that does not depend on temperature explicitly. Using (16.30), we express the explicit temperature dependence of the Auger current as

$$J_{CHCC}^{DH} = edn_{th}^3 \gamma_{CHCC} = J_{CHCC}^0 \left(\frac{k_B T}{E_g}\right)^3 \exp\left(-\frac{\Delta_{CHCC}}{k_B T}\right). \tag{16.35}$$

The CHHS recombination coefficient is sensitive to the relation between the bandgap and the spin-split energy. If $\Delta_{so} \approx E_g$, the CHHS recombination coefficient [11] takes the form,

$$\gamma_{CHHS} = \gamma_{CHHS}^0 \exp\left(-\frac{\Delta_{CHHS}}{k_B T}\right), \tag{16.36}$$

where γ_{CHHS}^0 is a weakly dependent function of temperature, Δ_{CHHS} is the threshold energy for the CHHS process. Corresponding Auger current carries temperature dependence shown below,

$$J_{CHHS}^{DH} = edn_{th}^3 \gamma_{CHHS} = J_{CHHS}^0 \left(\frac{k_B T}{E_g}\right)^{9/2} \exp\left(-\frac{\Delta_{CHHS}}{k_B T}\right). \tag{16.37}$$

Auger currents in bulk materials, (16.35) and (16.37), include exponential temperature factor due to the threshold nature of Auger transition. The energy released during electron–hole recombination can be transferred to another carrier only if both momentum and energy conservation conditions hold. These constraints require finite (threshold) momentum released to excite the third particle. The same considerations apply to QW structures with infinite barriers, where the conservation laws involve in-plane electron momenta; the Auger rates are similar to those calculated for bulk materials [2]. However, interfaces in actual QW structures have finite barriers that allow wave functions of electrons and holes to penetrate barrier materials. In this case, electron momentum in the growth direction is ill-defined, and momentum conservation breaks in the course of tunneling in the normal to QW-plane direction. In other words, for the part of electrons and holes which penetrate the barriers, momentum conservation is not required. This recombination channel requires no threshold energy. Thus, in a QW, there exist two channels of Auger recombination, one with bulk-like characteristics (with strict momentum conservation) and another one with no threshold.

The no-threshold-channel can be the dominant Auger mechanism below room temperature in thin QW of 1–3 nm width. Using recombination rates in type-I QW [8,12], at threshold injection level (16.31), one obtains temperature-dependent Auger currents in the form

$$J_{CHCC}^{QW} = j_{CHCC}^{0}\left(\frac{k_B T}{E_g}\right)^2,$$

$$J_{CHHS}^{QW} = j_{CHHS}^{0}\left(\frac{k_B T}{E_g}\right)^3. \tag{16.38}$$

The general conclusion is that the no-threshold Auger process results in the recombination current, which temperature dependence is much weaker than the exponential one in DH lasers. Below we use (16.38) to find the temperature dependence of threshold current in QWs.

16.7.4 Threshold Current

To determine intrinsic limitations to the threshold current, we analyze (16.18) neglecting Shockley-Read recombination and leakage. For DH laser, we use (16.32), (16.35), and (16.37) to get

$$J_{th}^{DH} = J_{R1}^{0}\left(\frac{k_B T}{E_g}\right)^{3/2} + J_{CHCC}^{0}\left(\frac{k_B T}{E_g}\right)^{3}\exp\left(-\frac{\Delta_{CHCC}}{k_B T}\right) + J_{CHHS}^{0}\left(\frac{k_B T}{E_g}\right)^{9/2}\exp\left(-\frac{\Delta_{CHHS}}{k_B T}\right). \tag{16.39}$$

The characteristic temperature reads as

$$T_0 = \left[d(\log)/dT\right]^{-1} = \frac{T(1+x+y)}{\frac{3}{2} + x\left(3 + \frac{\Delta_{CHCC}}{T}\right) + y\left(\frac{9}{2} + \frac{\Delta_{CHHS}}{T}\right)},$$

$$x = \frac{J_{CHCC}^{DH}}{J_R^{DH}}, \quad y = \frac{J_{CHHS}^{DH}}{J_R^{DH}}. \tag{16.40}$$

In the regime of high internal quantum efficiency ($x \ll 1$, $y \ll 1$), the maximum characteristic temperature is $T_0 \approx 2T/3$.

In mid-infrared lasers, $\lambda \geq 3\,\mu m$, providing CHCC Auger process dominates recombination,

$$T_0 \approx \frac{T}{3 + \frac{\Delta_{CHCC}}{k_B T}}. \tag{16.41}$$

Limit $\Delta_{CHCC} \ll k_B T$ gives theoretical maximum for DH lasers as

$$T_0 \approx T/3. \tag{16.42}$$

In the regime where CHHS Auger current is most relevant, one obtains

$$T_0 \approx T\left(\frac{9}{2}+\frac{\Delta_{CHHS}}{k_B T}\right)^{-1}. \tag{16.43}$$

If the bandgap is close to the spin split-off energy, the CHHS channel is resonant $(\Delta_{CHHS} \approx 0)$ and thus $T_0 \approx 2T/9$. At $T = 120\,\mathrm{K}$ it gives $T_0 \approx 27\,\mathrm{K}$, which is close to experimental results for mid-infrared lasers operating at $\lambda > 3\,\mu\mathrm{m}$ [13,14].

In QW lasers, one obtains the threshold current using (16.33) and (16.38):

$$J_{th}^{QW} = J_{R0}^{qW}\frac{k_B T}{E_g} + j_{CHCC}^{0}\left(\frac{k_B T}{E_g}\right)^2 + j_{CHHS}^{0}\left(\frac{k_B T}{E_g}\right)^3.$$

$$T_0 = \frac{T\,(1+x+y)}{1+2x+3y}, \quad x = \frac{J_{CHCC}^{QW}}{J_R^{QW}}, \quad y = \frac{J_{CHHS}^{QW}}{J_R^{QW}}. \tag{16.44}$$

In the regime of high internal quantum efficiency $(x \ll 1, y \ll 1)$, $T_0 \approx T$. If the CHCC process dominates, the upper limit $T_0 \approx T/2$ is higher than in DH lasers (16.42). The limit of $T_0 \approx T/2$ well fits room temperature experimental data in MQW InGaN / GaN lasers [15] and high-temperature data in GaAs and InGaAs QW lasers, $T_0 \approx (150 \sim 250)$ K [16,17].

To decrease the role of Auger recombination in laser performance, one uses the MQW structures. Suppression of the Auger recombination in MQW structures is discussed in Ref. [18].

16.8 Light-Emitting Diodes

The facets of the laser structure coated with reflection layers create a cavity thus forcing light to make many passes in the gain medium. That is the way to achieve the threshold condition. In the case of antireflection coating, light could pass once before going out through the facets. The threshold condition does not hold. Outgoing light comprises spontaneous emission or, if injection carrier density becomes high enough, *amplified spontaneous emission*. That is the regime the light-emitting diode works. In more detail, let us consider spectral distribution of the spontaneous emission given in (13.40), Chapter 13:

$$r_{sp}(\omega) \sim n_{ph}(\omega)\alpha(\omega), \tag{16.45}$$

where $n_{ph}(\omega)$ is the Bose function and $\alpha(\omega)$ is the spectral absorption coefficient.

Following arguments preceding (16.15), after one pass, the propagating spontaneous emission power acquires amplification factor $\exp[GL]$:

$$r_{sp}^{amp}(\omega) = r_{sp}(\omega)\cdot\exp[GL], \tag{16.46}$$

where L is the cavity length, $G(\omega) = g(\omega)\Gamma - \alpha_i$ is the net modal gain, α_i is the internal optical loss. Increase in injection carrier density to the point $G(\omega) > 0$ makes spontaneous emission amplified as illustrated in Figure 16.12.

As shown in Figure 16.12, amplification makes the line narrower than in an ordinary spontaneous emission spectrum. Line narrowing manifests that net gain is positive, and one can extract gain spectrum from $G(\omega) = \log\left[r_{sp}^{amp}(\omega)/r_{sp}(\omega)\right]/L$.

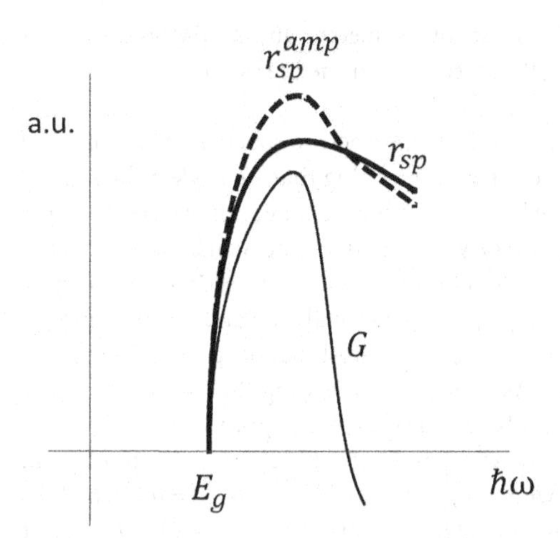

FIGURE 16.12 Spectral dependence of net gain (thin solid line), spontaneous emission (bold solid line), and amplified spontaneous emission (broken line).

16.9 Material Choice and Engineering in III–V Semiconductor Heterojunctions

Material combinations for DH and QW III–V semiconductor lasers should provide for effective electrical and optical confinements. Electrical confinement depends on the conduction and valence band offsets of constituent materials. The band offset can be engineered using computational data on valence band offsets between binaries [19–21] and then extrapolated to ternaries and quaternaries either by linear approximation (Vegard's law) or including bowing parameters [22]. In the next section, we give the example of engineering of laser structure using first-principle computational data.

Calculations of optical confinement are based on the model dielectric function that allows engineering a step in refractive index n_r at the interface. Usually, the smaller the bandgap, the larger the refractive index as contributions to n_r come from virtual interband electronic transitions. However, with the smaller bandgap, InAs has a lower refractive index than GaSb. This "anomaly" stems from the fact that n_r is determined by several energy gaps, including those located in L and X-points in the Brillouin zone. Contributions from L- and X-gaps are dominant due to a large joint density of states, and these gaps are lower in GaSb that explains the "anomaly". The model refractive index approach accounts for all energy gaps necessary for dielectric function calculation [23].

16.10 Quantum Cascade Laser

Quantum cascade laser (QCL) is the unipolar device. The optical power comes from electron transitions between conduction minibands in a biased superlattice. As optical transitions do not involve the valence band, the Auger recombination is out of the picture enabling the laser to operate at room temperature in a continuous-wave regime in the infrared spectral range up to 11 μm and up to 20 μm in a pulsed mode. The development of terahertz QCLs has also resulted in devices operating at room temperature [24]. The most recent achievement is the QCL operating up to 250 K and a frequency of 4 THz [25].

The QCL concept proposed in Ref. [26] followed by experimental implementation more than 20 years later [27]. To explain how QCL works, we consider an example of the device made of $Al_xGa_{1-x}As/GaAs$ superlattice starting from the choice of material parameters. The superlattice consists of many repeating periods [28]: each of them consists of an injector and MQW-active region. MQW comprises four barriers ($x = 0.33$) and three wells. Starting with the barrier, the MQW forms the thickness sequence ($\dot{A}$): 58 / 15 / 20 / 49 / 17 / 40 / 34.

In what follows, we assume the in-plane electron spectrum parabolic and use MQW conduction band profile in the flat-band approximation meaning that we neglect doping in the active region and so self-consistent corrections to the band edge alignment near interfaces. Within this simplified approach, the electron energy spectrum in MQW follows from the Schrödinger equation with potential determined by the electric field and the conduction band offset. To estimate the conduction band offset, we use the approach based on first-principles computational data and outlined in Section 16.7.

For alloy composition $x < 0.45$, the lowest bandgap in AlGaAs is in Γ-point in BZ, so QW $Al_xGa_{1-x}As / GaAs / Al_xGa_{1-x}As$ is of type I, meaning that on the energy scale, the bandgap of the well lies within the bandgap of the barrier as shown in Figure 16.13.

We assume natural valence band offset [21] to be linear function of alloy composition: $\Delta E_v(x) = 0.51\,x$, which gives $\Delta E_v = \Delta E_v(0.33) = 0.168$ meV. Bandgap in the alloy depends on Al content as [22]: $E_g(x) = 2.95x + (1-x)1.42 + 0.37x(x-1)$, which gives gaps $E_{gW} = 1.42$ eV, $E_{gB} = E_g(0.33) = 1.84$ eV. The conduction band offset follows from the band alignment and equal to $\Delta E_c = E_{gB} - E_{gW} - \Delta E_v = 0.255$ eV. Electron effective mass is given as [22]: $m(x) = 0.07(1+x)m_0$. The solution to the Schrödinger equation has been obtained assuming external electric field of 4.8×10^6 V/m. Eigenfunctions and eigenvalues are presented in Figure 16.14.

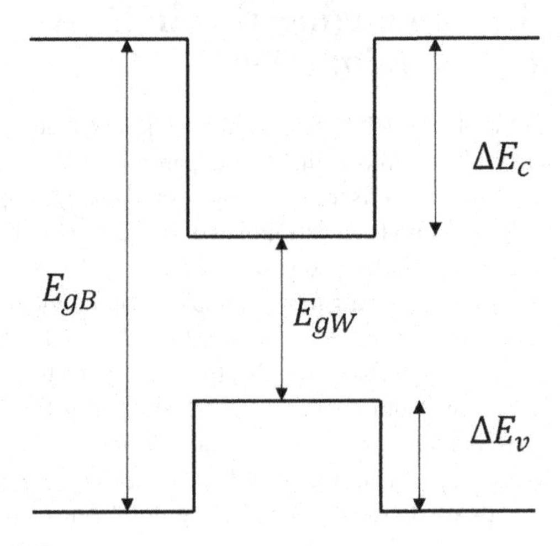

FIGURE 16.13 Type-I quantum well.

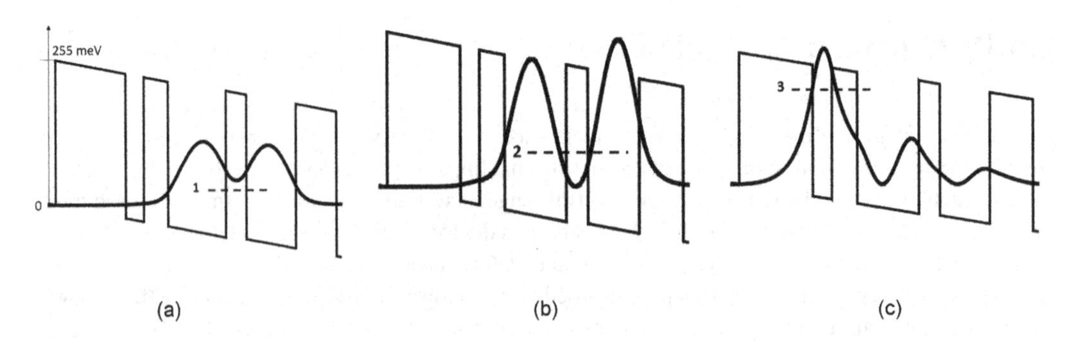

FIGURE 16.14 MQW band alignment and energy levels. (a) $E_1 = 27.3$ meV, (b) $E_2 = 60$ meV, (c) $E_3 = 188$ meV. Spatial distributions of $|\Psi(z)|^2$ along the growth direction render in bold lines.

The numerical example shown in Figure 16.14 corresponds to the lasing wavelength of 9.7 μm – close to experimentally observed in Ref. [28]. A small discrepancy may be attributed to our simplified assumptions discussed above. The wavelength does not depend on the bandgaps of constituent materials, and, with a particular choice of semiconductors, is tunable by the barrier alloy composition and layer thicknesses.

Wave functions corresponding to levels 1 and 2 are spread between two wide wells while electrons on level 3 are localized in the narrow well on the left to the active region. Optical transition $E_3 \rightarrow E_2$ shown in Figure 16.15 is accompanied by tunneling through the barrier and possible due to overlap of corresponding wave functions. The spatially diagonal transition increases the electron lifetime on level 3 (1.5 ps), as compared to that if a vertical transition had occurred. Fast escape of an electron out of level 2 occurs due to electron-phonon relaxation, which is most effective if energy difference E_{21} is close to the longitudinal optical (LO) phonon energy, $E_{21} \approx \hbar\omega_{LO}$. In our example $E_{21} \approx 33$ meV which is close to $\hbar\omega_{LO}$ in GaAs. As a result, the lifetime on level 2 (0.6 ps) is much shorter than on level 3, thus creating the population inversion.

One period in QCL comprises the injector and MQW active region. Injection of electrons (shown by the bold arrow in Figure 16.15) occurs from digitally graded alloy or superlattice where an electron loses the energy before being injected to the next period. The whole device consists of many periods and presents a "staircase", each step of which emits photons, as illustrated in Figure 16.16.

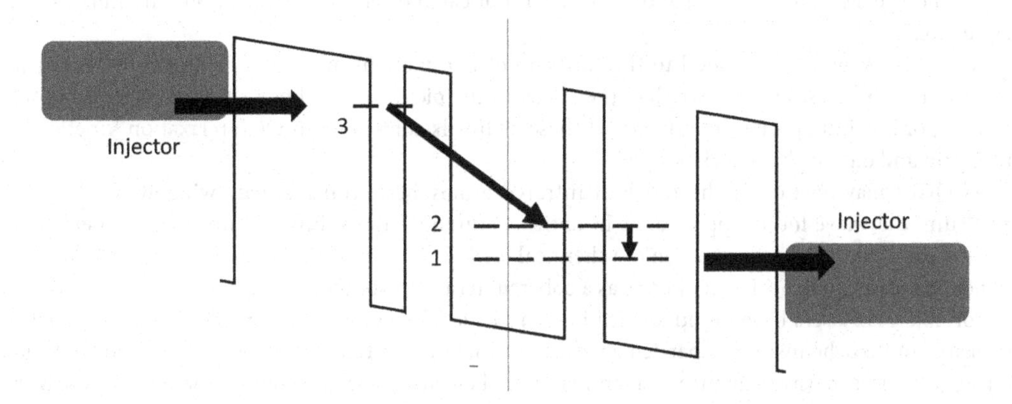

FIGURE 16.15 MQW active regions in QCL. Diagonal transition $3 \rightarrow 2$ followed by fast electron-phonon relaxation $2 \rightarrow 1$. The shaded area is the superlattice miniband.

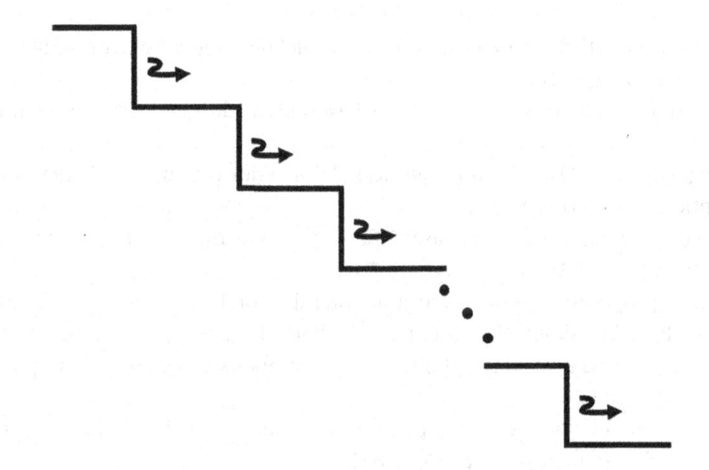

FIGURE 16.16 *N*-stage quantum cascade.

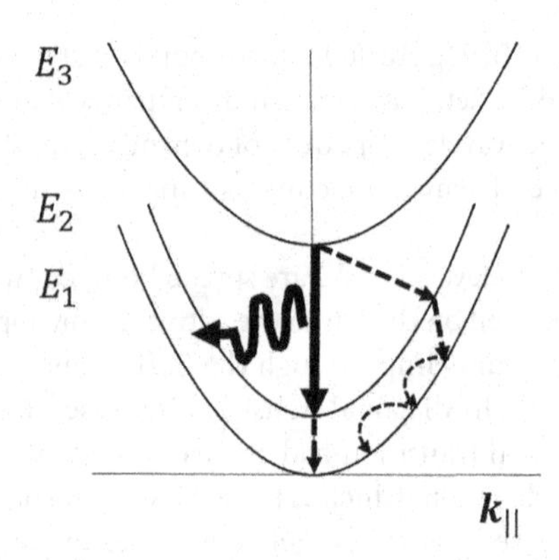

FIGURE 16.17 Intersubband transitions in QCL. Radiative and phonon relaxation transitions are shown as solid and broken arrows, respectively.

The cascade structure allows one electron to emit photons at many stages of the device, making the electron reusable, while in an interband laser, one electron can radiatively recombine once. In-plane electron dispersion on levels 1, 2, and 3, as well as optical and relaxation transitions, are illustrated in Figure 16.17.

Nonradiative processes related to the emission of LO-phonons are shown in Figure 16.17. Phonon reabsorption processes not shown, had they been in the picture, would have been depicted as broken arrows pointed in opposite directions. All these channels contribute to QCL relaxation kinetics and thus gain and output characteristics [29].

As QCLs may operate in the range of infrared atmospheric transparency windows of 3 – 5 and 8 – 12 μm, they have found applications in optical communications, infrared imaging, on-field atmospheric absorption spectroscopy, remote chemical testing, and complex gas testing for breath analysis. Tuned to far-infrared, QCL can operate as a coherent terahertz source.

Portable THz lasers have found applications in medical imaging, communications, quality control, security, and biochemistry. In particular, tunable single-mode terahertz sources are used in remote sensing of the explosive chemicals since many of them have spectroscopic signatures in terahertz. For more details on applications, see Ref. [30].

References

1. N. Holonyak Jr. and S.F. Bevacqua, "Coherent (visible) light emission from Ga($As_{1-x}P_x$) junctions", *Appl. Phys. Lett.* **1**, 82 (1962).
2. G.P. Agrawal and N.K. Dutta, *Semiconductor Lasers* (2nd ed., Van Nostrand Reinhold, New York, 1993).
3. S.H. Chuang, *Physics of Optoelectronic Devices* (John Wiley & Sons, Inc. Wiley Series in Pure and Applied Optics, New York, 1995).
4. Z. Alferov, "Double heterostructure lasers: Early days and future perspectives", *IEEE J. Sel. Top. Quant. Electron.* **6**, 832 (2000).
5. H. Kroemer, "A proposed class of hetero-junction injection lasers", *Proc. IEEE.* **51**(12), 1782 (1963).
6. D. Garbuzov, R.U. Martinelli, R.J. Menna, P.K. York, H. Lee, S.Y. Narayan, and J.C. Connolly, "2.7-μm InGaAsSb/AlGaAsSb laser diodes with continuous-wave operation up to −39 °C", *Appl. Phys. Lett.* **67**, 1346 (1995).
7. B.L. Gel'mont and G.G. Zegrya, "Temperature dependence of threshold current density for an injection laser", *Sov. Phys. Semicond.* **25**, 1216 (1991).

8. G.G. Zegrya, A.D. Andreev, N.A. Gun'ko, and E.V. Frolushkina, "Calculation of QW laser threshold currents in terms of new channels of nonradiative Auger recombination", *Proc. SPIE.* **2399**, 307 (1995).
9. B. Gelmont and Z. Sokolova, "Auger recombination in direct-gap semiconductors", *Sov. Phys. Semicond.* **16**, 1067 (1982).
10. V.N. Abakumov, V.I. Perel, and I.N. Yassievich, *Nonradiaive Recombination in Semiconductors* (North-Holland, Amsterdam, 1991).
11. B.L. Gelmont, Z.N. Sokolova, and V.B. Khalfin, "Inter-zone auger recombination in GaSb laser structures", *Sov. Phys. Semicond.* **18**, 1128 (1984).
12. G.G. Zegrya and V.A. Kharchenko, "New mechanism of Auger recombination of nonequilibrium current carriers in semiconductor heterostructures", *Sov. Phys. JETP.* **74**, 173 (1992).
13. H.K. Choi, S.J. Eglash, and G.W. Turner, "Double-heterostructure diode lasers emitting at 3 μm with a metastable GaInAsSb active layer and AlGaAsSb cladding layers", *Appl. Phys. Lett.* **64**, 2474 (1994).
14. J. Diaz, H. Yi, A. Rybaltowski, B. Lane, G. Lukas, D. Wu, S. Kim, M. Erdtmann, E. Kass, and M. Razeghi, "High power InAsSb/InPAsSb/InAs mid-infrared lasers", *Appl. Phys. Lett.* **70**, 40 (1997).
15. M. Razeghi, "GaN-Based Laser Diodes", In: *Advances in Semiconductor Laser and Applications*, pp. 161–234, Ed. by M. Dutta and M.A. Stroscio (World Scientific, Singapore, 2000).
16. G.P. Van der Ziel and N. Chand, "High-temperature operation (to 180°C) of 0.98 μm strained single quantum well $In_{0.2}Ga_{0.8}As$/GaAs lasers", *Appl. Phys. Lett.* **58**, 1437 (1991).
17. P. Derry, H. Hager, L. Chiu, D. Booher, E. Miao, and C. Hong, "Low threshold current high-temperature operation of InGaAs/AlGaAs strained-quantum-well lasers", *IEEE Photon. Technol. Lett.* **4**, 1189 (1992).
18. H. Mohseni, V. I. Litvinov, and M. Razeghi, "Interface-induced suppression of the Auger recombination in type-II InAs/GaSb superlattices", *Phys. Rev.* **B58**, 15378 (1998).
19. C.G. Van de Walle and R.M. Martin, "Theoretical study of band offsets at semiconductor interfaces", *Phys. Rev.* **B35**, 8154 (1987).
20. A. Qteish and R.J. Needs, "Improved model-solid-theory calculations for valence-band offsets at semiconductor-semiconductor interfaces", *Phys. Rev. B.* **45**, 1317 (1992).
21. H. Wei and A. Zunger, "Calculated natural band offsets of all II–VI and III–V semiconductors: chemical trends and the role of cation *d* orbitals", *Appl. Phys. Lett.* **72**, 2011 (1998).
22. M.P.C.M. Krijn, "Heterojunction band offsets and effective masses in III-V quaternary alloys", *Semicond. Sci. Technol.* **6**, 27 (1991).
23. S. Adachi, *Physical Properties of III-V Semiconductor Compounds* (John Wiley & Sons, New York, 1992).
24. M.A. Belkin and F. Capasso, "New frontiers in quantum cascade lasers: high performance room temperature terahertz sources", *Physica Scripta.* **90**, 118002 (2015).
25. A. Khalatpour, A. K. Paulsen, C. Deimert, Z.R. Wasilewski, and Q. Hu, "High-power portable terahertz laser systems", *Nat. Photon.* **15**, 6 (2021).
26. R. Kazarinov and R. Suris, "Possibility of the amplification of electromagnetic waves in a semiconductor with a superlattice", *Sov. Phys. Semicond.* **5**(4), 707 (1971).
27. J. Faist, F. Capasso, D.L. Sivco, C. Sirtori, A.L. Hutchinson, and A.Y. Cho, "Quantum cascade laser", *Science.* **264**, 553 (1994).
28. C. Sirtori, P. Kruck, S. Barbieri, P. Collot, J. Nagle, M. Beck, J. Faist, and U. Oesterle, "GaAs/$Al_xGa_{1-x}As$ quantum cascade lasers", *Appl. Phys. Lett.* **73**, 3486 (1998).
29. S. Slivken, V.I. Litvinov, M. Razeghi, and J.R. Meyer, "Relaxation kinetics in quantum cascade lasers", *J. Appl. Phys.* **85**, 665 (1999).
30. L. Zhang, G. Tian, J. Li, and B. Yu, "Applications of absorption spectroscopy using quantum cascade lasers", *Appl. Spectrosc.* 68, 1095 (2014).

17

Semiconductor Photodetectors

17.1 Photoconductors

A photodetector is a device that generates an electrical signal in response to optical illumination. If incident light excites nonequilibrium carriers in a semiconductor, the electrical conductivity, $\sigma = q(\mu_n n + \mu_p p)$, increases due to excess carrier densities $\Delta n(p)$; $\mu_n(\mu_p)$ is the mobility of electrons (holes), q is the elementary charge. The increase in conductivity, called *photoresponse*, turns a semiconductor into a photodetector. Virtually any semiconductor is a detector for incident light of wavelength corresponding to the energy gap for an optical transition. Intrinsic semiconductors are sensitive in spectral range $\lambda(\mu\text{m}) < 1.24/E_g(\text{eV})$, E_g is the bandgap (see Chapter 12). If band to impurity level transitions are involved in light absorption, the activation energy replaces E_g and determines the wavelength.

17.1.1 Generation Rate and Distribution of Carriers

Interband optical excitation in semiconductors creates electron–hole pairs, so in basic continuity equations (14.1) (Chapter 14), one uses generation rate G common for electrons and holes:

$$\frac{\partial \delta n}{\partial t} = G - \frac{\delta n}{\tau_n} + \frac{1}{q}(\nabla \cdot J_n),$$

$$\frac{\partial \delta p}{\partial t} = G - \frac{\delta p}{\tau_p} - \frac{1}{q}(\nabla \cdot J_p),$$

$$J_n = q\mu_n nE + qD_n(\nabla n),$$

$$J_p = q\mu_p pE - qD_p(\nabla p), \tag{17.1}$$

where G and $\tau_n(\tau_p)$ are the electron–hole generation rate and lifetimes in the conduction (valence) band, respectively. Finite lifetimes originate from interband and surface recombination, as well as capture by traps in the bandgap.

To establish a relation between generation rate and basic semiconductor parameters, we consider geometry typical for photoconductors in Figure 17.1. The energy flux in a semiconductor (Eq. (12.52), Chapter 12):

$$\Phi = \Phi_0 \exp(-\alpha z), \quad \left[\frac{\text{W}}{\text{m}^2}\right], \tag{17.2}$$

where z is the distance inside semiconductor from the illuminated surface, $\Phi_0 = \Phi_{in}(1-R)$, R is the reflectivity on the surface, Φ_{in} is the incident power flux, α is the absorption coefficient for optical intensity.

DOI: 10.1201/9780429285929-17

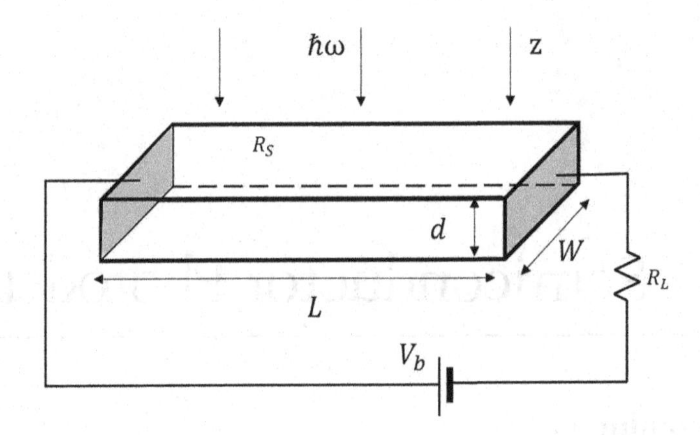

FIGURE 17.1 Semiconductor slab as a photoconductor, R_S and R_L are the semiconductor and load resistance, respectively.

The difference in photon flux at z and $z+dz$ is equal to the sheet density of absorbed photons per second:

$$dI = \frac{\Phi_0\alpha}{\hbar\omega}\exp(-\alpha z)\,dz. \tag{17.3}$$

Photon absorption rate per unit volume is $N = \Phi_0\alpha\exp(-\alpha z)/\hbar\omega$. Not each photon absorbed generates an e–h pair. A photon of energy $\hbar\omega > E_g$ may lose energy in different ways: to the creation of e–h pairs, transfer to lattice vibrations, or to electrons and holes in corresponding bands without new pair excitation.

The number ratio of generated e–h pairs to absorbed photons is called *internal quantum efficiency* η_i. Finally, the rate of e–h pair generation becomes

$$G(z) = \eta_i N = \eta_i\frac{\Phi_0\alpha}{\hbar\omega}\exp(-\alpha z),\ \left[\mathrm{m^{-3}s^{-1}}\right], \tag{17.4}$$

The internal quantum efficiency η_i is not necessarily less than one. If absorption occurs at optical energy much higher than the semiconductor bandgap, an e–h pair may produce one more pair by impact ionization, the process inverse to Auger recombination. If the process goes as an avalanche, pair multiplication may result in $\eta_i \gg 1$.

With the injection rate (17.4), we can solve basic equations (17.1) to get spatial distributions of excited carriers. Band-to-band optical excitation creates e–h pairs meaning that $\delta n = \delta p$ and in p-semiconductors, the distribution of minority carriers (electrons) under constant in time excitation follows from first and third equations (17.1):

$$D_n\frac{\partial^2\delta n}{\partial z^2} + \alpha\gamma\exp(-\alpha z) - \frac{\delta n}{\tau_n} = 0,\quad \gamma \equiv \eta_i\frac{\Phi_0}{\hbar\omega}. \tag{17.5}$$

The solution to (17.5) depends on the geometry and boundary conditions. To give an example, we find injected carriers distribution in a semi-infinite sample $(d \to \infty)$ with illuminated surface $z = 0$. Boundary conditions imply surface recombination at $z = 0$, $S\delta n = D_n\delta n/\partial z$, and $\delta n \to 0$, $z \to \infty$, where S is the surface recombination rate [m/s] (see Chapter 13). The solution has the form:

$$\delta n(z) = \frac{\alpha\gamma\tau_n}{\alpha^2L_n^2 - 1}\left[\frac{\alpha L_n^2 + S\tau_n}{L_n + S\tau_n}\exp\left(-\frac{z}{L_n}\right) - \exp(-\alpha z)\right],\quad L_n^2 = \tau_n D_n. \tag{17.6}$$

The density of injected carriers at the illuminated surface is

$$\delta n(0) = \frac{\alpha L_n \, \gamma \tau_n}{(\alpha L_n + 1)(L_n + S\tau_n)}. \tag{17.7}$$

Integrating distribution (17.6), one obtains

$$\int_0^\infty \delta n(z)\, dz = \frac{\gamma \tau_n}{\alpha^2 L_n^2 - 1}\left[\frac{\alpha L_n^2 + S\tau_n}{L_n + S\tau_n}\alpha L_n - 1\right], \tag{17.8}$$

the surface density of carriers generating in a thick sample and contributing to photoconductivity.

17.1.2 Uniform Illumination, Photoresponse, and Relaxation Times

When absorption is weak $(\alpha L_n \ll 1)$, carrier distribution (17.6) becomes

$$\delta n(z) = \alpha \gamma \tau_n \exp(-\alpha z) = \frac{\alpha \tau_n \Phi_{in}}{\hbar\omega}\, \eta_i\, (1 - R)\exp(-\alpha z), \tag{17.9}$$

where we assumed the surface ideal and thus $S = 0$.

Integrating (17.9) one finds the volume density of e–h pairs in a layer of thickness d from the illuminated surface:

$$\Delta n = \frac{1}{d}\int_0^d \delta n(z) = \eta \frac{\tau_n \Phi_{in}}{d\hbar\omega},$$

$$\eta \equiv \eta_i\, (1 - R)\left[1 - \exp(-\alpha d)\right], \tag{17.10}$$

where η is the *quantum efficiency*.

Incident power, partially reflecting from the illuminated surface, drops down in the course of pairs excitation on length d and then gets lost due to transmission toward the rest of the sample. Factors $\left[1 - \exp(-\alpha d)\right]$ and $(1 - R)$ account for transmission and reflection loss, respectively. If a sample were of finite thickness, one would have to account for corrected carrier distribution and backside reflection – the effects negligible in thick samples, $\alpha d > 1$.

The density of injected carriers (17.10) relates to generation rate as follows:

$$\Delta n = \eta \frac{\tau_n \Phi_{in}}{d\hbar\omega} = \eta \frac{\tau_n P_{in}}{\Omega\hbar\omega},$$

$$G = \frac{\Delta n}{\tau_n} = \eta \frac{P_{in}}{\Omega\hbar\omega}, \quad \left[\mathrm{s^{-1} m^{-3}}\right], \tag{17.11}$$

where P_{in} is the incident power $[W]$, $\Omega = dLW$ is the sample volume defined in Figure 17.1.

Photoresponse. Uniform illumination discussed above implies that diffusion of excited carriers is negligible, and thus basic equations simplify:

$$\frac{\partial \Delta n}{\partial t} = G - \frac{\delta\Delta}{\tau_n},$$

$$\frac{\partial \Delta p}{\partial t} = G - \frac{\delta\Delta}{\tau_p}, \tag{17.12}$$

As typical lifetimes are much longer than the energy and momentum relaxation times, non-equilibrium carriers thermalize fast and acquire almost equilibrium energy and momentum distribution in corresponding bands. Consequently, photoresponse depends on excess carriers only while mobilities remain at their equilibrium values: $\Delta\sigma = q(\mu_n\ \Delta n + \mu_p\ \Delta p)$. Kinetics of photoconductivity follows from (17.12) as

$$\frac{\partial\Delta\sigma}{\partial t} = Gq(\mu_n + \mu_p) - \frac{\Delta\sigma}{\tau_{pc}}, \quad \tau_{pc} = \frac{(\mu_p\Delta p + \mu_n\Delta n)}{\left(\frac{\mu_p}{\tau_p}\Delta p + \frac{\mu_n}{\tau_n}\Delta n\right)}. \tag{17.13}$$

To obtain (17.13), one multiplies rows in (17.12) by $q\mu_n$ and $q\mu_p$, respectively, and then adds to each other. Combination τ_{pc} is the *photoconductivity relaxation time.*

17.1.3 Steady-State Illumination. Photoconductive Gain and Responsivity

In the steady-state, (17.12) and (17.13) give

$$\begin{aligned} &\Delta n = G\tau_n, \quad \Delta p = G\tau_p, \\ &\tau_{pc} \to \tau'_{pc} = \frac{\mu_p\tau_p + \mu_n\tau_n}{\mu_p + \mu_n}, \\ &\Delta\sigma = Gq\tau'_{pc}(\mu_n + \mu_p) \end{aligned} \tag{17.14}$$

Time τ'_{pc} is often called a *photoconductivity lifetime.* As follows from (17.14), the larger τ'_{pc}, the stronger the photoresponse. At the same time, long relaxation time, as will be shown below, diminishes response to the alternating signal and thus slows down the speed of operation.

If under illumination an external voltage is applied, the short-circuit *photocurrent* ($R_L = 0$ in Figure 17.1), caused by photoresponse (17.14), has the form

$$\begin{aligned} \Delta I &= \Delta\sigma\frac{V_b}{L}Wd \\ &= qG(\mu_n\tau_n + \mu_p\tau_p)\frac{V_b}{L}Wd. \end{aligned} \tag{17.15}$$

Assuming excess carrier lifetime $\tau_{n,p} \approx \tau$ and introducing transit time τ_{tr}, one obtains

$$\begin{aligned} &\Delta I = qG\Omega g, \\ &g \equiv \frac{\tau}{\tau_{tr}}\left(1 + \frac{\mu_p}{\mu_n}\right), \\ &\tau_{tr} \equiv L^2/\mu_n V_b, \end{aligned} \tag{17.16}$$

where coefficient g is called *photoconductive gain.* In what follows, we suppose $\mu_n \gg \mu_p$ and thus $g \approx \tau/\tau_{tr}$. Factor $G\Omega$ equals the number per second of e–h pairs created in the whole sample. The longer the photocarriers dwell in the band, and the faster they are swept to electrodes by the applied voltage, the higher the photocurrent.

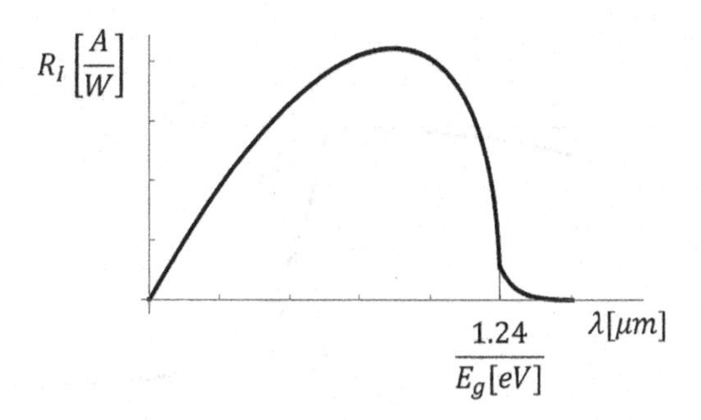

FIGURE 17.2 The spectral responsivity of a photoconductor.

Using (17.16) and the relation between generation rate and incident power (17.11), one can introduce *current responsivity* as

$$R_I = \frac{\Delta I}{P_{in}} = \frac{q\eta}{\hbar\omega} g, \quad \left[\frac{A}{W}\right]. \tag{17.17}$$

Often, as a detector device characteristic, Eq. (17.17) is written as $R_I = \eta g \lambda[\mu\text{m}]/1.24$, λ is the wavelength in microns corresponding to optical transition. Responsivity is proportional to gain and thus, in virtue of (17.16), reads as $R_I \sim g \approx \tau/\tau_{tr} = \tau\mu_n V_b/L^2$, so contains the product of intrinsic semiconductor parameters, $\tau\mu_n$, which is the figure of merit as for material choice.

After initial linear in λ dependence, the spectral responsivity drops down when absorption (and thus quantum efficiency) goes to zero for wavelengths close to the inverse bandgap, as shown in Figure 17.2.

The maximum R_I depends on quantum efficiency and the voltage applied to the photoconductor. If load resistance in Figure 17.1 is larger than the semiconductor resistance, $R_L \gg R_S$, it is convenient to read out the signal as the change of voltage across the load:

$$\Delta V_S = \frac{\partial}{\partial R_S}\left(\frac{R_L V_b}{R_L + R_S}\right)\Delta R_S \approx -\frac{V_b}{R_L}\Delta R_S = \frac{R_S V_b}{R_L}\frac{\Delta\sigma}{\sigma}. \tag{17.18}$$

Substituting $\Delta\sigma$ from (17.14), one obtains the *voltage responsivity*

$$R_V = \frac{\Delta V_S}{P_{in}} = \frac{V_b R_S}{R_L}\frac{\eta}{\Omega\hbar\omega}\frac{\tau_p\left(1+\mu_p/\mu_n\right)}{n_0 + p_0\mu_p/\mu_n}, \quad \left[\frac{V}{W}\right], \tag{17.19}$$

where n_0, p_0 is the equilibrium carrier density in the slab.

In a practical InAsSb detector, the maximum current responsivity is about 0.2 A/W at room temperature and a wavelength of 3.5 μm.

Responsivities characterize intrinsic properties that enable a semiconductor material to detect light. However, the response itself does not guarantee the proper performance of the detector due to various types of noise that could mask the signal. Ultimate characterization includes signal-to-noise ratio, noise-equivalent power, and detectivity – all to be discussed later.

17.1.4 Rectangular Pulse Illumination

If the light is on during time T, then off during the same time – the photoconductivity behaves as illustrated in Figure 17.3.

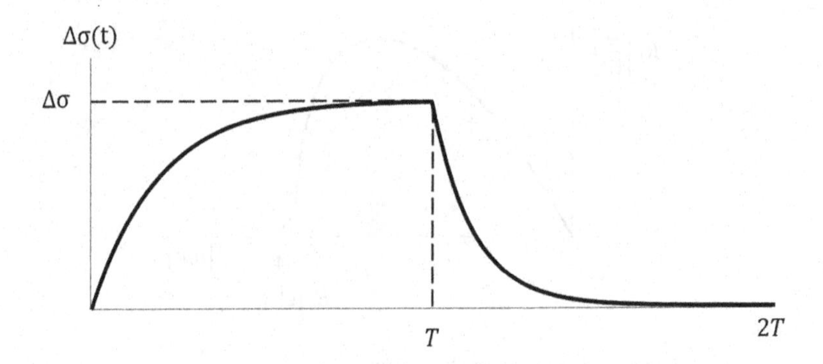

FIGURE 17.3 Time-dependent photoconductivity under rectangular-pulse-modulated illumination of duty cycle $T > \tau_{pc}$.

Illumination turned on at $t = 0$ causes the response shown Figure 17.3 ($0 < t < T$):

$$\Delta\sigma(t) = \Delta\sigma\left[1 - \exp\left(-\frac{t}{\tau_{pc}}\right)\right]. \tag{17.20}$$

If the duty cycle is large, $T \gg \tau_{pc}$, $\delta\sigma$ reaches the maximum value before it starts to decay when light is off ($T < t < 2T$ in Figure 17.3):

$$\Delta\sigma(t) = \Delta\sigma \exp\left(-\frac{t-T}{\tau_{pc}}\right), \quad t \geq T. \tag{17.21}$$

So, during time τ_{pc}, the photoconductivity rises after illumination is turned on and decays after the light is off. Similar solutions for excess carrier density are discussed already in Chapter 13.

If the duty cycle were short, $T < \tau_{pc}$, $\delta\sigma$ would not reach its stationary (maximum) value $\Delta\sigma$. The shorter the duty cycle as compared to τ_{pc}, the lower the magnitude of the response. So, τ_{pc} restricts the speed the photoconductor can detect the signal.

It is instructive to look at the photoresponse in early-stage $t \ll T$. From (17.13), one obtains

$$\frac{\partial\Delta\sigma}{\partial t} \approx Gq(\mu_n + \mu_p),$$

$$\Delta\sigma \approx Gq(\mu_n + \mu_p)\, t. \tag{17.22}$$

The temporal behavior of photoresponse at small t does not depend on lifetimes. Experimental measurement of the initial stage of photoconductivity (17.22) allows us to find G and use (17.11) to estimate quantum efficiency η.

17.1.5 Frequency Response

The widely used method of light detection is to modulate the light incident to the photodetector and then analyze the signal at the modulation frequency. For modulation, one uses some form of a light chopper, which periodically interrupts a continuous light source. Alternatively, one may measure the dc-current generated under steady-state illumination. The ac-method suppresses the noise current, being preferable in small-signal detection, while measurements of dc-current imply noise accumulation during a long integration time.

Excess carrier kinetics under time-periodic excitation with frequency f, $G(t) = G\sin(2\pi ft)$, follows from equation

$$\frac{\partial(\delta n)}{\partial t} = -\frac{\delta n}{\tau} + G\sin(2\pi ft). \tag{17.23}$$

Equation (17.23) has the exact solution given in (13.9) (Chapter 13). In the oscillation regime, after the initial relaxation process, $t \gg \tau$, excess carriers density oscillations follow the modulation with phase delay $\varphi \equiv arctg(2\pi f\tau)$:

$$\delta n = G\tau\cos(\varphi)\sin(2\pi ft - \varphi). \tag{17.24}$$

Correspondingly, from (17.11), (17.16), and (17.24), one obtains the photoresponse as

$$I_S(t) = \frac{I_0}{\sqrt{1+(2\pi f)^2\tau^2}}\sin(2\pi ft - \varphi), \quad I_0 = q\eta g\frac{P_{in}}{\hbar\omega}, \tag{17.25}$$

where P_{in} is the amplitude of incident optical power. Response signal cuts off high frequencies $f \gg f_c = (2\pi\tau)^{-1}$. The photoconductive gain (17.16) and the bandwidth $(0, f_c)$ depend on the lifetime of the excess (minority) carriers τ so that the gain-bandwidth product is a function of transit time only, $2\pi f_c g = 1/t_{tr}$, meaning a tradeoff – high gain suppresses maximum operation frequency.

17.1.6 The Noise in Photodetectors

To discuss basic parameters, which characterize a photoconductor, we consider steady-state illumination and detection as a measurement of signal (17.17):

$$i_S = R_I P_{in}. \tag{17.26}$$

Alternatively, one can take a voltage across the load resistance, u_S as a detectable signal. The signal goes to the receiver, which accepts ac-signal within the pass-band Δf and measures the mean-square voltage.

Even under steady-state conditions, the current through a semiconductor slab fluctuates in time. Fluctuations affect the performance of the detecting devices and will be discussed below in more detail.

The fluctuation (noise) is the deviation from the average $i(t) = I(t) - \langle I \rangle$, which is characterized by the spectral density [1,2]:

$$S(f) = 2\int_{-\infty}^{\infty} \exp(2\pi ift)\langle i(t')\, i(t+t')\rangle\, dt, \tag{17.27}$$

where $\langle \ldots \rangle$ means ensemble average, which equals the average over t' as we assume the process ergodic. The mean-square noise current in the frequency band B is $\langle i^2 \rangle = S(f)\, B$. Those part of the noise spectrum that falls into the measurement system frequency window Δf, $S(f)\,\Delta f$, enters the electronic circuit and amplify there along with the useful signal. The finite window restricts noise accumulation that is equivalent to averaging the fluctuations during integration time $t_{in} \approx 1/\Delta f$. Long averaging time t_{in} favors noise suppression. The spectral position of the window should not exceed the cut-off limit of photodetector f_c. Several noise sources are shown in Figure 17.4.

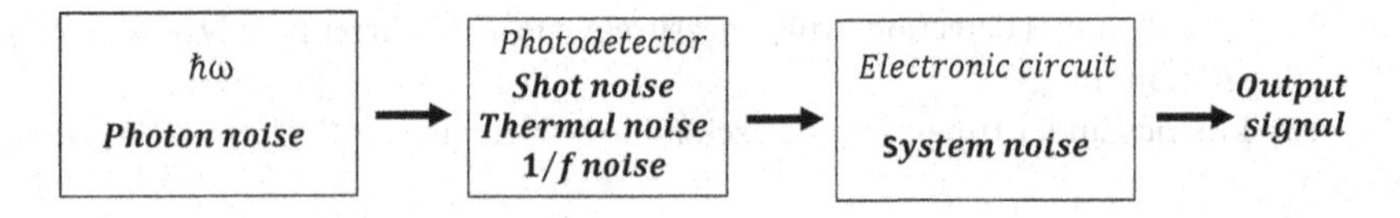

FIGURE 17.4 Noise in a photodetector.

Thermal noise. Thermal (*Nyquist-Johnson*) noise exists in thermal equilibrium and originates from the thermal motion of charges in a resistor at finite temperature. Fluctuation-dissipation theorem [3] gives the spectral density as

$$S_T(f) = \frac{4\hbar\pi f}{R_S}\left[1 + \frac{2}{\exp\left(\frac{2\hbar\pi f}{k_B T}\right) - 1}\right], \tag{17.28}$$

where the first term is the zero-point fluctuations. At low frequency, $2\hbar\pi f \ll k_B T$, $S_T(f)$ does not depend on the frequency and thus represents an example of *white noise*:

$$S_T(f) = \frac{4k_B T}{R_S}. \tag{17.29}$$

Noise power $\langle i_T^2 \rangle R_S = 4k_B T\ \Delta f$ depends on temperature and the spectral bandwidth of the measuring device. The voltage noise that follows from (17.29) is $\langle u_T^2 \rangle = 4k_B T R_S\ \Delta f$. The only way to suppress this noise within a fixed bandwidth is to cool all resistive elements of the circuit.

Flicker noise. One more type of equilibrium noise is related to fluctuating resistance and manifests itself as *flicker* noise, $S_F(f) \sim 1/f$, or $1/f$ *noise*. The physical processes that lead to fluctuations in the carrier mobility and (or) density are considered responsible for flicker noise. The way to reduce the noise is to increase the modulation frequency of the signal entering the detector.

Generation-recombination and shot noise. Noise in a nonequilibrium state is related to the direct current flowing through the photoconductor. The time a carrier dwells in the band and contributes the direct current is limited either by recombination act (for the excited carrier) or by the time it swept to ohmic contact by an applied voltage (for both excited and equilibrium carriers). The dwell time fluctuates, thus creating generation-recombination noise [4]:

$$\langle i_{GR}^2 \rangle = 4qgi_S\ \Delta f, \tag{17.30}$$

where i_S is the total dc-component of the current under illumination. The generation-recombination noise is a flavor of so-called *shot noise*, which originates from fluctuations of direct current of any origin. If recombination is ineffective due to the spatial separation of electrons and holes (this is the case in photodiodes, $g = 1$), the shot noise is the generation-only noise:

$$\langle i_G^2 \rangle = 2qi_S\ \Delta f. \tag{17.31}$$

In the absence of illumination, the shot noise contains dark current instead of i_S and often is called *dark noise*. Shot noise of any type reduces through lower both dc-current and spectral bandwidth.

Photon noise. Light incident to a photodetector comprises the optical signal of target and a broad spectrum of background emission coming as radiation from the environment. Even without signal, environment photons, absorbed by a semiconductor, create dc-current and then shot noise often called *photon noise*:

$$\left\langle i_{ph}^2 \right\rangle = 2uqI_{ph}\ \Delta f, \tag{17.32}$$

where, following (17.30) or (17.31), $u = 2g$ or 1 depending on photoconductor or photodiode, respectively, is under consideration. The background-induced current I_{ph} is the subject to calculation below.

In a photoconductor, single-mode photons (monochromatic illumination) generate current calculated in (17.11) and (17.16):

$$I = q\eta g \frac{P_{in}}{\hbar\omega}. \tag{17.33}$$

Coefficient $P_{in}/\hbar\omega$ is the number of photons per second incident to a photoconductor aperture of area $A = LW$ (see Figure 17.1). Alternatively, the coefficient reads as ΦA, where $\Phi\left[\mathrm{m^{-2}s^{-1}}\right]$ is the incident photon flux – the density of single-mode photons $n_{ph}(k)$ multiplied by their velocity component normal to the illuminated surface (see Figure 17.5):

$$\Phi(\boldsymbol{k}) = cn_{ph}(k)\cos\theta. \tag{17.34}$$

So, photocurrent (17.33) induced by single-mode photons takes the form,

$$I(\boldsymbol{k}) = Acqg\ \eta(k)\ n_{ph}(k)\cos\theta, \tag{17.35}$$

where $\eta(k)$ is the k-dependent quantum efficiency.

In an equilibrium background (blackbody radiation), the filling factor of photon modes is the Bose distribution function (see Chapter 4). The total current is a sum over all modes ($\omega = ck$):

$$I_{ph} = Acqg\sum_{k} n_{ph}(k)\eta(k)\cos\theta = \frac{Acqg}{(2\pi)^3}\int \frac{\eta(k)\cos\theta\, d^3k}{\exp\left(\dfrac{\hbar ck}{k_BT}\right)-1}. \tag{17.36}$$

Performing integration in (17.36), one obtains

$$I_{ph} = \frac{Acqg}{(2\pi)^3}\left\{4\pi\int_0^{\varphi/2}\sin\theta\cos\theta\, d\theta\int_{k_{\min}}^{\infty}\frac{\eta(k)k^2dk}{\exp\left(\dfrac{\hbar ck}{k_BT}\right)-1}\right\} = \frac{Aqgc}{4\pi^2}\sin^2\left(\frac{\varphi}{2}\right)\int_{k_{\min}}^{\infty}\frac{\eta(k)k^2dk}{\exp\left(\dfrac{\hbar ck}{k_BT}\right)-1}, \tag{17.37}$$

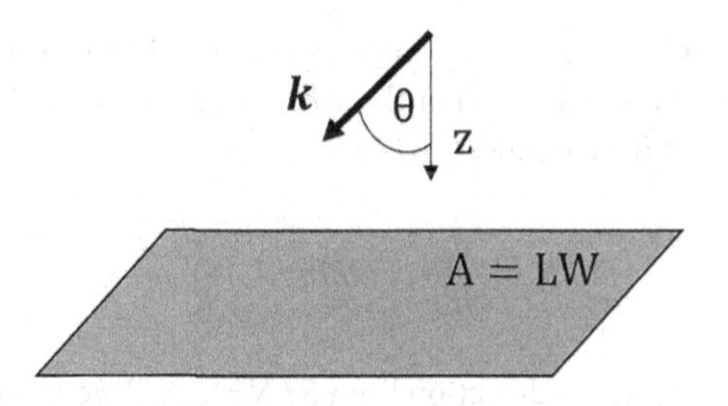

FIGURE 17.5 Incident photon flux, $-\varphi/2 \le \theta \le \varphi/2$; φ and $\boldsymbol{k}$ are the field-of-view angle and momentum of the background photon mode, respectively.

where $k_{\min} = 2\pi/\lambda_{\max}$ defines the infrared cut-off: only photons with energy larger than the bandgap $(\lambda < \lambda_{\max})$ can be absorbed by a semiconductor. Replacing variable to $\lambda = 2\pi / k$, one obtains a more commonly used expression:

$$I_{ph} = Aqg \sin^2\left(\frac{\phi}{2}\right) \int_0^{\lambda_{\max}} \frac{2\pi c\, \eta(\lambda)\, d\lambda}{\lambda^4 \left[\exp\left(\frac{2\pi\hbar c}{\lambda k_B T}\right) - 1\right]}. \tag{17.38}$$

If $\eta(\lambda) = 1$, the integrand equals $W / \hbar\omega$, where

$$W = \frac{(2\pi c)^2 \hbar}{\lambda^5 \left[\exp\left(\frac{2\pi\hbar c}{\lambda k_B T}\right) - 1\right]},$$

the power radiated at wavelength λ per unit wavelength interval by unit area of a blackbody at temperature T.

Noise mean-square current associated with background radiation then follows from (17.32) and (17.38):

$$\langle i_{ph}^2 \rangle = 2uqI_{ph}\, \Delta f. \tag{17.39}$$

The total noise current affects the detectivity of the device discussed in the next section.

17.1.7 Specific Detectivity

Total noise current comprises all relevant components as follows:

$$i_n = \sqrt{\langle i_{GR}^2 \rangle + \langle i_T^2 \rangle + \langle i_{Ph}^2 \rangle}. \tag{17.40}$$

In most photodetectors, the main sources of noise are thermal and generation-recombination currents. If these currents are small as compared to photon noise, we deal with background limited infrared performance (*BLIP*).

Not in a *BLIP* regime, to discern the signal from a noisy background, one defines the *signal-to-noise ratio as*

$$S / N = \frac{i_S}{i_n} \equiv \frac{i_S}{\sqrt{\langle i_{GR}^2 \rangle + \langle i_T^2 \rangle}} = \frac{R_I P_{in}}{\sqrt{4qgi_S \Delta f + 4k_B T \Delta f / R_S}} \approx \frac{R_I P_{in}}{\sqrt{4qgi_S \Delta f}}. \tag{17.41}$$

Approximation holds if thermal noise is negligible as compared to shot noise, in other words, if the voltage across the conductor exceeds $k_B T/q$. For the sake of convenience, (17.41) can be written as a ratio of incident power and *noise equivalent power* (*NEP*):

$$S / N = \frac{P_{in}}{NEP}, \quad NEP = \frac{\sqrt{4qgi_S \Delta f}}{R_I}. \tag{17.42}$$

For detectable signals, $S / N \geq 1$. At the detection limit $S / N = 1$, *NEP* equals the incident power.

Noise equivalent power quantifies noise and allows us to compare various materials and device configurations toward maximum sensitivity of the photodetector. However, as defined in (17.42), *NEP* contains the frequency window of the measurement device Δf and, through i_S, depends on the geometric parameters of the photoresistor. In detail, the dark current through the n-type slab, shown in Figure 17.1, is expressed as

$$i_d = q\left(n_0\mu_n + p_0\mu_p\right)EWd \approx qn_0\mu_n\frac{V}{L}Wd, \tag{17.43}$$

where the approximation is valid for n-type semiconductor, V is the voltage applied to semiconductor, n_0 is the equilibrium density of majority carriers. The gain follows from (17.16): $g = \tau\mu_n V/L^2$. Calculating *NEP* with (17.17), (17.42), and (17.43), one obtains

$$NEP = \frac{2\hbar\omega}{\eta}\sqrt{\frac{\Omega n_0\Delta f}{\tau}}, \tag{17.44}$$

Expression (17.44) can hardly be a figure-of-merit for material choice because it depends on bandwidth Δf and volume of semiconductor slab Ω. To avoid extrinsic parameters and quantify sensitivity, we divide *NEP* by $\sqrt{A\ \Delta f}$, $A = WL$ is the area of illuminated surface, and define *specific detectivity* as

$$D^* = \left(\frac{NEP}{\sqrt{A\Delta f}}\right)^{-1} = \frac{\eta}{2\hbar\omega}\sqrt{\frac{\tau}{n_{0S}}}, \quad \left[\frac{m\sqrt{Hz}}{W}\right] \tag{17.45}$$

where $n_{0S} = n_0 d$ is the sheet density of majority carriers in a semiconductor slab.

In a p-type photoconductor, following (17.43)–(17.45), one obtains

$$D^* = \frac{\eta}{2\hbar\omega}\sqrt{\frac{\tau\mu_n}{p_{0S}\mu_p}}. \tag{17.46}$$

Bearing in mind $\mu_n \gg \mu_p$, one may conclude that material with fast minority carriers favors higher detectivity.

Dependence $D^* \sim 1/\sqrt{n_0 d}$ limits detectivity in infrared at room temperature as infrared detectors have the bandgap of the order of $k_B T$, and thus high background carrier concentration. That is why long wavelength detection needs cooling of the device.

In virtue of (17.10), detectivity depends on the slab thickness as $D^* \sim \eta/\sqrt{d} \approx \left[1-\exp(-\alpha d)\right]/\sqrt{d}$. Coefficient $\eta/\sqrt{d}$ reaches the maximum value of $\approx 0.64\sqrt{\alpha}$ at the optimum thickness $d_{opt} \approx 1.26/\alpha$.

17.1.8 Background Limited Performance

Were noise in a semiconductor circuit suppressed, the device would operate in *BLIP* mode, which means that photonic noise is the only one that limits the detection capability at wavelength λ_S. Responsivity, *NEP*, and specific detectivity determined by (17.17), (17.39), and (17.45), respectively, become

$$R_I = \frac{qg\eta(\lambda_S)\lambda_S}{2\pi\hbar c}, \quad NEP = \frac{\sqrt{4qgI_{ph}\Delta f}}{R_I},$$

$$D^*_{BLIP} = \frac{\eta(\lambda_S)\lambda_S}{2\pi\hbar c\sin\left(\dfrac{\phi}{2}\right)\sqrt{8\pi c\displaystyle\int_0^{\lambda_{max}}\eta(\lambda)\lambda^{-4}\left[\exp\left(\dfrac{2\hbar\pi c}{\lambda k_B T}\right)-1\right]^{-1}d\lambda}} \tag{17.47}$$

Expression (17.47) sets a fundamental limit on detectivity at a certain temperature. In photodiodes, the numerical coefficient under the square root is twice as less as in (17.47). For details on detectors in the BLIP regime, see Ref. [5].

17.2 Photodiodes

Photoconductors, described in previous sections, are not the only option for light detection. Typical semiconductor impedance is low that makes it difficult to match the photoconductor to the rest of the circuit. It is more convenient to use a photodiode which is a light-sensitive p–n or metal-semiconductor junction (for details, see Chapter 14). The junction responds to incident optical power as its I–V characteristic is sensitive to illumination. The structure of the illuminated p–n junction is shown in Figure 17.6.

Photocurrent mode. As illustrated in Figure 17.6a, optically generated and separated by built-in electric field electrons and holes increase the current through the junction under an inverse bias and decrease it in forward-biased condition. Consequently, under illumination, the current downshifts for any voltage sign, as shown in Figure 17.7.

The I-V characteristic (Chapter 14) shifted by photocurrent density J_{ilm},

$$J = J_S\left[\exp\left(\frac{qU}{k_BT}\right)-1\right]-J_{ilm},$$

$$J_{ilm} = q\eta\Phi, \tag{17.48}$$

where Φ is the incident photon flux and η is the quantum efficiency of the photodiode.

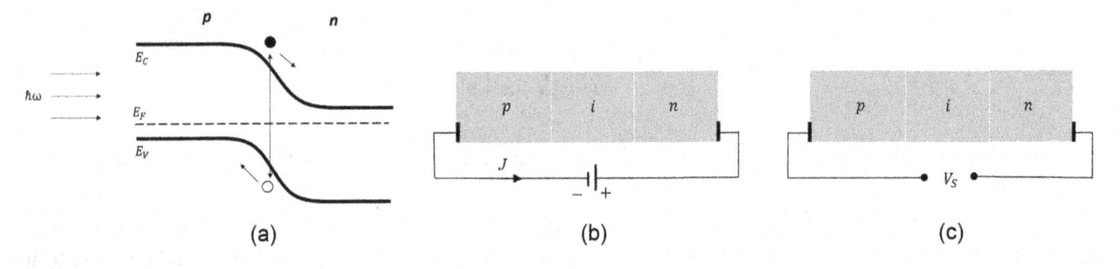

FIGURE 17.6 (a) Separation of optically generated e-h pair in a built-in electric field. (b) Photocurrent mode. (c) Photovoltaic (solar cell) mode.

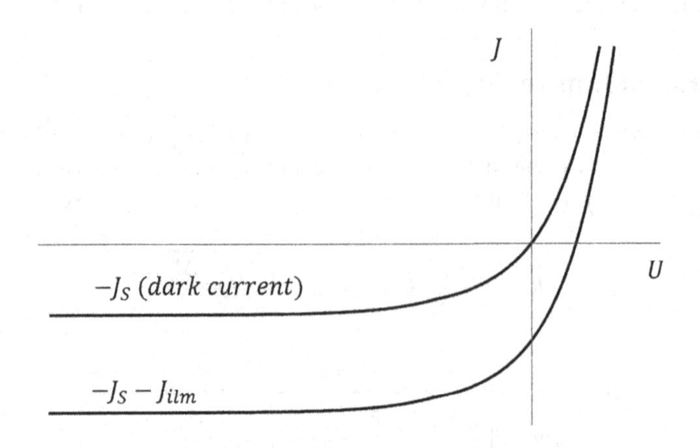

FIGURE 17.7 Current–voltage characteristics in p-n junction under illumination.

Detectivity in photovoltaic mode. If the junction is not biased (see Figure 17.6b), signal voltage V_S appears under illumination. The signal magnitude is equal to current AJ_{ilm} (illumination-induced short-circuit current) multiplied by the junction resistance at zero bias R_0:

$$V_S = q\eta\Phi AR_0, \quad R_0A = \left(\frac{\partial J}{\partial U}\right)^{-1}_{U=0} \tag{17.49}$$

where A is the area of the p–n junction.

At zero bias, there is no dark current and thus shot noise. Mean square noise current comprises thermal and photon noise, (17.29) and (17.39), respectively:

$$i_n = \sqrt{i_T^2 + i_{Ph}^2}, \quad NEP = \frac{\sqrt{2qI_{ph}\,\Delta f + 4k_BT\,\Delta f/R_0}}{R_I}. \tag{17.50}$$

If not in the BLIP regime, the thermal noise dominates and

$$NEP = \frac{1}{R_I}\sqrt{\frac{4k_BT\,\Delta f}{R_0}}, \quad R_I = \frac{q\eta\lambda_S}{2\pi\hbar c}, \quad D^* = \frac{\sqrt{A\Delta f}}{NEP} = \frac{q\eta\lambda_S}{2\pi\hbar c}\sqrt{\frac{R_0A}{4k_BT}}. \tag{17.51}$$

Following (17.51), one considers R_0A as a figure of merit for the thermally limited photodiode.

The diode feels incident light and thus may be used in a photocurrent mode in both forward and reverse bias regimes. However, the reverse bias operating point is preferable as the diode resistance is high (large R_0A), and the shot noise is limited as far as saturation current is small. Also, reverse bias favors lower junction capacitance that is preferable for high-frequency operation.

Absolute values of detectivity depend on wavelength and operation temperature and reach $(1 \div 8)\times 10^{11}$ $\text{cm}\sqrt{\text{Hz}}/\text{W}$ in the mid-infrared range of $(2.7 \div 6)\mu\text{m}$ [5].

17.3 Quantum Well Photodetectors

Spectral responsivity in bulk semiconductors reaches the maximum at photon energy close to semiconductor bandgap, as illustrated in Figure 17.2. The ultraviolet part of the spectrum is covered with AlGaN alloys: $\lambda \approx 300 - 400$ nm. The bandgap in II–VI materials CdS and CdSe corresponds to the absorption wavelength in the visible light region, $\lambda \approx 700$ nm. Semiconductors Si and Ge detect light at 1.1 μm and 1.85 μm, respectively. Alloys from the III–V material system, InGaAs-InSb, respond to $\lambda = (0.87 - 7.0)\mu\text{m}$. In mid- and far-infrared regions, narrow-gap II–VI and IV–VI alloys CdHgTe or PbSnTe, respectively, can be used for detection up to $\lambda \approx 20$ μm. For detectors in far-infrared, $(20 - 100)\mu\text{m}$, silicon, and germanium, doped with shallow donors, are used, and in this case, absorption is due to donor-conduction band optical transitions.

As seen from (17.46), the detectivity of photoconductors suffers from high background carrier concentration. That is the issue for long-wavelength detectors because as-grown II–VI and IV–VI binary and ternary narrow-gap crystals are highly non-stoichiometric and thus have high equilibrium carrier density. On the other hand, III–V materials are more stable, their growth methods well-developed, so doping and carrier density controllable. Artificial structures, such as quantum wells (QW), quantum dots, and superlattices made of wide-bandgap materials cover the mid-infrared range. By manipulating geometrical parameters and alloy content in quantum wells and barriers, one can engineer the energy distance between subbands involved in optical transitions irrespectively to the fundamental bandgap. So fabricated from AlGaAs material system which is easy to grow and process, it is possible to design quantum well photodetectors (QWIP) tuned to detect infrared light in $(6 \div 25)\mu\text{m}$ range. QWIPs made of InGaAs / GaAs QW's operate up to 15 μm [6,7].

Designing a quantum well to detect light of a particular wavelength becomes a simple matter of tailoring the potential depth and width of the well to produce two states separated by the desired photon energy.

17.3.1 Intersubband Absorption

In 3D-semiconductors, the interband absorption coefficient is calculated and given in (12.70) (Chapter 12). The basis wavefunctions used are the Bloch functions $\varphi_{nk}(\boldsymbol{r})=u_{nk}(\boldsymbol{r})\exp(i\boldsymbol{k}\cdot\boldsymbol{r})$, which are the product of Bloch amplitude and the envelope function $\exp(i\boldsymbol{k}\cdot\boldsymbol{r})$. Bloch amplitude and the electron effective mass both carry information on the periodic lattice potential, while the envelope is the wavefunction of a free-like electron.

As discussed in Chapter 2, in quasi-2D structures such as quantum wells, the envelope is modified such as it becomes a product of in-plane free-electron function $\exp\left(i\boldsymbol{k}_{\parallel}\cdot\boldsymbol{r}_{\parallel}\right)/\sqrt{A}$ and the solution of the one-dimensional Schrödinger equation in the confinement potential in the growth direction $f_{ni}(z)$:

$$\varphi_{nik}(\boldsymbol{r})=\frac{1}{\sqrt{A}}u_{n0}(\boldsymbol{r})\exp\left(i\boldsymbol{k}_{\parallel}\cdot\boldsymbol{r}_{\parallel}\right)f_{ni}(z),$$

$$\boldsymbol{k}_{\parallel}=\left(k_x,k_y\right),\quad \boldsymbol{r}_{\parallel}=\left(x,y\right), \tag{17.52}$$

where A is the area of the quantum well, n is the band index, $u_{n0}(\boldsymbol{r})$ is the Bloch amplitude at $\boldsymbol{k}=0$, i numerates quantum states in the confinement potential. Normalization conditions follow:

$$\int_{-\infty}^{\infty} f_{ni}^{*}(z)f_{nj}(z)dz=\delta_{ij},$$

$$\frac{1}{\Omega_0}\int_{\Omega_0} u_{n0}^{*}(\boldsymbol{r})u_{n'0}(\boldsymbol{r})d^3\mathrm{r}=\delta_{nn'}, \tag{17.53}$$

where Ω_0 is the volume of the unit cell.

Following (12.54) (Chapter 12), we present the optical transition rate $\left[\mathrm{s}^{-1}\right]$ as

$$w_{ij}=\frac{2\pi}{\hbar}\left|\langle\varphi_j\,|\,F\,|\,\varphi_i\rangle\right|^2\delta\left(E_j-E_i-\hbar\omega\right),$$

$$\langle\varphi_j|F|\varphi_i\rangle=\frac{\hbar qE}{2m\omega}\left\langle\varphi_{n'jk'}(\boldsymbol{r})\middle|e^{i\boldsymbol{q}\cdot\boldsymbol{r}}\,\boldsymbol{e}\cdot\nabla\middle|\varphi_{nik}(\boldsymbol{r})\right\rangle. \tag{17.54}$$

Neglecting the optical wavevector as compared to electron one, we obtain transition rate in the form,

$$w_{ij}=\frac{\pi\hbar q^2E^2}{2m^2\omega^2}\left|\left\langle\varphi_{n'jk}(\boldsymbol{r})\middle|\boldsymbol{e}\cdot\nabla\middle|\varphi_{nik}(\boldsymbol{r})\right\rangle\right|^2\delta\left(E_j-E_i-\hbar\omega\right), \tag{17.55}$$

and similarly to (12.70) (Chapter 12), the absorption coefficient in QW made of nonmagnetic materials becomes

$$\alpha(\omega)=\frac{2\pi\hbar^2q^2}{\Omega m^2\omega c\varepsilon_0 n_r}\sum_k\left|\left\langle\varphi_{n'jk}(\boldsymbol{r})\middle|\boldsymbol{e}\cdot\nabla\middle|\varphi_{nik}(\boldsymbol{r})\right\rangle\right|^2\delta\left(E_{jk}-E_{ik}-\hbar\omega\right)\left[f(E_{ik})-f\left(E_{jk}\right)\right], \tag{17.56}$$

where $\boldsymbol{k}$ is in-plane wavevector, E_{ik} and $f(E_{ik})$ are the $2D$-conduction subband and its Fermi occupation probability, respectively. The optical matrix element comprises three parts:

$$I = \langle \varphi_{n'jk} | \boldsymbol{e} \cdot \nabla | \varphi_{nik} \rangle = I_1 + I_2 + I_3, \tag{17.57}$$

$$I_1 = \frac{1}{A} \int d^3r \, f_{ni}(z) f_{n'j}^*(z) u_{n'0}^*(\boldsymbol{r}) \boldsymbol{e} \cdot \nabla u_{n0}(\boldsymbol{r})$$

$$I_2 = i\boldsymbol{e} \cdot \boldsymbol{k}_{\parallel} \frac{1}{A} \int d^3r \, u_{n'0}^*(\boldsymbol{r}) u_{n0}(\boldsymbol{r}) f_{n'j}^*(z) f_{ni}(z),$$

$$I_3 = \frac{1}{A} \int d^3r \, u_{n'0}^*(\boldsymbol{r}) \, u_{n0}(\boldsymbol{r}) \, f_{n'j}^*(z) e_z \nabla_z f_{ni}(z). \tag{17.58}$$

The integral in I_1 contains lattice-periodic factor $u_{n'0}^*(\boldsymbol{r}) \, u_{n0}(\boldsymbol{r})$ and slowly varying envelope $f_{n'j}^*(z) f_{ni}(z)$. Following the procedure used in Chapter 12, we represent the integral as a sum over unit cells numbered by their lattice vectors $\boldsymbol{R}$. Integration now runs over the unit cell:

$$I_1 = \frac{1}{A} \sum_{R} f_{ni}(R_z) \, f_{n'j}^*(R_z) \int_{\Omega_0} d^3r \, u_{n'0}^*(\boldsymbol{r}+\boldsymbol{R}) \boldsymbol{e} \cdot \nabla \, u_{n0}(\boldsymbol{r}+\boldsymbol{R}). \tag{17.59}$$

The envelope in (17.59) weakly depends on coordinates within the unit cell limits and thus is removed from the integral. The integral now contains lattice-periodic function and thus does not depend on $\boldsymbol{R}$:

$$I_1 = \frac{1}{A} \int_{\Omega_0} d^3r \, u_{n'0}^*(\boldsymbol{r}) \boldsymbol{e} \cdot \nabla u_{n0}(\boldsymbol{r}) \sum_{R_z} f_{ni}(R_z) f_{n'j}^*(R_z) \sum_{\boldsymbol{R}_{\parallel}} 1 = \frac{N_{\parallel}}{A} \int_{\Omega_0} d^3r \, u_{n'0}^*(\boldsymbol{r}) \boldsymbol{e} \cdot \nabla u_{n0}(\boldsymbol{r}) \sum_{R_z} f_{ni}(R_z) \, f_{n'j}^*(R_z), \tag{17.60}$$

where we used momentum conservation identity (Umklapp processes are neglected), $\sum_{\boldsymbol{R}_{\parallel}} 1 = N_{\parallel}$ is the number of unit cells in the QW area. After writing the coefficient in (17.60) as $N_{\parallel}/A = a_z/\Omega_0$, a_z is the lattice constant in the growth direction, and replacing $a_z \sum_{R_z} f_{ni}(R_z) \, f_{n'j}^*(R_z)$ by an integral over z, one obtains

$$I_1 = \frac{1}{\Omega_0} \int_{\Omega_0} d^3r \, u_{n'0}^*(\boldsymbol{r}) \boldsymbol{e} \cdot \nabla \, u_{n0}(\boldsymbol{r}) \int_{-\infty}^{\infty} f_{n'j}^*(z) \, f_{ni}(z) \, dz. \tag{17.61}$$

Calculating I_2 and I_3 similarly, we get

$$I_2 = i\boldsymbol{e} \cdot \boldsymbol{k}_{\parallel} \, \delta_{nn'} \delta_{ij},$$

$$I_3 = \delta_{nn'} \int_{-\infty}^{\infty} f_{nj}^*(z) e_z \nabla_z f_{ni}(z) dz. \tag{17.62}$$

Term I_1 is responsible for interband optical absorption. By selection rules, it equals zero for intraband transitions ($n = n'$). Term I_2 corresponds to non-resonant intraband transitions for which energy and momentum conservation conditions hold only for photons with $\hbar\omega \to 0$. If considering resonant intraband absorption only, one can neglect I_2. The remaining term I_3 describes intersubband optical transitions ($n = n', i \neq j$), which are at the origin of intraband resonant absorption in quantum wells.

In a geometrically symmetric QW, the subband wavefunctions in the conduction band $f_i(z)$ have definite parity to inversion $z \to -z$. Matrix element I_3 is nonzero if calculated on states of opposite parity. Also, I_3 depends on the angle of incidence of the optical wave. The angle of incidence should provide for z-directed electrical polarization in the wave. This means that linear polarized wave causes intersubband transitions only when the QW surface is illuminated by transverse magnetic (TM) wave in which polarization vector $\boldsymbol{e}$ lays in the plane of incidence, as illustrated in Figure 17.8.

Transverse electric wave $(\boldsymbol{e} \perp \boldsymbol{z})$ does not interact with n-type QW, meaning that the detector cannot work at normal incidence. To guarantee incidence tilted to the QW growth direction and thus achieve the effective light coupling, device design may include either a 45° edge facet geometry or a grating reflection mirror [7]. However, since normal incidence is required for the focal plane array, p-type QWIP is the preferred choice. In III–V p-type QWIP's, transitions occur between valence subbands, so the TE matrix element is nonzero due to the mixing of light- and heavy-hole wavefunctions, thus enabling detectors to work at normal incidence. Generic n-type QWIP with grating coupling is shown schematically in Figure 17.9.

Oscillator strength. Sometimes, matrix element I_3 is presented in an equivalent form using equality $p = mv$ in operator form: $mv_z = p_z$, where $v_z \equiv \dot{z}$ and $p_z \equiv -i\hbar\nabla_z$ are the operators of velocity and momentum in coordinate representation, respectively. The equality projected on states f_i, f_j takes the form,

$$\langle f_j | p_z | f_i \rangle = m \langle f_j | \dot{z} | f_i \rangle, \tag{17.63}$$

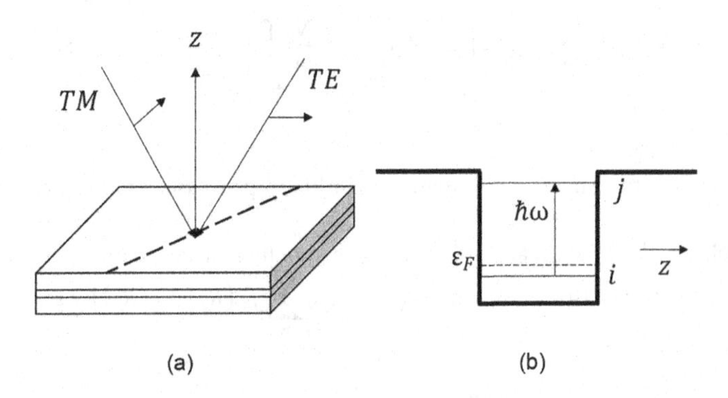

FIGURE 17.8 (a) Mutual orientation of incident linear polarized wave and the growth direction. The dashed line is the intersection of the incident plane with the surface of a sample. Arrows show polarization vector $\boldsymbol{e}$. (b) Intersubband optical transition in the conduction band.

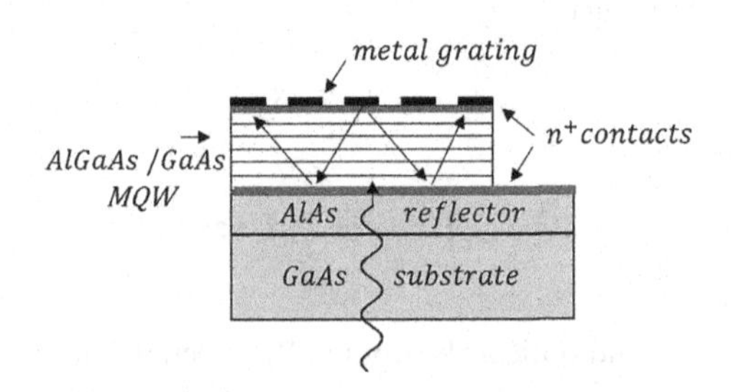

FIGURE 17.9 Back-side illuminated n-type QWIP. Metal grating designed for total internal reflection provides for high efficiency [6,7].

where $\dot{z}$ obeys Heisenberg equation of motion

$$\dot{z} = \frac{1}{i\hbar}[z, H], \quad [z, H] \equiv zH - Hz, \tag{17.64}$$

H is the Hamiltonian, which describes confinement in z-direction: $Hf_{i,j} = E_{i,j} f_{i,j}$, where eigenvalues $E_{i,j}$ are the edges of the conduction subbands. Substitution $\dot{z}$ in (17.63) gives

$$\hbar\langle f_j | p_z | f_i \rangle = i(E_j - E_i) m \langle f_j | z | f_i \rangle,$$

or

$$\left|\langle f_j | \nabla_z | f_i \rangle\right|^2 = \frac{\omega^2 m^2}{\hbar^2} z_{ij}^2, \tag{17.65}$$

where $\hbar\omega = E_j - E_i$, $z_{ij}^2 \equiv \left|\langle f_j | z | f_i \rangle\right|^2$.

It is instructive to note that the left- and right-hand sides of (17.65), being physically equivalent, correspond to different gauges for vector-potential (see Eqs. (12.20) and (12.27), Chapter 12).

Using (17.65), we present the absorption coefficient (17.56) as

$$\alpha(\omega) = \frac{2\pi q^2 \omega\, z_{ij}^2 e_z}{\Omega c \varepsilon_0 n_r} \sum_k \delta(E_{jk} - E_{ik} - \hbar\omega)\left[f(E_{ik}) - f(E_{jk})\right],$$

$$E_{ik} = E_i + \frac{\hbar^2 \mathbf{k}^2}{2m}. \tag{17.66}$$

Strict energy conservation, formally expressed by δ-function, implies that electron lives indefinitely on energy levels E_{ik} $(\tau = \infty)$. Scattering makes lifetime finite and thus broadens levels so that linewidth following from the uncertainty principle, $\Gamma = \hbar/\tau$. In a multiquantum well (MQW), there is another contribution to the linewidth related to tunneling through the barrier into a neighboring well. As a result, broadening relaxes conservation condition to

$$\delta(x) \to \frac{\Gamma/2\pi}{x^2 + (\Gamma/2)^2} \tag{17.67}$$

Taking the integral in (17.66) the way used in (4.16) (Chapter 4), one gets

$$\alpha(\omega) = \frac{q^2 k_B T e_z f_{osc}}{4\pi\hbar c \varepsilon_0 n_r L} \frac{\Gamma}{(E_j - E_i - \hbar\omega)^2 + (\Gamma/2)^2} \log \frac{1 + \exp\left(\dfrac{\mu - E_i}{k_B T}\right)}{1 + \exp\left(\dfrac{\mu - E_j}{k_B T}\right)}, \tag{17.68}$$

where $f_{osc} = 2\omega m z_{ij}^2 / \hbar$ is the dimensionless parameter that includes the optical matrix element and called *oscillator strength,* L is the effective QW-width. Writing the wave functions as $f_i = \varphi_i / \sqrt{L_i}$, one finds L using normalization condition: $L_i = \int_{-\infty}^{\infty} \varphi_i^2\, dz$. If wavefunctions do not penetrate the barriers (infinite barrier), L is the geometrical width of QW.

17.3.2 Responsivity and Gain

Generic QWIP consists of MQW so that the photocurrent is a sum of contributions generated by each illuminated QW, which can be designed deliberately to provide optical transitions between bound states or bound-to-continuum states. The band diagram in biased MQW and two types of optical transitions are shown in Figure 17.10.

Bound-to-bound state QWIP requires high applied voltage, as the higher the voltage, the thinner the barrier and thus higher tunneling probability $T(E)$ (see Figure 17.10b). Due to high voltage, the dark current is large and thus suppresses detectivity. For that matter, bound-to-continuum QWIPs are preferable as the photocurrent does not rely on tunneling but rather on over-barrier optical excitation, as shown in Figure 17.10a.

The photocurrent generated by an individual well can be expressed by (17.33) at $g = 1$:

$$i_{ph} = q\eta_{QW}\Phi A p_e, \tag{17.69}$$

where η_{QW} is the quantum efficiency of a single QW, p_e is the escape probability. In bound-to-bound state QWIP, the excited (hot) electron is involved in two competitive processes: energy relaxation on the way back to the ground state during time τ_{rel} and escape out the well in time τ_{esc} due to both tunneling and thermionic emissions. The escape probability is the combination of two characteristic times:

$$p_e = \frac{\tau_{rel}}{\tau_{rel} + \tau_{esc}}. \tag{17.70}$$

In bound-to-continuum QWIP, $p_e \approx 1$.

Inverse to escape, the process of electron trapping by the well is regulated by the capture to the ground level in time τ_c (lifetime) and transit time τ_{tr} for an electron across one period of MQW:

$$p_c = \frac{\tau_{tr}}{\tau_{tr} + \tau_c}. \tag{17.71}$$

As determined from experiments on AlGaAs-based QW's [7], the intersubband relaxation due to phonon emission is $\tau_{rel} \approx 1$ ps. Escape time is determined by tunneling in the electric field and estimated as $\tau_{esc} \ll 1$ ps. Lifetime $(\tau_c \approx 5\text{ ps})$ is much shorter than that in bulk interband photoconductors $(\approx 1\text{ ns})$. For an electron velocity of 10^7 cm/s and a QW width of $(30-50)$ nm, the estimated transit time is $(0.3 \div 0.5)$ ps.

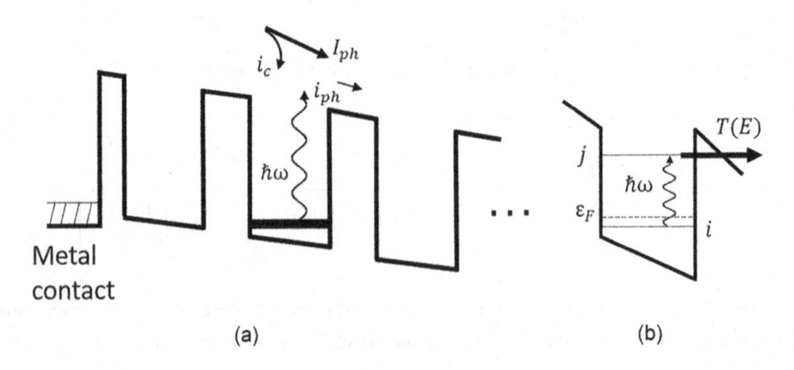

FIGURE 17.10 (a) Conduction band diagram of bound-to-continuum QWIP. (b) Bound-to-bound state transition and tunneling.

As presented in Figure 17.10a, conservation of the total current requires the capture (i_c) and escape $\left(i_{ph}\right)$ contributions to compensate each other. The capture portion of the total current, $i_c = p_c I_{ph}$. Then, condition $i_c = i_{ph}$ gives:

$$i_{ph} = p_c I_{ph}, \tag{17.72}$$

After assuming N_{QW} is the number of quantum wells in the detector, $\eta_{QW} \ll 1$, and overall efficiency $\eta = \eta_{QW} N_{QW}$, both total current and responsivity follow from (17.69) and (17.72):

$$I_{ph} = \frac{p_e}{p_c} q\eta_{QW}\Phi A = \frac{p_e \eta_{QW}}{p_c \eta} q\eta\Phi A \equiv qg\eta\Phi A,$$

$$R_I = \frac{qg\eta}{\hbar\omega}, \quad g \equiv \frac{p_e \eta_{QW}}{p_c \eta} = \frac{p_e}{p_c N_{QW}}, \tag{17.73}$$

where g is the overall photoconductive gain.

Responsivity, $R_I = q\eta_{QW} p_e / p_c \hbar\omega$, relies on properties of individual QW and does not depend on the number of quantum wells.

Under conditions $p_e \approx 1$, and $p_c \approx \tau_{tr}/\tau_c \ll 1$, the gain becomes similar to that in bulk photoconductors, which is proportional to the electron lifetime in a conduction band τ_c and inversely proportional to the transit time τ_{tr}.

17.3.3 The Noise and Detectivity

Like in other types of detectors, the main contribution to intrinsic noise is the shot noise that, similarly to (17.30), is expressed as [8]:

$$\left\langle i_S^2 \right\rangle = 2(2 - p_c) qgI\,\Delta f, \tag{17.74}$$

where I is the direct current, which could be either the dark current in a biased MQW, in which case it is generation-capture noise, or photocurrent induced by background illumination presenting the photon noise relevant under BLIP conditions.

Noise current (17.74) tends to the generation-only noise (17.31) at $p_c \to 1$ (photodiode mode). The capture process is not a random one and thus does not contribute to noise. In the opposite limit of $p_c \approx \tau_{tr}/\tau_c \ll 1$, noise current (17.74) tends to $\left\langle i_S^2 \right\rangle = 4I\,\Delta f \tau_c/\tau_{tr}$ that is equivalent to (17.32) photoconductive mode in which both generation and capture involved.

Assuming dark current $I_d = AJ_d$ as the primary noise source, for bound-to-continuum QWIP $\left(p_e \approx 1\right)$, NEP and detectivity follow from (17.42) and (17.73) as

$$NEP = \frac{\hbar\omega\sqrt{2\left(2 - p_c\right) AJ_d\,\Delta f}}{\eta\sqrt{gq}}, \quad D^* = \frac{\eta_{QW}}{\hbar\omega}\sqrt{\frac{qN_{QW}}{2\left(2 - p_c\right) p_c J_d}}. \tag{17.75}$$

Dark current density in (17.75) in a biased MQW originates from tunneling and thermionic emission out the ground subband i (see Figure 17.10) with subsequent drift in the electric field with velocity v_d:

$$J_d = \frac{q}{\Omega}\sum_k f\left(E_{ik}\right) T\left(E_{ik}\right) \mathrm{v}_d\left(E_{ik}\right). \tag{17.76}$$

One may calculate J_d assuming maximum barrier transparency $\left(T(E_{ik})=1\right)$ and energy independent drift velocity:

$$J_d = \frac{q v_d}{\Omega} \sum_k f(E_{ik}) = \frac{q v_d n_{2D}}{L_p},$$

$$n_{2D} = \frac{m k_B T}{\pi \hbar^2} \log\left[1 + \exp\left(\frac{\varepsilon_F - E_i}{k_B T}\right)\right], \tag{17.77}$$

where L_p is the length of one period in MQW, $\Omega = AL_p$, sheet density n_{2D} follows from $\boldsymbol{k}$-summation performed in (4.16) (Chapter 4).

In BLIP conditions, the dark current AJ_d in (17.75) to be replaced with photocurrent (17.38) generated by background illumination.

For more details on technical aspects of photodetectors, various figures of merit, and electrical parameters, see Ref. [9].

17.4 Concluding Remarks

As far as spectral properties of intersubband devices are not directly related to the fundamental semiconductor bandgap, the detector performance could be tuned to infrared or terahertz regions using various material systems and structure design options. The GaAs–InAs–AlAs material system allows us to grow lattice-matched heterojunctions and high structural quality MQWs. Well-elaborated III-As QWIPs cover a practically important mid-infrared range of (5–20) μm and present the mainstream in infrared detection technology.

Because of the small conduction band offset, GaAs / AlGaAs intersubband devices cannot work at $\lambda < 3\,\mu$m. Another III–V material family, the wide bandgap GaN / AlGaN junctions, has conduction band offsets large enough to allow an intersubband device operating in a communication band of $(1.3-1.55)\,\mu$m and further up to 4.5 μm. If designed as a quantum cascade structure, it covers terahertz (see Ref. [10] for a review). Single-layer AlGaN devices, which operate on fundamental valence-to-conduction band transitions, are sensitive from visible to ultraviolet and even X-ray region $\lambda < 350$ nm.

The search for new materials and structures is on-going: much effort focuses on III–V quantum cascade [11] and quantum dots photodetectors. High-quality quantum dot heterostructures become available, providing a wide spectral range of detection, high efficiency, and a high level of integration into mature Si-based electronics. [12,13].

Also, new materials attract attention in infrared optoelectronics: topological insulators [14] and graphene [15]. Both carry two-dimensional electrons with a zero-bandgap spectrum and high electron mobility. Because of that, they are useful to detect light in the infrared and terahertz. For a brief discussion of topological insulators $Bi_2(Te,Se)_3$ and their device applications, see the next chapter.

References

1. A. van der Ziel, *Noise in Solid State Devices and Circuits* (Wiley, New York, 1986).
2. S.L. Chuang, *Physics of Optoelectronic Devices* (Wiley, New York, 1995).
3. H.B. Callen and T.W. Welton, "Irreversibility and generalized noise", *Phys. Rev.* **83**, 34 (1951).
4. E. Rosencher and B. Vinter, *Optoelectronics* (Cambridge University Press, Cambridge, 2002).
5. A. Rogalski, *Infrared Detectors* (2nd ed., CRC Press, Boca Raton, FL, 2010).
6. B.F. Levine, "Quantum well infrared photodetectors", *J. Appl. Phys.* **74**, R1 (1993).
7. H.C. Liu, "Quantum Well Infrared Photodetector Physics and Novel Devices", in *Intersubband Transitions in Quantum Wells: Physics and Device Applications I*, eds. H.C. Liu and F. Capasso, pp. 129–196 (Academic Press, San Diego, CA, 2000).

8. W.A. Beck, "Photoconductive gain and generation-recombination noise in multiple quantum well infrared detectors", *Appl. Phys. Lett.* **63**, 3589 (1993).
9. J.D. Vincent, S.E. Hodges, J. Vampola, M. Stegall, and G. Pierce, *Fundamentals of Infrared and Visible Detector Operation and Testing* (Wiley, Hoboken, NJ, 2016).
10. M. Beeler, E. Trichas, and E. Monroy, "III-nitride semiconductors for intersubband optoelectronics: a review", *Semicond. Sci. Technol.* **28**, 074022 (2013).
11. F.R. Giorgetta, E. Baumann, M. Graf, Q. Yang, C. Manz, K. Köhler, H.E. Beere, D.A. Ritchie, E. Linfield, A.G. Davies, Y. Fedoryshyn, H. Jäckel, M. Fischer, J. Faist, and D. Hofstetter, "Quantum cascade detectors", *IEEE Quant. Electron.* **45**, 1029 (2009).
12. P. Martyniuk and A. Rogalski, "Quantum-dot infrared photodetectors: Status and outlook", *Prog. Quant. Electron.* **32**(3), 89–120 (2008).
13. A. Ren, L. Yuan, H Xu, J. Wu, and Z. Wang, "Recent progress of III–V quantum dot infrared photodetectors on silicon", *J. Mater. Chem. C.* **7**, 14441 (2019).
14. X. Zhang, J. Wang, and S.-C. Zhang, "Topological insulators for high-performance terahertz to infrared applications", *Phys. Rev.* **B82**, 245107 (2010).
15. F.H.L. Koppens, T. Mueller, Ph. Avouris, A.C. Ferrari, M.S. Vitiello, and M. Polini, "Photodetectors based on graphene, other two-dimensional materials and hybrid systems", *Nat. Nanotechnol.* **9**, 780 (2014).

18

Device Applications of Novel 2D Materials

New crystalline materials under study toward device applications are graphene and topological insulators (TI). The point is that electrons at a single TI surface and on a single sheet of graphene both have a linear-in-momentum energy spectrum, meaning that the effective mass is negligible and creating hope for high mobility and thus applications in high-speed electronics. Two-dimensional electrons in graphene and surface states in TI bring their specifics in Schottky diodes, photodetectors, field-effect transistors, and thermoelectric units. Apart from generic semiconductor properties, TI materials carry topological spin-momentum locked states not existing in ordinary semiconductors. These states make TI materials a template for various spintronic devices enabling electrical spin manipulation and applicable in random access memory, information storage, and microwave sources.

18.1 Graphene

The electron spectrum in single-layer graphene has the form discussed in Chapter 1:

$$E_{1,2}(k) = \pm \mathrm{v}\hbar k, \tag{18.1}$$

where $E_{1,2}(k)$ is the spectrum referenced to Dirac point in the corner $\boldsymbol{K}$ of the hexagonal Brillouin zone. The same spectrum is located in non-equivalent point $\boldsymbol{K}'$.

In bilayer graphene comprised of two identical layers of the same orientation, one upon the other, interlayer hopping integral γ generates the bandgap so that gapless Dirac cones (18.1) morph into a gapped semiconductor spectrum:

$$H_k = \begin{pmatrix} 0 & \hbar\mathrm{v}\left(k_x - ik_y\right) & 0 & 0 \\ \hbar\mathrm{v}\left(k_x + ik_y\right) & 0 & \gamma & 0 \\ 0 & \gamma & 0 & \hbar\mathrm{v}\left(k_x - ik_y\right) \\ 0 & 0 & \hbar\mathrm{v}\left(k_x + ik_y\right) & 0 \end{pmatrix}$$

$$E_{1,2} = \pm\frac{1}{2}\left(-\gamma + \sqrt{\gamma^2 + 4\hbar^2\mathrm{v}^2k^2}\right). \tag{18.2}$$

The presence of a second layer creates the energy gap, thus transforming the metallic type of medium into a semiconductor. The energy spectrum depends on a twist angle [1]. A small rotation of layers to each other at the "magic angle" of $\sim 1.1°$ makes the Fermi velocity of Dirac electrons vanishes. At this angle,

DOI: 10.1201/9780429285929-18

the electron spectrum in a bilayer becomes flat (not-dependent on k), manifesting Van Hove singularity in electron density of states. A large density of states is at the origin of unconventional superconductivity in twisted bilayer material [2].

18.1.1 Field-Effect Transistors

Due to high Fermi velocity of 10^6 m/s and carrier mobility at a room temperature of 10^5 cm^2/Vs, graphene as well as heterostructures Graphene/Boron Nitride or Graphene/Tungsten Chalcogenide, are promising materials for high-speed electronics.

A field-effect transistor is proven to work with graphene as a channel [3]. Schematically, the transistor is shown in Figure 18.1.

The channel has ambipolar conduction depending on the gate voltage, which shifts the Fermi level. By choosing gate voltage to minimize the drain current ($V_{GS} = V_{Dirac}$), we put the Fermi level in the Dirac point, as illustrated in Figure 18.2 The ac-signal applied to the gate would shift the Fermi level above and below the Dirac point switching conductivity from n - to p -type, respectively. The current–voltage characteristic is typical for the full-wave rectifier.

Graphene–field-effect transistor (FET) has found use in environmental sensing. Being exposed to molecules, which bind to graphene, the channel alters the conductivity, generating the electrical response. Among others, the device can detect molecules of glucose, hemoglobin, cholesterol, or hydrogen peroxide.

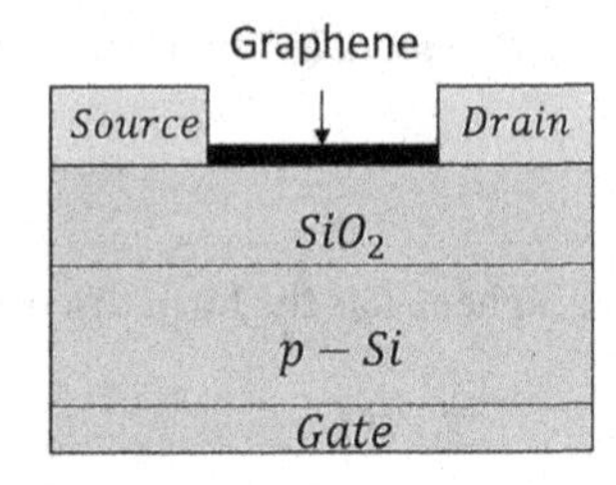

FIGURE 18.1 Cross-section of a field-effect transistor, G-FET.

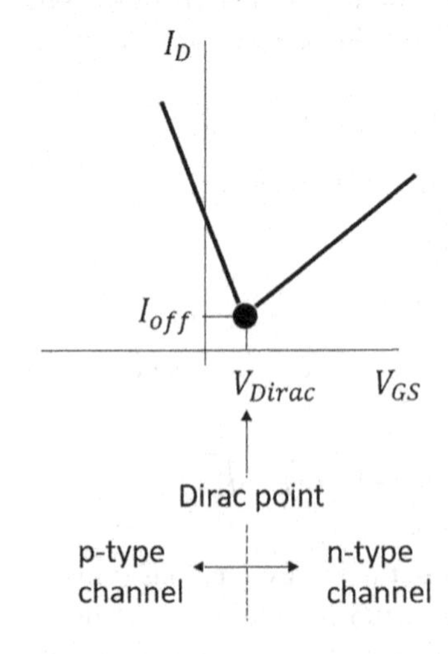

FIGURE 18.2 Transfer characteristics, I_D is the drain current, V_{GS} is the gate voltage.

18.2 Topological Insulators

The surface electron spectrum in $Bi_2(Te_xSe_{1-x})_3$ TI film is spin-resolved if vertical bias is applied:

$$E_{c,v\uparrow} = Dk^2 \pm R_\uparrow,$$

$$E_{c,v\downarrow} = Dk^2 \pm R_\downarrow,$$

$$R_{\uparrow\downarrow} = \sqrt{\left(\frac{\Delta}{2} - Bk^2\right)^2 + (Ak \pm V)^2}, \tag{18.3}$$

where $k^2 = k_x^2 + k_y^2$, V is proportional to the vertical bias that lifts spin degeneracy between two surfaces, $v = A/\hbar$ is the Fermi velocity, Δ and B are the parameters originated from top-bottom tunneling and thus relevant at a thickness less than six quintuple layers. In thick enough layers, electrons on both surfaces are independent and have gapless spectra linear in planar momenta $\boldsymbol{k}$: $E_{c,v,\uparrow,\downarrow} = \pm(Ak \pm V)$ (see Ref. [4] for review).

Any application implies access to 2D-electrons, gating to manipulate carrier density, and high-quality electrical connection. Since topological states are pinned to surfaces, the Fermi level position at the metal-TI contact affects the accessibility to 2D-states, mobility, and device characteristics.

18.2.1 Contacts and Gating

Native defects (mostly chalcogen vacancies) in $Bi_2(TeSe)_3$ create n-type doping and pin the Fermi level in the conduction band [5,6]. The band bending near the surface forms an accumulation layer, thus creating a 2D electron gas and making it difficult to distinguish between 2D electrons and surface topological states. The density of the native defects depends on the growth method, postgrowth treatment, as well as the choice of substrate and buffer layer. Proper postgrowth annealing can reduce the density of chalcogen vacancies that move the Fermi level in the bulk bandgap close to the Dirac point [7].

The carrier density in an accumulation layer can be controlled by electrostatic gating with metal contacts, which could be either Ohmic or Schottky type, depending on the choice of metal. The active regions in most optoelectronic devices comprise single or double heterojunctions, which have variable energy gaps, high carrier mobility, and the ability to confine nonequilibrium carriers. Several examples of contacts and heterojunctions are given below in the text.

First-principle calculations, performed for $Bi_2Se_3/(Au, Pt, Ni, Pd, graphene)$, contacts predict n-type Ohmic contacts only, regardless of the choice of metals from the group. Graphene and Au were found to have the weakest charge transfer to TI, thus providing the best Ohmic contacts which do not interfere with the topological states and preserve their spin-momentum locked character [8]. To prevent charge transfer to metals, it was proposed to use a thin large bandgap dielectric layer placed between a metal and a TI. This type of contact is normally used as a gate to achieve surface conduction tunable from n - to p - type by a gate voltage. Effective gating was reported for $(Bi_{1-x}Sb_x)_2Te_3$ in geometry with top [9,10] and back (through the dielectric substrate) gates [11]. To arrange ambipolar surface conduction, bulk conductivity should be suppressed as much as possible, for example, by choice of alloy composition, i.e. by the Bi/Sb or (and) Se/Te ratio. With the help of compositional tuning of the Fermi level, one makes the topological states accessible at $0.65 \le x \le 0.73$ in $(Bi_{1-x}Sb_x)_2Se_3$ [12] and $y = 0.94$ in $(Bi_{1-y}Sb_y)_2Te_3$ [13]. The p - to n -type conductivity transition around $y = 0.94$ is at the origin of the topological p - n junction proposal [14]. Manipulations of the Fermi level by alloy composition with subsequent observation of gate-controlled ambipolar conductivity were reported in $(Bi_{1-x}Sb_x)_2Te_3$ [15] and $(Bi_{1-x}Sb_x)_2Te_{1.25}Se_{1.75}$ [16].

18.2.2 Heterojunctions

The quality of a semiconductor heterojunction as a device depends on the structural quality of an interface and electrical parameters: band offsets, Schottky barrier height, and electron (hole) mobility. In the heterojunction TI/(ordinary semiconductor), the band offsets indicate if the topological states are accessible, thus helping to predict device performance.

The heterojunctions consisting of a TI and semiconductors of group-IV and III–V Si, Ge, (Al, Ga, In)As, (Al, In, Ga)N are of special interest as they help integrate TI's into well-developed semiconductor templates and expand the functionalities of optoelectronic devices and field-effect transistors. Several examples of such heterojunctions are given below.

Group IV semiconductors. The band diagram of the interface Si(111)/Bi_2Se_3 is illustrated in Figure 18.3. In the diagram, barrier height $\phi_B = 0.31$ eV, valence band offset of $\Delta E_v = 0.19$ eV, and Fermi energy, $E_F = 0.37$ eV [17].

For the surface orientation Si(001), a higher Schottky barrier of $\phi_B = 0.34$ eV was reported in Ref. [18]. The discrepancy between the values of ϕ_B found in Refs. [17] and [18] may be attributed to the different Si-face orientations used in the experiments, and also to the hydrogen-passivation of the Si(111) surface aimed at improving interface quality and exclusion of defects formation in the Schottky barrier. As the Fermi level in both examples is pinned deeply in the TI conduction band, the topological states would hardly be accessible, so they do not affect the Schottky device performance. The Schottky barrier height depends on the position of the Fermi level pinned at the interface. In n-Si(111)/p-Bi_2Te_3 heterojunctions used in photodetector devices the barrier height was estimated as 0.65 eV [19].

In a germanium contact to Sb_2Te_3, the conduction and valence band offsets were estimated as $\Delta E_c = 0.07$ eV and $\Delta E_v = 0.25$ eV, respectively. The band diagram is shown in Figure 18.4.

The Fermi level position in the TI bandgap allows topological states to take part in the in-plane transport. The spin-locked character of topological states is preserved as the Ge/Sb_2Te_3 interface introduces weak disturbance to the states [20].

Cubic III–V semiconductors. The GaAs substrates in the (001) [21,22] and (111)B [23,24] surface orientations are often used for TI epitaxial growth. GaAs is not the best substrate as it has a lattice mismatch

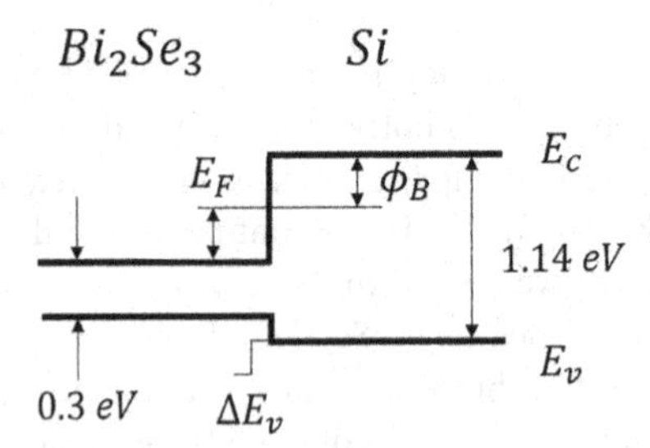

FIGURE 18.3 The band diagram at n-Si(111)/n-Bi_2Se_3 interface [17].

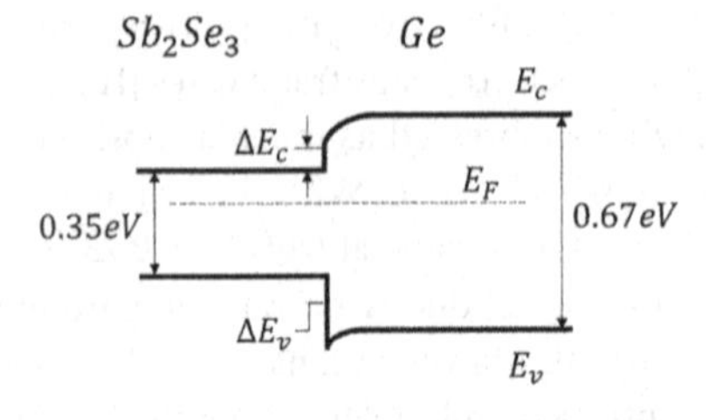

FIGURE 18.4 Band alignment in Ge/Sb_2Te_3 heterojunction [20].

as large as 3.4%, 5.9%, and 8.7% for Bi_2Se_3, Sb_2Te_3, and Bi_2Te_3, respectively [25]. Lattice mismatch causes compressive strain in coherently grown layers and may result in generating defects if the layer thickness exceeds the critical value. Despite this, the attractive side of using a GaAs substrate is that it favors the integration of a TI into a template suitable for high-speed electronic and optoelectronic devices.

The interface band structures of the GaAs/Bi_2Se_3 and GaAs/Bi_2Te_3 heterojunctions are shown in Figure 18.5a and b. The band alignment follows from a simplified consideration of conduction and valence level positions relative to vacuum. The type II heterojunction shown in Figure 18.5a was studied by first-principle calculations in Ref. [26]. It was shown that the hybridization between the topological and GaAs valence states at the interface adds momentum-dependent spin texture to nontopological states.

Circularly polarized light with interband photon energy $\hbar\omega \geq E_g\,(\text{GaAs})$ excites spin-polarized electrons on both sides of the junction shown in Figure 18.5b. If the TI film is thin enough, a considerable part of the optical power penetrates Bi_2Te_3 and is absorbed in the GaAs substrate. After spin-polarized hot electrons being injected across the junction into the Bi_2Te_3 layer, they create a spin imbalance. This kind of spin injection contributes to photocurrent due to the circular photogalvanic effect [27,28]. The photo-excited heterostructure GaAs/$(Bi_{0.5}Sb_{0.5})_2Te_3$ was studied as a possible broadband terahertz source which converts very fast transient spin current to charge current due to the inverse spin-Hall effect [29].

II–VI Semiconductors. In type-I heterointerface Bi_2Te_3/CdTe(111), the band diagram is similar to that rendered in Figure 18.3, where the conduction and valence band offsets being $\Delta E_c = 1.12$ eV and $\Delta E_v = 0.22$ eV, respectively. These offsets follow from the spectra of photoelectron spectroscopy in the X-ray and ultraviolet spectral regions [30].

Wurtzite III–V semiconductors. Wide bandgap III–V semiconductors are used widely in high-power field-effect transistors and various optoelectronic devices such as light-emitting diodes and lasers. The combination of two groups of materials – AlGaN-based semiconductors and bismuth chalcogenides – may open new possibilities in device applications, expanding their area of functionality. Optoelectronic devices grown on the common substrate would cover ultraviolet and infrared (IR) ends of the spectrum, providing surface states were not affected by the substrate. Wafers AlN/Si(111), well developed industry-grade substrates, can be used to integrate the epitaxial TI layers into existing electronic devices. In this context, substantial steps were a fabricaton of thin epitaxial Bi_2Se_3 layers directly on the AlN/Si(111) substrate, resulting in layer quality high enough to preserve the gapless topological surface states [31].

The electrical properties of the interface between AlN and TI are characterized by large conduction and valence band offsets, as illustrated in Figure 18.6. The band offsets were estimated using the conduction band affinity to vacuum with no regard to the charge transfer across the junction.

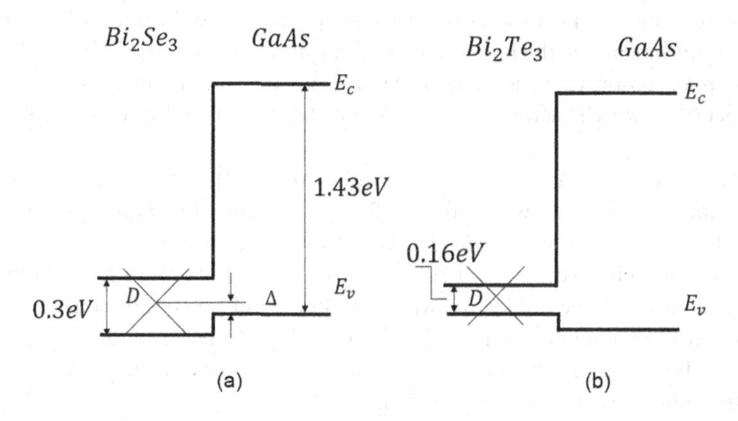

FIGURE 18.5 (a) Type II band alignment in a GaAs/Bi_2Se_3 heterojunction. $\Delta = 0.05$ eV [26]. (b) Type I heterojunction GaAs/Bi_2Te_3 [27]. Crossed lines mimic the linear surface spectrum in a TI, D is the Dirac point level.

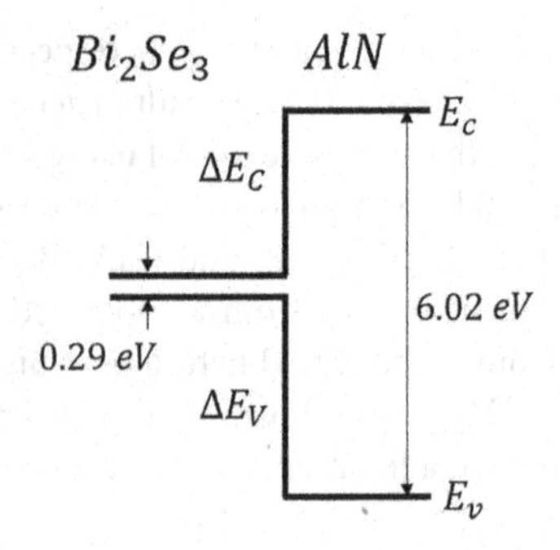

FIGURE 18.6 $\mathbf{Bi_2Se_3/AlN}$ heterojunction. $\Delta E_c = 2.86\,\text{eV}, \Delta E_v = 2.87\,\text{eV}$ [32].

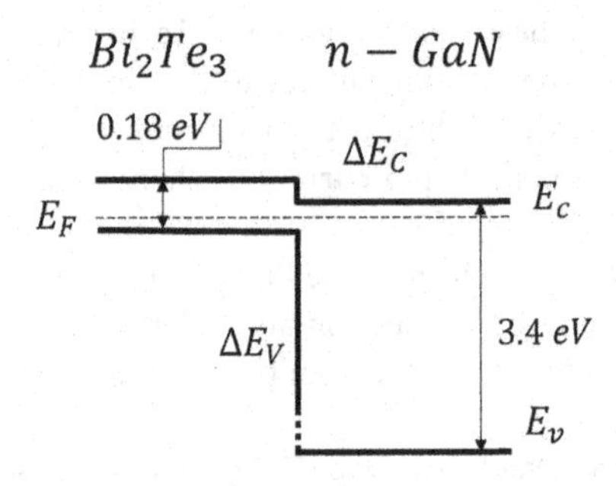

FIGURE 18.7 Type-II heterojunction Bi_2Te_3/GaN. $\Delta E_c = -0.05\,\text{eV}$, $\Delta E_v = 3.27\,\text{eV}$ [33].

In heterojunctions with $Al_{1-x}Ga_xN$, the band diagram is tuneable with composition x. In the Bi_2Te_3/GaN heterointerface, the conduction band offset is negative, manifesting type-II (staggered) alignment, as shown in Figure 18.7.

The GaN substrate by itself presents a template for various visible wavelength optoelectronic devices. Had a TI heterojunction been grown on the same substrate, it would have been an integrated IR photodetector with a Bi_2Te_3 active region. IR photodetectors based on Bi_2Te_3 are discussed in the next section.

18.2.3 Photodetectors

In topological insulators, the spectrum of electrons is gapped in the bulk and gapless or narrow-gapped at the surface, depending on the thickness. The small gap at the surface in $Bi_2(Te,Se)_3$ is at the origin of the broadband optical absorption and photosensitivity over the IR to terahertz spectral range. Various types of photodetectors with TI active regions have been fabricated on various substrates. Examples are given below.

The p - n heterostructure Bi_2Te_3/GaN is shown in Figure 18.7. The photoresponse upon illumination by the standard source of $\lambda = 1\,\mu\text{m}$ was studied in Ref. [34]. The quality of the p - n junction and the IR responsivity was found promising in the device illustrated in Figure 18.8.

In a semiconductor with a bandgap of 0.21 eV (the bandgap of bulk Bi_2Te_3), maximum performance would be expected in the range $\lambda < 5.9\,\mu\text{m}$, where the interband transitions come into effect. Gapless surface electrons in TI's extend the sensitivity of the device into the far-IR, and as far as light is absorbed by transitions in the gapless linear spectrum, the absorbance does not depend on wavelength [35]. A generic photoconductive detector is shown in Figure 18.9.

Molecular beam epitaxial growth of a Sb_2Te_3 photoconductive detector on an Al_2O_3 substrate shows the maximum responsivity and specific detectivity of $R = 21.7$ A/W and $D^* = 1.22\times10^{11}\,\text{cmHz}^{1/2}/\text{W}$, respectively, at $\lambda = 980$ nm and temperature of 150 K [36]. In a device with an active region made of

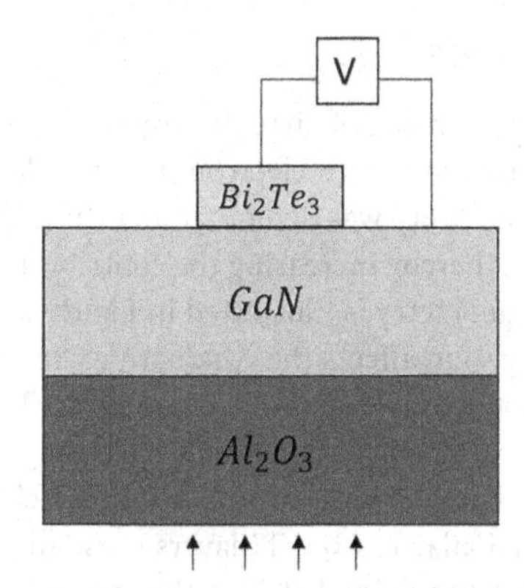

FIGURE 18.8 Back-side illuminated Bi_2Te_3/GaN IR detector [34]. The GaN layer and the substrate serve as windows for incident IR radiation.

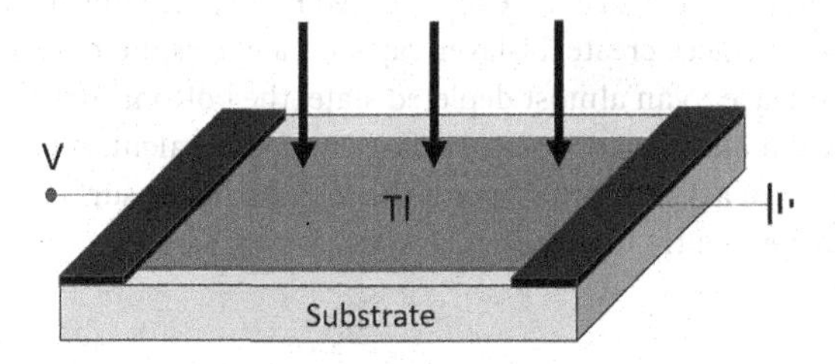

FIGURE 18.9 Schematic photodetector setting.

Sb_2SeTe_2 nanoflakes, room temperature responsivity was measured as high as 2,293 A/W at $\lambda = 532$ nm [37]. Si - based TI detectors attract considerable attention as they can be integrated with other electronic components using industry-grade templates of high structural quality. The high room-temperature responsivity of 300 A/W and high-speed operation were reported for a photoconductor made of Bi_2Se_3 nanowire on a Si/SiO_2 substrate. Improved characteristics were attributed to strong electron quantum confinement in the wire [38].

The Bi_2Te_3/Si photoconductive device demonstrated room temperature responsivity of 3.64 mA/W at $\lambda = 1.064\,\mu m$ [39]. The p - n junction in the heterostructure $(p)Bi_2Te_3/(n)Si$ spatially separates the nonequilibrium carriers, preventing fast recombination and increasing their lifetime. The broadband room temperature responsivity of the vertical *p-n* Bi_2Te_3/Si device has been experimentally observed in the spectral range from ultraviolet to terahertz, at $\lambda = (0.37, 1.55, 118.8)\,\mu m$ and maximum responsivity and specific detectivity of 1 A/W and 2.5×10^{11} $cmHz^{1/2}/W$, respectively [40]. So far, the best performance for a heterojunction device has been reported in Bi_2Se_3/Si [19]: a responsivity of 24.28 A/ W and detectivity of 4.39×10^{12} $cmHz^{1/2}/W$ in the range from ultraviolet to the optical communication band $(0.35-1.1)\mu$.

There are photodetector devices in which the topological insulators are used in combination with graphene [41] and WSe_2 [42]. Parameters of various TI-based photodetectors have been compared in the review article [43].

The insulating bulk and the conductive surface make TI a unique material for guiding light. The simultaneous presence of the high refractive index in the bulk and Dirac plasmon excitations on the surface allows the fabrication of new nanostructures, which increase light absorption in solar cells and photodetectors. In this respect, the surface of $Bi_{1.5}Sb_{0.5}Te_{1.8}Se_{1.2}$ patterned in nano-cones has improved the efficiency of solar cells due to the plasmonic resonance in the visible spectral range. It has proven to serve as an antireflection coating [44].

18.2.4 Field-Effect Transistors

Similar to photodetectors, FETs can be made of virtually any semiconductor. The use of TI as a channel material is driven by the hope to maintain high electron mobility due to suppressed backscattering of the surface modes. An increase in mobility was demonstrated when the electrostatic gating pushed the Fermi level close to the Dirac point, thereby increasing the contribution of topological states in the total carrier mobility [45]. Generic FET geometry is illustrated in Figure 18.10.

As mentioned in Section 18.2.1, gating affects the conductivity in the channel and this is enough for the structure to operate as a transistor. However, if the channel is a TI, the absence of an energy gap on the upper surface allows the source-drain current to flow, even if the gate bias is lower than the threshold value. This degrades the pinch-off characteristics of a FET (for FET characteristics and pinch-off conditions, see Chapter 15). For this particular reason, TI layers normally used in FETs are thin enough to create an energy gap due to electron tunneling between the top and the bottom surfaces. An ultrathin FET with a clear OFF state at negative gate voltage was demonstrated in a Bi_2Se_3 / SiO_2 / Si structure [46]. The performance of the Bi_2Se_3nanowire device shows the sharp pinch-off, which is attributed to the destruction of the gapless conduction channel by negative gate voltage [47].

To achieve the high-frequency performance of TI-based FET's, the issue to be addressed relates to the fact that upper and lower surfaces create TI-layer capacitance. Despite manipulating the gate voltage, one can tune the upper surface to an almost depleted state, the bottom surface may remain conductive as the vertical electric field makes two opposite surfaces non-equivalent. As Bi_2Se_3 has a high refractive index, the high capacitance of a TI layer may shunt the top and upper surfaces. In order to tune the top and bottom surfaces independently to a dielectric state, a scheme with two (top and bottom) gates is being used in Refs. [48,49].

18.2.5 Magnetic Devices

Current-induced magnetization switching is a key effect that underlies the operation of magnetoresistive random-access memory and memory-in-logic integrated circuits. The main blocks of these devices are magnetic tunnel junctions (MTJ) which comprises two ferromagnetic metal layers, one with fixed and the other with free magnetization, separated by a thin nonmagnetic dielectric. The spin-transfer torque (STT) in MTJ causes magnetization switching at a vertical current density on the order of 10^3 kA/cm^2.

Another type of torque is related to spin-orbit interaction and appears when the current flows parallel to an interface. That is why spin-orbit torque (SOT) requires a different device geometry compared to vertical STT devices. Various options for STT- and SOT-based switching geometry and modes of operation can be found in Refs. [50–55].

It was found that TI-based heterostructures demonstrate large SOT and allow switching at lower currents, so TI-based devices could replace MTJ's and other non-TI SOT- and STT-based units [56]. The ferromagnetic material in a bilayer TI/FM can also be made of a magnetically doped TI. As an example,

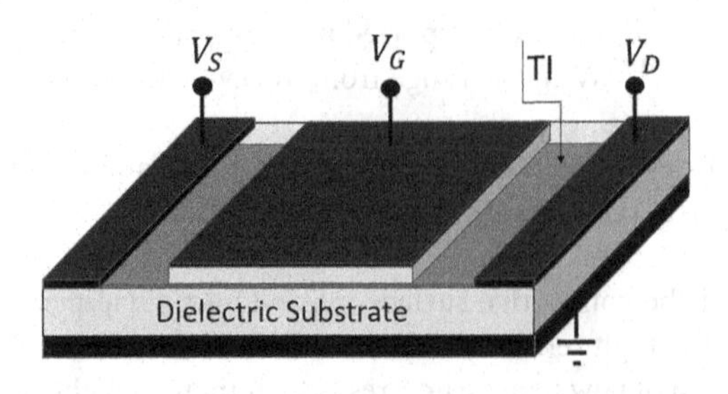

FIGURE 18.10 Metal-oxide-semiconductor field-effect transistor. A gate dielectric lays upon the TI layer.

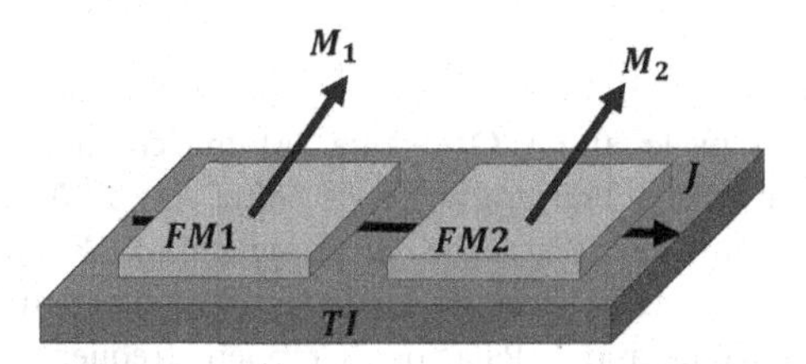

FIGURE 18.11 Phase-locked precession of magnetic moments in a TI-STO array.

the bilayer $(Bi_{0.5}Sb_{0.5})_2Te_3/(Cr_{0.08}Bi_{0.54}Sb_{0.38})_2Te_3$ reveals switching current reduced down to 89 kA/cm^2 at low temperature [57]. That is much lower than the current in heavy metal/FM bilayers. Large SOT also was observed in a Bi_2Se_3/NiFe heterostructure [58]. The Bi_2Se_3/NiFe switch operates at room temperature providing a threshold current density as low as 6×10^2 kA/cm^2 [59]. Both room-temperature operation and current density at least one order of magnitude lower than that in non-TI devices promise that TI-based switches will become an attractive option for logic units and memory cells because of their low power consumption.

In an MTJ, besides magnetization switching in a free magnetic layer, the steady-state precession of the magnetic moment occurs at a particular set of geometry and material parameters. It happens once magnetic damping being compensating by an STT. The precession generates ac current in the GHz frequency range through a giant magnetoresistance effect [60]. Such an MTJ is a spin-transfer oscillator (STO) that can be used as a microwave source. To increase the ac output power, it is possible to connect multiple STOs in series and synchronize. There is a domain of parameters where an ac signal produced by every single STO creates the feedback that leads to phase locking in the whole array. The synchronization does not require direct interaction between magnetic moments in neighboring STO. The process was studied for an array of two [61,62] and of N [63–66] STO devices, scaling the output ac power as N^2.

In a TI-STO, the anomalous Hall effect and spin-momentum locking play major roles in the nonlinear magnetic dynamics of the device. The self-consistent theoretical analysis predicts periodic and aperiodic oscillations depending on the input parameters [67,68].

Low power consumption in a TI-STO promises to facilitate highly energy-efficient arrays of phase-locked oscillators which can be used in phase-logic computing [69]. Theoretical analysis of the performance of the coupled oscillator system, shown in Figure 18.11, was carried out in Ref. [70].

An individual STO presents a nonlinear system in which current-induced spin polarization affects FM magnetization through exchange coupling and torque. In turn, the precessing FM magnetization causes an anomalous Hall effect in a TI and acts directly on electron spin polarization, thus affecting the TI current, maintaining the feedback from an FM. In a particular range of system parameters, the moment experiences steady-state precession. Moments $\boldsymbol{M}_1$ and $\boldsymbol{M}_2$ taken separately oscillate at different frequencies. However, a common current provides nonlinear coupling between them and synchronizes the oscillators. The operation of a phase-locked array is expected for a TI-STO made of the material combinations that follow: $Bi_2Se_3/YIG(Y_3Fe_5O_{12})$, Bi_2Te_3/GdN, Bi_2Se_3/EuS, , and $Bi_2Se_3/Cr_2Ge_2Te_4$.

18.2.6 Optoelectronics

As in the optical frequency range, the refractive indexes in the bulk and on the surface are different, a TI film can be utilized as a waveguide or an optical cavity [71].

The presence of almost gapless branches at the Bi_2Se_3 surface increases the number of absorption bands in the visible and mid-IR spectral ranges [72]. Saturable absorption makes TI a material of choice for ultrafast Q-switched lasers in communication and mid-IR bands. For a detailed review of TI applications in lasers, see Refs. [43,73].

References

1. K. Uchida, S. Furuya, J-I. Iwata, and A. Oshiyama, "Atomic corrugation and electron localization due to Moiré patterns in twisted bilayer graphenes", *Phys. Rev.* **B90**, 155451 (2014).
2. Y. Cao, V, Fatemi, S. Fang, K. Watanabe, T. Taniguchi, E. Kaxiras, and P. Jarillo-Herrero, "Unconventional superconductivity in magic-angle graphene superlattices", *Nature.* **556**, 43 (2018).
3. H. Wang, D. Nezich, J. Kong, and T. Palacios, "Graphene frequency multipliers", *IEEE Electron Device Lett.* **30**, 547 (2009).
4. V. Litvinov, *Magnetism in Topological Insulators* (Springer Nature, Switzerland AG, 2020).
5. C. E. ViolBarbosa, C. Shekhar, B. Yan, S. Ouardi, E. Ikenaga, G. H. Fecher, and C. Felser, "Direct observation of band bending in the topological insulator Bi_2Se_3", *Phys. Rev.* **B88**, 195128 (2013).
6. J. Suh, D. Fu, X. Liu, J.K. Furdyna, K.M. Yu, W. Walukiewicz, and J. Wu, "Fermi-level stabilization in the topological insulators Bi_2Se_3 and Bi_2Te_3---Origin of the surface electron gas", *Phys. Rev.* **B89**, 115307 (2014).
7. L.A. Walsh, A.J. Green, R. Addou, W. Nolting, C.R. Cormier, A.T. Barton, T.R. Mowll, R. Yue, N. Lu, J. Kim, M.J. Kim, V.P. LaBella, C.A. Ventrice, Jr., S. McDonnell, W.G. Vandenberghe, R.M. Wallace, A. Diebold, and C.L. Hinkle, "Fermi level manipulation through native doping in the topological insulator Bi_2Se_3", *ACS Nano.* **12**, 6310 (2018).
8. C.D. Spataru and F. Leonard, "Fermi-level pinning, charge transfer, and relaxation of spin-momentum locking at metal contacts to topological insulators", *Phys. Rev.* **B90**, 085115 (2014).
9. H. Steinberg J.-B. Laloe, V. Fatemi, J.S. Moodera, and P. Jarillo-Herrero, "Electrically tunable surface-to-bulk coherent coupling in topological insulator thin films", *Phys. Rev.* **B84**, 233101 (2011).
10. F. Yang, A.A. Taskin, S. Sasaki, K. Segawa, Y. Ohno, K. Matsumoto, and Y. Ando, "Top gating of epitaxial $(Bi_{1-x}Sb_x)_2Te_3$ topological insulator thin films", *Appl. Phys. Lett.* **104**, 161614 (2014).
11. X. He, T. Guan, X. Wang, B. Feng, P. Cheng, L. Chen, Y. Li, and K. Wu, "Highly tunable electron transport in epitaxial topological insulator $(Bi_{1-x}Sb_x)_2Te_3$ thin films", *Appl. Phys. Lett.* **101**, 123111 (2012).
12. Y. Satake, J. Shiogai, D. Takane, K. Yamada, K. Fujiwara, S. Souma, T. Sato, T. Takahashi, and A. Tsukazaki, "Fermi-level tuning of the Dirac surface state in $(Bi_{1-x}Sb_x)_2Se_3$ thin films", *J. Phys.---Condens. Matter.* **30**, 085501 (2018).
13. J. Kellner, M. Eschbach, J. Kampmeier, M. Lanius, E. Młynczak, G. Mussler, B. Holländer, L. Plucinski, M. Liebmann, D. Grützmacher, C.M. Schneider, and M. Morgenstern, "Tuning the Dirac point to the Fermi level in the ternary topological insulator $(Bi_{1-x}Sb_x)_2Te_3$", *Appl. Phys. Lett.* **107**, 251603 (2015).
14. J. Wang, X. Chen, B.-F. Zhu, and S.-C. Zhang, "Topological p-n junction", *Phys. Rev.* **B85**, 235131 (2012).
15. D. Kong, Y. Chen, J.J. Cha, Q. Zhang, J.G. Analytis, K. Lai, Z. Liu, S.S. Hong, K.J. Koski, S-K. Mo, Z. Hussain, and I.R. Fisher, Z.-X. Shen, and Y. Cui, "Ambipolar field effect in the ternary topological insulator $(Bi_{1-x}Sb_x)_2Te_3$ by composition tuning", *Nat. Nanotech.* **6**, 705 (2011).
16. A. Banerjee, A. Sundaresh, K. Majhi, R. Ganesan, and P.S.A. Kumar, "Accessing Rashba states in electrostatically gated topological insulator devices", *Appl. Phys. Lett.* **109**, 232408 (2016).
17. H. Li, L. Gao, H. Li, G. Wang, J. Wu, Z. Zhou, and Z. Wang, "Growth and band alignment of Bi_2Se_3 topological insulator on H terminated Si (111) van der Waals surface", *Appl. Phys. Lett.* **102**, 074106 (2013).
18. C. Ojeda-Aristizabal, M. S. Fuhrer, N. P. Butch, J. Paglione, and I. Appelbaum, "Towards spin injection from silicon into topological insulators---Schottky barrier between Si and Bi_2Se_3", *Appl. Phys. Lett.* **101**, 023102 (2012).
19. H. Zhang, X. Zhang, C. Liu, S.T. Lee, and J. Jie, "High-Responsivity, high-detectivity, ultrafast topological insulator Bi_2Se_3/Silicon heterostructure broadband photodetectors", *ACS Nano*, **10**, 5113 (2016).

20. B. Zheng, Y. Sun, J. Wu, M. Han, X. Wu, K. Huang, and S. Feng, "Group IV semiconductor Ge integration with topological insulator Sb_2Te_3 for spintronic application", *J. Phys. D---Appl. Phys.* **50**, 105303 (2017).
21. X. Liu, D.J. Smith, J. Fan, Y.-H. Zhang, H. Cao, Y.P. Chen, J. Leiner, B.J. Kirby, M. Dobrowolska, and J.K. Furdyna, "Structural properties of Bi_2Te_3 and Bi_2Se_3 topological insulators grown by molecular beam epitaxy on GaAs(100) substrates", *Appl. Phys. Lett.* **99**, 171903 (2011).
22. X. Liu, D.J. Smith, H. Cao, Y.P. Chen, J. Fan, Y.-H. Zhang, R.E. Pimpinella, M. Dobrowolska, and J.K. Furdyna, "Characterization of Bi_2Te_3 and Bi_2Se_3 topological insulators grown by MBE on (001) GaAs substrates", *J. Vac. Sci. Technol.* **B30**, 02B103 (2012).
23. A. Richardella, D.M. Zhang, J.S. Lee, A. Koser, D.W. Rench, A.L. Yeats, B.B. Buckley, D.D. Awschalom, and N. Samarth, "Coherent heteroepitaxy of Bi_2Se_3 on GaAs (111)B", *Appl. Phys. Lett.* **97**, 262104 (2010).
24. Z. Zeng, T.A. Morgan, D. Fan, C. Li, Y. Hirono, X. Hu, Y. Zhao, J.S. Lee, J. Wang, Z.M. Wang, S. Yu, M.E. Hawkridge, M. Benamara, and G.J. Salamo, "Molecular beam epitaxial growth of Bi_2Te_3 and Sb_2Te_3 topological insulators on GaAs (111) substrates---a potential route to fabricate topological insulator pn junction", *AIP Adv.* **3**, 072112 (2013).
25. L. He, X. Kou, and K.L. Wang, "Review of 3D topological insulator thin-film growth by molecular beam epitaxy and potential applications", *Phys. Status Solidi RRL.* **7**, 50–63 (2013).
26. L. Seixas, D. West, A. Fazzio, and S.B. Zhang, "Vertical twinning of the Dirac cone at the interface between topological insulators and semiconductors", *Nat. Commun.* **6**, 7630 (2015).
27. Y.Q. Huang, Y.X. Song, S.M. Wang, I.A. Buyanova, and W.M. Chen, "Spin injection and helicity control of surface spin photocurrent in a three dimensional topological insulator", *Nat. Commun.* **8**, 15401 (2017).
28. P. Hosur, "Circular photogalvanic effect on topological insulator surfaces---Berry-curvature-dependent response", *Phys. Rev.* **B83**, 035309 (2011).
29. D. Qu, "Spin-based broadband terahertz radiation from topological insulators", Lawrence Livermore National Laboratory, LDRD Annual Report (2017).
30. K.-K. Lee and T.H. Myers, "Band structure measurement and analysis of the Bi_2Te_3/CdTe (111)B heterojunction", *J. Vac. Sci. Technol.* **A33**, 031602 (2015).
31. P. Tsipas, E. Xenogiannopoulou, S. Kassavetis, D. Tsoutsou, E. Golias, C. Bazioti, G.P. Dimitrakopulos, P. Komninou, H. Liang, M. Caymax, and A. Dimoulas, "Observation of surface Dirac cone in high-quality ultrathin epitaxial Bi_2Se_3 topological insulator on AlN(0001) dielectric", *ACS Nano.* **8**, 6614 (2014).
32. E. Xenogiannopoulou, P. Tsipas, K.E. Aretouli, D. Tsoutsou, S.A. Giamini, C. Bazioti, G.P. Dimitrakopulos, P. Komninou, S. Brems, C. Huyghebaert, I.P. Raduc, and A. Dimoulas, "High-quality, large-area $MoSe_2$ and $MoSe_2/Bi_2Se_3$ heterostructures on AlN(0001)/Si(111) substrates by molecular beam epitaxy", *Nanoscale.* **7**, 7896 (2015).
33. P. Chaturvedi, S. Chouksey, D. Banerjee, S. Ganguly, and D. Saha, "Carrier and photon dynamics in a topological insulator Bi_2Te_3/GaN type II staggered heterostructure", *Appl. Phys. Lett.* 107, 192105 (2015).
34. M.Y. Pang, W.S. Li, K.H. Wong, and C. Surya, "Electrical and optical properties of bismuth telluride/gallium nitride heterojunction diodes", *J. Non-Crystal. Solids.* **354**, 4238 (2008).
35. X. Zhang, J. Wang, and S.-C. Zhang, "Topological insulators for high-performance terahertz to infrared applications", *Phys. Rev.* **B82**, 245107 (2010).
36. L. Luo, K. Zheng, T. Zhang, Y.H. Liu, Y. Yu, R. Lu, H. Qiu, Z. Li, and J.C.A Huang, "Optoelectronic characteristics of a near infrared light photodetector based on a topological insulator Sb_2Te_3 film", *J. Mater. Chem.* C3, 9154 (2015).
37. S.-M. Huang, S.-J. Huang, Y.-J. Yan, S.-H. Yu, M. Chou, H.-W. Yang, Y.-S. Chang, and R.-S. Chen, "Extremely high-performance visible light photodetector in the $Sb_2(SeTe)_2$ nanoflake", *Sci. Rep.* **7**, 45413 (2017).

38. A. Sharma, B. Bhattacharyya, A.K. Srivastava, T.D. Senguttuvan, and S. Husale, "High performance broadband photodetector using fabricated nanowires of bismuth selenide", *Sci. Rep.* **6**, 19138 (2016).
39. J. Liu, Y. Li, Y. Song, Y. Ma, Q. Chen, Z. Zhu, P. Lu, and S. Wang, "Bi_2Te_3 photoconductive detectors on Si", *Appl. Phys. Lett.* **110**, 141109 (2017).
40. J. Yao, J. Shao, Y. Wang, Z. Zhao, and G. Yang, "Ultra-broadband and high response of the Bi_2Te_3-Si heterojunction and its application as a photodetector at room temperature in harsh working environments", *Nanoscale.* **7**, 12535 (2015).
41. H. Qiao, J. Yuan, Z. Xu, C. Chen, S. Lin, Y. Wang, J. Song, Y. Liu, Q. Khan, H.Y. Hoh, C-X. Pan, S. Li, and Q. Bao, "Broadband photodetectors based on graphene- Bi_2Te_3 heterostructure", *ACS Nano.* **9**, 1886 (2015).
42. J. Yao, Z. Zheng, and G. Yang, "Layered-material WS2/topological insulator Bi_2Te_3 heterostructure photodetector with ultrahigh responsivity in the range from 370 to 1550 nm", *J. Mater. Chem.* **C4**, 7831 (2016).
43. W. Tian, W. Yu, J. Shi, and Y. Wang, "The property, preparation and application of topological insulators---a review", *Materials.* **10**, 814, (2017).
44. Z. Yue, B. Cai, L. Wang, X. Wang, M. Gu, "Intrinsically core-shell plasmonic dielectric nanostructures with ultrahigh refractive index", *Sci. Adv.* **2**, e1501536 (2016).
45. P. Wei, Z. Wang, X. Liu, V. Aji, and J. Shi, "Field-effect mobility enhanced by tuning the Fermi level into the band gap of Bi_2Se_3", *Phys. Rev. B.* **85**, 201402(R) (2012).
46. S. Cho, N.P. Butch, J. Paglione, and M.S. Fuhrer, "Topological insulator quantum dot with tunable barriers", *Nano Lett.* **12**(1), 469 (2012).
47. H. Zhu, C.A. Richter, E. Zhao, J.E. Bonevich, W.A. Kimes, H.-J. Jang, H. Yuan, H. Li, A. Arab, O. Kirillov, J.E. Maslar, D.E. Ioannou, and Q. Li, "Topological insulator Bi_2Se_3 nanowire high performance field-effect transistors", *Sci. Rep.* **3**, 1757 (2013).
48. S.K. Banerjee, L.F. Register, A. Macdonald, B.R. Sahu, P. Jadaun, J. Chang, Topological insulator-based field-effect transistor. Patent No. 8,629,427, 14 January 2014.
49. V. Fatemi, B. Hunt, H. Steinberg, S.L. Eltinge, F. Mahmood, N.P. Butch, K. Watanabe, T. Taniguchi, N. Gedik, R.C. Ashoori, and P. Jarillo-Herrero, "Electrostatic coupling between two surfaces of a topological insulator nanodevice", *Phys. Rev. Lett.* **113**, 206801 (2014).
50. S. Ikeda, J. Hayakawa, Y.M. Lee, F. Matsukura, Y. Ohno, T. Hanyu, and H. Ohno, "Magnetic tunnel junctions for spintronic memories and beyond", *IEEE Trans. Electron Devices.* **54**, 991 (2007).
51. I. M. Miron, K. Garello, G. Gaudin, P.-J. Zermatten, M. V. Costache, S. Auffret, S. Bandiera, B. Rodmacq, A. Schuhl, and P. Gambardella, "Perpendicular switching of a single ferromagnetic layer induced by in-plane current injection", *Nature.* **476**, 189 (2011).
52. K.L. Wang, J.G. Alzate, and P.K. Amiri, "Low-power non-volatile spintronic memory—STT-RAM and beyond", *J. Phys. D—Appl. Phys.* **46**, 074003 (2013).
53. R. Ramaswamy, J.M. Lee, K. Cai, and H. Yang, "Recent advances in spin-orbit torques—Moving towards device applications", *Appl. Phys. Rev.* **5**, 031107 (2018).
54. L. Liu, C.-F. Pai, D.C. Ralph, and R.A. Buhrman, "Magnetic oscillations driven by the spin Hall effect in 3-terminal magnetic tunnel junction devices", *Phys. Rev. Lett.* **109**, 186602 (2012).
55. L. Huang, S. He, Q.J. Yap, and S.T. Lim, "Engineering magnetic heterostructures to obtain large spin Hall efficiency for spin-orbit torque devices", *Appl. Phys. Lett.* **113**, 022402 (2018).
56. Y. Wang, P. Deorani, K. Banerjee, N. Koirala, M. Brahlek, S. Oh, and H. Yang, "Topological surface states originated spin-orbit torques in Bi_2Se_3", *Phys. Rev. Lett.* **114**, 257202 (2015).
57. Y. Fan, P. Upadhyaya, X. Kou, M. Lang, S. Takei, Z. Wang, J. Tang, L. He, L.-T. Chang, M. Montazeri, G. Yu, W. Jiang, T. Nie, R.N. Schwartz, Y. Tserkovnyak, and K.L. Wang, "Magnetization switching through giant spin-orbit torque in a magnetically doped topological insulator heterostructure", *Nat. Mater.* **13**, 699 (2014).

58. A.R. Melnik, J.S. Lee, A. Richardella, J.L. Grab, P.J. Mintun, M.H. Fischer, A. Vaezi, A. Manchon, E.-A. Kim, N. Samarth, and D.C. Ralph, "Spin-transfer torque generated by a topological insulator", *Nature.* **511**, 449 (2014).
59. Y. Wang, D. Zhu, Y. Wu, Y. Yang, J. Yu, R. Ramaswamy, R. Mishra, S. Shi, M. Elyasi, K.-L. Teo, Y. Wu, and H. Yang, "Room temperature magnetization switching in topological insulator-ferromagnet heterostructures by spin-orbit torques", *Nat. Commun.* **8**, 1364 (2017).
60. S.I. Kiselev, J.C. Sankey, I.N. Krivorotov, and N.C. Emley, "Microwave oscillations of a nanomagnet driven by a spin-polarized current", *Nature.* **425**, 380 (2003).
61. F.B. Mancoff, N.D. Rizzo, B.N. Engel, and S. Tehrani, "Phase-locking in double-point-contact spin-transfer devices", *Nature.* **437**, 393 (2005).
62. A.N. Slavin and V.S. Tiberkevich, "Theory of mutual phase locking of spin-torque nanosized oscillators", *Phys. Rev.* **B74**, 104401 (2006).
63. J. Grollier, V. Cros, and A. Fert, "Synchronization of spin-transfer oscillators driven by stimulated microwave currents", *Phys. Rev.* **B73**, 060409 (R) (2006); B. Georges, J. Grollier, V. Cros, and A. Fert, "Impact of the electrical connection of spin transfer nano-oscillators on their synchronization---an analytical study", *Appl. Phys. Lett.* **92**, 232504 (2008).
64. M. Elyasi, C.S. Bhatia, and H. Yang, "Synchronization of spin-transfer torque oscillators by spin pumping, inverse spin Hall, and spin Hall effects", *J. Appl. Phys.* **117**, 063907 (2015).
65. J. Turtle, P.-L. Buono, A. Palacios, C. Dabrowski, V. In, and P. Longhini, "Synchronization of spin torque nano-oscillators", *Phys. Rev.* **B95**, 144412 (2017).
66. T. Qu and R.H. Victora, "Phase-lock requirements in a serial array of spin transfer nano-oscillators", *Sci. Rep.* **5**, 11462 (2015).
67. T. Yokoyama, "Current-induced magnetization reversal on the surface of a topological insulator", *Phys. Rev.* **B84**, 113407 (2011).
68. X. Duan, X.-L. Li, Y.G. Semenov, and K.W. Kim, "Nonlinear magnetic dynamics in a nanomagnet–topological insulator heterostructure", *Phys. Rev.* **B92**, 115429 (2015); Y.G. Semenov, X. Duan, and K.W. Kim, "Voltage-driven magnetic bifurcations in nanomagnet–topological insulator heterostructures", *Phys. Rev.* **B89**, 201405(R) (2014).
69. R.A. Kiehl and T. Ohshima, "Bistable locking of single-electron tunneling elements for digital circuitry", *Appl. Phys. Lett.* **67**, 2494 (1995); T. Ohshima and R.A. Kiehl, "Operation of bistable phase-locked single-electron tunneling logic elements", *J. Appl. Phys.* **80**, 912 (1996).
70. C.-Z. Wang, H.-Y. Xu, N.D. Rizzo, R.A. Kiehl, and Y.-C. Lai, "Phase locking of a pair of ferromagnetic nano-oscillators on a topological insulator", *Phys. Rev. Appl.* **10**, 064003 (2018).
71. Z. Yue and M. Gu, "Resonant cavity-enhanced holographic imaging in ultrathin topological insulator films", Frontiers in Optics, Optical Society of America, 2016, FTu3F. 1.
72. L. Sun, Z. Lin, J. Peng, J. Weng, Y. Huang, and Z. Luo, "Preparation of few-layer bismuth selenide by liquid-phase-exfoliation and its optical absorption properties", *Sci. Rep.* **4**, 4794 (2014).
73. Z. Yue, X. Wang, and M. Gu, "Topological insulator materials for advanced optoelectronic devices", arXiv:1802.07841v1 [physics.optics] (2018).

Index

Printed in the United States
by Baker & Taylor Publisher Services

Printed in the United States
by Baker & Taylor Publisher Services